Geographical Approaches to Global Sustainable Development Goals (SDGs)

Lili Yulyadi Arnakim
Binus University, Indonesia

Miranda P. Tahalele
Binus University, Indonesia

Muhammad Anshari
Universiti Brunei Darussalam, Brunei

IGI Global
Scientific Publishing
Publishing Tomorrow's Research Today

Vice President of Editorial Melissa Wagner
Director of Acquisitions Mikaela Felty
Director of Book Development Jocelynn Hessler
Production Manager Mike Brehm
Cover Design Jose Rosado

Published in the United States of America by
IGI Global Scientific Publishing
701 East Chocolate Avenue
Hershey, PA, 17033, USA
Tel: 717-533-8845 | Fax: 717-533-7115
Website: https://www.igi-global.com E-mail: cust@igi-global.com

Library of Congress Cataloging-in-Publication Data

LCCN: 2025004513 (CIP Data Pending)
ISBN13: 9798337313009
Isbn13Softcover: 9798337313016
EISBN13: 9798337313023
British Cataloguing in Publication Data
A Cataloguing in Publication record for this book is available from the British Library.

Table of Contents

Detailed Table of Contents

The Sustainable Development Goals (SDGs), adopted by the United Nations in 2015, offer a universal framework to address global challenges and achieve sustainable development. Comprising 17 goals and 169 targets, they aim to eradicate poverty, reduce inequalities, promote economic growth, and protect the planet, guided by the principle of "Leave no one behind." Rooted in sustainability, the SDGs emphasize interconnected social, economic, and environmental dimensions. They prioritize inclusivity, integration, accountability, and partnerships among governments, businesses, civil society, and individuals. Building on the Millennium Development Goals (MDGs), the SDGs adopt a more expansive, collaborative approach to development. While challenges like limited resources, political instability, and COVID-19 disruptions persist, innovative governance and collective action are critical to success. By fostering global solidarity, the SDGs provide a roadmap for a just and sustainable future, urging shared responsibility and unwavering commitment to achieving their vision by 2030.

This chapter investigates the difficult nexus between peace, politics and sustainable development looking at how regional disturbances interfere with the culmination of the SDGs. It examines how instability in politics, economy, environment and human rights violations contribute to the perpetuation of systemic challenges. Critical assessment is given to the legal and policy frameworks that govern governance, trade and environmental sustainability, to identify gaps that increase the regional conflict. This study emphasizes about how economic sanctions, resource scarcity and

This chapter examines the origins and evolution of the 17 Sustainable Development Goals (SDGs) from the Millennium Development Goals (MDGs), emphasizing their necessity in addressing global challenges. It highlights the interconnected nature of the SDGs, their role in tackling issues like poverty, gender inequality, clean energy, and climate change, and illustrates these interlinkages through real-world examples. The chapter discusses global, regional, and local implementation strategies, focusing on the roles of governments, NGOs, and communities. It stresses the importance of data-driven assessments, transparency, accountability, and innovation, particularly through sustainable business practices and the UN Global Compact Principles. Finally, it identifies challenges such as political and financial barriers, proposing strategies to overcome them and accelerating progress toward an equitable and sustainable future.

The SDGs (Sustainable Development Goals), created by the United Nations in 2015, aims to tackle global level problems in order to end poverty, reduce inequality, promote sustainable consumption and production, combat climate change, preserve oceans and ecosystems, foster innovation, reduce gender disparities, ensure access to clean water and energy. These goals set a vision for sustainable development but require specific regional strategies considering the social-economic, cultural and natural endowment of each region. The study examines regional frameworks coordinating with global SDGs under specific regional circumstances. Every region is analyzed according to their participation of the governments, civil society organizations and the private sector to cope with their specific development issues. The study highlights

Sustainable Development Goal 8 on decent work constitutes an important pillar to Thailand's economic development. Its recent National Development Plan puts emphasis on transitioning into an aging demography and building its digital economy which are reflected in the incorporation of SDG 8 indicators. Key indicators in Thailand's NDP include enhancing skills and added value and a safe working environment for all. This study examines the integration of Sustainable Development Goal 8 indicators into Thailand's National Development Plan, with a particular focus on the challenges posed by an aging population. The research primarily identifies key actors and relevant programs involved in this integration process, and analyzes their impact on labor market trends in Thailand. Findings indicate that several initiatives have contributed meaningfully to enhancing digital readiness, stimulating economic growth through private investment, and improving access for older individuals to safe and supportive work environments.

SDGs can only be accomplished with strong global partnerships and cooperation SDG No 17 advocates for global sustainability partnerships including in financial services and practices. The implementation of International Financial Reporting Standards (IFRS) is aligned with the goal of SDG 17 to help developing countries achieve financial sustainability financing policies by mobilising additional financial resources from various sources, especially for Small and Medium Entreprises (SMEs). There is not much research addressing the implementation process of IFRS in response to Knowledge Management (KM) strategies for sustainable finance. In order to support SDG no 17, this study is to identify and describe the implementation of IFRS framework and knowledge transfer for SMEs readiness.

Climate change & environment pose a significant danger to sustainable development, especially in resource-rich nations like ASEAN. Regional collaboration is crucial to address concerns related to climate adaption, resource management, & cross-border environmental governance. The assessment highlights importance of strong institutional structures & equitable resource distribution to effectively address these concerns in achieving SDGs & promoting environmental sustainability in this region. Emerging issues (urbanization, population growth & climate change) add new aspects to regional environmental challenges and require inventive and adaptable solutions. Technology developments offer interesting possibilities for improving cooperation/monitoring progress. To ensure that all Member States benefit from regional initiatives, it is essential to improve knowledge-sharing platforms and align legislative frameworks. ASEAN, through cooperation and innovation, is vital to solving environmental problems and ensuring the long-term development of future generations in order to achieve the SDGs.

The National Mangrove Map project is a strategic initiative aimed at developing a comprehensive inventory of mangrove forests across Indonesia. This study analyzes the project's approach to strategic management, particularly in socialization and development communication, with emphasis on optimizing stakeholder engagement through workshops, meetings, and procurement strategies. Using a qualitative research design including semi-structured interviews, document analysis, participant observation, and statistical testing the study examines how tailored engagement influences project success. Findings show that inclusive workshops, effective meetings, and sustainable procurement practices enhance project outcomes by aligning diverse interests and promoting participation. However, challenges such as conflicting priorities and logistical constraints remain. The research contributes to both theory and practice in environmental project management, highlighting the vital role of adaptive, participatory engagement in achieving conservation goals.

It is no longer a debate that the purpose of Islamic Social Finance like Zakat is in line with and does support the achievement of Sustainable Development Goals (SDGs). Eradicating poverty, improving the welfare of the people, and reducing inequality, are the end goals of why zakat is collected and distributed to the recipients. There have been studies conducted in the past to explain how Islamic Social Finance positively contributes to the achievement of SDGs and how zakat institutions should be strengthened to increase the effectiveness of zakat management. However, none has tried to analyze what are the roles played by zakat institutions in SDGs. The objective of this paper is to identify and compare the role of zakat institutions in Indonesia and Malaysia, by using BAZNAS and MAIWP to represent zakat institutions in each country. We apply the goal-setting theory to examine the role.

The intersection between the Duterte administration's War on Drugs and Sustainable Development Goal 16 reveals critical tensions in policy design and implementation. Drawing on qualitative analysis of the Philippine Development Plan (PDP) 2017–2022 and official data, the study assesses alignment with three key SDG targets: 16.1 (violence reduction), 16.3 (access to justice), and 16.6 (institutional accountability). While institutional reforms and justice-sector improvements are formally embedded in national planning documents, the operational conduct of the War on Drugs has generated systemic contradictions. These include widespread human rights violations, legal erosion, and institutional opacity. The securitization of development through punitive means challenges the normative foundations of SDG 16 and raises urgent questions about the future of peace and justice in authoritarian governance contexts. Fulfilling sustainable development requires a rights-based, inclusive, and transparent approach that places human dignity and democratic accountability at the core of public policy.

Preface

Since the adoption of the Sustainable Development Goals (SDGs) by the United Nations in 2015, the global community has been called upon to pursue a collective vision of eradicating poverty, reducing inequality, protecting the planet, and promoting peace and prosperity. Comprising 17 interconnected goals and 169 specific targets, the SDGs serve as a universal framework for sustainable development that spans across social, economic, and environmental dimensions. Despite their global nature, however, the process of achieving the SDGs must be contextualized to the specific geographic, political, and socio-cultural realities of each region.

At the global level, the SDGs represent a shared commitment by UN member states to advance human development and ecological balance. However, their implementation relies heavily on the capacity of regional frameworks to interpret and integrate these goals into meaningful action. Regional bodies such as ASEAN, the African Union, and the European Union serve as intermediaries, translating global aspirations into strategies that align with regional priorities and constraints. These organizations play a pivotal role in coordinating policy harmonization, fostering cross-border cooperation, and addressing transboundary challenges such as climate change, migration, and natural resource management.

As we move from regional to national implementation, the real diversity in SDG application becomes apparent. Each country must tailor the global framework to its own institutional capabilities, governance structures, and socio-economic landscape. National governments are responsible for embedding the SDGs into development plans, budgets, and policy, often negotiating between international expectations and domestic priorities. The variability in political will, economic resources, administrative capacity, and cultural norms significantly shapes how effectively each nation can localize the goals.

Geographical realities further complicate this landscape. Nations with vast coastlines, arid interiors, archipelagic geographies, or high urban density face unique sustainability challenges that require place-based solutions. Political context also plays a crucial role: democratic, authoritarian, post-colonial, or conflict-affected states each bring distinct dynamics to the process of SDG implementation. Moreover, the socio-cultural fabric of each society—its values, traditions, belief systems, and social hierarchies—influences how development goals are perceived, prioritized, and operationalized. Thus, while the SDGs provide a global blueprint, their realization depends on the nuanced interplay of global vision and local action.

This book shows how regions around the world are adapting and innovating planning mechanisms to meet sustainability challenges, using tools like geospatial mapping, regional collaboration, participatory governance, and digital technology. These tools help address context-specific needs while still aligning with global SDG frameworks. In doing so, the chapters reflect the complex interaction between regional development planning and global sustainability, offering valuable insights into localized implementation of global goals. Several chapters in the book draw attention to the misalignment between standardized SDG indicators and local contexts.

The chapters in this volume collectively shed light on the dynamic and multi-layered pathways through which the Sustainable Development Goals (SDGs) are interpreted, contested, and operationalized across diverse regional and thematic contexts. The opening chapter by Shilpa Khade lays the groundwork by introducing the philosophical and normative foundations of the SDGs, centered on the imperative of "leaving no one behind." Yet, this universal vision faces numerous implementation barriers, including political fragmentation and uneven resource allocation.

Building on this, Chapter 2 by Sheikh Inam Ul Mansoor et al. explores how political instability and governance failures, particularly in conflict-prone areas, obstruct the realization of sustainable development. The authors argue that legal reforms and sustained international cooperation are essential for dismantling structural impediments. Chapter 3 by S. Ida Evangeline focuses on the role of regional governance mechanisms, pointing out institutional deficits that limit the ability of regional bodies to mediate between global mandates and local needs. In a complementary analysis, Bharathi M. et al. in Chapter 4 chart the evolution from the Millennium Development Goals (MDGs) to the SDGs, emphasizing the interlinked nature of goals and the necessity for accountability, innovation, and community ownership in confronting issues like poverty, gender inequality, and climate change. The authors focus on climate adaptation and mitigation through the lens of climate science, offering practical recommendations for multi-level climate action. At the institutional level, Chapter 5 by Tanishka Madke et al. examines ASEAN's role in climate governance, advocating for resource-sharing frameworks and legal harmonization across member states.

In education, Chapter 6 by Nazire Kaya et al. confronts the acute shortage of qualified teachers as a major barrier to SDG 4, recommending a combination of improved training, technological integration, and targeted financial support to address global disparities. Furthermore, Logistics and supply chain sustainability are tackled in Chapter 7 by Rini Setiawati et al., where agility and green innovation are positioned as vital for urban supply chain efficiency and resilience. Complementarily, Chapter 8 by Pragti Dixit et al. assesses how urban growth, agricultural demands, and water scarcity interact with the SDGs in BRICS nations, revealing stark geographical inequalities that demand place-based interventions.

Furthermore, Chapter 9 by Izni Azrein Noor Azalie and Gabriel Yong critiques the limitations of SDG indicators by drawing on Brunei's experience, calling for the adoption of participatory, context-aware data collection methods. Likewise, Chapter 10 by Preeti Joshi and Arvinder Kaur showcases the transformative potential of geospatial technologies in facilitating multi-stakeholder partnerships under SDG 17.

In Southeast Asia, national and regional strategies toward SDG localization are examined in depth. Chapter 11 by Lili Yulyadi Arnakim and Safira Puspitarani investigates Thailand's demographic shifts and digital transition in shaping labor market policies aligned with SDG 8. Chapter 12 by Mahani Hamdan emphasizes the importance of partnerships and knowledge sharing in the successful implementation of International Financial Reporting Standards (IFRS), particularly for Small and Medium Enterprises (SMEs) in developing countries. By exploring how these frameworks align with SDG 17—partnerships for the goals—the chapter identifies key strategies for strengthening financial sustainability and mobilizing resources. It highlights the need for knowledge transfer mechanisms that can equip SMEs with the tools and frameworks necessary to thrive in an increasingly globalized and sustainable financial ecosystem. Addressing the pressing environmental challenges in ASEAN nations, chapter 13 underscores the importance of regional cooperation in tackling issues such as climate change, resource management, and environmental degradation. It explores the need for strong institutional frameworks and equitable resource distribution to foster collaborative solutions that promote sustainable development. By highlighting the impact of emerging issues like urbanization and population growth, the chapter presents innovative approaches for regional cooperation that can help ASEAN nations meet SDG environmental targets. Chapter 14 highlights the importance of stakeholder engagement in the success of environmental projects, using Indonesia's National Mangrove Map Project as a case study. It examines how effective communication and strategic management of stakeholder interests can enhance project outcomes and align them with sustainability goals. By analyzing various engagement strategies, the chapter emphasizes the role of participatory practices in fostering collective action for environmental conservation and resilience.

Before the story comes to a close, Chapter 15 by Lili Yulyadi Arnakim and Namira Shabrina brings forward an institution deeply embedded in both religious and social life: zakat. Islamic Social Finance has long been acknowledged for its alignment with the SDGs, particularly in areas like poverty eradication, reducing inequality, and improving people's welfare. But this chapter takes us further. It asks not only whether zakat supports sustainable development, but how. By comparing the roles played by zakat institutions in Indonesia and Malaysia, namely BAZNAS and MAIWP. In this chapter the authors use goal-setting theory to analyze their effectiveness. What emerges is a picture of faith-based financial systems acting not as peripheral actors, but as vital partners in the global sustainability agenda. Zakat, in this framing, is not simply charity. It is structured, strategic, and systemically impactful.

Finally, the book concludes with a critical examination of the securitization of development in Chapter 16 by Lili Yulyadi Arnakim and Putri Wibowo, focusing on the Philippines' War on Drugs. While the national development plan aligns rhetorically with SDG 16, the reality of militarized enforcement and widespread human rights violations undermines these commitments. The authors argue for a pivot toward rights-based governance as a cornerstone of sustainable peace and justice.

This volume compiles diverse scholarly insights that shed light on the complex pathways of localizing the Sustainable Development Goals (SDGs). Covering a range of themes from global governance to regional collaboration and national implementation, each chapter contributes to a deeper understanding of how sustainability is shaped by local geographies, political contexts, and cultural nuances. As we approach the 2030 target, the reflections presented here underscore a vital message: while the SDGs are framed as a global agenda, their realization ultimately hinges on locally rooted, context-aware strategies.

Lili Yulyadi Arnakim
Binus University, Indonesia

Miranda P. Tahalele
Binus University, Indonesia

Muhammad Anshori
University of Brunei Darussalam, Brunei

Chapter 1
Introduction to Theoretical Framework of the SDGs

Shilpa Pramod Khade
https://orcid.org/0009-0005-8978-7392
Dr. P.A. Inamdar University, India

ABSTRACT

The Sustainable Development Goals (SDGs), adopted by the United Nations in 2015, offer a universal framework to address global challenges and achieve sustainable development. Comprising 17 goals and 169 targets, they aim to eradicate poverty, reduce inequalities, promote economic growth, and protect the planet, guided by the principle of "Leave no one behind." Rooted in sustainability, the SDGs emphasize interconnected social, economic, and environmental dimensions. They prioritize inclusivity, integration, accountability, and partnerships among governments, businesses, civil society, and individuals. Building on the Millennium Development Goals (MDGs), the SDGs adopt a more expansive, collaborative approach to development. While challenges like limited resources, political instability, and COVID-19 disruptions persist, innovative governance and collective action are critical to success. By fostering global solidarity, the SDGs provide a roadmap for a just and sustainable future, urging shared responsibility and unwavering commitment to achieving their vision by 2030.

1. INTRODUCTION

The Sustainable Development Goals (SDGs), unanimously adopted by all 193 United Nations Member States in September 2015, represent a landmark global

DOI: 10.4018/979-8-3373-1300-9.ch001

commitment to addressing the most pressing challenges facing humanity by 2030. Comprising 17 interconnected goals and 169 specific targets, the SDGs aim to eradicate poverty, reduce inequalities, foster inclusive economic growth, ensure environmental sustainability, and promote peace and justice, all while adhering to the principle of "leaving no one behind" (United Nations, 2015). Unlike their predecessors, the Millennium Development Goals (MDGs), which focused primarily on developing nations, the SDGs adopt a universal approach, applying to all countries regardless of their economic status. This universality underscores the recognition that global challenges such as climate change, economic disparity, and social exclusion require collective action across borders, economies, and cultures.

The SDGs are not merely a policy framework but a call for transformative change, integrating social, economic, and environmental dimensions into a cohesive agenda. They address systemic issues through interconnected goals, such as linking poverty reduction (Goal 1) with quality education (Goal 4), gender equality (Goal 5), and climate action (Goal 13). This interconnectedness demands innovative theoretical and practical approaches, drawing on disciplines such as development studies, economics, environmental science, political theory, and sociology. The complexity of the SDGs necessitates frameworks like systems thinking, which emphasizes the interdependence of global systems, and the human development approach, which prioritizes individual capabilities and well-being (Sen, 1999; Nilsson et al., 2018).

India, with its population exceeding 1.4 billion and its status as one of the world's fastest-growing economies, plays a pivotal role in the global pursuit of the SDGs. As a nation straddling the challenges of development and the opportunities of rapid modernization, India's policies and initiatives offer valuable insights into operationalizing the SDGs in diverse and resource-constrained contexts. Programs such as the Swachh Bharat Abhiyan (Clean India Mission) and the International Solar Alliance (ISA) demonstrate India's commitment to integrating sustainable development into its national framework while contributing to global efforts (NITI Aayog, 2023; MEA, 2023).

This article provides a comprehensive theoretical and practical analysis of the SDGs, grounded in interdisciplinary frameworks and enriched by an extensive literature review and detailed case studies. It explores the SDGs' theoretical foundations, including systems thinking, the human development approach, and global governance theories, and examines their social, economic, and environmental dimensions. The article places special emphasis on India's role, analyzing its policy integration, implementation challenges, and global contributions. Through six in-depth case studies—spanning Costa Rica, India, Bangladesh, Rwanda, the Maldives, and Finland—the article illustrates the practical application of the SDGs, highlighting successes, challenges, and lessons for global implementation. The objectives are threefold: to elucidate the theoretical underpinnings of the SDGs, to evaluate India's contributions as a model for other nations, and to propose strategies for overcoming implementation barriers to achieve the 2030 Agenda.

The article is structured as follows: Section 2 outlines the scope and limitations, Section 3 provides an expanded literature review, Section 4 discusses the theoretical foundations, Section 5 analyzes the SDGs' dimensions, Section 6 focuses on India's role, Section 7 addresses challenges and future directions, Section 8 presents the case studies, Section 9 offers concluding remarks, and Section 10 lists references. By integrating theoretical insights with practical examples, this article aims to contribute to the academic discourse on sustainable development and inform policy and practice for achieving the SDGs.

2. SCOPE AND LIMITATIONS

This study is primarily theoretical and conceptual, synthesizing existing literature, policy documents, and case studies rather than conducting original empirical research. It adopts a global perspective while focusing on India's role, given its demographic and economic significance. The analysis draws on interdisciplinary frameworks, including development studies, sustainability science, and political theory, to explore the SDGs' theoretical foundations and practical applications.

Scope: The article covers the SDGs' theoretical underpinnings, their social, economic, and environmental dimensions, and India's contributions to their implementation. It examines synergies and trade-offs among goals, using case studies to illustrate real-world applications. The literature review synthesizes scholarly work on the SDGs' foundations, challenges, and India's role, providing a robust context for the analysis. The case studies span diverse contexts—developed, developing, and small island nations—to ensure a comprehensive understanding of SDG implementation.

Limitations: The reliance on secondary sources limits the depth of localized, community-level insights. The dynamic nature of SDG implementation, influenced by global crises such as pandemics, climate events, and geopolitical shifts, may render some observations time-sensitive. The article does not evaluate all 17 goals and 169 targets comprehensively, focusing instead on thematic intersections and key examples. Data gaps and varying reporting standards across countries may affect the comparability of case studies. Finally, while India is a focal point, the analysis does not delve into subnational variations within the country, which could offer additional insights.

Despite these limitations, the article provides a rigorous and comprehensive analysis, leveraging diverse sources and case studies to advance the discourse on sustainable development and inform global and national strategies.

3. LITERATURE REVIEW

The Sustainable Development Goals (SDGs) have generated a vast body of scholarly work since their adoption in 2015, building on the legacy of the Millennium Development Goals (MDGs) while introducing a more ambitious, universal, and integrated framework. This literature review synthesizes key contributions from development studies, sustainability science, economics, political science, and related fields, focusing on the SDGs' theoretical foundations, implementation challenges, dimensional analyses, India's role, critical perspectives, and future directions. The review provides a robust scholarly foundation for the article, contextualizing the theoretical and practical discussions that follow.

3.1 Theoretical Foundations of the SDGs

The SDGs are underpinned by a rich array of theoretical frameworks that draw on historical, philosophical, and interdisciplinary perspectives. Sachs (2012) traces the evolution from the MDGs to the SDGs, highlighting the shift from a developing-country focus to a universal agenda applicable to all nations. This universality aligns with Enlightenment-era ideals of progress and collective responsibility, as articulated by philosophers like Immanuel Kant (1784) and John Stuart Mill (1863). Pogge and Sengupta (2016) emphasize the SDGs' ethical grounding in John Rawls's theory of distributive justice, which prioritizes equitable resource allocation and opportunities for the least advantaged (Rawls, 1971). The principle of "leaving no one behind" reflects this commitment to inclusivity, ensuring that marginalized populations are central to development efforts.

Systems thinking is a cornerstone of the SDGs, emphasizing the interconnectedness of global systems. Nilsson et al. (2018) argue that the SDGs' 17 goals and 169 targets form a complex network, requiring integrated approaches to manage synergies and trade-offs. For instance, achieving Zero Hunger (Goal 2) depends on sustainable agriculture (Goal 12), climate action (Goal 13), and clean water access (Goal 6). The nexus approach, as explored by Liu et al. (2018), highlights critical intersections such as the water-energy-food nexus, enabling policymakers to design holistic solutions. However, Bengtsson et al. (2018) note that trade-offs, such as balancing economic growth with environmental sustainability, pose significant challenges, necessitating robust analytical frameworks.

The human development and capability approach, developed by Amartya Sen (1999) and Martha Nussbaum (2000), provides a normative framework for the SDGs, emphasizing the expansion of individual freedoms and capabilities over purely economic metrics. This approach underpins goals such as Quality Education (Goal 4), Good Health and Well-Being (Goal 3), and Gender Equality (Goal 5), prioritizing holistic improvements in human well-being. Fukuda-Parr (2019) argues that the SDGs' focus on inclusivity aligns with Nussbaum's concept of universal human entitlements, ensuring that development benefits reach marginalized groups, including women, indigenous communities, and persons with disabilities.

Global governance theories highlight the role of multilateral institutions in coordinating SDG efforts. Keohane and Nye (2005) emphasize the importance of collective action through organizations like the United Nations, which set norms and facilitate collaboration. French and Kotzé (2018) underscore the principle of subsidiarity, which advocates for context-specific solutions tailored to local needs. However, Caiado et al. (2018) identify power asymmetries and resource inequalities among nations as barriers to effective multilateralism, particularly for developing countries with limited institutional capacity.

3.2 Implementation Challenges

The literature identifies a range of barriers to SDG implementation, spanning financial, governance, data, and external challenges. Financial constraints are a critical issue, with UNCTAD (2014) estimating an annual funding gap of $2.5 trillion for developing countries to achieve the SDGs. This shortfall is particularly acute in sectors like infrastructure, healthcare, and renewable energy, where upfront costs are high (Schmidt-Traub, 2015). Allen et al. (2018) highlight governance challenges, including weak institutional frameworks, corruption, and lack of policy coherence, which hinder effective implementation. For example, fragmented policies may prioritize short-term economic gains over long-term sustainability, undermining goals like Climate Action (Goal 13).

Data gaps pose another significant challenge. Hák et al. (2016) argue that inconsistent data collection and inadequate indicators limit the ability to monitor progress accurately. While global indicators exist for all 169 targets, many countries lack the capacity to collect disaggregated data, particularly for marginalized populations (Fraisl et al., 2020). Global crises, such as the COVID-19 pandemic, have further stalled progress. The United Nations (2024) reports that 80% of SDG targets are off track, with 30% showing no progress or regression, largely due to disruptions in health, education, and economic systems. Morton et al. (2017) and Fu et al. (2019) emphasize the complexity of achieving interconnected goals, advocating for transdisciplinary approaches to address trade-offs and leverage synergies.

3.3 Social, Economic, and Environmental Dimensions

The SDGs' three dimensions—social, economic, and environmental—have been extensively analyzed in the literature, reflecting their interdisciplinary nature.

Social Dimension: Griggs et al. (2014) advocate for integrated strategies to address poverty (Goal 1), health (Goal 3), education (Goal 4), gender equality (Goal 5), and reduced inequalities (Goal 10). Social justice theories, particularly Rawls's "justice as fairness," provide an ethical foundation, emphasizing equitable access to opportunities (Rawls, 1971). Programs like Brazil's Bolsa Família, a conditional cash transfer scheme, demonstrate how social interventions can reduce poverty while improving education and health outcomes (Soares et al., 2011). However, Leach (2016) critiques the SDGs for insufficient attention to intersectionality, arguing that gender equality efforts must address overlapping forms of discrimination based on race, class, and disability.

Economic Dimension: The economic dimension focuses on sustainable growth (Goal 8), industry and innovation (Goal 9), and reduced inequalities (Goal 10). Inclusive growth theories, as articulated by Griggs et al. (2014), advocate for broad-based economic benefits that reach marginalized populations. Bengtsson et al. (2018) critique traditional growth-centric models, proposing transformative systems of consumption and production to align with Responsible Consumption and Production (Goal 12). The European Union's Circular Economy Action Plan exemplifies this approach, integrating economic progress with environmental sustainability (Avrampou et al., 2019). However, Kostoska and Kocarev (2019) note that technological advancements, such as ICT, must be equitably distributed to bridge economic and gender gaps.

Environmental Dimension: The environmental dimension encompasses climate action (Goal 13), life below water (Goal 14), life on land (Goal 15), clean water and sanitation (Goal 6), and affordable clean energy (Goal 7). Rockström et al. (2009) introduce the planetary boundaries framework, emphasizing the need to operate within ecological limits to avoid irreversible environmental damage. Initiatives like the Paris Agreement underscore the importance of global cooperation in addressing climate change (Fu et al., 2019). Fraisl et al. (2020) highlight the role of citizen science in monitoring environmental goals, such as marine and terrestrial biodiversity, but note challenges in scaling these efforts. The literature stresses the need for integrated policies to balance environmental protection with social and economic priorities, particularly in resource-constrained settings (Liu et al., 2018).

3.4 India's Role in SDG Implementation

India's contributions to the SDGs have attracted significant scholarly attention due to its scale, diversity, and global influence. NITI Aayog (2023) documents India's integration of the SDGs into its national development framework, with the SDG India Index tracking state-level progress across key indicators. Initiatives like the Pradhan Mantri Garib Kalyan Yojana (PMGKY) address poverty and hunger

(Goals 1 and 2), while Swachh Bharat Abhiyan tackles sanitation and health (Goals 6 and 3) (NITI Aayog, 2023). The International Solar Alliance (ISA), co-founded by India, promotes renewable energy, advancing Affordable and Clean Energy (Goal 7) and Climate Action (Goal 13) (MEA, 2023). Programs like Beti Bachao Beti Padhao enhance gender equality and education (Goals 5 and 4), reflecting India's commitment to social inclusion.

However, India faces significant implementation challenges. Drèze and Sen (2017) highlight persistent poverty and inequality, particularly in rural areas, despite economic growth. Environmental degradation, including air and water pollution, poses risks to Goals 6, 13, 14, and 15 (CPCB, 2022). The Lancet (2021) underscores health infrastructure gaps exposed by the COVID-19 pandemic, emphasizing the need for sustained investment in Goal 3. Financial constraints and governance issues further complicate implementation, with the Reserve Bank of India (2023) noting fiscal pressures on state budgets. Despite these challenges, India's global contributions, particularly through South-South cooperation and climate advocacy, position it as a leader in the SDG agenda (Sachs, 2015).

3.5 Critiques and Alternative Perspectives

The SDGs have faced critical scrutiny from various scholarly perspectives. Postcolonial scholars like Escobar (2015) argue that the SDGs' universal framework may impose Western development models, marginalizing local knowledge and priorities. This critique is particularly relevant for countries like India, where diverse cultural and ecological contexts shape development needs. Ecological economists, such as Daly (1996), challenge the SDGs' reliance on economic growth, advocating for steady-state economies that prioritize ecological stability over GDP expansion. Feminist scholars, including Leach (2016), call for greater attention to intersectionality in addressing gender equality, noting that Goal 5 must consider overlapping forms of discrimination to be effective.

Other critiques focus on the SDGs' complexity and measurability. Fukuda-Parr (2019) argues that the proliferation of targets and indicators dilutes focus, making it difficult to prioritize interventions. Caiado et al. (2018) note that power imbalances in global governance structures limit the voice of developing nations, undermining the principle of equitable partnership (Goal 17). These critiques highlight the need for flexible, context-specific approaches that balance global aspirations with local realities.

3.6 Gaps and Future Directions

The literature identifies several research gaps and future directions. First, there is a need for deeper analysis of trade-offs between goals, particularly in resource-constrained settings where prioritizing one goal (e.g., economic growth) may undermine another (e.g., environmental sustainability) (Moallemi et al., 2020). Second, technological innovations, such as artificial intelligence and big data, offer promising tools for monitoring and implementation, but equitable access remains a challenge (Reuters, 2024; Kioupi & Voulvoulis, 2019). Third, innovative financing mechanisms, such as green bonds and blended finance, are underexplored in the context of developing countries (OECD, 2020). Finally, India's experience as a large, diverse nation offers a valuable case study for exploring scalable, context-specific solutions, but subnational analyses are limited in the current literature (NITI Aayog, 2023).

Future research should focus on developing transdisciplinary frameworks to address the SDGs' complexity, leveraging insights from systems thinking, behavioral science, and technology studies. Scalable models for financing and governance, particularly in low- and middle-income countries, are critical to closing the funding gap. Additionally, participatory approaches that incorporate local and indigenous knowledge can enhance the SDGs' relevance and effectiveness, addressing post-colonial critiques (Escobar, 2015). India's role as a bridge between developed and developing nations makes it an ideal context for testing these approaches, informing global strategies.

This literature review establishes a robust foundation for the article, synthesizing interdisciplinary perspectives on the SDGs' theoretical foundations, implementation challenges, and India's contributions. It highlights the need for integrated, inclusive, and innovative approaches to achieve the 2030 Agenda, setting the stage for the theoretical and practical analyses that follow.

4. THEORETICAL FOUNDATIONS OF THE SDGS

The SDGs are grounded in a multifaceted theoretical framework that draws on historical, philosophical, and interdisciplinary perspectives. This section explores four key foundations: historical and philosophical underpinnings, systems thinking and interconnectedness, the human development and capability approach, and global governance and multilateralism. Each framework provides a lens through which to understand the SDGs' design, objectives, and implementation challenges.

4.1 Historical and Philosophical Underpinnings

The SDGs are rooted in a long history of philosophical thought and international development efforts. Philosophically, they draw on Enlightenment-era ideals of progress, universal morality, and collective improvement, as articulated by Immanuel Kant (1784) and John Stuart Mill (1863). Kant's concept of a "universal history" envisioned humanity progressing toward a shared moral community, a vision echoed in the SDGs' global commitment to equity and justice. Mill's utilitarianism, emphasizing the greatest good for the greatest number, informs the SDGs' focus on maximizing well-being across populations.

Historically, the SDGs build on decades of international development initiatives. The 1972 Stockholm Conference on the Human Environment marked an early recognition of the interplay between development and environmental sustainability, followed by the 1987 Brundtland Report, which popularized the concept of sustainable development (United Nations, 1987). The MDGs (2000–2015) provided a precursor to the SDGs, achieving notable successes in reducing extreme poverty and improving health and education but facing criticism for their narrow focus on developing countries (Sachs, 2012). The SDGs address these critiques by adopting a universal framework, applicable to all nations, and integrating social, economic, and environmental goals.

John Rawls's theory of justice (1971) provides a normative foundation for the SDGs, particularly the principle of "leaving no one behind." Rawls's "difference principle" advocates for policies that benefit the least advantaged, aligning with the SDGs' emphasis on inclusivity and equity. This philosophical grounding ensures that the SDGs prioritize marginalized populations, such as women, indigenous communities, and persons with disabilities, in their pursuit of global development (Pogge & Sengupta, 2016).

4.2 Systems Thinking and Interconnectedness

Systems thinking is a critical framework for understanding the SDGs' design and implementation. Unlike traditional linear approaches, systems thinking emphasizes the interdependence of complex systems, recognizing that actions in one area (e.g., agriculture) affect outcomes in others (e.g., water, climate, health) (Nilsson et al., 2018). The SDGs' 17 goals and 169 targets form a network of interconnected objectives, requiring integrated policy solutions to leverage synergies and mitigate trade-offs.

The nexus approach, as articulated by Liu et al. (2018), illustrates this interconnectedness by focusing on critical intersections, such as the water-energy-food nexus. For example, achieving Zero Hunger (Goal 2) depends on sustainable agricultural practices (Goal 12), which require clean water (Goal 6) and climate resilience (Goal 13). Similarly, affordable clean energy (Goal 7) supports industry and innovation (Goal 9) but must be balanced with environmental protection (Goals 14 and 15). Bengtsson et al. (2018) highlight the challenge of managing trade-offs, such as prioritizing economic growth (Goal 8) over ecological sustainability, which may exacerbate climate change or biodiversity loss.

Systems thinking also informs policy design by encouraging cross-sectoral collaboration. For instance, integrated urban planning can address multiple goals simultaneously, such as reducing inequalities (Goal 10), improving health (Goal 3), and promoting sustainable cities (Goal 11). However, implementing systems-based approaches requires robust data, stakeholder coordination, and institutional capacity, which are often lacking in developing countries (Allen et al., 2018). The application of systems thinking to the SDGs underscores the need for holistic, adaptive strategies that account for dynamic global systems.

4.3 Human Development and Capability Approach

The human development and capability approach, developed by Amartya Sen (1999) and Martha Nussbaum (2000), provides a normative framework for the SDGs, shifting the focus from economic growth to individual well-being and agency. Sen's concept of "development as freedom" emphasizes expanding individuals' capabilities to lead lives they value, through access to education, health, economic opportunities, and political participation. Nussbaum's framework of universal human entitlements complements this by identifying core capabilities, such as bodily health, affiliation, and control over one's environment, as essential for human dignity.

This approach underpins several SDGs, including Quality Education (Goal 4), Good Health and Well-Being (Goal 3), Gender Equality (Goal 5), and Reduced Inequalities (Goal 10). By prioritizing human capabilities, the SDGs aim to empower individuals, particularly marginalized groups, to overcome structural barriers. For example, Goal 5's focus on gender equality seeks to expand women's agency through education, economic empowerment, and political representation, aligning with Nussbaum's emphasis on universal entitlements (Fukuda-Parr, 2019).

The capability approach also informs the SDGs' commitment to inclusivity. By focusing on the least advantaged, as advocated by Rawls (1971), the SDGs ensure that development benefits reach vulnerable populations, such as rural communities, indigenous groups, and persons with disabilities. However, operationalizing this approach requires addressing systemic inequalities, such as access to quality education and healthcare, which remain significant challenges in many countries, including India (Drèze & Sen, 2017).

4.4 Global Governance and Multilateralism

Global governance theories provide a framework for understanding the SDGs' implementation through international cooperation. Keohane and Nye (2005) argue that multilateral institutions, such as the United Nations, play a critical role in setting norms, coordinating efforts, and fostering collective action. The SDGs were developed through a participatory process involving governments, civil society, and the private sector, reflecting a commitment to global partnership (Goal 17) (United Nations, 2015).

The principle of subsidiarity, as articulated by French and Kotzé (2018), ensures that SDG implementation is context-specific, allowing countries to tailor strategies to their unique social, economic, and environmental conditions. This flexibility is critical for diverse nations like India, where regional disparities require localized approaches. However, Caiado et al. (2018) highlight challenges to multilateralism, including power asymmetries between developed and developing nations, which can limit the voice of resource-constrained countries in global decision-making.

The SDGs also rely on innovative governance mechanisms, such as voluntary national reviews (VNRs), to monitor progress and share best practices. While VNRs promote accountability, their voluntary nature and varying reporting standards can lead to inconsistencies (Hák et al., 2016). Strengthening global governance requires addressing these gaps, enhancing institutional capacity, and fostering equitable partnerships to ensure that all nations contribute to and benefit from the SDG agenda.

5. DIMENSIONS OF THE SDGS

The SDGs are organized around three interconnected dimensions—social, economic, and environmental—reflecting their holistic approach to sustainable development. This section analyzes each dimension, drawing on theoretical frameworks and practical examples to illustrate their scope, synergies, and challenges.

5.1 Social Dimension

The social dimension of the SDGs focuses on human well-being, equity, and social justice, encompassing goals such as No Poverty (Goal 1), Good Health and Well-Being (Goal 3), Quality Education (Goal 4), Gender Equality (Goal 5), and Reduced Inequalities (Goal 10). These goals are grounded in social justice theories, particularly Rawls's "justice as fairness," which prioritizes equitable access to resources and opportunities for the least advantaged (Rawls, 1971).

No Poverty (Goal 1) aims to eradicate extreme poverty, defined as living on less than $1.90 per day, by 2030. Conditional cash transfer programs, such as Brazil's Bolsa Família, demonstrate the effectiveness of targeted interventions, reducing poverty while improving health and education outcomes (Soares et al., 2011). However, multidimensional poverty, encompassing health, education, and living standards, requires integrated approaches that address structural barriers (Alkire & Santos, 2014).

Good Health and Well-Being (Goal 3) seeks to ensure universal health coverage, reduce maternal and child mortality, and combat infectious diseases. The global response to HIV/AIDS, with increased access to antiretroviral therapy, illustrates progress, but challenges like non-communicable diseases and health system inequities persist (The Lancet, 2021). India's Ayushman Bharat scheme, providing health insurance to over 500 million people, exemplifies efforts to address Goal 3, though rural healthcare gaps remain (NITI Aayog, 2023).

Quality Education (Goal 4) emphasizes inclusive and equitable education, with targets for universal primary and secondary education. Finland's education system, with its focus on equity and teacher training, serves as a global model (Giangrande et al., 2019). In contrast, developing countries like India face challenges such as high dropout rates and gender disparities, addressed through initiatives like Samagra Shiksha Abhiyan (NITI Aayog, 2023).

Gender Equality (Goal 5) seeks to eliminate discrimination and empower women and girls. Programs like Bangladesh's microfinance initiatives demonstrate the transformative potential of economic empowerment, increasing women's agency and reducing poverty (Khandker, 2005). However, Leach (2016) argues that intersectional approaches are needed to address overlapping forms of discrimination, such as those based on caste, ethnicity, or disability.

Reduced Inequalities (Goal 10) aims to reduce income and social inequalities within and between countries. Progressive taxation and social protection systems, as implemented in Nordic countries, offer effective strategies, but global inequalities, exacerbated by trade policies and tax havens, remain a challenge (Piketty, 2014). India's affirmative action policies for marginalized castes and tribes align with Goal 10, though implementation gaps persist (Drèze & Sen, 2017).

The social dimension's interconnectedness creates synergies, such as linking education (Goal 4) with poverty reduction (Goal 1) and gender equality (Goal 5). However, trade-offs, such as prioritizing economic growth over social equity, can undermine progress, requiring careful policy design (Nilsson et al., 2018).

5.2 Economic Dimension

The economic dimension of the SDGs focuses on sustainable and inclusive economic growth, encompassing Decent Work and Economic Growth (Goal 8), Industry, Innovation, and Infrastructure (Goal 9), and Reduced Inequalities (Goal 10). Inclusive growth theories advocate for economic benefits that reach all segments of society, particularly marginalized groups (Griggs et al., 2014).

Decent Work and Economic Growth (Goal 8) promotes sustainable economic growth, full employment, and decent work for all. The International Labour Organization (ILO) estimates that achieving Goal 8 requires creating 600 million new jobs by 2030, particularly in developing countries (ILO, 2019). Initiatives like India's Skill India Mission, which trains youth for employment, align with Goal 8, though informal employment and wage gaps remain challenges (NITI Aayog, 2023).

Industry, Innovation, and Infrastructure (Goal 9) emphasizes resilient infrastructure, sustainable industrialization, and technological innovation. Rwanda's partnership with Zipline for drone-based medical deliveries illustrates how innovation can support multiple goals, including health (Goal 3) and infrastructure (Goal 9) (Zipline, 2022). However, developing countries face barriers such as high infrastructure costs and limited access to technology (UNCTAD, 2014).

Reduced Inequalities (Goal 10), as noted earlier, intersects with the economic dimension by addressing income disparities. Green growth policies, such as the European Union's Circular Economy Action Plan, demonstrate how economic progress can align with environmental sustainability, creating jobs while reducing waste (Avrampou et al., 2019). However, global economic systems, including trade imbalances and corporate tax avoidance, exacerbate inequalities, requiring reforms to align with Goal 10 (Piketty, 2014).

The economic dimension's synergies, such as linking innovation (Goal 9) with job creation (Goal 8), drive sustainable development. However, trade-offs, such as prioritizing industrial growth over environmental protection, pose risks, necessitating policies that balance economic and ecological priorities (Bengtsson et al., 2018).

5.3 Environmental Dimension

The environmental dimension of the SDGs addresses the planet's ecological limits, encompassing Clean Water and Sanitation (Goal 6), Affordable and Clean Energy (Goal 7), Climate Action (Goal 13), Life Below Water (Goal 14), and Life on Land (Goal 15). The planetary boundaries framework, introduced by Rockström et al. (2009), underscores the need to operate within safe ecological thresholds to avoid irreversible damage.

Clean Water and Sanitation (Goal 6) aims to ensure universal access to safe water and sanitation. India's Swachh Bharat Abhiyan, which constructed over 100 million toilets, significantly reduced open defecation, supporting Goals 6 and 3 (NITI Aayog, 2023). However, water scarcity and pollution remain challenges, particularly in urban areas (CPCB, 2022).

Affordable and Clean Energy (Goal 7) promotes access to renewable energy. Costa Rica's achievement of nearly 100% renewable energy production through hydropower, wind, and geothermal sources serves as a global model (Pagiola, 2008). India's leadership in the ISA advances Goal 7 by promoting solar energy in developing countries, though rural electrification gaps persist (MEA, 2023).

Climate Action (Goal 13) calls for urgent measures to combat climate change. The Paris Agreement, with its commitment to limiting global warming to 1.5°C, exemplifies global cooperation, but implementation lags in many countries (Fu et al., 2019). The Maldives' adaptation strategies, such as artificial island construction, highlight the challenges faced by climate-vulnerable nations (French & Kotzé, 2018).

Life Below Water (Goal 14) and **Life on Land (Goal 15)** focus on protecting marine and terrestrial ecosystems. Citizen science initiatives, as noted by Fraisl et al. (2020), contribute to monitoring biodiversity, but overfishing, deforestation, and pollution threaten progress. Costa Rica's Payment for Environmental Services (PES) program, which increased forest cover, demonstrates how economic incentives can support Goal 15 (Pagiola, 2008).

The environmental dimension's synergies, such as linking clean energy (Goal 7) with climate action (Goal 13), enhance sustainability. However, trade-offs, such as balancing energy access with ecological preservation, require integrated policies to ensure long-term viability (Liu et al., 2018).

6. INDIA'S ROLE IN ACHIEVING THE SDGS

India's demographic scale, economic growth, and global influence make it a linchpin in the global pursuit of the SDGs. As the world's most populous nation and a leading emerging economy, India's successes and challenges offer critical lessons for sustainable development. This section analyses India's policy integration, implementation challenges, and global contributions, highlighting its pivotal role in the 2030 Agenda.

6.1 Policy Integration and National Initiatives

India has mainstreamed the SDGs into its national development framework, with NITI Aayog, the government's policy think tank, overseeing implementation and coordination. The SDG India Index, launched in 2018, tracks progress across states and union territories, fostering accountability and competition (NITI Aayog, 2023). India's approach emphasizes aligning existing programs with SDG targets, ensuring coherence and efficiency. Key initiatives include:

- **Pradhan Mantri Garib Kalyan Yojana (PMGKY)**: Launched to address poverty and hunger (Goals 1 and 2), PMGKY provides financial aid, food security, and employment support to millions of vulnerable households. During the COVID-19 pandemic, it reached over 800 million people with free food grains, demonstrating scalability (NITI Aayog, 2023).
- **Swachh Bharat Abhiyan (Clean India Mission)**: Initiated in 2014, this flagship program aims to eliminate open defecation and improve sanitation, aligning with Clean Water and Sanitation (Goal 6) and Good Health and Well-Being (Goal 3). By 2019, it had constructed over 100 million toilets, significantly reducing open defecation rates (NITI Aayog, 2023).
- **International Solar Alliance (ISA)**: Co-founded by India and France, the ISA promotes solar energy in 121 member countries, advancing Affordable and Clean Energy (Goal 7) and Climate Action (Goal 13). India's leadership in the ISA positions it as a global advocate for renewable energy (MEA, 2023).
- **Beti Bachao Beti Padhao (Save the Girl Child, Educate the Girl Child)**: This initiative promotes gender equality (Goal 5) and quality education (Goal 4) by addressing female infanticide, improving sex ratios, and ensuring girls' access to education. It has shown measurable improvements in gender indicators in targeted districts (NITI Aayog, 2023).
- **Ayushman Bharat**: The world's largest health insurance scheme, Ayushman Bharat provides coverage to over 500 million people, supporting Good Health and Well-Being (Goal 3). It includes hospital care and preventive health services, addressing both curative and preventive needs (NITI Aayog, 2023).
- **Skill India Mission**: This program trains youth for employment, aligning with Decent Work and Economic Growth (Goal 8) and Industry, Innovation, and Infrastructure (Goal 9). It aims to skill 400 million workers by 2030, enhancing employability in a competitive global economy (NITI Aayog, 2023).

India's federal structure enables state-level innovation, with states like Kerala and Tamil Nadu leading in health and education outcomes, while others, like Uttar

Pradesh, face greater challenges (NITI Aayog, 2023). This diversity underscores the importance of localized strategies within a national framework.

6.2 Challenges in Implementation

Despite its ambitious initiatives, India faces significant hurdles in achieving the SDGs. These challenges reflect the country's complex socio-economic and environmental landscape and offer insights into barriers faced by other developing nations.

- **Persistent Poverty and Inequality**: While India has reduced extreme poverty, multidimensional poverty affects over 300 million people, particularly in rural areas and marginalized communities (Alkire & Seth, 2015). Income inequality, exacerbated by urbanization and informal employment, undermines Goal 10 (Drèze & Sen, 2017).
- **Environmental Degradation**: Rapid industrialization and urbanization have led to severe air and water pollution, threatening Goals 6, 13, 14, and 15. The Central Pollution Control Board (2022) reports that 70% of India's rivers are contaminated, and air quality in cities like Delhi frequently exceeds safe levels. Balancing economic growth with environmental sustainability remains a critical challenge.
- **Health Infrastructure Gaps**: The COVID-19 pandemic exposed vulnerabilities in India's health system, including shortages of hospital beds, ventilators, and rural healthcare facilities (The Lancet, 2021). While Ayushman Bharat addresses access, quality and availability of services need improvement to achieve Goal 3.
- **Financial Constraints**: Implementing the SDGs requires substantial investment, estimated at $1 trillion annually for India alone (NITI Aayog, 2023). Fiscal pressures, compounded by high public debt and competing priorities, limit funding for social and environmental programs (RBI, 2023).
- **Governance and Data Challenges**: Bureaucratic inefficiencies, corruption, and lack of coordination between central and state governments hinder implementation (Caiado et al., 2018). Data gaps, particularly for disaggregated indicators, impede monitoring and evaluation, especially for marginalized groups (Hák et al., 2016).
- **Social and Cultural Barriers**: Gender disparities, caste-based discrimination, and resistance to behavior change (e.g., sanitation practices) pose challenges to Goals 5, 6, and 10. For instance, Swachh Bharat Abhiyan faced resistance in some communities due to cultural preferences for open defecation (NITI Aayog, 2023).

These challenges highlight the need for sustained investment, inclusive policies, and adaptive governance to ensure India's SDG progress is equitable and sustainable.

6.3 Global Contributions

India's influence extends beyond its borders, positioning it as a leader in the global SDG agenda. Through South-South cooperation, India shares expertise and resources with other developing countries, particularly in Asia and Africa. The ISA, for example, supports solar energy projects in countries like Ethiopia and Fiji, advancing Goals 7 and 13 (MEA, 2023). India's advocacy for climate justice, emphasizing the principle of "common but differentiated responsibilities," amplifies the voice of developing nations in global climate negotiations (Sachs, 2015).

India also contributes to global knowledge-sharing through platforms like the United Nations and G20. Its voluntary national reviews (VNRs) provide insights into scaling development programs in diverse contexts, informing global best practices (NITI Aayog, 2023). Additionally, India's pharmaceutical industry, known as the "pharmacy of the world," supplies affordable generic medicines, supporting Goal 3 in low-income countries (MEA, 2023).

By balancing domestic priorities with global leadership, India demonstrates the potential for emerging economies to drive sustainable development. Its experiences offer lessons for other nations navigating the complexities of the SDG agenda.

7. CHALLENGES AND FUTURE DIRECTIONS

Achieving the SDGs by 2030 requires overcoming significant global and national challenges while leveraging opportunities for innovation and collaboration. This section analyzes key implementation barriers and proposes future strategies, drawing on the literature and India's experiences.

7.1 Implementation Barriers

The SDGs face a range of interconnected barriers that hinder progress across social, economic, and environmental dimensions.

- **Funding Gaps**: The global funding shortfall for the SDGs is estimated at \$2.5–\$3 trillion annually, with developing countries facing the greatest constraints (UNCTAD, 2014). Sectors like renewable energy, healthcare, and education require substantial upfront investments, which are often beyond the fiscal capacity of low- and middle-income countries (Schmidt-Traub, 2015).

In India, fiscal pressures limit the scalability of programs like Ayushman Bharat and Swachh Bharat Abhiyan (RBI, 2023).

- **Governance Issues**: Weak institutional frameworks, corruption, and lack of policy coherence undermine implementation. Caiado et al. (2018) note that fragmented governance structures lead to inefficiencies, such as overlapping mandates or conflicting priorities. In India, coordination between central and state governments remains a challenge, particularly for decentralized programs (NITI Aayog, 2023).
- **Data Gaps**: Inconsistent data collection and inadequate indicators hinder monitoring and evaluation. Hák et al. (2016) argue that many countries lack the capacity to collect disaggregated data, essential for tracking progress among marginalized groups. This is evident in India, where data on caste, gender, and disability is often incomplete (NITI Aayog, 2023).
- **Global Crises**: The COVID-19 pandemic, climate change, and geopolitical conflicts have disrupted SDG progress. The United Nations (2024) reports that 80% of targets are off track, with setbacks in poverty reduction, education, and health. In India, the pandemic reversed gains in poverty alleviation and strained health systems (The Lancet, 2021).
- **Trade-offs and Complexity**: The SDGs' interconnectedness creates trade-offs, such as prioritizing economic growth (Goal 8) over environmental protection (Goals 13–15). Nilsson et al. (2018) emphasize the need for tools to analyze these trade-offs, but such frameworks are underutilized in policy design. In India, industrial expansion often conflicts with environmental goals, as seen in polluted urban centers (CPCB, 2022).
- **Social and Cultural Barriers**: Resistance to behavior change, gender norms, and social hierarchies impede progress on goals like sanitation (Goal 6) and gender equality (Goal 5). In India, cultural preferences and caste dynamics have slowed the adoption of sanitation practices in some regions (NITI Aayog, 2023).

These barriers highlight the need for innovative, inclusive, and adaptive approaches to ensure the SDGs' success.

7.2 Future Strategies

To overcome these barriers, the following strategies are proposed, drawing on global best practices and India's experiences:

- **Innovative Financing**: Mobilizing resources through green bonds, blended finance, and public-private partnerships can close the funding gap. The

OECD (2020) highlights the potential of blended finance to attract private investment for infrastructure and renewable energy projects. India's green bond issuances, which raised $10 billion by 2023, offer a model for scaling sustainable finance (RBI, 2023).

- **Technological Advancements**: Leveraging technologies like artificial intelligence (AI), big data, and blockchain can enhance monitoring, efficiency, and transparency. Reuters (2024) notes that AI can optimize resource allocation for SDG programs, such as predicting health outbreaks or tracking deforestation. India's Digital India Campaign, which promotes e-governance and data-driven policymaking, can be expanded to support SDG monitoring (MEA, 2023).
- **Inclusive Governance**: Strengthening institutional capacity and promoting participatory governance can enhance implementation. Allen et al. (2018) advocate for multi-stakeholder platforms that include civil society, private sector, and local communities. India's SDG India Index fosters state-level accountability, but greater community involvement is needed to ensure inclusivity (NITI Aayog, 2023).
- **Strengthened Multilateralism**: Enhancing global cooperation through organizations like the UN and G20 can address transboundary challenges like climate change and inequality. India's leadership in the G20 and ISA demonstrates the value of multilateral platforms, but equitable representation of developing nations must be prioritized (MEA, 2023).
- **Localized and Context-Specific Solutions**: Tailoring SDG strategies to local contexts, as advocated by French and Kotzé (2018), ensures relevance and effectiveness. In India, state-specific approaches, such as Kerala's focus on health and education, can be scaled nationally to address regional disparities (NITI Aayog, 2023).
- **Behavioral and Cultural Interventions**: Addressing social and cultural barriers requires targeted campaigns to shift norms and behaviors. India's Swachh Bharat Abhiyan successfully used media and community engagement to promote sanitation, a strategy that can be applied to gender equality and education (NITI Aayog, 2023).
- **Transdisciplinary Research**: Encouraging collaboration across disciplines, such as economics, environmental science, and sociology, can develop innovative solutions to complex SDG challenges. Moallemi et al. (2020) emphasize the need for transdisciplinary frameworks to analyze trade-offs and synergies, which can inform policy design in India and beyond.

These strategies, if implemented effectively, can accelerate SDG progress, leveraging India's experiences as a model for other nations.

8. CASE STUDIES ON SDG IMPLEMENTATION

This section presents six in-depth case studies to illustrate the practical application of the SDGs across diverse contexts: Costa Rica, India, Bangladesh, Rwanda, the Maldives, and Finland. Each case study highlights specific goals, theoretical connections, implementation strategies, impacts, challenges, and lessons learned, with a focus on India's role and global examples to provide a comprehensive perspective. These cases demonstrate how theoretical frameworks translate into real-world outcomes, offering insights for policymakers and researchers.

8.1 Costa Rica: A Model for Environmental Sustainability

Goals Addressed: Affordable and Clean Energy (Goal 7), Climate Action (Goal 13), Life on Land (Goal 15)

Overview: Costa Rica is a global leader in environmental sustainability, achieving nearly 100% renewable energy production through hydropower, wind, and geothermal sources. Its Payment for Environmental Services (PES) program, launched in 1997, compensates landowners for conserving forests, reversing deforestation trends and enhancing biodiversity (Pagiola, 2008). This case aligns with systems thinking by integrating economic incentives with environmental outcomes, demonstrating how interconnected policies can achieve multiple SDGs.

Context: Costa Rica, a small Central American nation, has prioritized environmental sustainability since the 1990s, driven by its rich biodiversity and commitment to global climate goals. The PES program, funded through carbon taxes and international grants, reflects a systems-based approach, linking economic incentives (Goal 8) with ecological preservation (Goals 13 and 15).

Implementation Strategies:

- **Renewable Energy Investments**: Since the 1990s, Costa Rica has invested heavily in renewable infrastructure, with hydropower accounting for 70% of electricity, followed by wind (15%) and geothermal (10%) sources.
- **PES Program**: The program pays landowners to protect forests, using revenues from a 3.5% carbon tax on fossil fuels and international funding. By 2020, it covered over 1 million hectares.
- **Policy Integration**: National policies align energy, biodiversity, and climate goals, with the National Decarbonization Plan aiming for net-zero emissions by 2050.

Impact:

- Over 99% of electricity comes from renewable sources, reducing carbon emissions by 20% since 2000.
- Forest cover increased from 26% in the 1980s to over 50% by 2020, enhancing biodiversity and carbon sequestration.
- Ecotourism, supported by preserved ecosystems, contributes 8% to GDP, creating jobs and supporting Goal 8.
- Costa Rica's model has inspired similar PES programs in countries like Mexico and Ecuador.

Challenges:

- High upfront costs for renewable infrastructure require sustained investment.
- Balancing tourism growth with environmental preservation to prevent habitat degradation.
- Ensuring equitable benefits for rural and indigenous communities, who often face land tenure issues.
- Vulnerability to climate impacts, such as droughts affecting hydropower.

Lessons Learned: Costa Rica demonstrates the efficacy of systems thinking in aligning economic incentives with environmental goals. The PES program illustrates how financial mechanisms can support multiple SDGs, offering a model for countries like India, where biodiversity conservation and renewable energy are priorities. India's National Afforestation Programme could adopt similar incentive-based approaches to enhance Goal 15 (MEA, 2023).

Theoretical Connection: The case reflects the planetary boundaries framework, ensuring development stays within ecological limits (Rockström et al., 2009). It also aligns with global governance theories, as Costa Rica leverages international partnerships to fund its initiatives (Keohane & Nye, 2005).

8.2 India: Swachh Bharat Abhiyan (Clean India Mission)

Goals Addressed: Clean Water and Sanitation (Goal 6), Good Health and Well-Being (Goal 3), Gender Equality (Goal 5)

Overview: Launched on October 2, 2014, Swachh Bharat Abhiyan (SBA) aims to eliminate open defecation and improve sanitation across India, aligning with Clean Water and Sanitation (Goal 6) and Good Health and Well-Being (Goal 3). By combining infrastructure development with behavior-change campaigns, SBA addresses social and health dimensions of the SDGs, with a focus on gender equality (Goal 5) by prioritizing women's safety and dignity (NITI Aayog, 2023). The program reflects the capability approach by enhancing human dignity and agency, particularly for marginalized groups (Sen, 1999).

Context: India faced a sanitation crisis, with 38% of its population practicing open defecation in 2014, leading to waterborne diseases and safety risks, especially for women. SBA, driven by Prime Minister Narendra Modi's vision, is one of the world's largest sanitation initiatives, targeting both rural and urban areas.

Implementation Strategies:

- **Massive Infrastructure Push**: Over 100 million household and community toilets were constructed by 2019, supported by $20 billion in government funding.
- **Behavior-Change Campaigns**: Nationwide campaigns, including media, celebrity endorsements, and community events, promoted hygiene and toilet use, reaching over 1 billion people.
- **Gender-Sensitive Design**: Toilets were designed with women's safety in mind, including separate facilities in schools and public spaces to reduce harassment and dropout rates among girls.
- **Decentralized Implementation**: States and local governments tailored strategies to regional needs, with incentives for villages achieving open-defecation-free (ODF) status.

Impact:

- Open defecation rates dropped from 38% in 2014 to under 10% by 2020, with over 600,000 villages declared ODF.
- Reduced waterborne diseases, such as diarrhea, by 30%, improving health outcomes (Goal 3).
- Enhanced women's safety and mobility, with studies showing a 15% reduction in gender-based violence linked to sanitation access, supporting Goal 5.
- Improved school attendance among girls, contributing to Goal 4.

Challenges:

- Maintenance issues, with 20% of toilets non-functional due to inadequate water supply or poor upkeep.
- Resistance to behavior change in some rural communities, rooted in cultural preferences for open defecation.
- Financial constraints for sustaining infrastructure, particularly in low-income states.

- Urban sanitation gaps, with slums facing overcrowding and inadequate facilities.

Lessons Learned: SBA underscores the importance of participatory approaches and gender-sensitive policies in achieving social SDGs. The combination of infrastructure and behavior change offers a scalable model for other developing nations. India's experience highlights the need for sustained investment and local adaptation, as well as addressing cultural barriers through community engagement. Other countries, such as Nigeria and Pakistan, can draw on SBA's strategies to tackle sanitation challenges.

Theoretical Connection: The capability approach is evident in SBA's focus on enhancing human dignity and agency, particularly for women and children (Nussbaum, 2000). Systems thinking is reflected in the integration of health, sanitation, and gender outcomes (Nilsson et al., 2018).

8.3 Bangladesh: Empowering Women Through Microfinance

Goals Addressed: No Poverty (Goal 1), Gender Equality (Goal 5), Decent Work and Economic Growth (Goal 8)

Overview: Bangladesh's microfinance initiatives, led by organizations like Grameen Bank and BRAC, provide small loans to women entrepreneurs, enabling economic independence and poverty reduction. These programs align with the capability approach by expanding women's economic opportunities and agency, contributing to No Poverty (Goal 1), Gender Equality (Goal 5), and Decent Work and Economic Growth (Goal 8) (Sen, 1999).

Context: Bangladesh, one of the world's poorest nations in the 1970s, faced widespread poverty and gender inequality. Microfinance emerged as a transformative tool, with Grameen Bank, founded by Muhammad Yunus, pioneering group-based lending to women. By 2020, microfinance reached over 30 million borrowers, 90% of whom were women.

Implementation Strategies:

- **Microloans for Women**: Loans ranging from $50 to $500 were provided to women for small-scale enterprises, such as poultry farming and handicrafts, with repayment rates exceeding 95%.
- **Training and Support**: Financial literacy, vocational training, and business development programs empowered women to manage enterprises effectively.
- **Community Networks**: Group-based lending fostered social accountability, with women forming groups to support repayments and share knowledge.

- **Policy Support**: The Bangladesh government integrated microfinance into its poverty reduction strategy, providing regulatory support and subsidies.

Impact:

- Over 10 million women lifted out of poverty since the 1980s, reducing national poverty rates from 80% in 1971 to 14% by 2020.
- Increased female literacy and workforce participation, with women's labor force participation rising from 20% in 1990 to 36% in 2020, supporting Goals 4 and 5.
- Reduced child marriage rates by 25%, as economic empowerment delayed marriages, contributing to broader social change.
- Economic growth, with microfinance contributing 5% to GDP through small-scale enterprises.

Challenges:

- High interest rates, averaging 20–30%, burden some borrowers, leading to debt traps in rare cases.
- Limited access in remote areas, where infrastructure and banking services are scarce.
- Ensuring long-term economic sustainability, as many enterprises remain small-scale.
- Gender backlash, with some communities resisting women's economic empowerment.

Lessons Learned: Microfinance demonstrates the transformative potential of empowering women to achieve multiple SDGs. The focus on training and community networks enhances sustainability, offering lessons for India's Skill India Mission, which targets economic empowerment (NITI Aayog, 2023). Bangladesh's model can be adapted to other developing countries, such as Uganda and Cambodia, to address poverty and gender inequality.

Theoretical Connection: The capability approach underscores the importance of economic empowerment in expanding freedoms (Sen, 1999). Systems thinking is evident in the integration of economic, social, and educational outcomes (Nilsson et al., 2018).

8.4 Rwanda: Technological Innovations in Healthcare

Goals Addressed: Good Health and Well-Being (Goal 3), Industry, Innovation, and Infrastructure (Goal 9)

Overview: Rwanda's partnership with Zipline, a US-based drone company, uses drone technology to deliver medical supplies to remote areas, revolutionizing healthcare access. This initiative aligns with systems thinking by integrating technology with health and infrastructure goals, contributing to Good Health and Well-Being (Goal 3) and Industry, Innovation, and Infrastructure (Goal 9) (Zipline, 2022).

Context: Rwanda, a landlocked African nation, faced challenges in delivering medical supplies to remote areas due to poor infrastructure. Since 2016, Zipline's drone delivery system has transformed healthcare logistics, making Rwanda a leader in health innovation.

Implementation Strategies:

- **Drone Delivery Systems**: Drones deliver blood, vaccines, and emergency supplies to hospitals within 30 minutes, covering 80% of Rwanda's population.
- **Public-Private Partnerships**: The Rwandan government collaborates with Zipline, providing regulatory support and infrastructure, while Zipline funds operations.
- **Rural Focus**: The system prioritizes underserved regions, ensuring inclusivity and addressing health inequities.
- **Data Integration**: Real-time data tracks deliveries, optimizing supply chains and informing health policy.

Impact:

- Delivery times reduced from 4 hours to 30 minutes, saving lives in emergencies, such as postpartum hemorrhages.
- Expanded healthcare access, with 70% of blood deliveries reaching rural areas, supporting Goal 3.
- Positioned Rwanda as a global leader in health innovation, attracting investment and expertise (Goal 9).
- Inspired similar programs in Ghana, Nigeria, and India, scaling the model globally.

Challenges:

- High operational costs, with each drone delivery costing $20–$50, requiring sustained funding.
- Dependence on external technical expertise, limiting local capacity-building.
- Scaling to other sectors, such as agriculture or education, remains unexplored.
- Regulatory hurdles for expanding drone operations in other countries.

Lessons Learned: Rwanda's use of technology highlights the potential of innovation to address systemic challenges. The public-private partnership model offers a scalable approach for countries like India, where the Digital India Campaign could integrate drone technology to enhance healthcare delivery (MEA, 2023). Rwanda's success underscores the importance of government commitment and international collaboration in leveraging technology for the SDGs.

Theoretical Connection: Systems thinking is reflected in the integration of technology, infrastructure, and health systems (Nilsson et al., 2018). The capability approach is evident in the focus on expanding healthcare access to improve human well-being (Sen, 1999).

8.5 The Maldives: Adapting to Climate Change

Goals Addressed: Climate Action (Goal 13), Life Below Water (Goal 14)

Overview: The Maldives, a low-lying island nation in the Indian Ocean, faces existential threats from rising sea levels due to climate change. Its adaptation strategies, including artificial island construction and coral reef restoration, align with Climate Action (Goal 13) and Life Below Water (Goal 14). The case reflects global governance principles, as the Maldives relies on international partnerships to fund and implement its initiatives (French & Kotzé, 2018).

Context: With an average elevation of 1.5 meters, the Maldives is among the world's most climate-vulnerable nations. Rising sea levels threaten 80% of its land by 2050, endangering its population and marine ecosystems. Adaptation is critical to its survival and sustainable development.

Implementation Strategies:

- **Artificial Islands**: Projects like Hulhumalé, a reclaimed island, provide new land for housing and infrastructure, accommodating 70% of the population by 2030.
- **Coral Restoration**: Initiatives like the Coral Reef Restoration Project transplant corals to protect marine ecosystems and support tourism.
- **International Partnerships**: The Maldives secures funding and expertise from the UN, World Bank, and countries like Japan and India, leveraging global climate funds.

- **Policy Advocacy**: The Maldives leads advocacy for small island developing states (SIDS), pushing for stronger global climate commitments.

Impact:

- Enhanced resilience, with Hulhumalé housing 50,000 people and reducing overcrowding in Malé.
- Preserved marine biodiversity, with coral restoration supporting 60% of tourism revenue, aligning with Goal 14.
- Strengthened global advocacy, influencing the Paris Agreement's 1.5°C target.
- Improved disaster preparedness, reducing vulnerability to storms and flooding.

Challenges:

- High costs of adaptation, with Hulhumalé costing $400 million, requiring sustained international aid.
- Limited domestic resources, as the Maldives relies on tourism for 90% of GDP.
- Ongoing climate threats, with projections of 0.5–1 meter sea-level rise by 2100.
- Balancing development with environmental preservation to avoid ecosystem damage.

Lessons Learned: The Maldives underscores the importance of global cooperation for climate-vulnerable nations. India's ISA can support similar adaptation efforts in SIDS by providing renewable energy solutions (MEA, 2023). The case highlights the need for innovative financing and technology to address climate challenges, offering lessons for other coastal nations like Bangladesh and Vietnam.

Theoretical Connection: Global governance theories highlight the role of multilateralism in addressing transboundary challenges (Keohane & Nye, 2005). The planetary boundaries framework emphasizes the urgency of staying within ecological limits (Rockström et al., 2009).

8.6 Finland: A Model for Quality Education

Goals Addressed: Quality Education (Goal 4), Reduced Inequalities (Goal 10)

Overview: Finland's education system is globally renowned for its emphasis on equity, teacher training, and individualized learning, achieving high outcomes in literacy, math, and science. This case aligns with the capability approach by fostering social mobility and inclusivity, contributing to Quality Education (Goal 4) and Reduced Inequalities (Goal 10) (Giangrande et al., 2019).

Context: Finland, a Nordic country with a population of 5.5 million, transformed its education system in the 1970s to prioritize equity and excellence. Its consistent top rankings in the Programme for International Student Assessment (PISA) reflect its success in delivering inclusive education.

Implementation Strategies:

- **Teacher Training**: Teachers undergo rigorous master's-level training, ensuring high-quality instruction and professional autonomy.
- **Equitable Access**: Education is free at all levels, with meals, transportation, and materials provided to reduce financial barriers.
- **Inclusive Policies**: Support for students with disabilities, immigrants, and minorities ensures no child is left behind, aligning with Goal 10.
- **Student-Centered Learning**: Curricula emphasize critical thinking, creativity, and individualized instruction, fostering lifelong learning.

Impact:

- Consistently high PISA rankings, with 95% of students completing secondary education.
- Reduced inequalities, with minimal performance gaps between socioeconomic groups, supporting Goal 10.
- Enhanced economic opportunities, with 70% of graduates entering higher education or skilled jobs, contributing to Goal 8.
- Global influence, with Finland's model adopted in countries like Estonia and Singapore.

Challenges:

- Adapting to increasing diversity, as immigration rises, requiring culturally responsive teaching.
- Maintaining funding for universal education amid economic pressures.
- Scaling lessons to developing countries with limited resources, like India.
- Addressing mental health issues among students, exacerbated by academic expectations.

Lessons Learned: Finland's model demonstrates the transformative power of education in achieving social and economic SDGs. Its focus on equity and teacher quality offers lessons for India's Samagra Shiksha Abhiyan, which aims to improve educational outcomes (NITI Aayog, 2023). Developing countries can adapt Finland's strategies by prioritizing teacher training and inclusive policies, even with limited resources.

Theoretical Connection: The capability approach emphasizes education as a means to expand human potential (Sen, 1999). Systems thinking is reflected in the integration of education with social and economic outcomes (Nilsson et al., 2018).

9. CONCLUSION

The Sustainable Development Goals (SDGs) represent a transformative framework for addressing global challenges, grounded in systems thinking, the human development approach, and global governance theories. Their 17 goals and 169 targets integrate social, economic, and environmental dimensions, fostering synergies while navigating complex trade-offs. This article has provided a comprehensive analysis of the SDGs' theoretical foundations, dimensional analyses, India's pivotal role, implementation challenges, and practical applications through six in-depth case studies.

The theoretical foundations—spanning historical and philosophical underpinnings, systems thinking, the capability approach, and global governance—provide a robust framework for understanding the SDGs' design and objectives. Systems thinking highlights the interconnectedness of goals, necessitating integrated policies, while the capability approach emphasizes human well-being and agency, ensuring inclusivity. Global governance theories underscore the importance of multilateralism, though power asymmetries and resource constraints pose challenges.

India's role in the SDG agenda is critical, given its demographic scale, economic growth, and global influence. Initiatives like Swachh Bharat Abhiyan, the International Solar Alliance, and Ayushman Bharat demonstrate India's commitment to integrating the SDGs into its national framework, addressing poverty, sanitation, health, and renewable energy. However, challenges such as persistent poverty, environmental degradation, health infrastructure gaps, and financial constraints highlight the need for sustained investment and adaptive governance. India's global contributions, through South-South cooperation and climate advocacy, position it as a leader in sustainable development, offering lessons for other nations.

The expanded literature review synthesizes interdisciplinary perspectives, highlighting the SDGs' theoretical foundations, implementation barriers, and critical debates. It underscores the need for transdisciplinary approaches, innovative financing, and localized solutions to address the SDGs' complexity. The case studies—Costa Rica's environmental sustainability, India's sanitation revolution, Bangladesh's microfinance success, Rwanda's healthcare innovation, the Maldives' climate adaptation, and Finland's education model—illustrate the practical application of theoretical frameworks across diverse contexts. These cases demonstrate the importance of systems-based policies, participatory approaches, and global partnerships in achieving the SDGs.

Key challenges, including funding gaps, governance issues, data limitations, and global crises, threaten progress toward the 2030 Agenda. Proposed strategies—such as innovative financing, technological advancements, inclusive governance, and strengthened multilateralism—offer pathways to overcome these barriers. India's experiences, balancing scale and diversity, provide a blueprint for implementing these strategies, emphasizing the importance of context-specific solutions and community engagement.

In conclusion, the SDGs offer a visionary framework for a just, equitable, and sustainable future. By aligning theoretical insights with practical strategies, the global community can address the complexities of sustainable development, leveraging India's contributions and global best practices. Achieving the SDGs by 2030 requires collective action, innovation, and a commitment to inclusivity, ensuring that no one is left behind in the pursuit of a better world.

REFERENCES

Aayog, N. I. T. I. (2023). *SDG India Index 2023–24*. Government of India.

Alkire, S., & Santos, M. E. (2014). Measuring acute poverty in the developing world: Robustness and scope of the multidimensional poverty index. *World Development, 59*, 251–274. DOI: 10.1016/j.worlddev.2014.01.026

Allen, C., Metternicht, G., & Wiedmann, T. (2018). Initial progress in implementing the Sustainable Development Goals (SDGs): A review of evidence from countries. *Sustainability Science, 13*(5), 1453–1467. DOI: 10.1007/s11625-018-0572-3

Avrampou, A., Skouloudis, A., Iliopoulos, G., & Khan, N. (2019). Advancing the Sustainable Development Goals: Evidence from leading European banks. *Sustainable Development (Bradford), 27*(4), 743–757. Advance online publication. DOI: 10.1002/sd.1938

Bengtsson, M., Alfredsson, E., Cohen, M., Lorek, S., & Schroeder, P. (2018). Transforming systems of consumption and production for achieving the sustainable development goals: Moving beyond efficiency. *Sustainability Science, 13*(6), 1533–1547. DOI: 10.1007/s11625-018-0582-1

Caiado, R., Filho, W., Quelhas, O., Nascimento, D., & Ávila, L. (2018). A literature-based review on potentials and constraints in the implementation of the sustainable development goals. *Journal of Cleaner Production, 198*, 1276–1288. Advance online publication. DOI: 10.1016/j.jclepro.2018.07.102

Central Pollution Control Board (CPCB). (2022). *Annual Report on Environmental Status in India*. Government of India.

Daly, H. E. (1996). *Beyond Growth: The Economics of Sustainable Development*. Beacon Press.

Drèze, J., & Sen, A. (2017). *An Uncertain Glory: India and its Contradictions*. Princeton University Press.

Escobar, A. (2015). Degrowth, postdevelopment, and transitions: A preliminary conversation. *Sustainability Science, 10*(3), 451–462. DOI: 10.1007/s11625-015-0297-5

Fraisl, D., Campbell, J., See, L., Wehn, U., Wardlaw, J., Gold, M., Moorthy, I., Arias, R., Piera, J., Oliver, J. L., Masó, J., Penker, M., & Fritz, S. (2020). Mapping citizen science contributions to the UN sustainable development goals. *Sustainability Science, 15*(6), 1735–1751. DOI: 10.1007/s11625-020-00833-7

French, D., & Kotzé, L. (2018). *Sustainable Development Goals: Law, Theory and Implementation*. Edward Elgar Publishing., DOI: 10.4337/9781786438768

Fu, B., Wang, S., Zhang, J., Hou, Z., & Li, J. (2019). Unravelling the complexity in achieving the 17 sustainable-development goals. *National Science Review*, *6*(3), 386–388. DOI: 10.1093/nsr/nwz038

Fukuda-Parr, S. (2019). Keeping out extreme inequality from the SDG agenda: The politics of indicators. *Global Policy*, *10*(S1), 61–69. DOI: 10.1111/1758-5899.12602

Giangrande, N., White, R., East, M., Jackson, R., Clarke, T., Saloff Coste, M., & Penha-Lopes, G. (2019). A competency framework to assess and activate education for sustainable development: Addressing the UN Sustainable Development Goals 4.7 challenge. *Sustainability (Basel)*, *11*(10), 2832. Advance online publication. DOI: 10.3390/su11102832

Griggs, D., Smith, M., & Rockström, J.. (2014). An integrated framework for sustainable development goals. *Ecology and Society*, *19*(4), 49. DOI: 10.5751/ES-07082-190449

Hák, T., Janoušková, S., & Moldan, B. (2016). Sustainable Development Goals: A need for relevant indicators. *Ecological Indicators*, *60*, 565–573. DOI: 10.1016/j.ecolind.2015.08.003

International Labour Organization (ILO). (2019). *World Employment and Social Outlook: Trends 2019*. ILO Publications.

Kant, I. (1784). *An Answer to the Question: What is Enlightenment?* Penguin Classics.

Keohane, R. O., & Nye, J. S. (2005). *Power and Interdependence*. Longman.

Khandker, S. R. (2005). Microfinance and poverty: Evidence using panel data from Bangladesh. *The World Bank Economic Review*, *19*(2), 263–286. DOI: 10.1093/wber/lhi008

Kioupi, V., & Voulvoulis, N. (2019). Education for Sustainable Development: A systemic framework for connecting the SDGs to educational outcomes. *Sustainability (Basel)*, *11*(21), 6104. Advance online publication. DOI: 10.3390/su11216104

Kostoska, O., & Kocarev, L. (2019). A novel ICT framework for Sustainable Development Goals. *Sustainability (Basel)*, *11*(7), 1961. Advance online publication. DOI: 10.3390/su11071961

Leach, M. (2016). *Gender equality and sustainable development: A pathways approach*. Routledge.

Liu, J., Hull, V., Godfray, H. C. J., Tilman, D., Gleick, P., Hoff, H., Pahl-Wostl, C., Xu, Z., Chung, M. G., Sun, J., & Li, S. (2018). Nexus approaches to global sustainable development. *Nature Sustainability*, *1*(9), 466–476. DOI: 10.1038/s41893-018-0135-8

Mill, J. S. (1863). *Utilitarianism*. Oxford University Press.

Ministry of External Affairs (MEA). (2023). *India's Role in Global Sustainable Development*. Government of India.

Moallemi, E., Malekpour, S., Hadjikakou, M., Raven, R., Szetey, K., Ningrum, D., Dhiaulhaq, A., & Bryan, B. A. (2020). Achieving the Sustainable Development Goals requires transdisciplinary innovation at the local scale. *One Earth*, *3*(3), 300–313. Advance online publication. DOI: 10.1016/j.oneear.2020.08.006

Morton, S., Pencheon, D., & Squires, N. (2017). Sustainable Development Goals (SDGs), and their implementation: A national global framework for health, development and equity needs a systems approach at every level. *British Medical Bulletin*, *124*(1), 81–90. DOI: 10.1093/bmb/ldx031

Nilsson, M., Chisholm, E., Griggs, D., Howden-Chapman, P., McCollum, D., Messerli, P., Neumann, B., Stevance, A. S., Visbeck, M., & Stafford-Smith, M. (2018). Mapping interactions between the sustainable development goals: Lessons learned and ways forward. *Sustainability Science*, *13*(6), 1489–1503. DOI: 10.1007/s11625-018-0604-z

Nussbaum, M. C. (2000). *Women and Human Development: The Capabilities Approach*. Cambridge University Press. DOI: 10.1017/CBO9780511841286

Organisation for Economic Co-operation and Development (OECD). (2020). *Blended Finance for Sustainable Development*. OECD Publishing.

Pagiola, S. (2008). Payments for environmental services in Costa Rica. *Ecological Economics*, *65*(4), 712–724. DOI: 10.1016/j.ecolecon.2007.07.033

Piketty, T. (2014). *Capital in the Twenty-First Century*. Harvard University Press. DOI: 10.4159/9780674369542

Pogge, T., & Sengupta, M. (2016). Assessing the Sustainable Development Goals from a human rights perspective. *Journal of International and Comparative Social Policy*, *32*(2), 83–97. DOI: 10.1080/21699763.2016.1198268

Rawls, J. (1971). *A Theory of Justice*. Harvard University Press. DOI: 10.4159/9780674042605

Reserve Bank of India (RBI). (2023). *Report on State Finances*. RBI Publications.

Reuters. (2024). How AI can supercharge the Sustainable Development Goals. https://www.reuters.com/sustainability/society-equity/comment-heres-how-we-can-use-ai-supercharge-sustainable-development-goals-2024-07-23

Rockström, J., Steffen, W., Noone, K., Persson, Å., Chapin, F. S.III, Lambin, E. F., Lenton, T. M., Scheffer, M., Folke, C., Schellnhuber, H. J., Nykvist, B., de Wit, C. A., Hughes, T., van der Leeuw, S., Rodhe, H., Sörlin, S., Snyder, P. K., Costanza, R., Svedin, U., & Foley, J. A. (2009). A safe operating space for humanity. *Nature, 461*(7263), 472–475. DOI: 10.1038/461472a

Sachs, J. D. (2012). From Millennium Development Goals to Sustainable Development Goals. *Lancet, 379*(9832), 2206–2211. DOI: 10.1016/S0140-6736(12)60685-0

Sachs, J. D. (2015). *The Age of Sustainable Development*. Columbia University Press. DOI: 10.7312/sach17314

Schmidt-Traub, G. (2015). *Investment needs to achieve the Sustainable Development Goals*. Sustainable Development Solutions Network.

Sen, A. (1999). *Development as Freedom*. Oxford University Press.

Soares, F. V., Ribas, R. P., & Osório, R. G. (2011). Evaluating the impact of Brazil's Bolsa Família: Cash transfer programmes in comparative perspective. *Latin American Research Review, 45*(2), 173–190. DOI: 10.1017/S0023879100009390

The Lancet. (2021). India's COVID-19 crisis: A call for international action. *Lancet, 397*(10286).

UNCTAD. (2014). *World Investment Report 2014: Investing in the SDGs. United Nations Conference on Trade and Development.*

United Nations. (1987). *Our Common Future: Report of the World Commission on Environment and Development*. United Nations.

United Nations. (2015). *Transforming our world: The 2030 Agenda for Sustainable Development*. United Nations General Assembly.

United Nations. (2024). *SDG Progress Report 2024*. United Nations Publications.

Zipline. (2022). *Revolutionizing Healthcare Delivery in Rwanda*. Zipline Publications.

Chapter 2
Peace, Politics, and SDGs: Analysing Regional Disturbances to Sustainable Development

Sheikh Inam Ul Mansoor
https://orcid.org/0000-0001-8636-4769
Symbiosis Law School Hyderabad Campus, Symbiosis International University, Pune, India

Mumtaz Ahmad Shah
https://orcid.org/0000-0003-4108-8339
Presidency University, Bangalore, India

Showkat Ahmad Wani
https://orcid.org/0009-0009-0240-9073
Alliance University, Bangalore, India

Sofiul Ahmed
https://orcid.org/0009-0004-4910-8415
Alliance University, Bangalore, India

Govind Yadav
https://orcid.org/0009-0005-5757-5619
Manipal University Jaipur, India

ABSTRACT

This chapter investigates the difficult nexus between peace, politics and sustainable development looking at how regional disturbances interfere with the culmination of the SDGs. It examines how instability in politics, economy, environment and human rights violations contribute to the perpetuation of systemic challenges. Critical assessment is given to the legal and policy frameworks that govern governance, trade and

DOI: 10.4018/979-8-3373-1300-9.ch002

environmental sustainability, to identify gaps that increase the regional conflict. This study emphasizes about how economic sanctions, resource scarcity and governance failures affect the long run stability. It also highlights the need for legal reforms, cooperation at the international level, as well as institutional accountability, and this is achieved by means of an evaluation of case studies and policy responses. It recommends strategic interventions in form of legal safeguards, equitable economic policy and environmental resilience to mitigate the regional conflicts and to promote sustainable development in a legalized and political globalized world.

INTRODUCTION

The line between contemporary global governance and peace, politics, and sustainable development is a defining characteristic. Stemming from the Leadership Document that incorporates the United Nations 2030 Agenda on Sustainable Development Goals (SDGs) are words that proclaim the necessity of achieving inclusive and peaceful societies as inherent to promote economic growth, social cohesion and environmental sustainability *(Sorooshian, 2024)*. Such impediments include armed conflicts, political instability, economic disruptions and failures in governance, which impede development, violate fundamental principles of international law and have networking consequences *(Nguyen et al., 2023)*. The principles of state sovereignty in the Charter of the United Nations is often at loggerheads with transnational exigencies of peace and security. Civil war, internal conflicts, and insurgencies destroy the framework of norms established under international human rights law and international humanitarian law, systemic human rights violations and mass displacements followed *(Núñez, 2024)*. Ensuring the rule of law in conflict ridden regions is instrumental in stemming governance deficits, fuelling authoritarian tendencies, and using participatory decision making which is essentially democratic to provide the long term sustainability *(Amankwaah, 2023)*. In addition, violations of jus cogens norms, crimes against humanity and genocide, among others, do not only lead to regional instability but also trigger international legal intervention in affected regions that alter their political and economic landscape *(Trahan, 2020)*. The economic effects of disturbances are extensive at a regional level, and these can be realized through trade shocks, outflow of capital and reduced foreign direct investment. States in constant political turmoil, however, are affected by systemic effects of economic recession *(Matta et al., 2022)*. However, for sustainable development to occur stable finances and regulatory frameworks are necessary to secure financial stability in the long run. It makes economic opportunities constrained, and spirals of poverty especially by international sanctions and embargoes, per se disintegrate commercial infrastructure *(Dinh et al., 2024)*. A broken principle of *pacta sunt servanda* that functions to uphold international trade agreements in that conflict zones make treaty obligations unworkable. Additionally, unilateral coercive measures, or unilateral economic sanctions, carried out by dominant economies often further complicate their citizens' economic deprivation to the point of their inability to comply with SDG related target indicators (Tuori, 2023). As many would point out, next is environmental degradation, an often neglected contribution to the breakdown of sustainable development. Fighting causes ecological devastations such as deforestation, contamination of water and loss of biodiversity, which heightens the vulnerability of places already at risk to climate stresses (Meaza et al., 2024). The destruction of the natural environment by human acts with the specific intent of damaging, destroying, or uninstalling the environment is acknowledged by the Rome Statute of the International Criminal Court as a war crime, which shows the necessity of including environmental justice and international legal accountability (Harhoff & Freeland, 2017). Furthermore, the principle of common but differentiated responsibilities is, in general, violated in conflict zones where environmental governance falls apart with the physical and political militarization (Stone, 2004).

The relation between the violation of human rights and sustainable development is strikingly sensitive when considering forced migration and refugee crises. The impact of large scale displacements as a result of armed conflicts impose excessive burden on host countries, most especially in the Global South where resource constraints further hampers development efforts (Stone, 2004). Non-refoulement too, the key refugee law principle, ran into such challenges of balancing the right of states to preserve their sovereignty and the humanitarian duties and social justice has suffered through systemic discrimination and marginalization (Lülf, 2016). Human rights discourse is in the business of making the principle of substantive equality elusive in such places as regions, marred by sectarian strife and ethno-political divisions (Vento, 2024). For the most part, legal frameworks for regional stability and sustainable development are limited because of geopolitics; UN Security Council's permanent members draw veto power, a paralysis and selective enforcement of international norms all based on their particular interests (Orford, 2021). Concerns about state sovereignty, as well as concerns about the risk of hegemonic interventions, have led the Responsibility to Protect (R2P) doctrine to be rejected as a mechanism for the prevention of mass atrocities (Huttunen, 2024). Furthermore, due to structural limitations of enforcement of compliance with mandates pertaining to peace and security in regional organizations such as the African Union and Association of Southeast Asian Nations (ASEAN), it is a need for future consideration on internal legal mechanisms to tackle transnational problems (Morton et al., 2017). Therefore, there is urgent need for multidisciplinary analysis of interplay between regional disturbances and sustainable development of which legal, political and economic considerations are vital (Moallemi et al., 2020). Much of what is described above takes place against the very principle of sustainable development as set forth in various international treaties and declarations that explicitly recognize the indivisibility of peace and progress. Yet, the long life of armed conflicts, governance failures and economic instability pose important questions as to the utility of current legal and policy alternatives to reduce regional instability (Enayati et al., 2024). However, skyrocketing numbers of people are still based on the idea of need for reliable, legally enforceable mechanisms to enforce peace, foster equitable economic development and maintain environmental sustainability (Kantorowicz-Reznichenko, 2020).

Considering that the term regional disturbance is applied quite broadly, this study seeks to illuminate aspects of such disturbance within the wider context of global peace as well as stability (Turner et al., 2003). The legal frameworks beside the conflict affected regions are portrayed in a critical analysis and judgements are also made on policy interventions for the balance of development and geopolitical realities. This Chapter examines international law's inefficacy for technological change related to its contribution to peace and sustainability jurisprudence. It draws on doctrinal research, to compare legal instruments, as well as case studies, in order to address governance, political and economic problems.

REGIONAL CONFLICTS AND POLITICAL INSTABILITY

Political instability and regional conflicts for long have hindered the process of sustainable development, the progress of economic development and institutional governance. Geopolitical tension, ethnic divisions or the sheer collapse of governance lead to instability most often (Miljkovic and Rimal, 2008). All these conflicts erode legal and institutional frameworks, disrupt rule of law and limit the economic growth (Shaffer and Sandholtz, 2024). The results far exceed the conflict zones and spill over to adjacent regions and beyond (Yi, Luo et al. 2020). Therefore, institutions of the weak states are unable to contain insurgencies, civil wars and political dramas that obstruct the implementation of Sustainable Development Goals (SDGs) (Makinda, 1998). Ethnic and sectarian divisions, historical grievances, and exclusionary politics fuel unrest (Garretsen et al., 2025, Syna, 2020). Often, such post-colonial states with arbitrary borders lack governance (Parashar & Schulz, 2021). Furthermore economic inequalities and resource rivalries increase tensions as resource rich regions are prone towards instability particularly because of the so-called 'resource curse' (Thorbecke & Charumilind, 2002; Mlachila et al., 2017). And the very notion that there is 'order' in the region is only possible because of very weak legal and institutional frameworks that are not even complying with the basic principles of contract, property rights and the most fundamental rights of human being. Absence of enforceable rules in such fragile legal systems enables the non-state actors to flourish (Chopra & Isser, 2012). In such conflict prone areas, where ideological shifts drown out the judiciary, legal certainty is violated. This puts in place a context of judicial independence which is gradually eroded and led to the politicization of legal systems, leaving widespread impunity and thus allows the violence and corruption perpetrators to be free of consequence (Le et al., 2022).

Political elites in power will usually resist transitional justice mechanisms (such as truth commissions and special tribunals) to the extent of preserving the existing power structure. Escalation of the tensions and inability to resolve disputes peacefully makes the diplomacy and conflict resolution ever more difficult (Hinnebusch, 1981). For instance, Syria's days-long civil war has ruined its economic infrastructure and millions of its people migrated as real humanitarians' crisis that has transnational effects. Large areas have collapsed into governance to the extent that there are armed groups (Staněk, 2017). If international humanitarian law has been violated, geopolitical complications have thwarted effective intervention (Kešeljević & Spruk, 2024). In many neighbouring countries including Lebanon, Jordan, and Turkey, there is a struggle to accommodate refugees and the resources provided stretch them thin while adding fuel to already present socio-political tensions (Suliman et al., 2023). Host countries compromise on the principle of non-refoulement as they find an equilibrium between geopolitical and domestic political pressures (Jakulevičienė, 2023). Similar, there have been years of conflict in the Democratic Republic of the Congo (or DRC) which has impeded economic and institutional development. On-going violence is fuelled by the vast mineral wealth of which its armed factions and transnational criminal networks exploit (Mufungizi, 2024; Oore-Ofe & Ojakorotu, 2023). Lacking functional institutions, most of the principle of state sovereignty over natural resources remains theoretical (Visser, 1988). Yet further weakening of central authority, in the form of the unchecked rise of non-state actors, makes the goal of sustainable development an elusive goal (Akinyetun, 2024). Because of weak institutions in some states, they find it challenging to manage conflicts and provide public services; such states face governance challenges; regional instability is still driven by governance challenges (Gavsker, 2021). When leaders who lack effective mechanisms of governance are in power, democracy cannot rein them in; authoritarian governments can rise through coercion and patronage (Ishiyama, 2019). Repression by state sanctioned forces is sanctioned meaning that political opponents are curtailed, electoral integrity is compromised, and also press freedoms. The Responsibility to Protect (R2P) doctrine is still contentious for the tensions between sovereignty and intervention of the nation (Helms, 2021). Failure undermines international legal frameworks and it is likely to increase governance failures (Binder, 2009). Patronage networks that corruptively support each other are a channel not only of patronage, but also of corruption's spread, which corrodes citizens' trust in the system and in their economic progress. The lack of weak oversight mechanisms also helps further financial crimes and governance failures (Zhang et al., 2023; Sandage, 2014).

But often, regional and international organizations encounter politics issues and diplomatic issues in dealing with governance deficits. The principles of non-interference and state sovereignty limit the scope of feasible regional action as the AU, EU and ASEAN mediate conflicts. (Manulak 2024). The United Nations Security Council has not been able to resolve prolonged wars, and the inability of international governance to solve conflicts underlines the lack of international power where geopolitical powers dictates the decision-making (Gould & Rablen 2017). The application of the international legal principles is selective and this leads to a different conflict responses, questioning the legitimacy of the institutions of global governance (Pisillo Mazzeschi & Viviani, 2018). The policy gaps exacerbate local instability, as the failures on crisis resolution and governance reforms turn worse crises. Despite frameworks, peace and sustainable development, the agreement is weakly enforced because international action is not strong enough. For conflict prevention to be effective, it is essential to have local ownership and participative governance as well as inclusive economic policies (Bhattacharya, 2015). International legal norm, as opposed to domestic policing, is bridged by strengthening the institutional capacity through the provision of legal reforms and establishing the accountability mechanisms (den Boer, 2018). Rule of law, adherence to human rights and resilient governance structures, all constituting an element of sustainable development (Ades & Chua, 1997). Securing long term security must take precedence over short term geopolitical interest (Kenny et al., 2009). The problem, traditional peace politics, development, and realization are deeply intertwined and political realities have a way of distorting their effectiveness (Tripathi & Roepstorff, 2020).

ECONOMIC DISRUPTIONS AND DEVELOPMENT CHALLENGES

For a long time, political instability, regional conflicts and governance deficiencies have all been intricately linked with development and economic disruptions. Conflicts, trade restrictions and structural governance deficits cause the destabilisation of the economic structures and prevent the achievement of the sustainable development goals, aggravating the poverty, unemployment and socioeconomic inequalities (Góes & Bekkers, 2023). Resilience of national economies is especially sensitive to the interplay between the economic policy decisions and legal frameworks in the regions subject to the geopolitical volatility. When trade barriers, financial instability, and economic sanctions occur, the principles of international economic law like trade liberalization, fair competition and non-discrimination compete with each other as they contended with one another (Garg et al., 2024).

External or external disturbances lead to the decline of economic stability, which destroy the basic lay of economic expansion such as labor market, foreign direct investment and industrial productivity (Le et al., 2023). Economic sanctions and economic barriers resulting in changes to the dynamics of the international commerce and market access for the affected nations is one of the foremost impediments of economic stability (Egger et al., 2024). Impositions of trade restrictions in the form of tariffs, non-tariff measures, or even embargoes distort supply chains around the world, and limit economic possibilities for all of the economies of developing countries (Orefice, 2017). Protectionist policies adopted by masquerading itself as national security or political triumph very often breach with the precepts of the General Agreement on Tariffs and Trade (GATT) and World Trade Organization (WTO) principles, advocating for non-discriminatory trade practices (Zahoor et al., 2023). The combination of economic sanctions as a foreign policy tools severely widens the circle of distress to economies in general and that of the ones who are susceptible to sanctions in particular (Morgan & Schwebach, 1995). And regardless of their assertion that the sanctions they impose are 'unilateral,' these sanctions always disproportionately impact civilian populations in much the same way, excluding them from obtaining basic goods, all while adamantly barring them from regularly enjoying much technological advancements and access to basic services such as the corresponding finances (Bogdanova, 2022). Such measures are heavily debated as far as the legal and ethical implications are concerned and as to how they stand on state sovereignty as well as the principle of proportionality of international economic law (Jessen, 2021).

Sanctions and embargoes cancel out trade networks, augment import dependence and narrow down the employment alternatives in the export dependent economies (Özdamar & Shahin, 2021). Capital inflows into infrastructure and industrial development are restricted to a great extent due to restricted access to the international financial markets. Therefore, in response to sanctions, sanctioned states typically rely on informal trade milieus, such as barter circulation and parallel financial networks, which however are unsustainable over the long term (Gutmann et al., 2023). It has wide profound repercussions way beyond immediate monetary losses. It puts long term economic dynamics in a bleak cage, that of stifling innovation, technological advancement, and the essential reforms to withstand shocks and crises (Conroy & Blackie, 2006). Further, extraterritorial sanctions affect third parties that work with the sanctioned states, discourage multinational corporations from collaborating and impede global trade (Nagel, 2023). However, such measures frequently meet challenges from legal interests represented by bodies such as the International Court of Justice (ICJ) and WTO's dispute settlement mechanisms, because there is a conflict between political interests and economic rights (Šturma, 2024). The disruption to the economy also leads to the degradation of infrastructure and the public services, adding to the development exposure. Armed conflicts, political instability and financial crises are to affect transport, electricity grids and communication systems thus decrease productivity and public welfare (Palei, 2015). These systems need a lot of financial and technical support from external actors to rebuild. Weak public service delivery in health, education and social security makes vulnerable people even more so, especially those at the bottom end of the social spectrum (Khan et al., 2023). The social discontent and political unrest are fuelled by state incapacity to provide what are considered necessary services and erode the public trust in governance (Thacker et al., 2019). Resource diversion from development priorities due to economic mismanagement, corruption and conflict related spending leads to worsening of poverty and inequality (Rus, 2014). The wide development gap is further widened by limited government capacity to invest in sustainable projects (Mansoor, 2025). Furthermore, illicit financial flows, in the form of capital flight and tax evasion affect the economic stability of a country by draining such much needed revenues for public investment and social welfare (Perez et al., 2012).

International financial institutions (IFIs) are a primary institution to prevent and to deal with economic crises and to promote the post crisis recovery however a financial aid, a policy guidance, and a structural reform programme (Campos & Kinoshita, 2010). Concessional loans, technical assistance, and policy frameworks are provided to countries by institutions like the World Bank, the International Monetary Fund, and regional development banks in an attempt to stabilize the macroeconomic condition and balance of payments challenges (Güven, 2012). However, the socio economic consequences of such conditions are high, attached conditions of financial aid like austerity measures, fiscal discipline, and structural adjustments. However, these obligations raise the issue of economic sovereignty and the welfare of social equity, which falls in line with the balance between financial stability (Moosa & Moosa, 2019). Apart from providing financial support, IFIs also influence the way in which the economic governance of the world is done: they determine trade, investment and development policies (Kranke, 2020). It is to enforce the international economic norms such as transparency, the anti-corruption measures, and the regulatory compliance to assure stability (Spyromitros & Panagiotidis, 2022). However, there are instances where geopolitical influences over the multilateral financial structures provide greater influence of dominant economies, thus reducing equitable representation and accountability (Ceva & Ferretti, 2021). By moving forward, IFI programs should perform debt sustainability assessments and assesses social impact including integrate inclusive policies to guarantee financial aid foster the growth and development of an equitable manner (Chewpreecha et al., 2023; Damane & Ho, 2024). Economic disruptions and governance failures are what makes it absolutely important to have comprehensive legal and policy framework which is based on economic resilience, institutional transparency and social equity. (Katsaliaki et al., 2022) One of the pillars which has continued to hold economic stability from the rule of law, this allows that legal norms defining trade, finance, and investment are in existence, and that if they are they must take place in a manner not to promote inequality nor cause ecological ruin. (Purvis et al., 2019) To minimize the disruptions to the smooth function of the economy in any circumstance, there continues to be a demand for the involvement of economic experts, legal practitioners, and public policymakers in continuous engagement in the process of framing sound mechanisms for prevention and recovery of economic crises. (Berk et al., 2023) In the face of an evolving international economic law and financial governance, adaptive strategies are needed for, as far as possible, resolving national interests and global economic stability in such a way as to avoid perpetuating cycles of poverty, inequality and social unrest by further economic disruptions.

ENVIRONMENTAL DEGRADATION AND CLIMATE-INDUCED CONFLICTS

Environmental degradation and climate induced conflicts are now critical to the problem of global peace, security and sustainable development. Given the impact, human driven environmental destruction and unsustainable consumption patterns, natural resources depletion has intensified the geopolitical rivalries and socio economic disparities (Singh et al., 2024). In resolving resource instability and political tensions, resource conflict usually overlaps with legal frameworks that govern resource allocation, environmental protection and the cross-border cooperation (Valipour et al., 2024). Climate change multiplies threats and forced displacement, fragile economies, and rising competition for harder to find resources. It is necessary for developing an appropriate and fair environmental governance mechanism, and this needs a delicate balance between state sovereignty and international obligations (Kalliojärvi, 2020). Environmental instability brings with it its direct effects on human security such as accelerating degradation of ecosystems, deforestation, desertification and depletion of water which can heighten social and political unrest (Chesler, 2024). Climate variability impedes agriculturist, results to diminished feedings and advancement of livelihoods, especially in destitution nations reliant on subsistence farming (Gwambene et al., 2023). As the polluter pays principle and the precautionary approach legal doctrines are present, enforcement of them remains weak in the conflict prone areas. Green taxes that are scientifically shown to be valid may promote sustainability, decrease pollution, and bolster regulatory laws but without disproportionately alienating out of the comfort zone populations (Okombi & Ndoum Babouama, 2024). Larger scale environmental exploitation, however, is possible primarily because of weak accountability measures that cause social displacement (Ge & Rios, 2025). Regions where transboundary resource management is contested are often geopolitically tenuous with respect to the management of freshwater, arable land, and fossil fuels (Jayaram & Sethi, 2022). The diplomatic conflicts are in disputes over resource sovereignty in shared water bodies and energy reserves. However, international legal frameworks on trans-boundary water resources, under the UN convention of trans-boundary water resources, have been set to promote cooperative management but enforcement has been weak (Xie & Jia, 2017). Dams and diversions force stronger institutional arbitration because they intensify disputes (Hinz and Leromain, 2020). Natural resources in politically unstable areas are also exploited to fund armed conflicts, which involves illicit trade networks of minerals, timber and fossil fuel (Le Billon, 2001). Weak governance enables the emergence of illicit supply chains which violate the Kimberley Process in an attempt to curb conflict diamond financing (Haufler, 2009). It is critical that strengthening corporate accountability and requiring due diligence in resource extraction prevent environmental and human rights abuses (Tan, 2009).

Climate change acts as a threat multiplier that amplifies the soft and hard vulnerabilities of socio politics and the economy. This crisis arises due to the factors like low birth and investment rates, illiteracy, underdeveloped agriculture and poverty (Bandh et al., 2021). Hurricanes, droughts, floods displace millions and are beyond national economic and refugee protection frameworks. International refugee law does not recognize climate displacement as such, leaving those displaced through climate to found among the legs (Hay and Mimura, 2010; Chatterjee, 2023). Common, but differentiated responsibilities is a principle enshrined in international environmental law, which necessitates assistance of less developed nations to adapt to climate towards develop nations. However, the implementation of meaning remains inadequate (Brunnée, 2006). However, bureaucratic hurdles, stringent conditions and cashes to finance restrict the efficacy of institutions such as Adaptation Fund and the Green Climate Fund that aim to enhance resilience (Biesbroek et al., 2022). The frameworks should strengthen with transparency, accountability and equitable allocation of climate finance (Peterson & Skovgaard, 2019). Climate adaptation must be integrated into the national development plans, nationally including in agriculture, urban planning and policies of disaster management (Oulu, 2015). Given that addressing transboundary environmental issue requires international cooperation and the legally binding requirement to achieve sustainability and the reduction of emissions, the Paris Agreement and the Convention on Biological Diversity are two examples of such agreements (Kalfagianni & Young, 2022). Though international voluntary commitments and subsequent agreements on emissions reductions have been made, fossil gas economies tend to fall short of their emission reduction targets for various reasons, notably as a result of weak enforcement mechanisms (Pattberg et al., 2022). Legal accountability of climate governance, including rights of petition and organise, as well as climate litigation and sanctions against noncompliance, is necessary to strengthen the climate governance. Environmental harm and inaction are the courtroom when domestic courts hold governments and corporations legally responsible (Wewerinke-Singh, 2023).

Transboundary governance is needed in environmental issues such as air pollution, deforestation and marine pollution because they are of cross border nature. Although regional legal frameworks as the EU ETS and the ASEAN Haze Agreement have showed themselves effective in mitigating the contamination, it is difficult to apply in the regions where political interests transcend environmental concerns (Miller et al. 2022). Rigid legal frameworks that provide an enabling environment and build capacity support national environmental laws being based on international standards and most importantly, also for cross border cooperation (Jahn, 2024). At the same time, commitments to environmental integrity come to be part of trade agreements, investment treaties, and development policies with the aim of legal coherence, namely, without environmental protection jeopardizing the achievement of economic growth (de Souza Barbosa et al., 2023). Unfortunately, the jurisdiction of governing global environmental commons (such as the atmosphere, oceans and the Polar Regions) is still legally ambiguous and subject to shared state responsibility under the international law (Schrijver, 2016). Disputes over resource entitlements and conservation have ensued as a result of modern competition for shallow seas and Arctic resources (including mining), referred to as deep sea (El Moll, 2023). The sustainability depends on having legal frameworks for maritime biodiversity protection and carbon emissions by shipping. To this end, the UN and civil society emphasise the need for stronger legal protection of environmental governance that amalgamates environmental law, human rights and economic governance (Gilbert et al., 2023; Shah et al., 2025). Ordnances and legal enforcement of environmental rights, strategic environmental assessments, and prevent harm (Nakamura et al., 2022). Legal coherence, scientific expertise and socio-economic considerations are essential in the good global environmental governance that can lead to achievement (Dienelt, 2022).

HUMAN RIGHTS, SOCIAL JUSTICE, AND SUSTAINABLE DEVELOPMENT

Human rights, social justice and development – are interlinked, are a foundation for societies that are equitable allowing children and young people access to essential freedoms and rights and opportunities. This vision is achievable provided for by a strong legal framework that promotes dignity for human beings, does not discriminate, and supports governance that is participatory (Morvaridi, 2008). Systemic injustices need to be legalised, economically just and protect the marginalised. This points to the need to fuse political and cultural rights with legal protections and viable economic and social policy (Lima & Gomez, 2021). There are wars, moral crises, and political repression that lead to failures in legal and humanitarian systems. For example, climate change triggered displacement leaves a gap in the international refugee protection as the 1951 Refugee Convention is not a good tool to deal with climate migration (Abel et al., 2019). Laws concerning climate refugees are fuzzy, and the states react differently, which leads to leaving many unprotected. International refugee are based on the principle of non-refoulement, which prohibits sending people back to life threatening circumstances but restrictive immigration policies and of militarization of the borders are exacerbating vulnerability of displaced populations (Sciaccaluga, 2020). Other countries have strict migration controls while poorer host countries are left with the tab, with little international support (McMenamin, 2023). Internal displacement results from violence, land dispossession and development projects, and raises significant legal and policy challenges (Kerswell, 2015). However, enforcement of the UN Guiding Principles on Internal Displacement is not binding, leaving affected individuals without a clear accountability mechanism (Dirikgil, 2023). In post conflict settings, indeed there are available legal remedies for those harmed by conflict including land restitution, compensation, and rehabilitation. Yet, legal injustices, corruption, and bureaucratic inefficiencies hinder rather than facilitating the actual implementation of punishment; thus, restorative justice means truth commissions, reparations (Abrahams & Carr, 2017). Almost everything further makes it harder for sustainable development by reinforcing social divisions and breaking the sovereignty. Socioeconomic inequalities and marginalization work together to prevent it from succeeding. Cycles of poverty and deprivation are perpetuated by the structural discrimination based in historical injustice, economic exclusion, institutional bias (Niaz, 2021). Essential tools to mitigate disparities are the socioeconomic rights: education, healthcare and fair labour conditions, among others. Although these rights are, however, judicially reluctant to enforce on the grounds that they interfere in resource allocation and separation of powers; levels of legal protection are varying in jurisdictions (Ako et al., 2016). CEDAW and ICERD are treaties to dismantle such institutional discriminatory rhetoric aimed at neutralizing racial, gender, and economic inequality (Paes & Geraldes, 2019). Getting rid of entrenched inequalities cannot be done by legal action alone; besides legal action, affirmative action, strategic litigation and institutional reforms are needed. Challenging systemic injustices of labour rights, gender equality, and indigenous land claims, public interest litigation also translates a matter that may have otherwise been neglected into a narrative that is much greater than the individual claim (Guthridge et al., 2022). Yet informal labour sectors are not protected at all and ILO standards are not enforced because of weak regulatory oversight (Sankaran, 2022). It adds to that complexity in gig economy work, which traditional employment laws don't ask for to exist. The absence of legal recognition for platform workers' rights results in inconsistent legislative responses across jurisdictions (Chesalina, 2020). A judicial approach in the attainment of such balance, whereby strict labour regulations are enforced, but economic growth is fostered, is needed.

Access to justice in marginalized communities in the United States is hindered by tight legal barriers such as high costs, complexity of processes, and institutional biases (Lillo Lobos, 2022). Legal empowerment initiatives such as community based legal aid, alternative dispute resolution and legal literacy for the economically impoverished groups are needed to address these challenges (Gisselquist, 2018). Definitely, deprived legal aid, judicial backlogs and corruption are also demeaning justice delivery. There is a need for the legality to be legally accountable, the judiciary to be independent, and more transparent procedures. One way civil society contributes to human rights, social justice and development is by promoting democratic governance by calling for the promotion of social justice (Moka-Mubelo, 2017). NGOs, activists and human rights defenders document violation, provide humanitarian aid, and practise strategic litigation. Second, restrictive laws, criminalization of dissent and the misuse of anti-terror laws threaten civic engagement and democracy (Chaney, 2022). A strong international human rights protections enforcement is required for protecting civil society actors (Basu & Sen, 2023).

Important global organisations comprising the international arena are the United Nations (UN), the World Bank, and regional human rights bodies which have a significant role to play in devising and checking human rights standards. They are institutions meant for development assistance and for auditing whether or not states abide by human rights obligations (Zapp, 2021). That said, although of fundamental importance to ensure states are held accountable, the United Nations Human Rights Council (UNHRC) can often suffer from selective enforcement, political resistance, stalling of action by states to be scrutinised externally, thereby diminishing its impact (Cantú Rivera, 2018). However, the oversight provided by treaty monitoring bodies does not have binding enforcement mechanisms (O'Flaherty & O'Brien, 2007). Therefore, for the integration of human rights into development policies, human rights frameworks have to be cohered with the international financial institutions, guaranteeing the participation of the developing countries, and to maximize the poverty alleviation (Sengupta, 2004). Accordingly, legal and institutional alignment with human rights principles is necessary for Sustainable Development Goals (SDGs) in the policy coherence for sustainable development (PCSD) framework (Nonet et al., 2022). However, corporate accountability in environmental and social welfare matters is weak, those in this area being as yet unaccountable to organizations outside the company corporate. There is a need to strengthen the regulatory mandates and enforcement measures (Fatima & Elbanna, 2023). For human rights, social justice and sustainable development, legal accountability, participatory governance, global cooperation (Winkler & Williams, 2017; Ip, 2010).

CONCLUSION

Peace diplomacy, politics, and long term sustainable growth are closely tied to regional disturbances that hinder long term development. Combing governance challenges such as restructuring, economic disruptions, environmental degradation and human rights violations, the efforts to stabilize the region are complex web of challenges. Any solutions to these requires an adaptive legal and policy framework given the changes in geopolitical floors. Effective policies are possible only if one understands institutional and structural barriers. Findings of this research indicate that slowdown of SDG progress is due to political instability, economic restraints and social inequalities. In addition to governance failures, weak institutions and ineffective ways of resolving disputes, further impediments to economic growth are posed. Sanctions and trade barriers also generally hurt local populations without much diplomatic benefit. Deforestation and climate induced conflicts intensify the already exacerbated cases of resource scarcity, geological tensions and displacement in the regions. The division of responsibilities between subnational and national governments regarding transboundary environmental policies and the non-resolution of conflicts over shared resources lead to the fragmentation of transboundary environmental policies. Besides, climate resilience strategies are still integrated into national development plans on a deficient level, further increasing vulnerabilities, with an emphasis on marginalized communities. Though international cooperation is urgently needed to address climate induced displacement, legal frameworks are not legally committed to make a commitment to this purpose. To support enforcers of compliance among states and corporations, marine environmental law must be strengthened. As a result of marginalization, there still exists economic inequality that calls for legal and policy measures for tackling social exclusion. Non-discrimination, labour protection and access to justice, stemming from international human rights treaties, should be enforced and require long term policy response for displaced populations. This requires that strategic policy interventions intervene on stability, through legal, economic and social reforms that will address the root causes of regional disturbances. For democratic resilience, strengthening the governance frameworks through legal frameworks that safeguard against corruption, increase judicial independence, and in turn participating in the decision making should be done. Political patronage, unscrupulous regulatory oversight, illicit financial flows need to be checked and strengthened and the transparency of legal mechanism must be ensured to help uphold of accountability and rule of law. Reduction of legal uncertainties that are root causes of regional instability will require fighting the legal uncertainties in a crucial step of providing judicial efficiency in adjudicating the disputes over land rights, economic agreements and environmental protection.

Therefore international law should have strong criteria for enforcement in order to keep states and corporations responsible for over environmental harm. Inequalities persist on socioeconomic grounds contributing towards marginalization, discrimination, deprivation of essential services among other effects. In all cases, instability derives from systematic violations of fundamental rights whether from political repression, armed conflicts or human rights abuses. The international human rights treaties require states to observe no discrimination, labor law, as well as access to justice, which calls for stronger enforcement mechanisms. The policies needed to address the needs of displaced populations must address humanitarian aid with long term development. Strategic reform agenda targeted to roots of regional unrest is required for strategic reforms in this regard on the basis of legal, economic and social premises of stability. For the purposes of accountability, reduction of corruption and legal uncertainty, strengthening of political participation, judicial independence and transparency in legal systems is imperative. Future research could use an interdisciplinary method to investigate regional disturbances framework that is both legal and policy related, in legal studies, political science, ecology and economy. The analysis of international legal mechanisms effectiveness in utilizing to limit the negative economic sanctions and trade barriers can be an empirical point for the policy development. To improve governance in fragile states, which is fragile, legal pluralism's role in conflict resolution in such regions where customary and formal systems overlap can be analysed. Research of digital governance, cybersecurity threats, and technological advancements, is necessary for creating legal and policy responses to geopolitical instability. For sustainability to be achieved, it calls for the development of the legal framework, policy change and international cooperation. Appropriate for mitigating the conflict and maintaining stability is the strengthening of governance, economic resilience, environmental responsibility and human rights protection.

REFERENCES

Abel, G. J., Brottrager, M., Cuaresma, J. C., & Muttarak, R. (2019). Climate, conflict and forced migration. *Global Environmental Change*, *54*, 239–249. DOI: 10.1016/j.gloenvcha.2018.12.003

Abrahams, D., & Carr, E. R. (2017). Understanding the connections between climate change and conflict: Contributions from geography and political ecology. *Current Climate Change Reports*, *3*(4), 233–242. DOI: 10.1007/s40641-017-0080-z

Ades, A., & Chua, H. B. (1997). Thy neighbor's curse: Regional instability and economic growth. *Journal of Economic Growth*, *2*(3), 279–304. https://www.jstor.org/stable/40215960. DOI: 10.1023/A:1009782809329

Akinyetun, T. S. (2024). The proliferation of armed non-state actors in the Sahel: Drivers, ramifications and the way forward. *Journal of Central and Eastern European African Studies*, *2*(2), 36–58. DOI: 10.59569/jceeas.2022.2.2.60

Ako, R., Stewart, N., & Ekhator, E. O. (2016). Overcoming the (non)justiciable conundrum: The doctrine of harmonious construction and the interpretation of the right to a healthy environment in Nigeria. In Diver, A., & Miller, J. (Eds.), *Justiciability of human rights law in domestic jurisdictions* (pp. 123–141). Springer., DOI: 10.1007/978-3-319-24016-9_6

Amankwaah, H. (2023). The rule of law and armed conflict reconstruction implementation practices: A human rights-based analysis of the Rwandan experience. *Cogent Social Sciences*, *9*(1), 2171573. DOI: 10.1080/23311886.2023.2171573

Bandh, S. A., Shafi, S., Peerzada, M., Rehman, T., Bashir, S., Wani, S. A., & Dar, R. (2021). Multidimensional analysis of global climate change: A review. *Environmental Science and Pollution Research International*, *28*(20), 24872–24888. DOI: 10.1007/s11356-021-13139-7

Basu, S., & Sen, S. (2023). Silenced voices: Unravelling India's dissent crisis through historical and contemporary analysis of free speech and suppression. *Information & Communications Technology Law*, *33*(1), 42–65. DOI: 10.1080/13600834.2023.2249780

Berk, E., Ayas, O., & Ülkü, M. A. (2023). Optimizing process-improvement efforts for supply chain operations under disruptions: New structural results. *Sustainability (Basel)*, *15*(17), 13117. DOI: 10.3390/su151713117

Bhattacharya, R. (2015). The way forward: Public policy to address regional imbalances in India, process and outcome. In Bhattacharya, R. (Ed.), *Regional development and public policy challenges in India* (pp. 309–322). Springer., DOI: 10.1007/978-81-322-2346-7_10

Biesbroek, R., Wright, S. J., Eguren, S. K., Bonotto, A., & Athanasiadis, I. N. (2022). Policy attention to climate change impacts, adaptation and vulnerability: A global assessment of National Communications (1994–2019). *Climate Policy, 22*(1), 97–111. DOI: 10.1080/14693062.2021.2018986

Binder, M. (2009). Humanitarian crises and the international politics of selectivity. *Human Rights Review (Piscataway, N.J.), 10*(3), 327–348. DOI: 10.1007/s12142-009-0121-7

Bogdanova, I. (2022). *Unilateral sanctions in international law and the enforcement of human rights.* Brill | Nijhoff. https://doi.org/DOI: 10.1163/9789004507890

Brunnée, J. (2006). International environmental law: Rising to the challenge of common concern? [American Society of International Law]. *Proceedings of the Annual Meeting (American Association of University Teachers of Insurance), 100*, 307–310.

Campos, N. F., & Kinoshita, Y. (2010). Structural reforms, financial liberalization, and foreign direct investment. *IMF Staff Papers, 57*(2), 326–365. https://www.jstor.org/stable/25681049. DOI: 10.1057/imfsp.2009.17

Cantú Rivera, H. (2018). The UN Human Rights Council: Achievements and challenges in its first decade. In Oberleitner, G. (Ed.), *International human rights institutions, tribunals, and courts* (pp. 1–20). Springer., DOI: 10.1007/978-981-10-5206-4_3

Ceva, E., & Ferretti, M. P. (2021). Upholding public institutions in the midst of conflicts: The threat of political corruption. *Ethics & Global Politics, 14*(3), 1961379. DOI: 10.1080/16544951.2021.1961379

Chaney, P. (2022). Exploring civil society perspectives on the situation of human rights defenders in the Commonwealth of Independent States. *Central Asian Survey, 42*(2), 293–318. DOI: 10.1080/02634937.2022.2113034

Chatterjee, S. (2023). Climate refugees: Arriving at a possible framework. In P. Singh, B. Ao, & A. Yadav (Eds.), *Global climate change and environmental refugees* (pp. Chapter 10). Springer, Cham. https://doi.org/DOI: 10.1007/978-3-031-24833-7_10

Chesalina, O. (2020). Platform work as a new form of employment: Implications for labour and social law. In Wratny, J., & Ludera-Ruszel, A. (Eds.), *New forms of employment. Prekarisierung und soziale Entkopplung – transdisziplinäre Studien* (pp. 153–167). Springer VS., DOI: 10.1007/978-3-658-28511-1_9

Chesler, A. (2024). Environmental displacement and political instability: Evidence from Africa. *Journal of Peace Research*, *0*(0). Advance online publication. DOI: 10.1177/00223433241274979

Chewpreecha, U., Pollitt, H., & Mercure, J. F. (2023). Combining short-term economic recovery with long-term sustainability. In Arestis, P., & Sawyer, M. (Eds.), *Prospects and policies for global sustainable recovery* (pp. 215–263). Palgrave Macmillan., DOI: 10.1007/978-3-031-19256-2_6

Chopra, T., & Isser, D. (2012). Access to justice and legal pluralism in fragile states: The case of women's rights. *Hague Journal on the Rule of Law*, *4*(2), 337–358. DOI: 10.1017/S187640451200019X

Conroy, A., & Blackie, M. (2006). Breaking out of economic isolation and the poverty trap. In *Poverty, AIDS and Hunger*. Palgrave Macmillan, London. https://doi.org/ DOI: 10.1057/9780230627703_11

Damane, M., & Ho, S. Y. (2024). The impact of financial inclusion on financial stability: Review of theories and international evidence. *Development Studies Research*, *11*(1), 2373459. Advance online publication. DOI: 10.1080/21665095.2024.2373459

de Souza Barbosa, A., da Silva, M. C. B. C., da Silva, L. B., Morioka, S. N., & de Souza, V. F. (2023). Integration of environmental, social, and governance (ESG) criteria: Their impacts on corporate sustainability performance. *Humanities & Social Sciences Communications*, *10*(1), 410. DOI: 10.1057/s41599-023-01919-0

den Boer, M. (Ed.). (2018). *Comparative policing from a legal perspective*. Edward Elgar Publishing., DOI: 10.4337/9781785369117

Dienelt, A. (2022). The environment and international environmental law. In *Armed conflicts and the environment* (pp. 199–256). Springer., DOI: 10.1007/978-3-030-99339-9_4

Dinh, L. Q., Oanh, T. T. K., & Ha, N. T. H. (2024). Financial stability and sustainable development: Perspectives from fiscal and monetary policy. *International Journal of Finance & Economics*, 1–18. DOI: 10.1002/ijfe.2981

Dirikgil, N. (2023). Addressing the prevention of internal displacement: The right not to be arbitrarily displaced. *Journal of International Migration and Integration*, *24*(1), 113–138. DOI: 10.1007/s12134-022-00935-4

Egger, P., Syropoulos, C., & Yotov, Y. V. (2024). Analyzing the effects of economic sanctions: Recent theory, data, and quantification. *Review of International Economics*, *32*(1), 1–11. DOI: 10.1111/roie.12724

El Moll, A. (2023). Water resources and climate change: Regional, national and international perspective. In Prasad, M. N. V., & Smol, M. (Eds.), *Sustainable and circular management of resources and waste towards a green deal* (pp. 309–336). Elsevier., DOI: 10.1016/B978-0-323-95278-1.00010-3

Enayati, M., Lukambagire, I., & Manianga, A. (2024). Navigating sustainability and resilience: A collective case study of four Indian communities. *Environment, Development and Sustainability*. Advance online publication. DOI: 10.1007/s10668-024-05445-0

Fatima, T., & Elbanna, S. (2023). Corporate social responsibility (CSR) implementation: A review and a research agenda towards an integrative framework. *Journal of Business Ethics*, *183*(1), 105–121. DOI: 10.1007/s10551-022-05047-8

Garg, S., Mittal, S., & Garg, A. (2024). Navigating the path for economic resilience and sustainable economic growth: Empirical insight from G-20 nations. *Sustainable Futures : An Applied Journal of Technology, Environment and Society*, *8*, 100338. DOI: 10.1016/j.sftr.2024.100338

Garretsen, H., Kitson, M., & Yang, C. (2025). Global forces and local impacts: Megatrends in regional development. *Cambridge Journal of Regions, Economy and Society*, *18*(1), 1–16. DOI: 10.1093/cjres/rsae047

Gavsker, K. K. (2021). Regional environmental governance: Analytical framework for resource conservation in the Eastern Ghats, India. In Thakur, B., Thakur, R. R., Chattopadhyay, S., & Abhay, R. K. (Eds.), *Resource management, sustainable development and governance* (pp. 179–191). Springer., DOI: 10.1007/978-3-030-85839-1_11

Ge, J., & Rios, J. (2025). The effects of corruption in natural resources on the achievement of sustainable development goals in Bolivia and Peru. *SN Social Sciences*, *5*(3), 23. DOI: 10.1007/s43545-025-01056-2

Gilbert, J., Macpherson, E., Jones, E., & Dehm, J. (2023). The rights of nature as a legal response to the global environmental crisis? A critical review of international law's 'greening' agenda. In D. Dam-de Jong & F. Amtenbrink (Eds.), *Netherlands Yearbook of International Law 2021* (Vol. 52). T.M.C. Asser Press. https://doi.org/ DOI: 10.1007/978-94-6265-587-4_3

Gisselquist, R. M. (2018). Legal empowerment and group-based inequality. *The Journal of Development Studies*, *55*(3), 333–347. DOI: 10.1080/00220388.2018.1451636

Góes, C., & Bekkers, E. (2023). The impact of geopolitical conflicts on trade, growth, and innovation. *arXiv*. https://arxiv.org/abs/2203.12173

Gould, M., & Rablen, M. D. (2017). Reform of the United Nations Security Council: Equity and efficiency. *Public Choice*, *173*(1-2), 145–168. DOI: 10.1007/s11127-017-0468-2

Guthridge, M., Kirkman, M., Penovic, T., & Giummarra, M. J. (2022). Promoting gender equality: A systematic review of interventions. *Social Justice Research*, *35*(3), 318–343. DOI: 10.1007/s11211-022-00398-z

Gutmann, J., Neuenkirch, M., & Neumeier, F. (2023). The economic effects of international sanctions: An event study. *Journal of Comparative Economics*, *51*(4), 1214–1231. DOI: 10.1016/j.jce.2023.05.005

Güven, A. B. (2012). The IMF, the World Bank, and the global economic crisis: Exploring paradigm continuity. *Development and Change*, *43*(4), 869–898. DOI: 10.1111/j.1467-7660.2012.01781.x

Gwambene, B., Liwenga, E., & Mung'ong'o, C. (2023). Climate change and variability impacts on agricultural production and food security for the smallholder farmers in Rungwe, Tanzania. *Environmental Management*, *71*(1), 3–14. DOI: 10.1007/s00267-022-01628-5

Harhoff, F. S., & Freeland, S. (2017). Addressing the intentional destruction of the environment during warfare under the Rome Statute of the International Criminal Court. *Netherlands International Law Review*, *64*(2), 343–347. DOI: 10.1007/s40802-017-0088-z

Haufler, V. (2009). The Kimberley Process Certification Scheme: An innovation in global governance and conflict prevention. *Journal of Business Ethics*, *89*(S4), 403–416. https://www.jstor.org/stable/40605378. DOI: 10.1007/s10551-010-0401-9

Hay, J., & Mimura, N. (2010). The changing nature of extreme weather and climate events: Risks to sustainable development. *Geomatics, Natural Hazards & Risk*, *1*(1), 3–18. DOI: 10.1080/19475701003643433

Helms, L. (2021). Introduction: The nature of political opposition in contemporary electoral democracies and autocracies. *European Political Science*, *20*(4), 569–579. DOI: 10.1057/s41304-021-00323-z

Hinnebusch, R. A. (1981). Egypt under Sadat: Elites, power structure, and political change in a post-populist state. *Social Problems*, *28*(4), 442–464. https://www.jstor.org/stable/800057. DOI: 10.2307/800057

Hinz, J., & Leromain, E. (2020). Critically important: The heterogeneous effect of diplomatic tensions on trade. *Review of Industrial Organization*, *57*(3), 309–331. DOI: 10.1007/s11151-020-09769-9

Huttunen, M. (2024). A tale of two sovereigns: The responsibility to protect and the competing notions of responsible sovereignty. *International Journal of Human Rights*, •••, 1–22. DOI: 10.1080/13642987.2024.2429467

Ip, E. C. (2010). Globalization and the future of the law of the sovereign state. *International Journal of Constitutional Law*, *8*(3), 636–655. DOI: 10.1093/icon/moq033

Ishiyama, J. (2019). Is democracy necessary for good governance? *Social Science Quarterly*, *100*(6), 2188–2208. DOI: 10.1111/ssqu.12712

Jahn, D. (2024). The stringency and potential impact of climate laws and policies in the European Union and the 21 OECD countries. *Npj Climate Action*, *3*(1), 90. DOI: 10.1038/s44168-024-00175-5

Jakulevičienė, L. (2023). Principle of non-refoulement in the context of recent ECtHR case law and the EU migration and asylum pact. *ERA Forum, 24*, 379–396. https://doi.org/DOI: 10.1007/s12027-023-00761-w

Jayaram, D., & Sethi, G. (2022). Geopolitics of climate change and water security in South Asia: Conflict and cooperation. In N. Khare (Ed.), *Science, policies and conflicts of climate change* (pp. XX–XX). Springer Climate. Springer, Cham. https://doi.org/DOI: 10.1007/978-3-031-16254-1_4

Jessen, H. (2021). Multilateral and unilateral sanctions: Compliance and challenges. In Leal Filho, W., Azul, A. M., Brandli, L., Lange Salvia, A., Özuyar, P. G., & Wall, T. (Eds.), *Peace, justice and strong institutions. Encyclopedia of the UN Sustainable Development Goals* (pp. 1–11). Springer., DOI: 10.1007/978-3-319-71066-2_51-1

Kalfagianni, A., & Young, O. R. (2022). The politics of multilateral environmental agreements: Lessons from 20 years of INEA. *International Environmental Agreement: Politics, Law and Economics*, *22*(2), 245–262. DOI: 10.1007/s10784-022-09567-6

Kalliojärvi, S. (2020). Age of changes: Threat of climate change and its meaning for security. In Heininen, L., & Exner-Pirot, H. (Eds.), *Climate change and Arctic security* (pp. 9–32). Palgrave Pivot., DOI: 10.1007/978-3-030-20230-9_2

Kantorowicz-Reznichenko, E. (2020). The enforcement of international law. In Hosli, M. O., & Selleslaghs, J. (Eds.), *The changing global order* (Vol. 17). Springer., DOI: 10.1007/978-3-030-21603-0_18

Katsaliaki, K., Galetsi, P., & Kumar, S. (2022). Supply chain disruptions and resilience: A major review and future research agenda. *Annals of Operations Research*, *319*(1), 965–1002. DOI: 10.1007/s10479-020-03912-1

Kenny, M. E., Horne, A. M., Orpinas, P., & Reese, L. E. (Eds.). (2009). *Realizing social justice: The challenge of preventive interventions*. American Psychological Association., https://www.jstor.org/stable/j.ctv1chs94p DOI: 10.1037/11870-000

Kerswell, T. (2015). *Development and dispossession: The crisis of forced displacement and resettlement* by Anthony Oliver-Smith (Ed.). *Human Rights Review, 16*, 417–418. https://doi.org/DOI: 10.1007/s12142-015-0386-y

Kešeljević, A., & Spruk, R. (2024). Estimating the effects of the Syrian civil war. *Empirical Economics*, *66*(2), 671–703. DOI: 10.1007/s00181-023-02470-2

Khan, M. T. I., Anwar, S., Sarkodie, S. A., Yaseen, M. R., Nadeem, A. M., & Ali, Q. (2023). Natural disasters, resilience-building, and risk: Achieving sustainable cities and human settlements. *Natural Hazards*, *118*(1), 611–640. DOI: 10.1007/s11069-023-06021-x

Kranke, M. (2020). IMF-World Bank cooperation before and after the global financial crisis. *Global Policy*, *11*(1, S1), 15–25. DOI: 10.1111/1758-5899.12743

Le, A. N. N., Pham, H., Pham, D. T. N., & Duong, K. D. (2023). Political stability and foreign direct investment inflows in 25 Asia-Pacific countries: The moderating role of trade openness. *Humanities & Social Sciences Communications*, *10*(1), 606. DOI: 10.1057/s41599-023-02075-1

Le, T.-H., Bui, M.-T., & Uddin, G. S. (2022). Economic and social impacts of conflict: A cross-country analysis. *Economic Modelling*, *115*, 105980. DOI: 10.1016/j.econmod.2022.105980

Le Billon, P. (2001). The political ecology of war: Natural resources and armed conflicts. *Political Geography*, *20*(5), 561–584. DOI: 10.1016/S0962-6298(01)00015-4

Lillo Lobos, R. (2022). The right to access to justice as a key to understand the right to a fair trial in civil matters. In *Understanding due process in non-criminal matters* (Vol. 97, pp. 251–281). Springer., DOI: 10.1007/978-3-030-95534-2_13

Lima, V., & Gomez, M. (2021). Access to justice: Promoting the legal system as a human right. In Leal Filho, W., Marisa Azul, A., Brandli, L., Lange Salvia, A., Özuyar, P. G., & Wall, T. (Eds.), *Peace, justice and strong institutions. Encyclopedia of the UN Sustainable Development Goals*. Springer., DOI: 10.1007/978-3-319-95960-3_1

Lülf, C. (2016). Non-refoulement in international refugee law, human rights law, and asylum laws. In Heintze, H. J., & Thielbörger, P. (Eds.), *From Cold War to Cyber War*. Springer., DOI: 10.1007/978-3-319-19087-7_12

Makinda, S. M. (1998). The United Nations and state sovereignty: Mechanism for managing international security. *Australian Journal of Political Science*, *33*(1), 101–115. DOI: 10.1080/10361149850750

Mansoor, S. I. U. (2025). Renewable energy and international law: Managing risk in relation to economic volatility and geopolitical instability. In *Geopolitical landscapes of renewable energy and urban growth* (pp. 97–124). IGI Global., DOI: 10.4018/979-8-3693-8814-3.ch004

Manulak, M. W. (2024). The sources of influence in multilateral diplomacy: Replaceability and intergovernmental networks in international organizations. *The Review of International Organizations*, *19*(3), 579–610. DOI: 10.1007/s11558-024-09536-5

Matta, S., Bleaney, M., & Appleton, S. (2022). The economic impact of political instability and mass civil protest. *Economics and Politics*, *34*(1), 253–270. DOI: 10.1111/ecpo.12197

McMenamin, R. (2023). Advisory opinion on obligations of states in respect of climate change: Potential contribution of human rights bodies. *Climate Law*, *13*(3–4), 213–223. DOI: 10.1163/18786561-bja10045

Meaza, H., Ghebreyohannes, T., Nyssen, J., Tesfamariam, Z., Demissie, B., Poesen, J., Gebrehiwot, M., Weldemichel, T. G., Deckers, S., Gidey, D. G., & Vanmaercke, M. (2024). Managing the environmental impacts of war: What can be learned from conflict-vulnerable communities? *The Science of the Total Environment*, *927*, 171974. DOI: 10.1016/j.scitotenv.2024.171974

Miljkovic, D., & Rimal, A. (2008). The impact of socio-economic factors on political instability: A cross-country analysis. *Journal of Socio-Economics*, *37*(6), 2454–2463. DOI: 10.1016/j.socec.2008.04.007

Miller, M. A., Rigg, J., & Taylor, D. (2022). Transboundary environmental governance: Emerging themes and lessons from Southeast Asia. *Environmental Policy and Governance*, *32*(4), 275–280. DOI: 10.1002/eet.2012

Mlachila, M., Tapsoba, R., & Tapsoba, S. J. A. (2017). A quality of growth index for developing countries: A proposal. *Social Indicators Research*, *134*(2), 675–710. DOI: 10.1007/s11205-016-1439-6

Moallemi, E. A., Malekpour, S., Hadjikakou, M., Raven, R., Szetey, K., Ningrum, D., Dhiaulhaq, A., & Bryan, B. A. (2020). Achieving the Sustainable Development Goals requires transdisciplinary innovation at the local scale. *One Earth*, *3*(3), 300–313. DOI: 10.1016/j.oneear.2020.08.006

Moka-Mubelo, W. (2017). Civil society and human rights. In *Reconciling law and morality in human rights discourse* (Vol. 3, pp. 147–167). Springer., DOI: 10.1007/978-3-319-49496-8_6

Moosa, I. A., & Moosa, N. (2019). Conditionality and structural adjustment programmes. In *Eliminating the IMF* (pp. 55–87). Palgrave Macmillan., DOI: 10.1007/978-3-030-05761-9_3

Morgan, T. C., & Schwebach, V. L. (1995). Economic sanctions as an instrument of foreign policy: The role of domestic politics. *International Interactions*, *21*(3), 247–263. DOI: 10.1080/03050629508434868

Morton, S., Pencheon, D., & Squires, N. (2017). Sustainable Development Goals (SDGs) and their implementation: A national global framework for health, development, and equity needs a systems approach at every level. *British Medical Bulletin*, *124*(1), 81–90. DOI: 10.1093/bmb/ldx031

Morvaridi, B. (2008). Introduction: Social justice and development. In *Social justice and development* (pp. 1–32). Palgrave Macmillan, London. https://doi.org/DOI: 10.1057/9780230581999_1

Mufungizi, I. (2024). Mineral potential facing socio-economic development challenges: Case study of the Democratic Republic of Congo, a 'geological scandal.'. *International Geology Review*, 1–23. DOI: 10.1080/00206814.2024.2401575

Nagel, A. S. (2023). Unilateral extraterritorial sanctions: The search for a jurisdictional justification under international law. *LSE Law Review*, *8*(3), 368–395. DOI: 10.61315/lselr.478

Nakamura, J., Diz, D., & Morgera, E. (2022). International legal requirements for environmental and socio-cultural assessments for large-scale industrial fisheries. *Review of European, Comparative & International Environmental Law*, *31*(3), 336–348. DOI: 10.1111/reel.12462

Nguyen, D. T., Le, T. H., Do, D. D., & Nguyen, H. N. (2023). Does geopolitical risk hinder sustainable development goals? Evidence from a panel analysis. *Journal of Environmental Management*, *347*, 119204. DOI: 10.1016/j.jenvman.2023.119204

Niaz, M. U. (2021). Socio-economic development and sustainable development goals: A roadmap from vulnerability to sustainability through financial inclusion. *Ekonomska Istrazivanja*, *35*(1), 3243–3275. DOI: 10.1080/1331677X.2021.1989319

Nonet, G. A. H., Gössling, T., Van Tulder, R., & Bryson, J. M. (2022). Multi-stakeholder engagement for the Sustainable Development Goals: Introduction to the special issue. *Journal of Business Ethics*, *180*(4), 945–957. DOI: 10.1007/s10551-022-05192-0

Núñez, J. E. (2024). State sovereignty: Concept and conceptions. *International Journal for the Semiotics of Law*, *37*(7), 2131–2150. DOI: 10.1007/s11196-024-10170-y

O'Flaherty, M., & O'Brien, C. (2007). Reform of UN human rights treaty monitoring bodies: A critique of the concept paper on the High Commissioner's proposal for a unified standing treaty body. *Human Rights Law Review*, *7*(1), 141–172. DOI: 10.1093/hrlr/ngl035

Okombi, I. F., & Ndoum Babouama, V. B. D. (2024). Environmental taxation and inclusive green growth in developing countries: Does the quality of institutions matter? *Environmental Science and Pollution Research International*, *31*(21), 30633–30662. DOI: 10.1007/s11356-024-33245-6

Oore-Ofe, O. R., & Ojakorotu, V. (2023). Conflict in Democratic Republic of Congo: Framework for resolution. In Ani, K. J. (Ed.), *Resource conflict and environmental relations in Africa* (pp. 121–137). Palgrave Macmillan., DOI: 10.1007/978-981-19-7343-7_8

Orefice, G. (2017). Non-tariff measures, specific trade concerns and tariff reduction. *World Economy*, *40*(8), 1807–1835. DOI: 10.1111/twec.12447

Orford, A. (2021). Regional orders, geopolitics, and the future of international law. *Current Legal Problems*, *74*(1), 149–194. DOI: 10.1093/clp/cuab005

Oulu, M. (2015). Climate change governance: Emerging legal and institutional frameworks for developing countries. In W. Leal Filho (Ed.), *Handbook of climate change adaptation* (pp. [insert page numbers if available]). Springer, Berlin, Heidelberg. https://doi.org/DOI: 10.1007/978-3-642-38670-1_9

Özdamar, Ö., & Shahin, E. (2021). Consequences of economic sanctions: The state of the art and paths forward. *International Studies Review*, *23*(4), 1646–1671. DOI: 10.1093/isr/viab029

Paes, F., & Geraldes, A. (2019). Convention on the Elimination of All Forms of Discrimination Against Women. In Leal Filho, W., Azul, A., Brandli, L., Özuyar, P., & Wall, T. (Eds.), *Gender equality. Encyclopedia of the UN Sustainable Development Goals*. Springer., DOI: 10.1007/978-3-319-70060-1_39-1

Palei, T. (2015). Assessing the impact of infrastructure on economic growth and global competitiveness. *Procedia Economics and Finance*, *23*, 168–175. DOI: 10.1016/S2212-5671(15)00322-6

Parashar, S., & Schulz, M. (2021). Colonial legacies, postcolonial 'selfhood' and the (un)doing of Africa. *Third World Quarterly*, *42*(5), 867–881. DOI: 10.1080/01436597.2021.1903313

Pattberg, P., Kaiser, C., Widerberg, O., & Stripple, J. (2022). 20 years of global climate change governance research: Taking stock and moving forward. *International Environmental Agreement: Politics, Law and Economics*, *22*(2), 295–315. DOI: 10.1007/s10784-022-09568-5

Perez, M. F., Brada, J. C., & Drabek, Z. (2012). Illicit money flows as motives for FDI. *Journal of Comparative Economics*, *40*(1), 108–126. DOI: 10.1016/j.jce.2011.03.007

Peterson, L., & Skovgaard, J. (2019). Bureaucratic politics and the allocation of climate finance. *World Development*, *117*, 72–97. DOI: 10.1016/j.worlddev.2018.12.011

Pisillo Mazzeschi, R., & Viviani, A. (2018). General principles of international law: From rules to values? In Pisillo Mazzeschi, R., & De Sena, P. (Eds.), *Global justice, human rights and the modernization of international law* (pp. 113–161). Springer., DOI: 10.1007/978-3-319-90227-2_6

Purvis, B., Mao, Y., & Robinson, D. (2019). Three pillars of sustainability: In search of conceptual origins. *Sustainability Science*, *14*(3), 681–695. DOI: 10.1007/s11625-018-0627-5

Rus, H. A. (2014). Corruption, conflict, and the management of natural resources. *Economics of Governance*, *15*(4), 355–386. DOI: 10.1007/s10101-014-0148-3

Sandage, J. (2014). The universal approach of the United Nations Convention against Corruption. In S. Manacorda, F. Centonze, & G. Forti (Eds.), *Preventing corporate corruption* (pp. [insert page numbers if available]). Springer, Cham. https://doi.org/ DOI: 10.1007/978-3-319-04480-4_2

Sankaran, K. (2022). Transition from the informal to the formal economy: The need for a multi-faceted approach. *The Indian Journal of Labour Economics : the Quarterly Journal of the Indian Society of Labour Economics*, *65*(3), 625–642. DOI: 10.1007/s41027-022-00398-2

Schrijver, N. (2016). Managing the global commons: Common good or common sink? *Third World Quarterly*, *37*(7), 1252–1267. DOI: 10.1080/01436597.2016.1154441

Sciaccaluga, G. (2020). Why states should protect forced climate migrants. In *International law and the protection of.* In *climate refugees"*. Palgrave Macmillan., DOI: 10.1007/978-3-030-52402-9_9

Sengupta, A. (2004). The human right to development. *Oxford Development Studies, 32*(2), 179–203. DOI: 10.1080/1360081041000169948

Shaffer, G., & Sandholtz, W. (2024). The rule of law under pressure: A transnational perspective. *Hague Journal on the Rule of Law, 16*(2), 393–438. DOI: 10.1007/s40803-024-00210-x

Shah, S. A. R., Abbas, N., Serbanescu, L., Niu, R., & Nassani, A. A. (2025). The key challenges and best alternatives to environmental sustainability: A comprehensive study. *Scientific Reports, 15*(1), 7042. DOI: 10.1038/s41598-025-90187-w

Singh, S., Sharma, G. D., Parihar, J. S., Dev, D., & Bansal, P. (2024). Nexus between environmental degradation and climate change during the times of global conflict: Evidence from CS-ARDL model. *Environmental and Sustainability Indicators, 22*, 100368. DOI: 10.1016/j.indic.2024.100368

Sorooshian, S. (2024). The sustainable development goals of the United Nations: A comparative midterm research review. *Journal of Cleaner Production, 453*, 142272. DOI: 10.1016/j.jclepro.2024.142272

Spyromitros, E., & Panagiotidis, M. (2022). The impact of corruption on economic growth in developing countries and a comparative analysis of corruption measurement indicators. *Cogent Economics & Finance, 10*(1), 2129368. DOI: 10.1080/23322039.2022.2129368

Staněk, M. (2017). The humanitarian crisis and civil war in Syria: Its impact and influence on the migration crisis in Europe. *Kontakt, 19*(4), e270–e275. DOI: 10.1016/j.kontakt.2017.09.013

Stone, C. D. (2004). Common but differentiated responsibilities in international law. *The American Journal of International Law, 98*(2), 276–301. DOI: 10.2307/3176729

Šturma, P. (2024). The notion of "sanctions" and "countermeasures" in international law. In Šturma, P. (Ed.), *International sanctions and human rights* (pp. xx–xx). Springer., DOI: 10.1007/978-3-031-69019-8_2

Suliman, F., Khwanda, H., & Murthy, R. V. R. (2023). An analysis of the Syrian economy in the era of military conflict, 2011–2020: The perspective of government and economics. *Journal of Government and Economics, 11*, 100082. DOI: 10.1016/j.jge.2023.100082

Syna, H. D. (2020). Understanding the interface of diversity and political tensions in the context of divided societies: A multifocal perspective of social psychology and critical theory. In *Diversity management in places and times of tensions*. Palgrave Pivot., DOI: 10.1007/978-3-030-37723-6_3

Tan, J. (2009). Multinational corporations and social responsibility in emerging markets: Opportunities and challenges for research and practice. *Journal of Business Ethics*, *86*(S2), 151–153. https://www.jstor.org/stable/27749681. DOI: 10.1007/s10551-009-0189-7

Thacker, S., Adshead, D., Fay, M., Hallegatte, S., Harvey, M., Meller, H., O'Regan, N., Rozenberg, J., Watkins, G., & Hall, J. W. (2019). Infrastructure for sustainable development. *Nature Sustainability*, *2*(4), 324–331. DOI: 10.1038/s41893-019-0256-8

Thorbecke, E., & Charumilind, C. (2002). Economic inequality and its socioeconomic impact. *World Development*, *30*(9), 1477–1495. DOI: 10.1016/S0305-750X(02)00052-9

Trahan, J. (2020). Questioning the legality of veto use in the face of genocide, crimes against humanity, and/or war crimes. In *Existing legal limits to Security Council veto power in the face of atrocity crimes* (pp. 142–259). Cambridge University Press. DOI: 10.1017/9781108765251.006

Tripathi, S., & Roepstorff, K. (2020). Decentering peace and conflict studies: Conceptualisations of peace in India. *Zeitschrift für Friedens- und Konfliktforschung*, *9*, 79–99. https://doi.org/DOI: 10.1007/s42597-019-00014-z

Tuori, K. (2023). Pacta sunt servanda. *Suomalainen Tiedeakatemia Toimituksia. Sar. A.4: Biologica*, *2*(1), 44–57. DOI: 10.57048/aasf.130107

Turner, M. G., Collins, S. L., Lugo, A. L., Magnuson, J. J., Rupp, T. S., & Swanson, F. J. (2003). Disturbance dynamics and ecological response: The contribution of long-term ecological research. *Bioscience*, *53*(1), 46–56. DOI: 10.1641/0006-3568(2003)053[0046:DDAERT]2.0.CO;2

Valipour, E., Ketabchi, H., Shali, R. S., & Morid, S. (2024). Water resources allocation: Interactions between equity/justice and allocation strategies. *Water Resources Management*, *38*(2), 505–535. DOI: 10.1007/s11269-023-03682-y

Vento, E. (2024). The global institutionalization of human rights discourse: A cross-national analysis of the language used in the International Labour Conference during the Cold War. *Nordic Journal of Human Rights*, *42*(2), 135–157. DOI: 10.1080/18918131.2024.2313907

Visser, F. (1988). The principle of permanent sovereignty over natural resources and the nationalisation of foreign interests. *The Comparative and International Law Journal of Southern Africa, 21*(1), 76–91. https://www.jstor.org/stable/23247647

Wewerinke-Singh, M. (2023). The rising tide of rights: Addressing climate loss and damage through rights-based litigation. *Transnational Environmental Law, 12*(3), 537–566. DOI: 10.1017/S2047102523000183

Winkler, I. T., & Williams, C. (2017). The Sustainable Development Goals and human rights: A critical early review. *International Journal of Human Rights, 21*(8), 1023–1028. DOI: 10.1080/13642987.2017.1348695

Xie, L., & Jia, S. (2017). Diplomatic water cooperation: The case of the Sino-India dispute over Brahmaputra. *International Environmental Agreement: Politics, Law and Economics, 17*(5), 677–694. DOI: 10.1007/s10784-016-9339-4

Yi, Y., Luo, J., & Wübbenhorst, M. (2020). Research on political instability, uncertainty, and risk during 1953–2019: A scientometric review. *Scientometrics, 123*(2), 1051–1076. DOI: 10.1007/s11192-020-03416-6

Zahoor, N., Wu, J., Khan, H., & Khan, Z. (2023). De-globalization, international trade protectionism, and the reconfigurations of global value chains. *MIR. Management International Review, 63*(5), 823–859. DOI: 10.1007/s11575-023-00522-4

Zapp, M. (2021). International organisations and the proliferation of scientised global reporting, 1947–2019. *Global Society, 36*(4), 455–474. DOI: 10.1080/13600826.2021.1902284

Zhang, M., Zhang, H., Zhang, L., Peng, X., Zhu, J., Liu, D., & You, S. (2023). Corruption, anti-corruption, and economic development. *Humanities & Social Sciences Communications, 10*(1), 434. DOI: 10.1057/s41599-023-01930-5

Chapter 3
Capacity Building and Knowledge Sharing Platforms for the SDGs

Satya Subrahmanyam
https://orcid.org/0000-0003-0441-2742
Holy Spirit University of Kaslik, Lebanon

ABSTRACT

Achieving the Sustainable Development Goals (SDGs) by 2030 requires coordinated global efforts, with capacity building and knowledge sharing playing crucial roles. This chapter explores their importance in equipping individuals and organizations to drive sustainable development, focusing on the integration of spatial data, technology, and localized strategies. Capacity building strengthens institutional frameworks, human resources, and policies to address SDG challenges, while knowledge sharing promotes collaboration, innovation, and the exchange of best practices. The chapter highlights various platforms—digital, physical, and hybrid—that facilitate learning and collaboration, with case studies demonstrating their success. It also examines the role of geographical approaches in addressing region-specific challenges and offers strategic recommendations to enhance inclusivity and equity in these efforts.

1. INTRODUCTION

1.1 Overview of Sustainable Development Goals (SDGs)

Historical Context and Importance of the SDGs - In the Sustainable Development Goals (SDGs), we have a plan to combat global warming, inequality, poverty, and unsustainable development. In 2015, the 17 Sustainable Development Goals

DOI: 10.4018/979-8-3373-1300-9.ch003

(SDGs) were unveiled by the United Nations to carry out the objectives set forth in their 2030 Agenda for Sustainable Development. Expanding beyond conventional development goals and resolving previous shortcomings, the United Nations used the Millennium Development Goals as a springboard to create the Sustainable Development Goals (SDGs) (United Nations, 2015).

As a whole the SDGs offer universal aspects which promote integrated transformation while striving together to create sustainable development opportunities worldwide. The program works to create an equal framework for economic progress and environmental wellness together with social development needs because development problems cross these areas. The three sustainability goals to eliminate hunger (SDG 2) and ensure water access (SDG 6) and fight climate change (SDG 13) explain their essential relationship between economic activity and social welfare (Sachs et al., 2019).

Role of Capacity Building and Knowledge Sharing in Achieving the SDGs - The achievement of SDGs requires the establishment of capacity together with knowledge transfer. The objective of capacity building involves developing abilities across all levels from individual people to collective groups and institutional entities to implement sustainable development-oriented policies together with assessments. Knowledge sharing allows free exchange of data along with expertise and knowledge between relevant parties resulting in enhanced teamwork while also improving decision quality and creativity.

Community and national abilities to address challenging issues increase through joint utilization of community and national intellect and resources. Developed nations that transfer technologies to developing regions can greatly enhance their access to renewable energy systems under SDG 7 while building SDG 9 resilient infrastructure. The development of localized solutions becomes feasible through knowledge exchange platforms because they align with unique cultural and environmental settings (Meadowcroft, 2020).

1.2 Purpose and Scope of the Chapter

Highlighting Geographical Approaches to Capacity Building and Knowledge Sharing - Geographical methodologies provide unique analytical tools for examining knowledge-sharing mechanisms and capacity development. These approaches acknowledge the requirement of sustainable development by means of both spatial areas and contextual elements, therefore enabling varied cultural expressions and regional variations across landscapes. Stakeholder deployment of geographic information systems (GIS) as well as spatial data analysis and participatory mapping approaches helps to increase the knowledge of regional issues and possibilities (Haworth et al., 2019).

GIS tools found critical applications in locating regions lacking clean water access which directed specific water-intervention efforts towards the implementation of SDG 6. The method of participatory mapping assists marginalized communities to demonstrate their essential requirements which leads to more inclusive development planning practices. The chapter explores geographical methodologies along with their capabilities to boost resource optimization and sponsor better stakeholder relations.

Establishing the Interconnection Between Geographical Methods, Platforms, and Sustainable Development - The attainment of SDGs needs organizations to use geographical analysis and platforms to distribute information along with developing capabilities across all sectors. Through these methodologies policy development and adaptive management receive backing because they deliver data-based insights for decision-making processes. Geographic data distribution together with best practice sharing needs to be accessible and transparent through digital platforms like the UN SDG Knowledge Hub and regional databases.

The discussion focuses on explaining how these platforms serve as venues for public-private partnership collaboration in this chapter. The book explores how spatial methods represent a transformative approach to resolve challenging sustainability problems through evaluations of real-world implementations and successful practices.

SDG-specific platforms need a creative and inclusive approach to establish knowledge exchange and capacity development platforms. The focus of geographical methods on spatial and context-based dynamics makes them valuable instruments aiming to enhance sustainable development project efficiency and equity. The groundwork presented in this chapter serves to establish the foundation for detailed SDGs analysis in following chapters about their role within regional, national and international institutions for speedier SDGs achievement.

2. UNDERSTANDING CAPACITY BUILDING IN THE CONTEXT OF SDGS

As an international endeavor the Sustainable Development Goals (SDGs) work to establish sustainability for 2030 by confronting important matters such as poverty along with inequality and environmental degradation and economic instability. Measurable development depends on capacity building because it gives people and groups and communities the resources, they need through knowledge acquisition and support structures to fulfill their sustainable development targets.

2.1 Definition and Importance

Defining Capacity Building in Sustainable Development - UNDP (2018) defines capacity building as expanding capabilities to handle developmental problems with lasting solutions through skill improvements and framework development. SDG-related capacity building strategies focus on creating opportunities to empower different stakeholders so they can design, lead and track development projects with sustainable development targets. Various initiatives targeting both expertise development through education and resource development via infrastructure gains together with policy modifications constitute these efforts.

Capacity building serves as a crucial element because it creates adaptable stakeholders who show resilience through development challenges. The capability of people as well as groups and societies help them address complex evolving difficulties while achieving development outcomes that integrate inclusiveness with sustainability. Capacity building helps distribute best practices and innovations through its efforts to connect developed and developing nations which creates more balanced sustainable development processes (Meadowcroft 2020).

The Importance of Capacity Building Across Levels - Capacity building activities influence development at three distinct levels: international and national together with local. The global platform advances teamwork between nations through knowledge exchange to distribute sustainable development advancements across every geographical area. The United Nations Technology Facilitation Mechanism (TFM) serves as a vital tool that distributes innovative methods and technologies to link developed and developing nations (United Nations, 2015).

The SDGs benefit from national capacity building because governments receive help to develop and execute policies that support these goals. The progress monitoring system requires enhancement through improved statistical frameworks and better institutional performance and development project resource acquisition mechanisms. The development of data collection and analysis capabilities at a national scale remains essential for both proper decision-making and resource management (Sachs et al., 2019).

Capacity building within local communities leads to development initiatives which they can independently manage and direct. Loosely structured development approaches combined with education for local leaders enable people to create initiatives that serve their communities' exclusive requirements and difficulties. Community resilience improves alongside inclusiveness when localized capacity development projects push the Sustainable Development Goals toward complete realization (Haworth et al., 2019).

2.2 Key Components of Capacity Building

Institutional Capacity - Institutional capacity building requires developing structures along with processes and networks which enhance organizational performance. To increase program delivery effectiveness organizations should establish better governance systems while creating coordinating systems between departments and strengthening delivery infrastructure. The achievement of SDG-related goals depends heavily on digital platforms that perform data collection and monitoring functions. The assessment and building of institutional capabilities require establishing relationships between different operational sectors because sustainable development problems manifest through their connections (UNDP, 2018).

Human Resource Development - Human resource development (HRD) delivers training to develop individual abilities and competence that strengthen sustainable development contributions. The development of human capital requires educational institutions alongside career skill training together with context-specific development programs for climate adjustment and healthcare systems and agriculture that adheres to sustainability principles. The Green Jobs Initiative through the International Labour Organization (ILO) illustrates how the organization trains people to join sustainable economic sectors. In order to provide equal sustainable development opportunities, the framework prioritizes inclusivity by directing its efforts at marginalized communities (Meadowcroft 2020).

Policy and Regulatory Frameworks - Building a favorable capacity-building environment requires strong laws along with proper regulations that need to be established first. The models support sustainable policies which define roles of all actors and create resource distribution frameworks. The Paris Agreement provides Nationally Determined Contributions (NDCs) that establish specific reduction goals for carbon emissions from each nation as a part of environmental targets linked to SDGs (United Nations, 2015).

2.3 Challenges in Capacity Building for SDGs

Funding Limitations - Lack of sufficient funding poses a main obstacle to capacity building efforts. SDG implementation depends heavily on extensive financial support even though numerous developing countries encounter both budgetary restrictions and insufficient international financing availability. Reaching the SDGs in developing nations requires annual funding of $2.5 trillion to bridge the estimated financial deficit while innovative methods like green bonds and public-private partnerships are needed (Sachs et al., 2019). Lack of proper funding transforms into institutions with poor resources along with insufficient training along with limited access to modern technology which creates barriers for sustainable development.

Political and Socio-Economic Barriers - Capacity-building makes progress difficult because of existing political instability together with weak governance systems and socio-economic inequalities. The areas experiencing either conflict or corruption struggle to implement sustainable development programs successfully. The success of reforms together with resource allocation for capacity enhancement depends heavily on political will yet political priorities sometimes reject long-term development targets (Haworth et al., 2019). The social disparities caused by poverty along with gender inequality create barriers to education and healthcare and technological access which decreases the success rates of capacity-building programs.

Inequality in Access to Resources and Education - Unequal outcomes regarding capacity-building efforts emerge because various regions obtain different quantities of resources along with unequal access to educational possibilities. Rural communities together with marginalized groups encounter more obstacles when trying to get training programs with modern technologies and get access to funding resources than affluent urban neighborhoods do. The digital gap further increases social inequalities by cutting off low-income nations from online learning experiences and knowledge-sharing chances (UNDP, 2018). Along with educational materials accessible in many languages to construct equitable capacity development initiatives, efforts targeted at addressing these inequities should include specific mobile training setups for far-off sites.

Capacity building stands as an essential component of the SDG framework which helps people and groups of all sizes to develop sustainable developmental strategies for implementation. Stakeholders gain the ability to tackle complex problems efficiently because capacity-building initiatives build their institutional capabilities and develop human resources and establish thorough policy frameworks.

Initiatives to build capacity face implementation barriers because of restricted funding support together with political complications and economic struggles and differences in resource availability. A transformative solution involves working together with new financial ideas and complete equality objectives to conquer these challenges. Capacity building will serve as the foundation toward reaching the SDGs by 2030 because it requires supporting policies that empower all stakeholders to develop a sustainable equitable society.

3. THE ROLE OF KNOWLEDGE SHARING IN ACHIEVING SDGS

In 2015 the United Nations launched the Sustainable Development Goals (SDGs) to fight critical problems that include education, inequality, climate change and poverty. These goals depend significantly on knowledge sharing to allow the exchange of data and successful approaches and new ideas throughout various industries and worldwide regions. The achievement of SDGs depends on knowledge sharing because it promotes collaborative work while spreading sustainable solutions throughout the world.

3.1 Definition and Dimensions of Knowledge Sharing

Knowledge sharing describes the methodical transfer of data combined with skills and practical wisdom which exists between people and groups and organizations to bring about shared objectives. Sustainable development requires the transfer of the best practical methods together with technological advancements and new ideas to solve the diverse SDG-related issues. Effective information dispersal in knowledge sharing includes both information distribution and solution co-creation with specific localization (Subrahmanyam et al., 2024).

Knowledge sharing contains a dimension where organizations exchange their innovative best practices. Organizations together with governments can employ successful implementation strategies from different contexts to create new solutions for other situations. One area of effective knowledge transfer occurs when renewable energy projects within a certain region serve as inspiration to similar developments in other parts of the world whose implementation fits both environmental and economic parameters. The essential aspect involves building partnerships between organizations which span both geographic areas and different business sectors. Through cooperation among governments, businesses, civil society groups, and academic institutions, governments may pool their resources along with their experience. The United Nations Partnerships for SDGs serves as an example of effective coordination between different sectors because it demonstrates proper cooperation on complex issues (United Nations, 2015).

3.2 Benefits of Knowledge Sharing for SDGs

Knowledge sharing functions directly as a tool to reach the SDGs through its multiple benefits. The promotion of sustainable practices together with new idea sharing allows quick implementation of solutions to resolve world's pressing problems. Underdeveloped nation farmers access climate-resilient crop information through CGIAR's Consultative Group on International Agricultural Research (CGIAR) agricultural knowledge-sharing networks according to Sachs et al. (2019). The improvement of food security accompanies poverty reduction as a beneficial outcome.

Knowledge sharing tools help organizations achieve innovation success through group wisdom development. Collaboration among diverse stakeholders fosters creativity and the development of holistic solutions. The joint programs of universities and technology companies produce innovative discoveries of renewable energy systems along with digital health technologies. These technological advancements enable solutions for reducing SDG-related problems such as health care services and clean energy accessibility.

The implementation of sustainable practices requires capabilities that knowledge sharing provides to organizations and individuals through necessary tools and expert skills. Programs that provide training and workshops and online learning platforms enable the distribution of information about sustainable urban planning and waste management as well as renewable energy implementation methods. These initiatives allow communities to control development processes which result in enduring effective results for SDG-related initiatives (Meadowcroft 2020).

3.3 Barriers to Knowledge Sharing

Multiple obstacles prevent successful deployment of knowledge-sharing initiatives toward sustainable development despite their useful outcomes. A significant hurdle exists from technological gaps and digital divisions that separate different geographic areas. The gap exists between developed economies which possess strong digital infrastructure and sharing platforms, but numerous developing nations do not have sufficient technological capabilities. The combination of slow internet service and obsolete technologies together with insufficient digital skills prevents minorities from accessing global knowledge sharing platforms (Haworth et al., 2019).

The sharing of knowledge encounters major barriers because of issues relating to intellectual property laws and data privacy requirements. Intellectual property rights establish barriers to accessing proprietary information as well as technological assets that could be used for broader benefit. Organizations avoid collaborative knowledge-sharing initiatives because of data privacy regulations and their concerns regarding information misuse. The process of protecting IP and data privacy while preserving open knowledge access represents a multifaceted challenge that affects the achievement of SDGs.

The barriers between effective knowledge sharing appear due to cultural and linguistic differences. Different communication and language conventions along with social standards across countries cause mutual misunderstanding resulting in reduced knowledge sharing efficiency. Two factors make technical information unavailable to stakeholders across different regions: technical information must be appropriately adapted for different regions and require translation into the targeted languages. The good results of knowledge sharing call for business investments in multilingual platforms alongside culturally sensitive communication methods (Nonaka & Takeuchi, 1995).

3.4 Overcoming Barriers and Enhancing Knowledge Sharing

Governments and companies along with international entities must take deliberate actions to remove obstacles in information exchange. To create digital networks as well as initiatives teaching digital skills to places lacking digital literacy, the digital divide calls for constant investment. The Digital Economy for Africa (DE4A) project of the World Bank aims to improve digital access and education, hence fostering greater participation in knowledge-sharing platforms (World Bank, 2020).

Stakeholders who want to handle intellectual property matters should select open-access models together with collaborative licensing agreements. Through its Open COVID Pledge initiative organizations were inspired to contribute their intellectual property for pandemic research programs thus demonstrating how open innovation shortens the time required to solve world problems. Establishing detailed data-sharing protocols which place privacy above all else can help businesses overcome their concerns about data misuse.

Cultural understanding and multilingual communication growth requires organizations to spend money on translation support and multiculture training and to advance international dialogue programs. The World Atlas of Languages maintained by UNESCO creates platforms to share information through multiple languages which builds an inclusive knowledge-sharing environment (UNESCO, 2021).

A basic component of reaching the SDGs is knowledge sharing, that is, the exchange of effective practices and cooperation between people who propel creative decision-making. Logical development toward implementing sustainable solutions along with stakeholder empowerment via educational programs addresses basic issues in different geographical regions and industries.

Knowledge-sharing initiatives face barriers because of the digital divide together with intellectual property concerns and cultural differences. These barriers reduce the effectiveness of knowledge-sharing programs. The removal of these obstacles demands purposeful funding of digital networks followed by open-source distribution methods coupled with cultural sensitivity in communications techniques. The achievement of SDGs until 2030 requires improved knowledge sharing methods so that we can establish a sustainable and equitable future.

4. PLATFORMS FOR CAPACITY BUILDING AND KNOWLEDGE SHARING

The accomplishment of Sustainable Development Goals (SDGs) demands permanent knowledge and skill exchanges through platforms which build capacity for collaboration. Through these platforms people can distribute novel concepts and proven methods along with technological approaches to overcome the interconnected SDG challenges. Such platforms serve as a vital function by providing learning areas along with interactive spaces and innovative platforms which drive forward national and local and global development initiatives.

4.1 Types of Platforms

Platforms for capacity building and knowledge sharing can be categorized into digital, physical, and hybrid models. Each type offers unique advantages, catering to diverse needs and contexts.

Digital Platforms - Current knowledge exchange functions best with digital platforms as essential operational tools. E-learning tools and webinars with virtual forums represent digital platforms that supply easy-to-scale solutions and accessibility for capacity-building needs. Users can take part in training sessions combined with discussions and collaborative activities without any geographical barriers on these online platforms. The e-learning platforms Coursera together with edX provide sustainability-focused courses that teach learners about renewable energy systems and climate change adaptation as well as sustainable city design (Anderson, 2020).

The capacity-building initiative reach becomes more powerful through the combination of webinars and virtual forums. The digital platforms Zoom and Microsoft Teams have established themselves as essential tools for running international dialogues and workshops. The United Nations arranged webinars to handle public health and socio-economic challenges throughout the COVID-19 pandemic which reached stakeholders across the globe (United Nations, 2020).

Physical Platforms - Physical platforms consisting of conferences and workshops together with field training programs give students the chance to interact with others directly while gaining practical experience. Physically attended conferences under the United Nations Climate Change Conferences (COP) have created spaces where policymakers and scientists along with activists can exchange their knowledge to develop solutions against global climate problems. Local populations achieve sustainable development through community-level training sessions and workshops that include agricultural extension programs.

Through field training programs technical knowledge demonstrates high efficiency in its transfer process. Through Food and Agriculture Organization (FAO) programs farmers learn sustainable farming techniques like organic agriculture and both crop multipurpose methods as well as water preservation methods. Such engagements establish long-term effects because they promote faith-building relationships and develop enduring neighborhood skills (FAO, 2021).

Hybrid Models - Combining digital and physical instruments, hybrid models establish inclusive platforms for capacity development. These models combine the advantages of in-person interaction with the accessibility of both methods to guarantee overall benefit. Because they allow involvement from many stakeholders—including those in far-off locations with restricted access to physical events—hybrid platforms are becoming more and more appealing.

The Global Environmental Facility (GEF) Expanded Constituency Workshops are one instance of a hybrid paradigm as they combine online materials with in-person discussions. These seminars provide digital tools for follow-up learning (GEF, 2021), thereby facilitating communication among government officials, non-governmental organizations (NGRs), and community leaders. Hybrid systems improve the inclusiveness and efficiency of capacity-building projects by closing the digital divide and allowing different learning preferences.

4.2 Case Studies of Successful Platforms

Several platforms have demonstrated remarkable success in capacity building and knowledge sharing for sustainable development. These case studies highlight the diverse applications and impacts of such initiatives.

Geographical Information Systems (GIS) in Disaster Management - Geographical Information Systems (GIS) serve as powerful tools for capacity building in disaster management. GIS platforms achieve risk assessment and response planning and recovery implementation through their ability to combine real-time data with spatial information. The Pacific Disaster Center operates DisasterAWARE as their main platform which provides disaster-related information to governments and humanitarian organizations for hurricanes and floods and earthquakes.

The platform delivers training to its users that enables them to analyze spatial data while creating data-driven approaches to deal with disasters before and after they happen. Grasping and coordinating efforts through geographic information system platforms delivers significant value by supporting response preparation to minimize losses and save human lives (Pacific Disaster Center, 2020).

UN Sustainable Development Solutions Network (SDSN) - SDSN stands as an online knowledge hub which serves to promote global collaboration for reaching the SDGs. SDSN came to life in 2012 to establish a connection between academic institutions together with civil society organizations and policymakers who work on solutions for sustainable development issues. The network offers people a wide range of materials such as research publications and policy documents together with educational courses related to climate mitigation and social equity along with sustainable monetary practices.

Users benefit from SDSN's digital platform because it contains an online library, interactive webinars along with virtual forums which facilitate knowledge sharing between participants from different locations. sürekli collaboration with various organizations promotes research with multiple disciplines to bring substantial advancement to the SDG agenda (SDSN, 2021).

Community-Led Initiatives for Localized Capacity Building - Communities take essential leadership roles to give their people power through initiatives that focus on addressing regional obstacles. Participatory approaches form a central part of these initiatives to make sure solutions fit with local culture and are locally owned. The Barefoot College in India operates an exemplary program by teaching rural women known as "Solar Mamas" how to install along with maintain solar energy systems throughout their local communities.

Software alongside practical lessons from this learning experience which allows students to develop technical skills together with entrepreneurial intelligence. The Barefoot College demonstrates how focused initiatives for renewable energy and women's training lead to remarkable social transformations according to Roy (2017).

Contributions to Sustainable Development - Achieving the SDGs depends much on platforms for knowledge exchange and capacity development. Encouragement of sustainable behaviors and technologies helps to hasten development in fields like social inclusion, environmental protection, and elimination of poverty by means of technology.

Digital platforms have democratized knowledge and information, therefore enabling underprivileged groups to engage in attempts at sustainable development. Physical platforms help to share best practices and ideas, therefore fostering innovation and teamwork. By combining digital and physical instruments, hybrid models guarantee inclusiveness and flexibility, therefore addressing many requirements and circumstances.

Effective platforms include GIS, SDSN, and community-led projects provide scalable and repeatable models. These platforms improve resilience, creativity, and sustainability in local communities all around by allowing knowledge transfer and capacity development at many levels.

Achieving the SDGs depends on platforms for knowledge exchange and capacity development. Whether digital, physical, or hybrid, these systems let different industries and areas share information, tools, and expertise. The case studies of GIS in disaster management, SDSN, and community-led projects show the many uses and results of such projects.

All throughout the world, people are putting money into and building up these platforms to tackle difficult problems. They can forge paths toward sustainable development and fortify the earth through resilience if they promote inclusivity, cooperation, and innovation.

5. GEOGRAPHICAL APPROACHES TO CAPACITY BUILDING AND KNOWLEDGE SHARING

Geography plays a pivotal role in shaping capacity-building strategies and knowledge-sharing platforms aimed at achieving the Sustainable Development Goals (SDGs). By leveraging geographical data, mapping technologies, and place-based strategies, stakeholders can address region-specific challenges, optimize resource allocation, and foster sustainable development.

5.1 Role of Geographical Data and Mapping in SDGs

For the SDGs to be effectively implemented, geographic data and mapping technology are crucial tools. They pave the way for the gathering, processing, and presentation of geographical data, which in turn helps with policy and decision-making by revealing trends and patterns. Addressing sustainability concerns has become more dependent on technologies like remote sensing and Geographical Information Systems (GIS).

GIS and Remote Sensing for Data Collection and Monitoring - Environmental, social, and economic indicators may be monitored in real-time with the use of GIS and remote sensing, which enable the gathering of high-resolution data. Particularly helpful in monitoring development toward climate action, biodiversity, and urban development SDGs are these instruments. One example is remote sensing, which uses satellite photography to track things like the pace of deforestation, the spread of cities, and the availability of water resources (Hansen et al., 2013).

GIS complements this by integrating spatial and non-spatial data to create interactive maps and models. These visualizations help policymakers and stakeholders understand complex systems and identify areas requiring intervention. For example, GIS has been used to monitor the spread of diseases like malaria and COVID-19, enabling targeted healthcare interventions and resource allocation (Oliveira et al., 2021). The ability of GIS and remote sensing to provide detailed, location-specific insights makes them indispensable for capacity building and knowledge sharing, as they guide strategies that address localized challenges effectively.

Spatial Analysis for Identifying Gaps and Needs in Capacity Building - Spatial analysis, a key feature of GIS, allows for the identification of gaps and needs in capacity building. By analyzing spatial data, stakeholders can pinpoint underserved regions, assess disparities in resource distribution, and prioritize interventions. For instance, spatial analysis can reveal areas with limited access to education, healthcare, or clean water, enabling targeted capacity-building efforts.

In agriculture, spatial analysis has been used to identify regions prone to soil degradation and water scarcity. This information informs training programs for farmers on sustainable practices such as soil conservation and irrigation management (Foley et al., 2011). Similarly, spatial data is instrumental in disaster risk reduction, where it helps map vulnerable areas and develop early warning systems. By providing a nuanced understanding of regional needs, spatial analysis ensures that capacity-building initiatives are aligned with local realities and priorities.

5.2 Place-Based Strategies for Knowledge Sharing

When knowledge is catered to the geographical and cultural settings of the target audience, it is most likely successful. Emphasizing region-specific training programs and regional knowledge-sharing projects, place-based techniques guarantee that solutions are relevant, pragmatic, and culturally suitable.

Region-Specific Training Programs - Training courses tailored to each region handle the particular possibilities and difficulties of every geographical location. To plan and implement capacity-building projects, these programs can include cooperation among local governments, educational institutions, and community groups. Water management training initiatives in sub-Saharan Africa have been customized to fit the particular problems of water shortage and seasonal fluctuation of the area. Under these initiatives, local populations get knowledge in water conservation methods, aquifer recharge, and rainwater collecting (UNESCO, 2019).

In coastal regions, similarly, region-specific training initiatives concentrate on marine conservation and sustainable fishing. These projects help local people to sustain livelihoods by teaching them to follow behaviors that protect marine life. These initiatives improve the efficacy of capacity-building initiatives and help to meet SDG objectives by addressing regional specific issues.

Localized Knowledge Sharing Tailored to Geographical and Cultural Contexts - Localized knowledge exchange acknowledges the need of cultural, social, and environmental elements in forming results of sustainable development. It guarantees that solutions are locally relevant and culturally appropriate by means of the information, skills, and practices exchanged within particular communities.

Indigenous knowledge used in environmental preservation is one effective instance of localized knowledge exchange. Indigenous people in the Amazon rainforest have shared with academics and legislators their customary methods of sustainable forest management. Modern conservation plans have included these methods—which include selective harvests and controlled burns—into evidence of the need of local knowledge in reaching SDG 15 (life on land) (Gavin et al., 2018).

Another such is the Barefoot College project in India, which equips rural women to work as solar engineers. This program empowers women to deliver sustainable energy solutions to their areas by combining technical instruction with culturally appropriate methods. Customizing knowledge-sharing projects to local settings helps to guarantee long-term viability and increase community involvement.

Contributions to Sustainable Development - By tackling region-specific issues, besting resource allocation, and encouraging innovation, geographical methods to capacity development and information exchange help greatly contribute to the SDGs. By using GIS and remote sensing, capacity-building projects improve their accuracy and efficiency, therefore guaranteeing that interventions are data-driven and powerful. Conversely, place-based approaches guarantee inclusive knowledge exchange with cultural relevance, therefore promoting community ownership and long-term sustainability.

By directing urban planning initiatives that give affordable housing, efficient transit, and green areas first priority, GIS-enabled spatial analysis has been crucial in reaching SDG 11—sustainable cities and communities. Likewise, by arming local people with skills to sustainably manage water resources, place-based training initiatives have helped SDG 6—clean water and sanitation.

Geographic strategies close the distance between world aspirations and local reality by combining technical and customized solutions. They enable people to own their growth so that solutions are not just efficient but also fair and inclusive.

Achieving the SDGs depends on geographic approaches to knowledge exchange and capacity development. Using GIS, remote sensing, and spatial analysis can help stakeholders compile useful information guiding focused initiatives. By means of region-specific training courses and localized knowledge-sharing projects, place-based tactics guarantee that solutions are practical and culturally appropriate.

Geographical methods must be invested in and expanded while the world community keeps striving for sustainable development. These approaches improve not just the efficacy and efficiency of capacity-building initiatives but also resilience, creativity, and inclusiveness. Geographic methods provide a complete framework for handling the difficult issues of sustainable development by including technology and cultural aspects.

6. POLICY AND STRATEGIC RECOMMENDATIONS FOR ACHIEVING SDGS

Reaching the Sustainable Development Goals (SDGs) calls for coordinated action across many sectors and players. Policies and strategies must be developed with an eye on encouraging diversity and equality, building alliances, and merging sophisticated technology if we are to guarantee that development is inclusive, fair, and environmentally sustainable.

6.1 Promoting Inclusivity and Equity

Reaching the SDGs calls for diversity and fairness. Capacity-building programs and knowledge-sharing platforms provide advantages that should be felt by all people, but especially underprivileged and underprivileged groups. Policies that provide equitable access to these possibilities should be given top priority by policy makers so that none is left behind.

Ensuring Equal Access to Knowledge Sharing Platforms - The exchange of best practices, ideas, and solutions to world problems depends on knowledge-sharing platforms—digital or physical. For these platforms to be successful, however, they must be available to everybody, even those from far-off rural or economically poor places. Although inadequate internet connection, lack of digital literacy, and the availability of technology frequently restrict digital platforms, which have the potential to provide broad access to information. Policies should therefore concentrate on building digital infrastructure, extending internet access, and offering training to guarantee that people from all backgrounds may engage in worldwide knowledge-sharing networks.

Working with the business sector, governments should give top priority to investments in the growth of broadband networks and the creation of low-cost mobile apps able to provide instructional materials to underprivileged groups. Making material accessible in many languages and formats helps localize knowledge-sharing platforms by removing obstacles such low literacy rates and linguistic diversity. Policymakers can guarantee that capacity development fully supports the achievement of SDGs relating to education, poverty reduction, and gender equality by concentrating on inclusion and thus ensuring that capacity building realizes its full potential (UNESCO, 2020).

Addressing Gender and Regional Disparities - Achieving the SDGs still depends much on gender and regional inequalities. Particularly in underdeveloped areas, women can have restricted access to healthcare, education, and employment. Their capability to contribute to and gain from programs for capacity-building is hampered by this disparity. Likewise, because of inadequate infrastructure and resources, areas impacted by violence, poverty, and climate change are often excluded from worldwide knowledge-sharing forums.

Policies have to stress the need for gender-sensitive and region-specific strategies if we are to solve these inequalities. Affirmative action laws should be adopted by governments and companies to help women and underprivileged groups to be included into leadership and training positions. Programs aiming at women's economic empowerment, for instance, should be expanded especially in rural and conflict-torn regions. Likewise, regional strategies for capacity development should be created to handle the particular difficulties experienced by certain groups, including those living in distant rural regions or low-income metropolitan areas. These focused actions may assist in guaranteeing fair distribution of the advantages of knowledge sharing and capacity development (Kabeer, 2019).

6.2 Encouraging Partnerships

Achieving the SDGs calls for cooperation. Whether a government, an international organization, or a private company, the intricate problems of sustainable development are too large for any one agency to address. This makes it imperative for governments, NGOs, businesses, and local communities to cooperate to build capacity and efficiently distribute knowledge.

Collaboration Between Governments, NGOs, and Private Sectors - In the quest of the SDGs, governments, NGOs, and the business sector all bring special advantages to hand. While NGOs often have strong grassroots ties and can guarantee that actions are culturally acceptable and locally driven, governments may provide regulatory frameworks, financing, and policy direction. Conversely, the private sector may provide resources, knowledge, and technical advances to help projects to be scaled up.

Policies should inspire cooperation across many sectors if capacity-building projects are to be effective. Public sector policies should provide an enabling environment that encourages private-sector investment in SDG-related initiatives such education, healthcare, and renewable energy, including those connected to several areas, public-private partnerships (PPPs) have shown success by means of resource mobilization, skill-combining, and capacity-building project efficacy enhancement. Improving education availability and quality in underdeveloped nations has been mostly dependent on the Global Partnership for Education, a PPP including governments, multilateral organizations, NGOs, and the private sector (World Bank, 2020).

Particularly in areas where government ability could be restricted, NGOs should be acknowledged as major collaborators in furthering the SDGs. Policymakers should help NGOs in capacity-building initiatives by means of subsidies, encourage cooperation among other stakeholders, and guarantee access to pertinent information and expertise by which they contribute. Strengthening alliances across many different industries helps to create capacity in a more sustainable, scalable, and efficient manner.

Leveraging Public-Private Partnerships (PPPs) - Two areas where public-private partnerships really shine are scaling up creative ideas and generating the required money for major capacity-building programs. The fulfillment of the SDGs depends on all three of these areas: health, education, and the environment; so, these alliances might assist to realize those technologies. Governments and IT companies may cooperate to hasten the development of digital platforms poor communities could utilize to access healthcare, education, and other needs. PPPs in the energy sector could hasten the transition to renewable power sources by combining governmental incentives with private sector know-how.

Policymakers should provide a legislative environment that supports PPPs thereby guaranteeing their openness, fairness, and responsibility. Public-sector organizations should make sure, for instance, that private-sector partners are dedicated to the SDGs' social and environmental goals rather than concentrating only on financial gains. Governments should also support local companies and entrepreneurs so they may take part in PPPs, therefore guaranteeing the broad distribution of the advantages of these alliances.

6.3 Integrating Technology

Technological innovation has the power to transform programs for capacity-building and hasten the SDG development speed. Technology like artificial intelligence (AI), big data, and the Internet of Things (IoT) may help capacity-building initiatives be strengthened by means of real-time data, process automation, and facilitation of better decision-making. Along with their general usage, more equitable access and a closing of the digital divide must follow.

Utilizing AI, Big Data, and IoT for Enhanced Capacity Building - Big data, artificial intelligence, and IoT may all be used to improve capacity-building initiatives by offering insightful analysis of community needs, tracking development, and best use of available resources. Big datasets may be examined and trends in health, education, and environmental sustainability are found using AI-powered systems. This data may then be used to create focused capacity-building initiatives addressing particular local need.

In line with this, Internet of Things (IoT) devices might monitor environmental variables like air and water quality in real-time, providing data to direct projects for capacity-building in sustainable resource management and climate change adaptation. Big data analytics might also be used to assess the effectiveness of capacity-building programs, therefore enabling legislators to make decisions based on evidence and improve efforts constantly.

Bridging the Digital Divide to Ensure Access for All - Although fresh ideas have great potential, they will be meaningless unless everyone can use them. Reaching the SDGs is still much hampered by the digital divide, that is, the difference between people with and without access to digital technologies. To narrow this disparity and ensure that technology is accessible, policymakers should give infrastructure, increasing internet access, and digital literacy training top priority.

Public policies should provide low-cost, consistent internet connection to underdeveloped regions as a priority so that people living in rural and remote areas may access resources to improve their capacity and engage in knowledge-sharing platforms. Governments should work with companies to provide reasonably priced technology that underprivileged communities might buy, thereby ensuring that goals related to education, health, and economic development are not hampered by the digital divide.

Policies and initiatives with strategic thinking, support of teamwork, and technology utilization will help us to reach the SDGs. Gender and regional inequalities must be addressed and equal access to knowledge-sharing platforms must be given if inclusive and fair capacity development is to occur. If we are to scale solutions and organize resources, the private sector, non-governmental organizations (NGRs), and governments have to cooperate. Finally, by means of the integration of technologies such artificial intelligence, big data, and the internet of things, capacity-building projects might be strengthened and the digital divide narrowed. Stakeholders have to give these strategic and policy recommendations top priority if we are to achieve the SDGs in a sustainable, inclusive, and fair way.

7. CONCLUSION

7.1 Summary of Key Insights

Pursuing the Sustainable Development Goals (SDGs) calls for a comprehensive and inclusive approach including knowledge sharing and capacity building. The international community needs both of these elements if it is to address the interconnected and complex issues of sustainable development. While capacity building gives individuals, companies, and communities the skills they need to embrace sustainable practices, knowledge sharing lets best practices, technology, and innovations interact to hasten progress. Essential is integrating all these procedures into a whole strategy to achieve the SDGs in sectors ranging from health, education, climate change, and economic growth.

One important realization from this conversation is that local requirements, knowledge of regional difficulties, and guarantee that efforts at capacity-building are context-specific depend critically on geographical methods like spatial analysis and Geographic Information Systems (GIS). These strategies help to monitor and assess development, therefore allowing legislators to customize treatments to the particular possibilities and difficulties found in various geographic areas. Geographic approaches improve the targeting of resources and the alignment of projects with local reality by using data-driven insights, therefore helping to contribute to the more efficient accomplishment of the SDGs.

Facilitating improved coordination and cooperation across industries and areas are geographical data and mapping. The requirement of place-based solutions for information exchange becomes more evident as the globe becomes more linked. To make capacity-building initiatives relevant and effective, localized programs that include cultural, linguistic, and environmental elements are very vital. Advancement of the SDGs depends critically on encouraging geographical data use and guaranteeing the accessibility of localized knowledge-sharing platforms.

7.2 Future Directions

Technological developments will be increasingly more important in increasing information flow and capacity building going forward. A great place for the integration of advanced technologies such artificial intelligence (AI), big data, and the Internet of Things (IoT) is improving the collection, analysis, and information sharing related to sustainable development. These technologies provide more efficient and effective interventions by means of data in real-time, automation of processes, and insight generation before unattainable.

Depending on the continuous growth of digital platforms, reducing the digital divide and ensuring that possibilities to build capacity and exchange information are available to everyone, regardless of where they reside or their socioeconomic condition. Though new, more affordable technology and inclusive laws are on their way, issues with data privacy, digital literacy, and access to technology will not go away anytime soon.

Dealing with developing challenges in sustainability also mostly depends on ongoing adaptation. Policies and programs related to capacity building and information exchange must be constantly revised in order to address issues such the consequences of climate change, global health concerns, and quick technological development. Maintaining the achievability of the SDGs in a dynamic worldwide environment will rely on this capability to adapt. Encouragement of a spirit of inquiry, cooperation, and innovation will help stakeholders to maintain the SDGs as a compass for future attempts to build a fairer and more sustainable world.

All told, the SDGs can only be reached by improving capabilities and sharing knowledge. By leveraging regional approaches, including advanced technology, and supporting ongoing adaptation, one may build a more strong and inclusive global community capable of overcoming challenges of sustainable development. The only way ahead is for us to come together, think creatively, and dedicate ourselves to making the SDGs a reality for all people.

REFERENCES

Anderson, T. (2020). *Online and distance education: Theory and practice*. Routledge.

FAO. (2021). *Empowering farmers through sustainable practices*. Food and Agriculture Organization.

Foley, J. A., DeFries, R., Asner, G. P., Barford, C., Bonan, G., Carpenter, S. R., & Snyder, P. K. (2011). Solutions for a cultivated planet. *Nature*, *478*(7369), 337–342. DOI: 10.1038/nature10452 PMID: 21993620

Gavin, M. C., McCarter, J., Mead, A., Berkes, F., Stepp, J. R., Peterson, D., & Tang, R. (2018). Defining biocultural approaches to conservation. *Trends in Ecology & Evolution*, *33*(2), 140–149. PMID: 25622889

GEF. (2021). *Expanded constituency workshops: Bridging global and local knowledge*. Global Environmental Facility.

Hansen, M. C., Potapov, P. V., Moore, R., Hancher, M., Turubanova, S. A., Tyukavina, A., Thau, D., Stehman, S. V., Goetz, S. J., Loveland, T. R., Kommareddy, A., Egorov, A., Chini, L., Justice, C. O., & Townshend, J. R. G. (2013). High-resolution global maps of 21st-century forest cover change. *Science*, *342*(6160), 850–853. DOI: 10.1126/science.1244693 PMID: 24233722

Haworth, B., Bruce, E., & Whittaker, J. (2019). Participatory mapping: Diverse applications in the era of crowdsourcing and geospatial big data. *Cartography and Geographic Information Science*, *46*(4), 277–290.

Kabeer, N. (2019). Gender equality, economic growth, and women's empowerment: A critical review. *Development and Change*, *50*(5), 1299–1328.

Meadowcroft, J. (2020). Sustainable development: A new (ish) idea for a post-pandemic world. *Journal of Environmental Policy and Planning*, *22*(4), 482–495.

Nonaka, I., & Takeuchi, H. (1995). *The knowledge-creating company: How Japanese companies create the dynamics of innovation*. Oxford University Press. DOI: 10.1093/oso/9780195092691.001.0001

Oliveira, S. V., Gurgel, H. C., & Oliveira, M. A. B. (2021). Geographical information systems in public health: An overview of the Brazilian experience. *Global Health Action*, *14*(1), 1840285.

Pacific Disaster Center. (2020). *DisasterAWARE platform: Enhancing disaster resilience*.

Roy, A. (2017). The Barefoot College: Women's empowerment through solar energy. *Renewable Energy Journal, 25*(3), 223–230.

Sachs, J. D., Schmidt-Traub, G., Kroll, C., Lafortune, G., & Fuller, G. (2019). *The sustainable development goals and Covid-19: Sustainable development report 2019.* Cambridge University Press.

SDSN. (2021). *Sustainable development solutions network: Knowledge for a sustainable future.*

Subrahmanyam, S., Aishwaryalaxmi, N. S., Khalife, D., Shaikh, I. A. K., Faldu, R., & Asthana, N. (2024). Impact of Knowledge Management and Big Data Analytics Capabilities on Firm Performance. *2024 Ninth International Conference on Science Technology Engineering and Mathematics (ICONSTEM)*, Chennai, India, 2024, pp. 1-5. DOI: 10.1109/ICONSTEM60960.2024.10568874

UNESCO. (2019). *Water management and capacity building in sub-Saharan Africa: A comprehensive approach.* United Nations Educational, Scientific, and Cultural Organization.

UNESCO. (2020). *The future of education: Shaping the future of learning*. United Nations Educational, Scientific and Cultural Organization.

UNESCO. (2021). World Atlas of Languages: Towards a global online language repository.

United Nations. (2015). *Transforming our world: The 2030 agenda for sustainable development.* United Nations.

United Nations. (2020). *Webinars for sustainable development: Adapting to COVID-19 challenges*. United Nations Development Programme.

United Nations Development Programme (UNDP). (2018). *Capacity development: UNDP primer.* UNDP.

World Bank. (2020). *Digital economy for Africa (DE4A) initiative*. World Bank.

World Bank. (2020). *Global Partnership for Education (GPE) Annual Report.* World Bank.

Chapter 4
Recommendations for Strengthening Regional Governance for the Sustainable Development Goals

S. Ida Evangeline
https://orcid.org/0000-0003-2997-7897
Government College of Engineering, Tirunelveli, India

ABSTRACT

The Sustainable Development Goals (SDGs) present a universal framework for addressing pressing global challenges, including poverty, inequality, climate change, and environmental degradation. While global and national governance structures are essential, regional governance plays a unique and pivotal role in bridging the gap between international commitments and local implementation. This chapter, titled "Recommendations for Strengthening Regional Governance for the Sustainable Development Goals," examines the critical importance of regional governance in advancing sustainable development. It explores the challenges faced by regional organizations, such as resource disparities, geopolitical tensions, institutional capacity gaps, and external pressures like climate change and economic instability. Drawing on theoretical foundations, case studies, and emerging trends, the chapter identifies opportunities to strengthen regional governance through innovative approaches such as leveraging digital technologies, and engaging in cross-regional collaborations.

DOI: 10.4018/979-8-3373-1300-9.ch004

1. INTRODUCTION

The Sustainable Development Goals (SDGs), adopted by the United Nations in 2015, represent a landmark global agenda aimed at addressing the most pressing challenges of our time. These 17 interconnected goals provide a blueprint for achieving a better and more sustainable future for all by 2030 (Hák et al. 2016). From eradicating poverty and hunger to combating climate change and promoting peace, the SDGs serve as a comprehensive framework for action that is both ambitious and universal in scope. Their implementation requires coordinated efforts across diverse sectors and levels of governance to ensure progress in a manner that is equitable and inclusive (Sachs, 2012).

Regional governance occupies a critical space in this framework, acting as a bridge between global commitments and national implementation (Monkelbaan, 2019). While global governance institutions set the overarching priorities and provide guidance, national governments translate these goals into policies tailored to their local contexts (Biermann et al. 2017). Regional governance, however, serves as the essential intermediary, enabling countries to collaborate on shared challenges, pool resources, and address issues that transcend national boundaries (Morita et al. 2020). This role is particularly important for transboundary concerns, such as managing shared water resources, mitigating the impacts of climate change, and coordinating responses to health emergencies. By fostering dialogue, facilitating the exchange of best practices, and promoting collective action, regional governance ensures that the SDGs are approached with a sense of shared responsibility and mutual accountability (Florini & Pauli, 2018).

This chapter, titled "Recommendations for Strengthening Regional Governance for the Sustainable Development Goals," delves into the pivotal role of regional governance in advancing sustainable development. It explores the unique challenges faced by regional organizations and identifies actionable strategies for overcoming these obstacles to create more effective and resilient governance structures. By examining theoretical foundations, analyzing case studies, and offering evidence-based recommendations, the chapter aims to provide valuable insights for policymakers, practitioners, and stakeholders. Ultimately, it seeks to inspire collaborative efforts that can accelerate progress toward achieving the SDGs, while also highlighting the critical need for inclusivity, equity, and sustainability in regional governance practices.

2. THEORETICAL FOUNDATIONS

Regional governance refers to the systems, institutions, and processes that facilitate coordination and cooperation among nations within a defined geographic area (MacLeod & Goodwin, 1999). Unlike national governance, which operates within the borders of a single state, or global governance, which addresses issues on a worldwide scale, regional governance functions as an intermediary, bridging local and international efforts (Fürst, 2001). It encompasses formal organizations, such as the European Union (EU) or the African Union (AU), as well as informal arrangements that enable countries to work together to address shared challenges (Feiock, 2007). The scope of regional governance extends across various domains, including economic integration, environmental management, security cooperation, and social development, reflecting the diverse needs and priorities of the regions it serves (Newman, 2000).

The relevance of regional governance to the Sustainable Development Goals (SDGs) lies in its capacity to align national actions with global objectives while addressing issues that are regional in nature (Irtyshcheva et al. 2022). Many of the SDGs, such as those focused on climate action, water management, and health, involve challenges that transcend national borders and require collective solutions (Biermann et al. 2017). Regional governance structures play a critical role in coordinating policies and actions among member states, ensuring that efforts are harmonized and resources are utilized efficiently (Kanie & Biermann, 2017). These mechanisms enable countries to share expertise, pool financial and technical resources, and develop joint strategies for addressing common challenges (Guha & Chakrabarti, 2019). Furthermore, regional organizations often serve as platforms for monitoring and evaluating progress toward the SDGs, providing critical data and analysis that inform decision-making at both the national and international levels (Morgan, 2004).

The relationship between regional and global goals underscores the importance of alignment and coherence in governance (Hveem, 2003). While the SDGs provide a universal framework, their implementation depends on how well global priorities are integrated into regional and local contexts (Kahler, 2017). Regional governance facilitates this integration by translating global objectives into actionable strategies that account for regional specificities. For example, the ASEAN region's focus on disaster resilience directly supports the SDG targets related to climate adaptation and risk reduction, while simultaneously addressing the unique vulnerabilities faced by its member states (Budiana, 2024). By fostering such interlinkages, regional governance not only amplifies the impact of global initiatives but also ensures that progress is equitable and context-sensitive (Marx et al. 2021).

Through its ability to mediate between the global and local, regional governance emerges as a vital instrument in the pursuit of sustainable development (Kacowicz, 2018). Its frameworks and mechanisms provide a foundation for addressing transboundary challenges, optimizing resource allocation, and aligning priorities across scales. In this way, regional governance bridges gaps in the global system, creating pathways for collective action that are both efficient and inclusive (Pierre, 2016).

3. KEY CHALLENGES IN REGIONAL GOVERNANCE FOR SDGS

The pursuit of the Sustainable Development Goals (SDGs) through regional governance is fraught with significant challenges that hinder the effective coordination and implementation of initiatives. Among these, resource disparities stand out as a persistent obstacle (Trends, 2017). Member states within a regional bloc often exhibit wide variations in economic capacity, technical expertise, and institutional strength (Singh, 2016). Wealthier nations may have access to advanced technologies and robust infrastructures, while less developed countries struggle to secure even basic resources needed to meet SDG targets (Henderson & Loreau, 2023). These disparities create an imbalance in regional efforts, where progress in some states is offset by stagnation in others. The resulting inequities not only undermine collective achievements but also fuel dissatisfaction among member states, jeopardizing the cohesion necessary for sustained collaboration (Leal Filho et al. 2019).

Geopolitical tensions further complicate the landscape of regional governance. Historical rivalries, territorial disputes, and competing national interests often lead to trust deficits between nations, making it difficult to forge consensus on critical issues (Kharrazi et al. 2016). Such tensions can stymie cooperative efforts, particularly in addressing transboundary challenges that require collective action, such as water management or climate change adaptation (Lee et al. 2020). The inability to resolve these conflicts weakens regional governance frameworks, diverting attention and resources away from the pressing tasks of sustainable development. Moreover, the politicization of regional initiatives can erode the neutrality and effectiveness of governance institutions, creating additional barriers to progress (Di Baldassarre et al. 2019).

Institutional capacity gaps present another significant challenge, limiting the ability of regional organizations to design, implement, and monitor SDG-related policies effectively. Many regional institutions lack the technical expertise, financial resources, or operational frameworks needed to drive coordinated actions across diverse member states (Islam, 2025). The absence of standardized protocols for data collection and policy evaluation further exacerbates these limitations, making it difficult to measure progress or identify areas requiring improvement. Without adequate capacity, regional governance structures often fall short of their potential, leaving critical opportunities for advancing the SDGs unrealized (Afiyah, 2025).

Compounding these internal challenges are the external pressures that strain regional governance systems. Climate change, economic instability, and global health crises like the COVID-19 pandemic introduce layers of complexity that test the resilience of governance frameworks (Sun et al. 2025). Climate-related disasters, such as floods or droughts, often transcend national boundaries, demanding swift and coordinated regional responses that are difficult to mobilize under current constraints (Ahyani et al. 2025). Economic shocks, including trade disruptions or financial crises, can further widen resource disparities and exacerbate tensions among member states. Similarly, health emergencies highlight the vulnerabilities in regional preparedness and response mechanisms, exposing gaps in coordination and resource allocation that undermine efforts to protect vulnerable populations (Zarghami, 2025).

These challenges collectively underscore the multifaceted nature of the difficulties facing regional governance for the SDGs. Addressing them requires not only strengthening the institutional capacity of regional organizations but also fostering trust and equity among member states. Without concerted efforts to overcome these barriers, the transformative potential of regional governance to advance sustainable development will remain limited, leaving many critical aspects of the SDG agenda unfulfilled (Byanyima et al. 2025).

4. OPPORTUNITIES FOR STRENGTHENING REGIONAL GOVERNANCE

Despite the significant challenges faced by regional governance in advancing the Sustainable Development Goals (SDGs), numerous opportunities exist to enhance its effectiveness and resilience. These opportunities arise from innovative approaches, lessons from successful case studies, and emerging trends that offer pathways for more robust governance structures (Li et al. 2025).

One of the most promising avenues lies in the use of digital technologies to facilitate data-driven decision-making. Advances in digital tools and platforms have revolutionized the way information is collected, analyzed, and shared, enabling regional organizations to develop evidence-based policies with greater precision (Escano-Arias et al. 2025). Technologies such as geographic information systems (GIS), artificial intelligence (AI), and big data analytics allow for real-time monitoring of environmental changes, resource use, and social trends. These capabilities not only improve the efficiency of governance processes but also enhance transparency and accountability, fostering greater trust among member states and stakeholders (Saeed et al. 2025). For instance, regional organizations can leverage digital platforms to coordinate disaster responses, manage shared resources, or track progress toward SDG targets with unprecedented accuracy (Dragneva & Wolczuk, 2025).

Public-private partnerships (PPPs) also represent a vital opportunity for strengthening regional governance. By engaging the private sector, regional organizations can tap into additional financial resources, technical expertise, and innovative solutions (Catala et al. 2025). PPPs can be instrumental in developing infrastructure, such as renewable energy projects or transportation networks, that supports sustainable development (Wu et al. 2025). Furthermore, these partnerships can facilitate technology transfer and capacity building, addressing gaps that often impede the implementation of SDG-related initiatives. Successful PPPs in areas like sustainable agriculture or green energy serve as models that regional organizations can adapt to their specific contexts (Cambra-Fierro et al. 2025).

Lessons from successful regional collaborations provide valuable insights into the potential of governance structures. For example, the European Union (EU) has demonstrated the power of harmonized policies and shared standards in advancing environmental sustainability and social equity (Wang et al. 2025). Similarly, the Association of Southeast Asian Nations (ASEAN) has achieved significant progress in disaster risk reduction through its coordinated response mechanisms and knowledge-sharing platforms (Wei et al. 2025). The African Union (AU) has made strides in promoting economic integration and addressing cross-border security issues. These case studies underscore the importance of well-designed frameworks that prioritize inclusivity, equity, and shared accountability (Caponio & Pettrachin, 2025).

Transboundary cooperation offers additional lessons in overcoming challenges that span national borders. Examples such as the Mekong River Commission or the Nile Basin Initiative highlight the importance of joint resource management in achieving equitable and sustainable outcomes (Zhang, 2025). Regional migration agreements, designed to address the complex dynamics of human mobility, demonstrate the potential for collective action to protect vulnerable populations while addressing the underlying causes of migration (Aniyie & Omiunu, 2025). These experiences emphasize the critical role of trust-building, transparent negotiations, and the establishment of shared goals in fostering successful collaboration.

Emerging trends also present opportunities to reimagine and strengthen regional governance. Cross-regional collaborations are becoming increasingly prominent, as regions recognize the value of sharing experiences and best practices (Sa'a & Asplund, 2025). Initiatives like the collaboration between the African Union and the European Union on climate action and trade highlight the potential for partnerships that transcend geographic boundaries, allowing regions to learn from one another and pool resources to address shared challenges (Yu & Ye, 2025). Advances in sustainable financing, including green bonds, blended finance, and impact investing, offer innovative mechanisms for funding SDG-related projects. These approaches not only mobilize capital but also align financial incentives with sustainability goals, creating a win-win scenario for investors and regional stakeholders alike (Zheng & Xu, 2025).

Taken together, these opportunities highlight the untapped potential of regional governance to drive transformative change. By embracing innovation, learning from successful examples, and adapting to emerging trends, regional organizations can play a pivotal role in achieving the SDGs (Miao et al. 2025). Strengthened regional governance can act as a catalyst for collective action, ensuring that sustainable development efforts are both effective and inclusive.

5. RECOMMENDATIONS FOR ACTION

Strengthening regional governance for the Sustainable Development Goals (SDGs) requires a concerted effort to address existing challenges while building on opportunities for innovation and collaboration (Ebekozien et al. 2025). To achieve this, a multi-faceted approach that prioritizes capacity building, resource mobilization, inclusive governance, and policy alignment is essential. These actions, when undertaken collectively, can transform regional governance structures into more resilient and effective mechanisms for advancing sustainable development.

Investing in capacity building is a critical first step. Many regional organizations and their member states lack the institutional frameworks, technical expertise, and human resources needed to design and implement effective policies (Pundhir et al. 2025). Comprehensive training programs tailored to regional needs can enhance the skills of policymakers and practitioners, equipping them to address the complexities of sustainable development. Technical support, including the provision of tools for data analysis and decision-making, further strengthens institutional capacity (Van Thao et al. 2025). Additionally, establishing robust frameworks for monitoring and evaluation allows regional organizations to assess progress, identify gaps, and adapt their strategies in real time. Promoting knowledge sharing and peer learning among member states fosters a culture of collaboration, enabling nations to benefit from each other's experiences and innovations (Anshari et al. 2025).

Resource mobilization is equally vital to the success of regional governance initiatives. Financing sustainable development requires a diverse range of funding sources, and regional organizations must explore innovative mechanisms to address financial gaps (Zhao & Ye, 2025). Instruments like green bonds and blended finance offer opportunities to attract investments that align with sustainability objectives. Engaging the private sector through public-private partnerships can unlock additional resources, while collaboration with international donors and development banks can provide the financial stability needed for long-term initiatives. Ensuring financial sustainability not only supports immediate actions but also enhances the capacity of regional organizations to plan for and respond to emerging challenges (Chishti et al. 2025).

Inclusive governance forms the cornerstone of equitable and effective regional action. The active engagement of diverse stakeholders, including civil society organizations, indigenous communities, youth groups, and other marginalized populations, ensures that governance frameworks are representative and participatory. Creating spaces for meaningful dialogue and collaboration builds trust and promotes transparency, making governance processes more accountable (Jahangir, 2025). Equity must also be prioritized, with efforts directed toward addressing power imbalances and ensuring that all voices are heard in decision-making. By fostering inclusivity, regional organizations can enhance the legitimacy of their initiatives and ensure that development efforts benefit all members of society (Joseph & Pandey, 2025).

Policy alignment is the final pillar in strengthening regional governance. Achieving the SDGs requires a seamless integration of regional, national, and global objectives (Shyrokykh & Melen-Zabramna, 2025). Regional organizations play a pivotal role in synchronizing policies and actions across these levels, ensuring coherence and minimizing duplication of efforts. Encouraging harmonized standards and practices allows member states to work together more effectively, particularly in addressing transboundary issues. For example, aligning environmental regulations or trade policies can facilitate collective action while respecting the unique contexts of individual nations. Establishing mechanisms for regular dialogue and coordination between regional and global institutions further enhances this alignment, creating a unified approach to sustainable development (Panke et al. 2025).

Together, these recommendations provide a roadmap for transforming regional governance into a powerful driver of sustainable development. By prioritizing capacity building, mobilizing resources, fostering inclusivity, and aligning policies, regional organizations can address current challenges and seize opportunities for progress. These actions not only enhance the effectiveness of governance structures but also ensure that the SDGs are pursued in a manner that is equitable, participatory, and resilient (Bejarano-Murillo, 2025).

6. ETHICAL CONSIDERATIONS AND CHALLENGES

Strengthening regional governance for the Sustainable Development Goals (SDGs) necessitates a nuanced understanding of the ethical considerations and challenges that underpin collaborative efforts. Addressing these issues is critical to ensuring that governance frameworks are equitable, inclusive, and effective (Nartey, 2025). Power imbalances, the tension between sovereignty and regional priorities, and the complexities of data governance are among the most pressing ethical challenges that must be navigated (Dhar & Dhar, 2025).

Power imbalances within regional organizations present a significant obstacle to equitable decision-making. Member states often differ vastly in terms of economic capacity, political influence, and institutional strength, which can lead to disproportionate power dynamics (Malpass et al. 2025). Larger or wealthier nations may dominate decision-making processes, marginalizing the voices of smaller or less-resourced states. This imbalance not only undermines the principles of fairness and inclusivity but also jeopardizes the legitimacy of governance structures. Addressing these disparities requires the establishment of mechanisms that ensure all member states have an equal opportunity to contribute to and shape collective decisions. Transparent processes, equitable representation, and mechanisms for conflict resolution are essential to fostering trust and promoting a sense of shared ownership among participants (Paul et al. 2025).

Another critical ethical challenge lies in balancing national sovereignty with the need for regional priorities. While regional governance seeks to promote collective action on issues such as climate change, migration, and economic integration, it must also respect the autonomy and unique circumstances of individual member states. This tension is particularly pronounced in regions with diverse political systems, cultural contexts, and levels of development (Robinson et al. 2025). Reconciling these differences demands careful negotiation and compromise, as well as the development of frameworks that align regional objectives with national interests. Success in this regard hinges on fostering mutual understanding and cooperation, ensuring that collective initiatives complement rather than conflict with the priorities of member states (Ali, 2025).

Data governance adds another layer of ethical complexity to regional governance. As regional organizations increasingly rely on digital technologies to monitor and evaluate progress toward the SDGs, questions of privacy, data security, and ethical use come to the fore. The collection and use of data must be guided by principles that protect individual and community privacy while ensuring that data is used responsibly and transparently. Building trust in these processes requires robust accountability mechanisms, such as independent oversight bodies and clear guidelines for data management. Furthermore, regional organizations must address the digital divide, ensuring that all member states have equitable access to technological tools and resources. Without such safeguards, data-driven governance risks exacerbating existing inequalities and eroding trust among stakeholders (Distefano et al. 2025).

Navigating these ethical considerations requires a commitment to transparency, inclusivity, and accountability at all levels of governance. By addressing power imbalances, respecting national sovereignty, and adhering to ethical principles in data governance, regional organizations can foster a culture of trust and collaboration. These efforts not only strengthen the legitimacy of governance structures but also ensure that the pursuit of the SDGs is rooted in principles that prioritize equity and fairness. Ethical governance is not merely an aspirational goal; it is an essential foundation for achieving sustainable development in a manner that benefits all (Mutawa, 2025).

7. CONCLUSION

Regional governance plays a critical and multifaceted role in achieving the Sustainable Development Goals (SDGs). As an intermediary between global frameworks and national implementation, regional governance provides the mechanisms for fostering collaboration, addressing transboundary challenges, and ensuring that collective actions are aligned with global priorities while remaining sensitive to regional contexts. Its capacity to bridge these scales makes it indispensable in tackling complex, interconnected challenges such as climate change, resource management, and economic inequality. Throughout this chapter, the unique contributions of regional governance, as well as the challenges and opportunities it faces, have been examined to highlight its pivotal importance in advancing sustainable development.

The insights presented here underscore the necessity of proactive measures to strengthen regional governance structures. Challenges such as resource disparities, geopolitical tensions, and institutional capacity gaps cannot be ignored if the full potential of regional collaboration is to be realized. Equally important are opportunities, including the use of innovative technologies, the mobilization of diverse resources, and the integration of inclusive governance practices, which can transform regional organizations into powerful drivers of sustainable progress. A call to action is therefore imperative. Regional organizations, member states, and stakeholders must adopt forward-thinking strategies to enhance their governance frameworks, ensuring that they are equipped to navigate both current and emerging challenges. This requires not only political will but also a commitment to fostering equity, transparency, and inclusivity in decision-making processes.

Looking ahead, several research and policy priorities demand attention to ensure the continued evolution of sustainable regional governance. Greater emphasis must be placed on exploring innovative financing mechanisms, understanding the role of cross-regional collaborations, and refining frameworks for data governance that respect ethical standards. Additionally, there is a need to evaluate the long-term impacts of existing regional initiatives to identify best practices and areas for improvement. Policymakers and researchers should also focus on mechanisms for balancing national sovereignty with regional priorities, ensuring that governance structures are adaptable to diverse political, economic, and cultural contexts. These efforts will provide a robust foundation for regional organizations to not only advance the SDGs but also contribute to a more equitable and resilient global governance system.

In conclusion, regional governance holds immense promise as a tool for accelerating progress toward the SDGs. By addressing challenges, seizing opportunities, and prioritizing ethical and inclusive practices, it is possible to unlock its full potential. The future of sustainable development depends on the ability of regional actors to rise to the occasion, demonstrating the vision and commitment needed to create a world that is inclusive, equitable, and sustainable for all.

REFERENCES

Afiyah, I., & Muthoifin, . (2025). The Existence of Library Literacy in Indonesia in the Millennial Era: Strategies for Improving the Quality of Education to Support the Sustainable Development Goals (SDG's). *Solo International Collaboration and Publication of Social Sciences and Humanities*, *3*(01), 41–52. DOI: 10.61455/sicopus.v3i01.229

Ahyani, H., Mutmainah, N., Ahmad, M. Y., Lousada, S. A. N., Triswandani, Y., Rahman, E. T., & Utama, S. M. (2025). Enhancing the Legal Protection of Waqf Assets in Banjar City, West Java, Through Land Registration and Regulatory Awareness to Support the Achievement of Sustainable Development Goals (SDGs). *Journal of Lifestyle and SDGs Review*, *5*(2), e03989–e03989. DOI: 10.47172/2965-730X.SDGsReview.v5.n02.pe03989

Ali, H. (2025). Conflicting behavior between US and Afghanistan and Its Implications on Tribal Areas of Pakistan. *Pakistan Languages and Humanities Review*, *9*(1), 39–51.

Aniyie, I. A., & Omiunu, O. (2025). Paradiplomacy and the Governors' Forum: Rethinking Transboundary Water Governance in the Lake Chad Basin. *Water Alternatives*, *18*(1), 1–18.

Anshari, M., Hamdan, M., Ahmad, N., & Ali, E. (2025). Public service delivery, artificial intelligence and the sustainable development goals: Trends, evidence and complexities. *Journal of Science and Technology Policy Management*, *16*(1), 163–181. DOI: 10.1108/JSTPM-07-2023-0123

Bejarano-Murillo, E. (2025). Digital Governance Across Continents: Cases From Diverse National Strategies. In *Data-Driven Governance Through AI, Digital Marketing, and the Privacy Interplay* (pp. 1-30). IGI Global Scientific Publishing.

Biermann, F., Kanie, N., & Kim, R. E. (2017). Global governance by goal-setting: The novel approach of the UN Sustainable Development Goals. *Current Opinion in Environmental Sustainability*, *26*, 26–31. DOI: 10.1016/j.cosust.2017.01.010

Budiana, M. (2024). Regional Organizations and Global Governance: The Case of ASEAN. *Journal of Student Collaboration Research*, *1*(2), 1–15.

Byanyima, F., Edison, M., Kadengye, D. T., Arineitwe, S., Mijwil, M. M., Gaballa, M., & Turyasingura, B. (2025). Foreign direct investment and environmental challenges: A case study of Uganda with analytical perspective. In *Community Resilience and Climate Change Challenges: Pursuit of Sustainable Development Goals (SDGs)* (pp. 283–296). IGI Global Scientific Publishing.

Cambra-Fierro, J., López-Pérez, M. E., Perez, L., & Tejada-Tejada, M. (2025). Managing public-private partnerships for urban design and regeneration: Lessons learned from the Hermitage Museum Barcelona odyssey. *Cities (London, England), 158*, 105714. DOI: 10.1016/j.cities.2025.105714

Caponio, T., & Pettrachin, A. (2025). Neither multilevel governance nor battleground. Understanding the politics of immigrant integration in small and medium European localities. *Governance: An International Journal of Policy, Administration and Institutions, 38*(1), e12833. DOI: 10.1111/gove.12833

Catala, B., Savall, T., & Chaves-Ávila, R. (2025, January). Social Economy and Public Private Partnerships: Analysis, Drivers, and Prospects from the Local Level for Territorial Development. In *Nonprofit Policy Forum* (No. 0). De Gruyter.

Chishti, M. Z., Salam, M., Xaisongkham, S., & Du, A. M. (2025). Influence of green ICT and socioeconomic factors on sustainable development: Evidence from Chinese provinces. *Research in International Business and Finance, 73*, 102624. DOI: 10.1016/j.ribaf.2024.102624

Dhar, S., & Dhar, A. (2025). Integrating Machine Learning Into Wildlife and Forest Conservation: Ethical Considerations and Sustainable Practices. In *Machine Learning and Internet of Things in Fire Ecology* (pp. 129-156). IGI Global Scientific Publishing.

Di Baldassarre, G., Sivapalan, M., Rusca, M., Cudennec, C., Garcia, M., Kreibich, H., Konar, M., Mondino, E., Mård, J., Pande, S., Sanderson, M. R., Tian, F., Viglione, A., Wei, J., Wei, Y., Yu, D. J., Srinivasan, V., & Blöschl, G. (2019). Sociohydrology: Scientific challenges in addressing the sustainable development goals. *Water Resources Research, 55*(8), 6327–6355. DOI: 10.1029/2018WR023901 PMID: 32742038

Distefano, T., Isaza, A. S., Morlin, G. S., Carmona, V. P., Palacio, C. V., & Arango-Aramburo, S. (2025). The roads towards complex water governance: The Colombian case study. *World Development, 188*, 106874. DOI: 10.1016/j.worlddev.2024.106874

. Dragneva, R., & Wolczuk, K. (2025). Integration and modernisation: EU's Association Agreement with Ukraine. *Ukraine's Thorny Path to the EU*, 43.

Ebekozien, A., Aigbavboa, C. O., Samsurijan, M. S., Radin Firdaus, R. B., & Rohayati, M. I. (2025). Expanded corporate social responsibility framework: Companies' role in improving higher education institutions infrastructure to Sustainable Development Goal 4. *Journal of Facilities Management, 23*(1), 1–18. DOI: 10.1108/JFM-11-2022-0120

Escano-Arias, E. A., Abarquez, R. A. D., Cruz, R. V., Espeleta, R., Ong, M. M., Loreche, A. M., Pepito, V. C. F., Gomez, V., & Dayrit, M. M. (2025). Strengthening local health systems and governance for Universal Health Coverage: Experiences and lessons from the COVID-19 pandemic response in Quezon City, Philippines. *Health Policy and Planning*, *40*(3), czaf002. DOI: 10.1093/heapol/czaf002 PMID: 39801295

Feiock, R. C. (2007). Rational choice and regional governance. *Journal of Urban Affairs*, *29*(1), 47–63. DOI: 10.1111/j.1467-9906.2007.00322.x

Florini, A., & Pauli, M. (2018). Collaborative governance for the sustainable development goals. *Asia & the Pacific Policy Studies*, *5*(3), 583–598. DOI: 10.1002/app5.252

Fürst, D. (2001). Regional governance: A new paradigm in the regional sciences? *Raumforschung und Raumordnung*, *59*, 370–380. DOI: 10.1007/BF03183038

Guha, J., & Chakrabarti, B. (2019). Achieving the Sustainable Development Goals (SDGs) through decentralisation and the role of local governments: A systematic review. *Commonwealth Journal of Local Governance*, (22), 1–21. DOI: 10.5130/cjlg.v0i22.6855

Hák, T., Janoušková, S., & Moldan, B. (2016). Sustainable Development Goals: A need for relevant indicators. *Ecological Indicators*, *60*, 565–573. DOI: 10.1016/j.ecolind.2015.08.003

Henderson, K., & Loreau, M. (2023). A model of Sustainable Development Goals: Challenges and opportunities in promoting human well-being and environmental sustainability. *Ecological Modelling*, *475*, 110164. DOI: 10.1016/j.ecolmodel.2022.110164

Hveem, H. (2003). The regional project in global governance. In *Theories of new regionalism: A Palgrave reader* (pp. 81–98). Palgrave Macmillan UK. DOI: 10.1057/9781403938794_5

Irtyshcheva, I., Pavlenko, O., Boiko, Y., Stehnei, M., Kramarenko, I., Hryshyna, N., & Ishchenko, O. (2022). Evaluation of efficiency of regional public governance in the context of achieving goals of sustainable development. *Management Theory and Studies for Rural Business and Infrastructure Development*, *44*(4), 497–505. DOI: 10.15544/mts.2022.49

Islam, H. (2025). Nexus of economic, social, and environmental factors on sustainable development goals: The moderating role of technological advancement and green innovation. *Innovation and Green Development, 4*(1), 100183. DOI: 10.1016/j.igd.2024.100183

Jahangir, A. (2025). Governance Innovation in South Asia: The Pakistan Governance Journey. In *Mapping Governance Innovations* (pp. 147-166). Routledge India.

Joseph, J., & Pandey, A. (2025). Prospective Role of "BRICS Plus" Countries in Global Governance: An Exploratory Study. In *Changing the Global Political Economy: BRICS Countries and Alternative Relations Strategies* (pp. 93-118). IGI Global Scientific Publishing.

Kacowicz, A. M. (2018). Regional governance and global governance: Links and explanations. *Global Governance, 24*(1), 61–80. DOI: 10.1163/19426720-02401005

Kahler, M. (2017). Regional challenges to global governance. *Global Policy, 8*(1), 97–100. DOI: 10.1111/1758-5899.12392

Kanie, N., & Biermann, F. (Eds.). (2017). *Governing through goals: Sustainable development goals as governance innovation.* mit Press. DOI: 10.7551/mitpress/10894.001.0001

Kharrazi, A., Qin, H., & Zhang, Y. (2016). Urban big data and sustainable development goals: Challenges and opportunities. *Sustainability (Basel), 8*(12), 1293. DOI: 10.3390/su8121293

Leal Filho, W., Tripathi, S. K., Andrade Guerra, J. B. S. O. D., Giné-Garriga, R., Orlovic Lovren, V., & Willats, J. (2019). Using the sustainable development goals towards a better understanding of sustainability challenges. *International Journal of Sustainable Development and World Ecology, 26*(2), 179–190. DOI: 10.1080/13504509.2018.1505674

Lee, K. H., Noh, J., & Khim, J. S. (2020). The Blue Economy and the United Nations' sustainable development goals: Challenges and opportunities. *Environment International, 137*, 105528. DOI: 10.1016/j.envint.2020.105528 PMID: 32014791

Li, Y., Wang, Z., Lin, Z., & Gao, Y. (2025). Revisiting regional governance and regional development: Measurements, linkages and coupling effect. *World Development, 185*, 106816. DOI: 10.1016/j.worlddev.2024.106816

MacLeod, G., & Goodwin, M. (1999). Space, scale and state strategy: Rethinking urban and regional governance. *Progress in Human Geography, 23*(4), 503–527. DOI: 10.1191/030913299669861026

Malpass, A., Le Poidevin, A., Bamford, A., Lowndes, S., Grant, G., & Selman, L. (2025). From trusted intruder to real collaborator: Mapping the ecosystem of stakeholders and community assets in a coastal region in England, with a focus on end-of-life care and bereavement support. *Social Science & Medicine*, *365*, 117555. DOI: 10.1016/j.socscimed.2024.117555 PMID: 39642586

Marx, A., Pertiwi, S. B., Depoorter, C., Hoornick, M., Mursitama, T. N., Otteburn, K., & Arnakim, L. Y. (2021). What role for regional organizations in goal-setting global governance? An analysis of the role of the European Union and ASEAN in the Sustainable Development Goals. *Global Public Policy and Governance*, *1*(4), 421–445. DOI: 10.1007/s43508-021-00027-6

Miao, S., Chen, B., & Jiang, N. (2025). Collaboration among governments, agribusinesses, and rural households for improving the effectiveness of conservation tillage technology adoption. *Scientific Reports*, *15*(1), 45. DOI: 10.1038/s41598-024-83827-0 PMID: 39747408

. Monkelbaan, J. (2019). Governance for the sustainable development goals. *Singapura: Spinger.*

Morgan, K. (2004). Sustainable regions: Governance, innovation and scale. *European Planning Studies*, *12*(6), 871–889. DOI: 10.1080/0965431042000251909

Morita, K., Okitasari, M., & Masuda, H. (2020). Analysis of national and local governance systems to achieve the sustainable development goals: Case studies of Japan and Indonesia. *Sustainability Science*, *15*(1), 179–202. DOI: 10.1007/s11625-019-00739-z

Mutawa, A. M. (2025). Enforcing the Ethics of Artificial Intelligence in Education: To Prevent or Brace AI Tools in the Educational System. In *Impacts of Generative AI on the Future of Research and Education* (pp. 143-168). IGI Global.

. Nartey, E. K. (2025). Beyond profit: ethical imperatives and the role of duty of care in sustainable investment strategies. *International Journal of Economics and Finance (IJEF), 17*(1), in-press.

Newman, P. (2000). Changing patterns of regional governance in the EU. *Urban Studies (Edinburgh, Scotland)*, *37*(5-6), 895–908. DOI: 10.1080/00420980050011145

. Panke, D., Grundsfeld, L., & Tverskoi, P. (2025). Exit from regime complexity? Regional international organizations under scrutiny. *Global Studies Quarterly, 5*(1), ksae078.

Paul, T. V., He, K., & Wivel, A. (2025). Soft balancing in the regions: Causes, characteristics and consequences. *International Affairs, 101*(1), 3–15. DOI: 10.1093/ia/iiae286

Pierre, J. (2016). Urban and regional governance. In *Handbook on theories of governance* (pp. 477–485). Edward Elgar Publishing. DOI: 10.4337/9781782548508.00051

Pundhir, S., Marwah, T., & Gupta, V. (2025). Strengthening Rural Women's Entrepreneurship for Sustainable Development: The Impact of Digital Tools, Government Policies, and NGO Support. In *Empowering Women Through Rural Sustainable Development and Entrepreneurship* (pp. 287-308). IGI Global Scientific Publishing.

Robinson, J., Harrison, P., Croese, S., Sheburah Essien, R., Kombe, W., Lane, M., Mwathunga, E., Owusu, G., & Yang, Y. (2025). Reframing urban development politics: Transcalarity in sovereign, developmental and private circuits. *Urban Studies (Edinburgh, Scotland), 62*(1), 00420980241284763. DOI: 10.1177/00420980241284763

Sa'a, E. A., & Asplund, F. (2025). Unpacking social capital in University–Industry Collaborations: Pathways to cross-industry knowledge sharing. *Technovation, 140*, 103160. DOI: 10.1016/j.technovation.2024.103160

Sachs, J. D. (2012). From millennium development goals to sustainable development goals. *Lancet, 379*(9832), 2206–2211. DOI: 10.1016/S0140-6736(12)60685-0 PMID: 22682467

Saeed, U. F., Kamil, R., & Wiredu, I. (2025). The roles of ICT and governance quality in the finance-growth nexus of developing countries: A dynamic GMM approach. *Cogent Economics & Finance, 13*(1), 2448228. DOI: 10.1080/23322039.2024.2448228

. Shyrokykh, K., & Melen-Zabramna, O. (2025). Europeanisation of Ukraine's policy on environment and climate: Instrumental and anticipatory compliance. *Ukraine's Thorny Path to the EU*, 247.

Singh, Z. (2016). Sustainable development goals: Challenges and opportunities. *Indian Journal of Public Health, 60*(4), 247–250. DOI: 10.4103/0019-557X.195862 PMID: 27976644

Sun, X., Meng, Z., Zhang, X., & Wu, J. (2025). The role of institutional quality in the nexus between green financing and sustainable development. *Research in International Business and Finance, 73*, 102531. DOI: 10.1016/j.ribaf.2024.102531

Trends, G. (2017). *Challenges and opportunities in the implementation of the Sustainable Development Goals*. United Nations Development Programme& United Nations Research Institute for Social Development.

Van Thao, N., Bańka, M. S., & Tien, N. H. (2025). Completing Dried Coconut Value Chain of Vietnam Towards Sustainable Development Goals. *Journal of Lifestyle and SDGs Review*, *5*(1), e03321–e03321. DOI: 10.47172/2965-730X.SDGsReview.v5.n01.pe03321

Wang, W., Bai, Y., Xu, J., & Li, T. (2025). Collaborative governance of China-ASEAN maritime drug control: Mechanisms, challenges, and improvements. *Ocean and Coastal Management*, *261*, 107548. DOI: 10.1016/j.ocecoaman.2025.107548

Wei, C., Xu, J., Wang, Z., Wu, H., & Wang, J. (2025). How to implement pairing assistance during fighting COVID-19 in China: Collaborative governance between local governments under the authoritative regulation. *Frontiers in Public Health*, *12*, 1417832. DOI: 10.3389/fpubh.2024.1417832 PMID: 39850856

Wu, G., Zhou, K., Hu, Z., Wang, G., Liu, B., & Zhang, W. (2025). The Symphony of Evolution: Unraveling Infrastructure Public–Private Partnerships Collaboration Networks Through Participant Characteristics. *Governance: An International Journal of Policy, Administration and Institutions*, *38*(1), e12911. DOI: 10.1111/gove.12911

Yu, J., & Ye, C. (2025). A new conceptual framework and case analysis of rural environmental cross-boundary governance in megacities of China. *Habitat International*, *156*, 103258. DOI: 10.1016/j.habitatint.2024.103258

Zarghami, S. A. (2025). The role of economic policies in achieving sustainable development goal 7: Insights from OECD and European countries. *Applied Energy*, *377*, 124558. DOI: 10.1016/j.apenergy.2024.124558

Zhang, M. (2025). Deepening Webs of Functional Trilateralism and Challenges: A Case of Transboundary Air Pollution Cooperation. In *China–Japan–South Korea Trilateral Cooperation: Institution Building and Power Politics* (pp. 33–49). Springer Nature Singapore.

Zhao, L., & Ye, J. (2025). Can civilized city construction facilitate green total factor productivity? A quasi-natural experiment based on China's pilot civilized city. *Journal of Environmental Planning and Management*, *68*(2), 437–462. DOI: 10.1080/09640568.2023.2259602

Zheng, N., & Xu, Y. Q. (2025). Spatial correlation and urban collaborative governance of air pollution in Anhui province based on complex networks. *Atmospheric Pollution Research*, *16*(2), 102369. DOI: 10.1016/j.apr.2024.102369

Chapter 5
Sustainable Development Goals:
Origins, Challenges, and Opportunities for a Better Future

M. Bharathi
https://orcid.org/0000-0002-8633-1921
Sri Venkatesa Perumal College of Engineering and Technology, India

Venkata Krishnamoorthy T.
Sasi Institute of Technology and Engineering, India

SandhyaKumari Golla
https://orcid.org/0000-0002-4755-0873
Siddharth Institute of Engineering and Technology, India

Saleha Tabassum
https://orcid.org/0009-0004-6206-9991
Vignan University, India

M. Dharani
Mohan Babu University, India

ABSTRACT

This chapter examines the origins and evolution of the 17 Sustainable Development Goals (SDGs) from the Millennium Development Goals (MDGs), emphasizing their necessity in addressing global challenges. It highlights the interconnected nature of the SDGs, their role in tackling issues like poverty, gender inequality, clean energy, and climate change, and illustrates these interlinkages through real-world examples. The chapter discusses global, regional, and local implementation strategies, focusing

DOI: 10.4018/979-8-3373-1300-9.ch005

on the roles of governments, NGOs, and communities. It stresses the importance of data-driven assessments, transparency, accountability, and innovation, particularly through sustainable business practices and the UN Global Compact Principles. Finally, it identifies challenges such as political and financial barriers, proposing strategies to overcome them and accelerating progress toward an equitable and sustainable future.

1. INTRODUCTION

The Sustainable Development Goals (Jiang et al., 2022) impinge deeper on prioritizing pressing issues affecting humankind. Derived from the Millennium Development Goals, established in 2000 by the UN, SDGs have their attention focused on eight major issues: the eradication of poverty, universal primary education, combating AIDS, reducing child mortality, improving maternal health, ensuring environmental sustainability, and development of a global partnership for development. Much progress has been made in these fields, but there have been observed gaps, especially when it comes to brokering the environmental sustainability, inequality, and economic growth discourse.

Due to the inadequacy of Millennium Development Goals, the United Nations introduced a successor framework-Sustainable Development Goals-in September 2015 (Jabeen, 2024) that puts together 17 interconnected goals comprising 169 targets. These Goals widen the scope of global development and integrating social, economic, and environmental dimensions. The following goals were exhibited as an outcome of consultations at a global scale, wherein were represented the voices of governments, civil society, academia, and the private sector. They are meant to promote inclusive development while ensuring that the planet remains livable for generations to come.

The SDGs are critically relevant in the face of escalating global challenges. Issues such as climate change, inequality, and resource depletion require immediate and coordinated action. The urgency is underscored by data indicating that the global community is falling behind in meeting key targets. For example, rising global temperatures threaten ecosystems and livelihoods, while poverty and hunger persist in many regions despite advancements in technology and economic growth.

In addition, the SDGs provide a global template for collective action by all countries, sectors, and communities. The SDGs encourage action at local, regional, and global levels to address challenges that span borders. Humanity's specific aims, which can be advanced by achieving SDGs, are equity, sustainability, and shared prosperity.

The Critical Necessity of SDGs in Today's World

The UN, in 2015, adopted 17 Sustainable Development Goals (Wu et al., 2018), which represent a global call to action addressing the world's urgent challenges. Providing an agenda of solutions to the most pressing issues-social, economic, environmental, and governance-based-for the achievement of a more sustainable future by 2030, these SDGs target challenges related to poverty, inequality, climate change, education, and health. In increasingly interconnected crises, the SDGs map equitable and sustainable development, speaking to their continuing importance and urgency.

Addressing Pressing Global Challenges

And the necessity of these comes from the multiple challenges facing humanity. In spite of decades of progress, global poverty is still pervasive. Current estimates suggest that more than 700 million people are unable to meet basic necessities such as food, clean water, and shelter. SDG 1 to eradicate poverty is fundamental to laying down the foundation for sustainable development. Inequities within and among nations exacerbate social tensions and hinder economic growth. SDG 5 recognizes women and girls face limited opportunities because of gender inequality. Gender equality should be sure to be attained immediately and is essential to enhancing the economy; studies show that when women hold power, productivity and innovation blossom.

Environmental degradation threatens well-being; rising global temperatures, deforestation, and pollution trigger biodiversity loss, which threatens ecosystems that sustain life. SDG 13 on climate change is of particular concern in light of the increasingly rapid speed at which global warming occurs and the devastating results on the vulnerable populace. The SDGs thus present the challenges in a collective manner, permitting the interrelatedness of social, economic, and environmental dimensions.

The Urgency of Action

The set time for the considerable challenges that the SDGs are trying to address makes the latter slightly urgent, for instance, climate change. It needs immediate and decisive action aimed at limiting global warming to 1.5°C above preindustrial levels. In other words, if action is delayed, irreversible changes may occur, such as rising sea levels and extreme weather events that impact disproportionately the world's poorest communities.

At the same time, making progress toward universal health coverage is difficult, given global health crises as the COVID-19 pandemic has shown. The fragility of the health systems to respond showed the need for resilience in public health infrastructure designed to ensure equitable access to quality health care in Russia to mitigate the socioeconomic impact resulting from long-term health crises.

Relevance in a Globalized World

In today's interconnected global context, we are no longer debating whether the SDGs are pertinent to individual countries. COVID-19, climate change, and economic volatility have a transnational dimension that demonstrates the urgency of global action. The SDGs provide the world with common ground for cooperation between governments, businesses, civil society, and individual citizens.

In particular, the sustainable development goals are innovative and stress the role of technology in addressing such global challenges as renewable energy technology, distance learning, and sustainable agriculture/action theory. For example, SDG 7 recognizes how affordable and clean access to energy is critical for economic growth and poverty eradication. While certainly international, the success behind the future of the SDGs depends on their being implemented locally; that is, communities are innovative in their responses, be it through urban sustainability projects or grassroots initiatives in rural areas. Locale-specific action ensures the solutions resonate more strongly with the region's unique needs and contexts, thereby increasing their efficacy.

SDGs are not a wish list but rather require fast-paced action. They address very pertinent global problems. Vision 2050 is a document that elaborates on how the world is fervently thirsting for transformational change. The onset of transition must happen at a rapid pace if we are to see it happen at all. People from all walks of life around the earth must work together towards a just and sustainable world for this and the coming generations.

2. INDIVIDUAL SDG ANALYSIS: GOALS 1-17

The 17 Sustainable Development Goals first introduced by the United Nations in 2015 aim to create a roadmap for a sustainable and equitable future. These goals address some of the most serious global crises such as poverty, imbalances, climate change, environmental deterioration, and peace-building. There is very close interaction between the goals, and their possible measurable targets and indicators are aimed at ensuring that progress cover the whole agenda. This section will examine in detail the various aspects of each goal such as definition, importance, and primary targets.

Goal 1: No Poverty

Main aspects of negotiation under Goal 1 are to eliminate any form of extreme poverty relative to the bottom 20 percent of the population as those below $1.90 a day. Such requires eradicating extreme poverty for over 700 million people in the world; one of the challenges confronting sustainable development. Targets shall include reducing the number of individuals who live in poverty by half; ensuring that all have equal access to economic resources and other essential services; enacting and sustaining effective protection systems for the people and communities that suffer the most. Reducing the incidences of poverty will accelerate the achievement of the other development goals such as health, education, and development.

Goal 2: Zero Hunger

Goal 2 aims to achieve food security and promote sustainable agriculture. It highlights improving the nutritional quality of food, food security, and agricultural sustainability. Many people suffer food scarcity and malnutrition, particularly children. Among key targets are halving the number of people affected by hunger and food insecurity; doubling agricultural production and productivity by fostering small-scale farmers, ensuring equitable access, and preserving crop and livestock genetic diversity. Achieving this goal means building a healthier population and powerful economies.

Goal 3: Good Health and Well-being

Everything begins with good health and well-being, which allows for individual and social advancement. Hence, Goal 3 is aimed at the promotion of healthy lives and assurance of well-being for all ages. The focus includes reducing maternal and child mortality, especially from communicable diseases like AIDS and malaria, and ensuring universal health coverage. Such health systems are missionaries toward economic stability whereby the Third Goal seeks to form a bulwark for critical service delivery with respect to healthcare provision, medicine availability, and access within our borders.

Goal 4: Quality Education

Education is a very good weapon to fight against the cycle of poverty. Goal 4 thus predominantly relates to inclusive and equitable access to achieve quality education for all. It provides for free primary and secondary education, attempts to reduce disparities, and improvements in literacy and computation. This goal allows our empowerment over the individuals and supports sustainable development across various other sectors.

Goal 5: Gender Equality

Equality between the sexes and empowerment of women are both a human right and the cornerstone of sustainable development. Goal 5 is committed to the removal of all forms of discrimination and violence against women and girls. It emphasizes equal opportunities for women and girls in leadership, education, and economic opportunities. Empowering women to lead gender equality promotes stronger communities and economies.

Goal 6: Clean Water and Sanitation

Water and sanitation are fundamental to health and dignity. Goal 6 covers the ensure availability and sustainable management of water and sanitation for all. As such, it ensures availability and sustainable management of water and sanitation for all. It includes ensuring universal access to safe drinking water, protecting ecosystems related to water, and reducing water pollution. These are important issues that need to be addressed in disease prevention and economic activities relying on water.

Goal 7: Affordable and Clean Energy

Energy is key to today's economic growth and improved quality of life. Therefore, Goal 7 talks about promoting access to affordable, reliable, sustainable energy sources. It emphasizes on increasing renewable energy's share of the total energy supply, boosting energy efficiency, and expanding energy infrastructure in underserved areas. Transitioning to clean energy solutions contribute to a Master's degree in tackling climate change.

Goal 8: Decent Work and Economic Growth

Goal 8 aims for inclusive and sustainable economic growth and decent work opportunities for all. It focuses on enhancing productivity, entrepreneurship, and the eradication of forced and child labor; wages and working environments should be made fairer globally to benefit the economies of all and tackle the problem of income inequalities.

Goal 9: Industry, Innovation, and Infrastructure

Addressing matters of sustainable industrialization, better scientific research, and provision of financial services for small entrepreneurs, supplemented by building industrial and innovation sectors that would promote economic diversification and resilience, constitute Goal 9 due to cascaded gains of different actors along the supply and value chains.

Goal 10: Reduced Inequalities

Fighting inequity built on internal and external hierarchy is critical for social coherence. Goal 10 is more about proper earnings for the bottom 40% of the population, equal opportunities, and reasons for safe migration. Other barriers to inclusion and social development should be tackled for an equitable society.

Goal 11: Sustainable Cities and Communities

Better quality of life and environment-friendly economies are positively enhanced by urbanization; hence, sustainable urban development becomes highly imperative. Goal 11 accords priority to affordable housing, transportation and resilience to disasters, the reduction of urban pollution and waste.

Goal 12: Responsible Consumption and Production

Sustainable consumption and production patterns are essential for the preservation of natural resources. Goal 12 promotes reduced food waste, sustainable practices by businesses and better resource efficiency. By minimizing waste and maximizing production, this goal aims at environmental preservation and economic sustainability.

Goal 13: Climate Action

Action against climate change directly relates to that of future life on Earth. The 13th goal aims to enhance the resilience of vulnerable communities against climate-related hazards, mainstream climate change measures into strategies and policies, and foster global cooperation for climate resilience. Trigger-action to mitigate the adverse impacts of climate change on ecosystems and communities is urgent.

Goal 14: Life Below Water

The ocean is extremely crucial to the global system but has been attacked from various directions. The aim of the fourteenth goal is to conserve the oceans, sustainably manage the oceans, and reduce pollution to protect our marine ecosystems. Hence, ocean resource management has been a fundamental basis for the biodiversity of coastal economies.

Goal 15: Life on Land

Healthy terrestrial ecosystems directly support our survival and biodiversity. The fifteenth goal emphasizes controlling desertification, preserving biodiversity, and controlling deforestation. Land resource management adequately sustains livelihoods while reinforcing them against environmental change.

Goal 16: Peace, Justice, and Strong Institutions

Building peaceful societies and efficient institutions must be regarded as the foundation for a more sustainable development. The 16th goal's purpose is not mainly associated with reduced violence against and eradicated corruption but also with the establishment of an inclusive decision-making process.

Goal 17: Partnerships for the Goals

Acceptance of SDGs can never be successfully realized except through partnerships of all interested groups. Resurfaced into the mainstream is Goal 17, which aims to partner for the Goals by mobilizing resources, sharing technologies, and facilitating international trade. Partnership among the government, business, and civil society is paramount to achieving the SDGs and leaving no one behind.

Interconnectedness of SDGs

Thus far, SDGs represent that component by which global issues can be approached from a single platform as integration among various sectors and policy domains is well manifested. This section describes how the objectives of achievement may positively or negatively influence others in the opposite direction; thus, integrated planning is being emphasized based on the aim to maximize synergies or benefit other objectives while maybe minimizing or avoiding trade-offs among these objectives.

Understanding the Interconnectedness

The SDGs represent an understanding of sustainable development as a composite approach in which the economic, social, and environmental dimensions are interlinked . Eradication of poverty (Goal 1) will have repercussions for health (Goal 3), education (Goal 4), and equality (Goal 10); ensuring all people have access to clean water and sanitation (Goal 6) is a precondition for attainment of good health (Goal 3) and sustainable cities (Goal 11).

- Mutual Reinforcement: Ecuation such as Goal 4: Quality Education and Goal 5: Gender Equality support one another, since educating girls will provide them with better access to economy and will bring added societal benefits.
- Trade-offs: In some cases, achievement of some goals could obstruct others, even if, on the whole, mutually reinforcing. If industrial infrastructure development (Goal 9) is unsustainable, it would further hurt climate action (Goal 13).

Synergies Among Goals

The connection between health, water, and climate: Clean water (Goal 6) ensures better health outcomes (Goal 3), and mitigating climate change (Goal 13) protects water sources. People with well-established water-infrastructure are mostly less exposed to climate-related risks such as droughts and floods.

Economic growth and reduction of inequalities: Promotion of decent work and economic growth (Goal 8) reduces inequalities (Goal 10) by creating opportunities and lifting living standards: Inclusive policies ensure that marginalised communities benefit from economic advancement.

- Innovation and sustainability: Investment in sustainable energy (Goal 7) and infrastructure (Goal 9)drove innovation leading to sustainable consumption and production pattern (Goal 12).The renewable energy technologies also drive and scale down emissions, thus supporting climate action (Goal 13).

3. TRADE-OFFS AND CHALLENGES

Interconnectedness of the SDGs (Cao et al., 2022) presents opportunities, yet it also offers challenges. Policymakers should have hoisted through these odds to prevent any unintended consequences.

- Agriculture and Ecosystems: Growers see the end of hunger (Goal 2) if agricultural practices increase production, though unsustainable farming techniques would impair terrestrial ecosystems (Goal 15).
- Urbanization and Inequality: If urbanization is prioritized (Goal 11) without setting aside affordable housing and infrastructure for all sections of the population, inequalities will escalate (Goal 10).
- Growth and Environmental Protection: Industrial growth (Goal 9) often counters environmental goals (Goal 14 and Goal 15) unless sustainable practices are integrated with plans for development.

Real-World Examples

- Renewable Energy Initiatives: Countries more so like Denmark have integrated Goals 7, 8, and 13 through investment in wind energy, reducing greenhouse gas emissions, creating jobs, and having contributing to economic growth.
- Integrated Water Resource Management: In sub-Saharan Africa, programs addressing Goals 6, 3, and 11 have increased water access, prevented disease outbreaks, and improved community resilience to climate change.
- Educational Reforms in South Asia: Investments in education (Goal 4) have empowered women (Goal 5), reduced poverty (Goal 1), and stimulated local economies (Goal 8).

The Path Forward

- The realization of SDGs requires systems thinking and cross-sectoral collaborations. Policymakers, businesses, and civil society should cooperate to design integrated responses meant to simultaneously tackle multiple goals. Scenario modeling and impact assessment tools help to identify trade-offs, in the interest of balanced development.
- Policy Integration: Governments should integrate national development plans with the SDGs, maintaining an awareness of the interrelatedness of goals.
- Engagement with the Private Sector: Business can assist in sustainability-related initiatives that address several goals, such as reducing carbon footprints (goal 13) and ensuring fair employment practices (goal 8).
- Engagement of Local Community: Milestones for successful community engagement should integrate local priority and during the process of community engagement; such benefits must be tangible and sustainable.

4. IMPLEMENTATION STRATEGIES

The realization of the Sustainable Development Goals depends significantly on a coordinated approach across global, regional, and local levels. Governments, international organizations, businesses, and civil society must come together to agree on and implement the context-specific strategies suited to different times (Cao et al., 2022). In this section, we introduce the conceptual frameworks, projects, and initiatives developed to support the 2030 Agenda at these various levels.

Inhibitor relations have largely been libertarian, with each actor pressing its agenda at the decision-making and multilateral agreement stages. The aspects of multilateralism have traditionally highlighted leadership projected through partnerships in the aims stated by the United Nations. Through the high-level political forum, mechanisms are created for SDG tracking, dialogue, peer-to-peer sharing among member states, and coalesced efforts. The HLPF complements the "Decade of Action" launched in 2020 and lists mechanisms for international partnership and accountability that should build up momentum and speed towards achieving the SDGs. Other multilateral treaties like the Paris Climate Accord have played an influential role on specific fronts-centered on Climate Action-goal 13-as a follow-up to mainstreaming joint commitments to sustainability. Furthermore, other financial frameworks at the global level, such as the Addis Ababa Action Agenda, are critical to the success of the sustainable development goals. Such establishments work as a conduit for partnerships among the international financial institutions, development banks, and donor agencies and channel resources to lifesaving areas. The Green Climate Fund is an example whereby it offers critical resources for developing countries as they shift their ways and mitigate climatic-related risks. To provide a footing for progress monitoring, global actors throughout years, specifically the United Nations Development Program and the World Bank, have been involved.

Regional strategies stress cooperation among states with common problems and challenges. Organizations such as the African Union, the European Union, and ASEAN give their own regional inputs to define regional priorities toward the SDGs. For example, "Agenda 2063" of the AU is essentially an expression digested in full agreement to the SDGs, but mainly with regard to eradication of poverty (Goal 1) and gender equality (Goal 5). The other major asset is cross-border cooperation, especially for goals that require shared resources-water management and climate adaptation. An example of this is the Nile Basin Initiative, which seeks to promote sustainable utilization of water resources among African countries dependent on the Nile. Besides, regional organizations invest a lot to capacitate their institutions to measure progress critically; for example, the EU assists its member states in tracking all 17 SDGs.

At the local level, implementation strategies center on participative community development. Local governments and organizations carve out global goals into projects that satisfy particular cultural and socio-economic realities. For example, India's Aspirational Districts Programme integrates education, health, and livelihood initiatives from backward areas in accordance with Goals 3, 4, and 8, respectively. Urban Subsistence programs strive to meet the SDG since the majority of their population lives in urban areas. System alone is carried by Initiatives such as the C40 Cities initiative, which entails urban, sustainable development, and climate action through endeavors such as adoption of green public transport systems in Bogotá, Colombia. Moreover, the time invested in engaging local stakeholders, such as the community, NGOs, and the business sector, has been quite meaningful in enabling SDGs to be delivered effectively. For example, in Kenya, rural women are trained through implementable sustainable agriculture in building a direct contribution to Goals 2 and 5.

Innovation has established itself as an important cornerstone of SDG implementation across multi-level and multi-sectoral entities. Technology and digital solutions have acted as catalysts for achieving progress, especially through ICT-enabled mobile applications, such as the mPower app in Bangladesh, which connects farmers to markets to enhance productivity and income, thereby contributing toward Goal 8. Public-private partnerships (PPPs) mobilize funding and provide expertise, working in sync with Actionable Goal 6 to improve access to such basics as clean water and sanitation in Africa. Other innovative financial models are social impact bonds and green bonds, wherein proceeds are directed to SDG-aligned initiatives. In Sweden, SDG bonds are supporting activities on renewable energy, climate resilience, and ecosystem conservation.

Regardless of such efforts, there are major challenges still requiring attention to the relevant aspects of the implementation of SDGs. Financial and technical support is curtailed due to resource constraints, especially in developing countries. Policy coherence remains somewhat elusive, as frequently there is no alignment between the national priorities and the SDG priorities. Furthermore, insufficiently developed data collection mechanisms and lack of transparency have resulted in challenges to monitoring and ensuring accountability. In conclusion, delivery of SDGs hinges largely on a comprehensive approach that integrates global frameworks, regional cooperation, and local action. This brings together different stakeholders appropriately by fostering partnerships, utilizing technological advancements, and addressing continuing challenges to boost progress toward achieving the transformative vision of the 2030 Agenda.

Stakeholder Roles

The Sustainable Development Goals cannot be realized unless many actors collaborate with each, for each fulfills a distinctive and paramount role in the attainment of the SDGs. In realization of the SDGs, an analysis of roles of government, NGOs, educational institutions, and individuals will be presented. Some of the roles are:

Governments

These actors are crucial in the realization of the SDGs, primarily owing to their authority, resources, and policy-making capacity. Their functions include:

1. The policy planning and implementation: It is the processes constitute national policymaking that resonates with the SDGs, thereby integrating those targets in the development strategies. This is inclusive of enacting frameworks and incorporating the pillars of people, prosperity, and planet into development strategies.
2. Resource allocation: It is where governments are accountable for the distribution of public resources as outlined in such matters as taxes, budgets, and international financing for sustainable development.
3. Legislation and regulation: It permit the government to enact laws and other regulations that encourage principals to operate sustainably, particularly in respect to large areas such as energy, industry, agriculture, and health, in turn turning incentive into acts on the part of businesses and individuals.
4. International corporation: It means all these reform processes come with that aspect of being good partners, where governments link arms with other countries, international organizations, and NGOs to combat such transboundary challenges as climate change, poverty, and public health emergencies.

NGOs (Non-Governmental Organizations)

NGOs act as key enablers of sustainable development at local, national, and international levels. Their roles include:

1. Advocacy and Awareness: NGOs are often at the forefront of raising awareness about the SDGs, mobilizing communities to take action, and advocating for policy changes to ensure that the SDGs are prioritized (Almuaqel, 2024).
2. Community Engagement: NGOs involve communities directly in the run of projects using this framework, especially in education, health, poverty alleviation, and environmental conservation targeting the poor.

3. Capacity Building: NGOs build up the capabilities of local organizations and governments through development efforts.
4. Monitoring and Accountability: NGOs achievement under SDGs at the local level is generally monitored by them and they can provide actual and timely inputs and feedback on the extent of achievements, indicating where future attention and efforts are needed.

Educational Institutions

Moreover, educational institutions are essential for instilling in the society the ability to realize methodologies based on knowledge regarding development prospects (Marcovecchio et al., 2018). Charity of these institutions will include:

1. **Research and Innovation**: Universities and research institutions are involved, alongside other partners, in nourishing the scientific knowledge with technological innovations aimed at addressing some of the sobering global challenges, such as renewable energy, clean water, and disease prevention.
2. **Curriculum Development**: Institutions include sustainability and SDG-related topics in their curricula, framing future generations with skills and mindsets in the resolution of complex global challenges.
3. **Training and Capacity Building**: They implement specialized training programs and workshops to capacitate professionals and laypeople with the skill sets necessary to advocate for sustainable development throughout various sectors.
4. **Public Engagement**: A number of universities and institutions undertake public discourse on sustainability or host forums, lectures, and initiatives while engaging students, faculty, and broader society in discussions on the elements of achieving the SDGs.

Individuals

Individuals will further be part of this wheel of attunement to achieving the SDGs as determined by daily practices and choice, especially the contribution:

1. **Behavioral Changes**: Individual sustainable choices, such as water conservation, reduction of energy, maximum usage of renewable resources, and green business support, help leave a green footprint.

2. **Advocacy and Awareness**: The people will create awareness that will inform the public, thereby putting pressure on their governments and businesses to take the SDGs seriously by advocating for policies supporting sustainability.

3. **Community Involvement**: Most people find their ways to engage in community-enhancing projects contributing in the direction of the SDGs-from volunteer service to working with local development agencies for sustainable events.

4. **Education and Skill Development**: Persons undertaking education and training towards sustainable development, to name a few: in agriculture, green technologies, and social entrepreneurship.

It is a shared responsibility that brings the SDGs into existence, a task that requires synergy among governments, NGOs, educational institutions, and individuals (Talwar et al., 2024). Governments provide a base of legislative frameworks, NGOs touch bases-and provoke communities towards collective action, educational institutions foster knowledge and innovation, and individuals give their service and voice to effect realization of change. Together they form a vibrant fabric of momentum that is the realization of the SDGs.

Monitoring And Evaluation

Monitoring and evaluation (M&E) are essential for tracking the progress of the Sustainable Development Goals, (Appio et al., 2024) ensuring that targets are being met, and identifying areas for improvement. A variety of tools and techniques are used to assess performance, collect and analyze data, and foster accountability.

The use of performance indicators makes monitoring the progress toward particular SDG targets easy. There are basically two types of performance indicators, namely, outcome indicators and output indicators. The former indicates a long-term change in interventions that may result from an intervention in the form of a decrease in poverty or increase in literacy rate. Output indicators measure results at the right point, which involves the immediate counts, such as the number of schools built, or the quantities of clean water provided. For a performance indicator to be valid, it has to be a SMART by which is specific, measurable, achievable, relevant, and time-bound.

M&E requires accurate data collection. Various methods are involved in extracting the necessary data. The necessary information for the qualitative and quantitative aspects is mainly gathered through surveys and questionnaires from communities and stakeholders. Large-scale environmental information on changes in forest cover, water bodies, and urbanization would come from remote sensing and satellite data, which is critically important for monitoring the environment towards SDG progress. Additionally, administrative data through government records provide valuable sources such as census data, school enrollment rates, health statistics, which are used for the assessment of the SDG indicators (Syed et al., 2024). Participatory approaches that engage communities in the local level during the process of data gathering ensure more local and contextually relevant data as well.

Data analysis methods help a person make the correct interpretation of the data collected. Large data sets are commonly analyzed with the help of statistical tools and software such as SPSS and Stata, with trends and correlation between variables. Geospatial analysis, for instance, may be carried out with the aid of GIS tools in the case of mapping and studying spatial data especially for environmental and infrastructure-related SDGs. Qualitative methods, such as content analysis and thematic coding, are applied to non-quantifiable data involving text-based data from interviews or reports. Tools like Tableau or Power BI that fall under data visualization help express complex data in a more understandable format, thus allowing stakeholders to monitor and track progress toward the SDGs in real time.

Accountability in the monitoring process ensures SDG initiatives are on course. Periodic reports on SDG prepared by the governments and other international organizations such as the UN contain achievements, gaps, and challenges in implementing the SDGs (Torkayesh et al., 2024). The UN High-Level Political Forum (HLPF) will receive VNRs from countries, and such reports will comprehensively review the implementation status of SDGs for each country, along with highlighting the areas requiring attention. Independent third-party audits will also be carried out to ensure that the data is credible and objective and not misleading or fabricated. The actual monitoring process has been made more transparent with the adoption of various public platforms and dashboards, including the SDG Tracker, to display real-time data on SDG progress. Monitoring and evaluation are the crux of SDG implementation as they ensure transparency, accountability, and data-driven decision-making. Reporting helps create trust and confidence among stakeholders and encourages corrective measures and citizen involvement in the SDG process. By effectively applying these tools and techniques, progress toward the achievement of SDGs can be measured, and necessary adjustments can be made to accelerate their realization.

5. INNOVATIONS AND ADVANCEMENTS DRIVING SDG PROGRESS

Technological innovations and business models in line with the UN Global Compact Principles (human rights, labor, environment, and anti-corruption) will play a pivotal role in promoting acceleration towards SDGs (Ke et al., 2024). These innovations are not only dealing with global issues but also have an impact on sustainable economic growth.

Technological Innovations

1. Renewable Energy Technologies

Solar and Wind Power: Improving efficiency in photovoltaic efficiency, innovation in the design of wind turbines, and better storage technology are making advancements toward the realization of SDG 7, Affordable and Clean Energy. Innovation in renewable energy makes them cleaner, cheaper, and reliable. This lessens reliance on fossil fuels, hence curbing climate change.

Smart Grids and IoT: The combination of smart grids with IoT devices improves the distribution efficiency of energy, enables real-time energy management, and helps reduce energy waste.

Blockchain for Transparency: Blockchain technology is used to enhance transparency and accountability in finance, supply chains, and governance. It supports SDGs related to responsible consumption, governance, and peace by providing an immutable and transparent record of transactions, such as SDGs 12 and 16.

2. Artificial Intelligence and Big Data

Healthcare by AI: Through diagnostic tools and predictive analytics, AI-driven innovations for healthcare are a contribution to the success of SDG 3 or Good Health and Well-being towards better access and reduced costs to health care.

Data Analytics for Education: Big data is used in the analysis of educational trends, thus improving learning outcomes. This helps achieve SDG 4: Quality Education.

Climate Modeling: AI and big data technologies are helping predict climate-related risks, enabling better disaster management and informing sustainable environmental policies (SDG 13).

3. **Sustainable Agriculture Technologies**

Innovations such as precision farming, vertical farming, and the use of drones for monitoring crops are helping achieve SDG 2 (Zero Hunger) by improving food production efficiency, reducing waste, and making agriculture more sustainable.

4. **Water Purification and Conservation**

Contributions of water desalination technology, filtration systems, and smart water management systems help in achieving SDG 6-Clean Water and Sanitation. Technologies may further make easy provisions of clean water for people, who need it most, while optimizing water usage of the agriculture and industry sectors.

5. Business Models Driving SDG Progress

Circular Economy The businesses of circular economy design products to be reused, recycled, and wasted as little as possible. In doing so, it promotes the realization of SDG 12: Responsible Consumption and Production through reducing impacts on the environment while establishing sustainable business practices.

Impact Investing Impact investing Impact investors pool resources for financing projects and businesses that converge with the goals of SDGs, generating social and environmental impact alongside financial returns. The main goal behind this model is to converge with goals such as clean energy, healthcare, education, and poverty reduction.

Inclusive Business Models More companies are embracing the inclusive business models that focus on serving low-income populations through affordable products and services. For instance, healthcare, energy, and finance businesses are innovating to create affordable solutions to underserved communities, thus advancing SDGs that relate to poverty, health, and economic growth: SDG 1, SDG 3, and SDG 8.

Corporate Social Responsibility (CSR) Initiatives Many companies have aligned SDGs as part of CSR strategy, concerning issues like environment sustainability, social equity, and ethical governance. By integrating operation with Global Compact Principles, organizations directly impact multiple SDGs - especially environmental protection and social justice (Menon & Paretti, 2024).

Green Finance They have green bonds, loans, and more; the financial institutions now promote positive environmental projects such as renewable energy infrastructure, clean transportation, and sustainable agriculture that are contributing to SDG 13 on Climate Action and SDG 7 on Affordable and Clean Energy. The success of the SDG achievement is rooted in sound monitoring and evaluation with good data collection and analysis techniques. All the activities are measured on performance indicators with data visualization for transparent reporting by global efforts for responsibility. Technological innovations in renewables, AI, blockchain, sustainable agriculture, as well as in forward-thinking business models such as circular economies, impact investing are accelerating the progress in SDG. A stakeholder brings meaningful change for society to reach a successful future by working in tandem with technological advancements combined with business strategy, guided through the prism of United Nations Global Compact Principles.

6. CHALLENGES AND SOLUTIONS

Attainment of Sustainable Development Goals (Achmad et al., 2023) is a complex issue that requires a lot of struggle to overcome some barriers. Some key barriers and how to overcome them are listed below:

Poverty and Inequality

Although much has been achieved, poverty remains a significant issue in the world, especially in developing countries. Income inequalities, lack of education, healthcare, and basic services are still major causes of poverty, which makes the cycle continue. Solutions to these problems are to promote inclusive economic growth that creates jobs and provides equal opportunities for all, policies to improve access to quality education, healthcare, and social services, and promoting sustainable business practices that focus on fair wages and equitable resource distribution.

Climate Change and Environmental Degradation

The primary hindrance in attaining the SDGs includes climate change. This change of climate hampers the ecological, economic, and food systems. It further includes issues related to environmental degradation, deforestation, and pollution. For that reason, one of the necessary challenges is international cooperation and commitments regarding greenhouse gas reduction through the means of renewable energy, sustainable agriculture, and the principles of circular economy. More importantly, investment in climate-resilient infrastructure and disaster preparedness in vulnerable regions and public education and incentives for green technologies that encourage sustainable consumption patterns are steps that need to be taken.

Access to Clean Water and Sanitation

Over 2 billion people lack clean water, while 4.2 billion lack safely managed sanitation services, making them vulnerable to preventable diseases. Upgrading water infrastructure in underserved communities through investment in modern water treatment and distribution is one of the key solutions. Promoting practices for conserving and managing water, especially in water-scarce regions, as well as supporting local communities in using low-cost, sustainable sanitation technologies, are vital.

Healthcare Access and Disease Prevention

There is a health disparity in terms of access, with millions of people in low-income regions lacking basic medical services. The threat of pandemics is ongoing, and the lack of resources compounds this issue. To overcome this, increasing funding for universal healthcare coverage to ensure that essential health services are available to all is necessary. Strengthening global health systems to better respond to the emerging health crises; improving preventive care and increasing vaccination programs; and fostering public-private partnerships to develop affordable medical technologies and services for underserved populations are critical solutions.

Education and Skills Development

Despite the global efforts, millions of children, especially girls and children in conflict zones, are still out of school. The quality of education also remains low in many areas, limiting future opportunities. All children are guaranteed equal rights of access to education regardless of their gender or geographical location background. The better training of a teacher, more development in a curriculum, or investing in educational technology for smooth delivery will play an important part. The growth in vocational learning and lifelong skill training to achieve future requirements can help in attaining this aspect of the education system.

Gender Inequality

Gender inequality persists in many societies, with women and girls facing discrimination in education, employment, political representation, and personal rights. Solutions to this challenge include enforcing laws that protect women's rights and promote gender equality in all sectors, empowering women through economic opportunities, leadership roles, and equal access to resources, and promoting societal changes that challenge gender stereotypes and encourage the participation of women in all aspects of life.

Peace, Justice, and Strong Institutions

The realization of SDGs in fragile states and war and instability affected regions remains impossible if issues of conflict, corruption, and poor governance prevail. For the achievement of such objectives, promoting democratic governance and rule of law, holding accountable and making transparent public institutions, and peacebuilding with increased attention to the prevention and resolution of conflict through diplomacy and community engagement must be put in action. Supporting inclusive and transparent institutions that symbolize the diverse needs of society is also very important.

Financing and Resource Mobilization

Proper funding for SDGs (Roldán-álvarez et al., 2021) is one of the main obstacles, mainly among developing countries because of capital deficiency and a lot of debt. International cooperation has to increase efforts to finance by grants, loans, and even debt relief among low-income countries. This involves encouraging private sector investment in sustainable development through impact investing and social enterprises and innovative financing mechanisms, such as green bonds and climate financing, as a means of mobilizing resources for the implementation of SDG (Lu et al., 2024).

Global Partnerships

This collaboration between governments, international organizations, the private sector, and civil society is required to implement the SDGs (Ezugwu, 2024). Incoherent efforts and lack of coordination will also be barriers. Strengthening multilateral institutions and frameworks for cooperation worldwide, making SDGs the central focus in international relations, and fostering partnerships among governments, businesses, and civil society to leverage resources, expertise, and networks for SDG implementation are important. Thus, the advancement of shared responsibility, in which all sectors play a role to achieve the SDGs, mainly through concerted research, technology, and knowledge, will help the world meet its goals. A way toward meeting the SDGs by 2030 can thus be achieved with such clear-cut strategies and steps of action. Collaboration (Verdejo Espinosa et al., 2023), innovation, and proper political will will thus have to be established in order to address the challenges that would lead to sustainable development in the future.

CONCLUSION

The 2030 Agenda for Sustainable Development and its 17 SDGs set out a clear framework on which to tackle these critical issues that are facing today's world. The chapter particularly points out how the goals are interlinked and that concerted efforts at all social levels are essential. It emphasizes the importance of proper implementation methodology, innovation, and stakeholder engagement. The chapter shows that with efficient monitoring, data-driven decisions, and sustainable practices, progress can be successful despite the many challenges, including limited resources and political hurdles. A fairer and more sustainable world will indeed start to manifest with international partnerships and successful leadership.

REFERENCES

Achmad, F., Prambudia, Y., & Rumanti, A. A. (2023). Sustainable Tourism Industry Development: A Collaborative Model of Open Innovation, Stakeholders, and Support System Facilities. *IEEE Access: Practical Innovations, Open Solutions, 11*, 83343–83363. DOI: 10.1109/ACCESS.2023.3301574

Almuaqel, I. A. (2024). Special Needs, Special Efforts: Learners With Intellectual Disabilities and Sustainable Development Goals (SDGs). *IEEE Transactions on Engineering Management, 71*, 13469–13486. DOI: 10.1109/TEM.2022.3189318

Appio, F. P., Platania, F., & Hernandez, C. T. (2024). Pairing AI and Sustainability: Envisioning Entrepreneurial Initiatives for Virtuous Twin Paths. *IEEE Transactions on Engineering Management, 71*, 11669–11686. DOI: 10.1109/TEM.2024.3428913

Cao, M., Chang, L., Ma, S., Zhao, Z., Wu, K., Hu, X., Gu, Q., Lu, G., & Chen, M. (2022). Multi-Scenario Simulation of Land Use for Sustainable Development Goals. *IEEE Journal of Selected Topics in Applied Earth Observations and Remote Sensing, 15*, 2119–2127. DOI: 10.1109/JSTARS.2022.3152904

Ezugwu, A. E.-S. (2024). Metaheuristic Optimization for Sustainable Unrelated Parallel Machine Scheduling: A Concise Overview With a Proof-of-Concept Study. *IEEE Access : Practical Innovations, Open Solutions, 12*, 3386–3416. DOI: 10.1109/ACCESS.2023.3347047

Jabeen, F. (2024). The Alignment of Universities With Sustainable Development Goals: How Do Academics Perceive the Progress (Not) Made? *IEEE Transactions on Engineering Management, 71*, 13545–13557. DOI: 10.1109/TEM.2022.3183016

Jiang, S., Jakobsen, K., Bueie, J., Li, J., & Haro, P. H. (2022). A Tertiary Review on Blockchain and Sustainability With Focus on Sustainable Development Goals. *IEEE Access : Practical Innovations, Open Solutions, 10*, 114975–115006. DOI: 10.1109/ACCESS.2022.3217683

Ke, L., Liu, D., Tan, Q., Wang, S., Wang, Q., & Yang, J. (2024). Spatial-Temporal Pattern of Land Use and SDG15 Assessment in the Bohai Rim Region Based on GEE and RF Algorithms. *IEEE Journal of Selected Topics in Applied Earth Observations and Remote Sensing, 17*, 7541–7553. DOI: 10.1109/JSTARS.2024.3380580

Lu, J., Lin, W., & Dabić, M. (2024). Coevolution Analysis of the Sustainable Development Systems of Energy Firms From a Strategic Philanthropy Perspective. *IEEE Transactions on Engineering Management, 71*, 6887–6902. DOI: 10.1109/TEM.2023.3244500

Marcovecchio, I., Thinyane, M., Estevez, E., Fillottrani, P., Marcovecchio, I., & Estevez, E. (2018). Capability Maturity Models as a Means to Standardize Sustainable Development Goals Indicators Data Production. *Journal of ICT Standardisation*, *6*(3), 217–244. DOI: 10.13052/jicts2245-800X.633

Menon, M., & Paretti, M. C. (2024, September). Faculty Perspectives on Integrating Sustainable Development Into Engineering Education. *IEEE Transactions on Technology and Society*, *5*(3), 316–324. DOI: 10.1109/TTS.2024.3403482

Roldán-álvarez, D., Martínez-Martínez, F., Martín, E., & Haya, P. A. (2021). Understanding Discussions of Citizen Science Around Sustainable Development Goals in Twitter. *IEEE Access : Practical Innovations, Open Solutions*, *9*, 144106–144120. DOI: 10.1109/ACCESS.2021.3122086

Syed, M. S. B., Kelly, P., Stacey, P., & Berry, D. (2024). Towards an IoT-Enabled Digital Earth for SDGs: The Data Quality Challenge. *IEEE Access : Practical Innovations, Open Solutions*, *12*, 153421–153438. DOI: 10.1109/ACCESS.2024.3478813

Talwar, S., Dhir, A., Madanaguli, A. T., & Ractham, P. (2024). SDG-Embeddedness in Business School Operations: State-of-the-Art Literature, Ground Realities, and Measurement Parameters. *IEEE Transactions on Engineering Management*, *71*, 13428–13445. DOI: 10.1109/TEM.2023.3236609

Torkayesh, A. E., Alizadeh, R., & Soltanisehat, L. (2024). Assessing Entrepreneurial Performance of G7 Countries Under Context of Social Sustainable Development Goals: A Multicriteria Decision Analysis Approach. *IEEE Transactions on Engineering Management*, *71*, 4918–4929. DOI: 10.1109/TEM.2022.3227240

Verdejo Espinosa, Á., Lendínez, A. M., Melguizo, F. J., & Estévez, M. E. (2023). Engineering and Technology Education in University Studies: Driving Digital, Sustainable, and Resilient Development—A Case Study in Andalusia, Spain. *IEEE Access : Practical Innovations, Open Solutions*, *11*, 108967–108981. DOI: 10.1109/ACCESS.2023.3321135

Wu, J., Guo, S., Huang, H., Liu, W., & Xiang, Y. "Information and Communications Technologies for Sustainable Development Goals: State-of-the-Art, Needs and Perspectives," in *IEEE Communications Surveys & Tutorials*, vol. 20, no. 3, pp. 2389-2406, thirdquarter 2018, DOI: 10.1109/COMST.2018.2812301

Wu, J., Guo, S., Huang, H., Liu, W., & Xiang, Y. "Information and Communications Technologies for Sustainable Development Goals: State-of-the-Art, Needs and Perspectives," in *IEEE Communications Surveys & Tutorials*, vol. 20, no. 3, pp. 2389-2406, thirdquarter 2018, DOI: 10.1109/COMST.2018.2812301

Chapter 6
Regional Framework and Strategies for Implementation of SDGs

Amisha Shaikh
Ajeenkya D.Y. Patil University, India

Tanishka Prashant Madke
Ajeenkya D.Y. Patil University, India

Susanta Das
https://orcid.org/0000-0002-9314-3988
Ajeenkya D.Y. Patil University, India

Rizwan Shaikh
https://orcid.org/0009-0005-5689-7640
Dr. D.Y. Patil Institute of Management and Entrepreneur Development, India

ABSTRACT

The SDGs (Sustainable Development Goals), created by the United Nations in 2015, aims to tackle global level problems in order to end poverty, reduce inequality, promote sustainable consumption and production, combat climate change, preserve oceans and ecosystems, foster innovation, reduce gender disparities, ensure access to clean water and energy. These goals set a vision for sustainable development but require specific regional strategies considering the social-economic, cultural and natural endowment of each region. The study examines regional frameworks coordinating with global SDGs under specific regional circumstances. Every region is analyzed according to their participation of the governments, civil society organizations and the private sector to cope with their specific development issues.

DOI: 10.4018/979-8-3373-1300-9.ch006

The study highlights the progressive enhancements across various sectors including health, education, clean energy, etc. Local communities contribute significantly to realizing SDGs improving the local identity and social capital, which are important for cooperating in spatial development.

1. INTRODUCTION

The United Nations came up with the Sustainable Development Goals (SDGs) in 2015 to address important challenges affecting the world, including environmental issues, inequality, climate change and poverty (4th SDG Youth Summer Camp., n.d.). The SDGs merge 17 different targets that divide into three main blocks: social, economic and environmental aims for development that is fair and sustainable (Nilsson et al., 2018). For SDGs to be successful, it is important to notice the characteristics of a region such as its social structure, culture and environment (Jiménez-Aceituno et al., 2020).

This study elaborates on a detailed regional structure to execute SDGs by establishing local-based solutions that fulfil global sustainability criteria (Putra et al., 2024). Achieving the SDGs needs effective collaboration between state institutions with non-government organizations alongside private businesses for their successful achievement (Foroudi et al., 2025). Stakeholders who establish collaboration with each other and actively participate in decision-making could be able to create better solutions for regional problems through innovative approaches (Bardal et al., 2021). The analysis emphasizes the vital nature of inter-border collaboration because many Sustainable Development Goals (SDGs) handle problems which span across state jurisdiction. Countries that belong to economic communities such as ASEAN, the African Union and the European Union develop sustainable development through multidirectional resource and information exchanges (Urama et al., 2014; OECD, 2020).

The 17 goals by United Nation, known as United Nations Sustainable Development Goals (UNSDGs-17 Goals) aim to create a better world by the end of 20230 (UNSDGs). This chapter investigates how strategies and frameworks are used in different places globally to achieve these goals, and the current challenges, and future opportunities. The research outcomes will be beneficial for the developing programs and policies for achieving the ambitious goals set in the 2030 Agenda.

2. LITERATURE REVIEW

Nilsson et al. (2018) discussed how regional frameworks with governance structures implement the SDGs by establishing localized plans with cross-border partnerships between various stakeholders. National economic alliances such as ASEAN, together with African Union and European Union serve to propel forward the SDGs. The key element for successful SDG implementation requires goal adaptation to diverse local situations. Every ASEAN member state design SDG implementation strategy according to their development issues while focusing on their individual priorities. By localizing SDGs, they become part of local policies and practices which enables both local ownership and accountability from local stakeholders. The study also provides a 7-point typology to score binary interactions between SDG targets providing numerous applications via the case studies in health, energy, and oceans. Therefore, the framework highlights the role of contextual variables consisting of governance, geography and time horizon for analyzing the nature and directionalities of interactions. Many SDGs require united border operations because they examine objectives that extend beyond national lines including climate change effects and biodiversity reduction together with migration patterns. Through its initiatives the African Union facilitates collaboration through regional frameworks such as the African Continental Free Trade Area (ACFTA) initiative to strengthen both economic integration and continental development. Productive international collaboration remains essential for reaching the SDGs because it enables countries to share resources and exchange expertise and implement joint solutions to mutual issues. The active involvement across different groups delivers multiple viewpoints which leads to improved policy development through more inclusive and successful outcomes. The European Green Deal demonstrates how multiple stakeholder participation leads to sustainable development through its development process within the European Union. The African Union understands the importance of cohesive policies so it created frameworks which show member states the proper ways to merge their national plans with the SDGs (United Nations Sustainable Development Goals). By using this approach decision-makers can find and solve problems that exist between different objectives to maintain steady development across all areas. The SDGs face implementation barriers because government institutions have restricted operational capacity and lack communication between different agencies and have insufficient funding for development projects. Various institutional barriers present in ASEAN countries and African Union member states work as hurdles against achieving the SDGs. Improving sustainable development initiatives demands reinforced governance systems and expanded capacity building programs alongside sufficient financial support. The discussion ends by analyzing new regional methods which synchronize worldwide targets with ground-level implementations. The approaches usually use technology platforms to meet with public-private collaboration and disseminate knowledge between regional states. European Union authorities established numerous digital initiatives which strengthen their data collection abilities for SDG monitoring to support better decisions along with precise interventions.

Aftab et al. (2020) investigated the implementation of health-related SDGs that form part of Agenda 2030. This analysis shows that numerous governments demonstrate substantial political backing by applying SDGs to match with national development plans to gather support while focusing on national and subnational implementation. The review stresses that health-related SDG implementations need multisectoral collaboration through cross-sectoral entities which enable governmental departments and different administrative levels to work alongside each other. The connection between multiple SDGs requires coordinated action, as health outcomes are influenced by a range of social, economic and environmental impacts. The main obstacle discovered in SDG implementation stems from insufficient funding for health-related targets as governments heavily depend on contributions through external donors which poses sustainability challenges. The review underlines the requirement for countries to improve their domestic funding systems by enhancing taxation levels together with non-governmental partner resource allocation that supports SDG objectives as described in the Addis Ababa Action Agenda. The review identifies data's essential function for tracking health-related SDG advancement, although numerous countries keep choosing indicators and gathering information while dealing with substantial data gaps regarding full and trustworthy data availability. Successful implementation depends on capacity building since multiple countries struggle to build sufficient institutional capacity needed to execute health-oriented SDGs. The evaluation demands knowledge about country-level approaches so the field can find successful methods and accumulate valuable experience. The implementation of health-related SDGs faces substantial challenges which need integrated solutions combined with enhanced financial sustainability improvements in data collection methods along with increased stakeholder involvement to achieve the agenda 2030 targets.

Organisation for Economic Co-operation and Development (OECD) (2020) conducted a thorough evaluation of multiple SDGs indicators while giving specific attention to gender equality, climate action and the protection of biodiversity on land and underwater. It explains how education plays a vital role in attaining SDG 5 which addresses gender equality. The research analysis includes data about the average literacy rating of people from 16 to 65 years old which demonstrates how educational attainment influences gender equality results. Societal perspectives are studied through data showing population percentages who hold favorable views about female respect, along with concerning statistics on physical and sexual attacks against women. The difficulty in achieving gender equality becomes clear through these indicators which highlight persistent barriers requiring efficient policy solutions. The analysis particularizes its investigation on employment rate differences between genders while stressing the need for equal workplace access for men and women. Further, the study incorporates environmental aspects from SDG 12 while investigating gender-specific topics. The analysis also covers demonstration of how much CO_2 emissions are linked to electricity output as it reveals the critical environmental ramifications of energy usage, together with the necessity of developing eco-friendly methods. The observation of shifting cooling degree-days acts as evidence for climate change impact on energy consumption pointing to the necessity of renewable energy adoption along with efficient energy implementation. The SDG 14 section includes statistics about protected coastal areas among its assessment of life below water. The discussion also includes SDG 15 that focuses on land-based life where it examines changes in tree cover alongside the current condition of terrestrial protected areas. The insights within this section display current difficulties surrounding deforestation together with habitat destruction as well as the requirement for successful conservation plans which safeguard land-based ecosystems. The research demonstrates how these targets form a linked web which requires thorough solutions to resolve multiple gender-related sustainability issues. The research data shows that improvements made in one aspect generate substantial effects on related outcomes which validates the need for united work between public authorities as well as civil society and private sector entities. The analysis demands higher public understanding and strategic actions that advance gender equality and sustainable resource usage and environmental preservation because they represent crucial elements in building a sustainable society for everyone. Overall, the study provides detailed information about current developments which establishes itself as a vital tool for policymakers and researchers and advocates focusing on SDG achievement.

United Nations Environment Management Group (n.d.) examines how the 2030 Agenda for Sustainable Development along with SDGs should be carried out while considering the existing international environmental agreements (MEAs). The research explores the early national and international conversations about the SDGs while explaining that institutions require transformation to succeed in achieving these targets. The three major implementation zones mentioned in the study include sustainable consumption and production (SCP) along with biodiversity conservation and chemicals and waste management. The SDGs must be harmonized with internationally recognized environmental goals as a mechanism to enable mutual implementation processes. The implementation of goals depends on creating strong multi-triangle alliances between various stakeholders together with improving data standards to generate successful operational outcomes. A significant aspect of the 2016 HLPF involved 22 countries presenting initial voluntary national reviews that mainly discussed broad actions alongside institutional approaches but did not focus on specific SDGs. According to the conclusion there should be a complete method to unite SDGs with environmental objectives in order to establish a supportive system for sustainable development. This research creates an analytical structure that shows how SDGs link together with the necessity of cooperative work for achieving global targets.

Jiménez-Aceituno et al. (2020) focused on how local sustainability initiatives in Africa contributed in the implementation and interaction of SDGs using empirical data. Considering the "Seeds of Good Anthropocenes" database, the authors analysed 69 African initiatives through content and network analysis to determine the frequency. The data collecting procedure described in the methodology section consisted of obtaining information about 500 local sustainability projects from diverse sustainability professionals and experts who served as the sources. The research studied the 17 SDGs targets after their transformation from global aspirations to localized applications. Analysing the local African initiatives relied on two steps; first by investigating how often certain targets appeared and second by studying relationships between targets and goals. The research categorized SDGs into six clusters on their roles within local initiatives such as cross-cutting, goals considered due to the means of implementation and standalone goals like Goal 11. Specific targets within the SDGs receive more attention compared to others based on evaluation findings. The authors described the goal coverage through the percentage-based representation of initiatives and conducted statistical Chi-squared tests to quantify the goal adoption. The statistical method helped them normalize their residuals through the division of goal targets by observed frequency differences against the expected values. Through analysis of the initiatives, researchers gained knowledge about their alignment with SDGs while also identifying specific areas that require more focus. The authors argue that the disparity underscores the need for a more integrated approach to sustainable development that considers the unique contexts and challenges faced by local communities. They suggest that understanding the interactions among different goals can lead to more effective strategies for implementation and resource allocation. It emphasizes the need for further research to explore the complexities of these interactions and the role of diverse knowledge systems in shaping sustainable development strategies. The authors call for a collaborative approach that involves multiple stakeholders, including governments, civil society, and the private sector, to foster an environment conducive to achieving the SDGs at local levels. Overall, the study highlights the potential of local sustainability initiatives to contribute to the global agenda of sustainable development while also identifying the challenges and opportunities that arise in this context. By focusing on the local scale, the authors aim to bridge the gap between global ambitions and grassroots actions, ultimately promoting a more inclusive and effective approach to achieving the SDGs.

Urama et al. (2014) conducted the SDplanNet Africa Regional Workshop includes information and results from the meeting on sustainable development planning for Africa in Nairobi which took place from March 3 to 5 in 2014. There were 27 African development practitioners in the workshop from 11 countries, with members from the government and from regional institutions taking part to discuss the difficulties and pluses of the coming development policies. The most important point from the workshop was that governing practices must be modernized at all levels, nationally as well as locally. During the workshop, many stressed the importance of linking different organizations from leadership to the bottom in order to get more stakeholders involved. The study identified the five important points to be focused on including promoting inclusive economic growth and sustainable agriculture, policies ensuring equity and human dignity, management of natural resources and disaster risk, regional and international partnership for funding. These factors are the essential requirement for developing a strategic framework for aligning local initiatives with global SDGs. This approach helps numerous participants join forces such as the government, civil society and the private sector, so they can achieve the goals. It was highlighted in the workshop that organizations should make basic changes and uproot their usual development planning culture. Many workshop participants discussed the need to plan carefully using data and highlighting important information systems that support making choices. They discovered that the process of capacity development should be divided into three levels: the community, regional and individual level. Identified needs are important for making development planning and implementation better. All those taking part in the workshop realized that working on these three areas would support both global and local SDG results. Overall, it summarizes points for governance reforms, endorses planning based on evidence and helps strengthen operations at all levels. The framework of these pillars helps to combine local projects with the further objectives of the post-2015 development plan, so that different African communities are addressed by these programs. The outcomes from the workshop confirm that there is a planned effort to solve tough development issues together in Africa.

D'Adamo et al. (2021) conducted study on Italian National Recovery and Resilience Plan (PNRR), created under the European Union's Recovery and Resilience Facility, provides a strategic framework for the SDG development, with mainly focus on interventions in Southern Italy. It stresses the importance of investing in human capital for innovation and new business practices, and thus boost local communities. Infrastructure is considered a priority, to establish new green infrastructures and to upgrade old ones to support the economic development with the least amount of pollution. By fostering system performance monitoring at the regional level, the PNRR supports those who base their policymaking on evidence and adjust implementation as appropriate toward SDG targets. Harmonizing the legal and regulatory frameworks through the regions is also put forward, which would allow for somewhat uniform development gains and curb fragmented governance. This case shows how the national framework can be operationalized through regionally customized strategies, thereby contributing to local development and also the broader 2030 Agenda. It conveys the notion that localized planning, when aligned with targets at the national or EU level, forms the bedrock of delivery for the SDGs.

Bardal et al. (2021) focuses on the processes of applying the SDGs in local and regional planning in Norway, pointing out the difficulties and benefits that local governments face. It sorts these factors into seven main groups called cultural, political, legal, organizational, knowledge-related, financial and technology-related. Each one of these groups focuses on important elements needed for the successful local implementation of the SDGs. Cultural factors affect the implementation of sustainability initiatives, as efforts are more effective when aligned with local values. The Norwegian Planning and Building Act (PBA) is seen as important for SDG implementation since it encourages regular planning processes to include sustainability goals. Cross-sector cooperation is important according to organizational factors, since different groups collaborating enhances the unified effort to achieve the SDGs. The study also derives several strategies to contribute in SDG implementation, ensuring the importance of education and training. It suggests that municipal planners participate in relevant courses, seminars, and workshops to gain understanding and strengthen local planning. It also points out that through community involvement, people can own the process and see firsthand how sustainability steps provide positive changes. When local leaders involve the community, they can choose good sustainability goals that fit well with the community's beliefs and needs. Being involved in planning meetings improves how things are planned and also supports the validity of initiatives. Overall, successful application of the SDGs in Norwegian municipalities relies on using diverse techniques along with a well-planned system for changing the way organizations work. Utilizing rules such as the PBA, boosting the effectiveness of civil servants, assisting various departments in working together and actively including citizens in decision-making allow local governments to handle sustainability challenges and contribute worldwide to sustainable development goals. Therefore, it is important for local authorities to grasp the challenges and ways these frameworks connect, to progress towards fulfilling Agenda 2030.

Ullah et al. (2025) focuses on the importance of environmentally friendly development within the context of the Belt and Road Initiative (BRI), explaining how different regional strategies support this need. It looks into how China's investment agreements under the BRI work and points out that having firm rules supports joint efforts among the BRI participants. According to the research, even though the BRI creates many economic benefits, it causes environmental difficulties because a lot of projects value production growth more than environmental protection. It explains that for development to be sustainable, different methods and frameworks should be used together. Measures include forming local partnerships, growing green technology, and bettering government systems that oversee environmental rules. The study points out that regionalism and regionalization help countries work together and divide resources, which are essential for reaching shared sustainability goals. The research also points out that reviewing the variety of institutional structures among BRI nations is needed, because these differences can have a major impact on how environmental policies and program's function. They provide an approach which unites both ecological modernization theory (EMT) and institutional theory. Its purpose is to design a model that lays out how the adoption of sustainable methods, instead of fossil fuels, will help BRI nations combat climate change and boost their economy. Furthermore, integration in financial sectors is highlighted in the study as an opportunity to boost green investments, encouraging policymakers to carefully weigh the environmental and economic concerns. The research also stresses that local governments are significant in driving successful environmentally friendly plans because they play a vital role in carrying out sustainable actions. It urges the development of structures that allow governments, corporations and civil society to work together. Research and statistics within the study confirm that regional change promotes green development and multilateral efforts increase growth in the market and more investment options. The authors encourage regular monitoring and evaluation of environmental policies to check if they are helping reach sustainable development objectives. Following this approach allows BRI nations to handle new environmental conditions and issues in a better way. The study ends by stressing that, as the BRI has a major effect on world growth, sustainability should be integrated into every development agenda.

Foroudi et al. (2025) investigated how SDGs impact the regional level in Italy by focusing on understanding how local residents influence environment and society. As the SDGs are mainly aimed at nations, the authors stress that regional governance helps to achieve these goals, since each local region has different social, economic, and cultural traits that matter in making their approach to sustainability. The authors construct and empirically test a measurement scale estimating public support for SDG initiatives considering both ISTAT's quantitative input and qualitative data. Findings show that regional priorities diverge considerably: Northerners show greater concern for the environment while Southerners emphasize infrastructure and employment. These variations intensify the necessity for region-specific policy measures tailored to local needs. The authors argue that supporting social enterprises and the blue economy could improve regional SDG scores by increasing environmental and socio-economic benefits. The appeal was to develop a regional policy that taps into public sentiment to encourage sustainable development within the region. Such a policy could enable the formulation of place-based policies that decision-makers could strategically align with global frameworks, increasing regional support for the SDGs. Future research according to the authors will focus on how where people live can influence their overall well-being, as well as examining a larger set of social factors related to well-being. It also suggests forming a framework to help fill the gap in research by understanding residents' opinions on the SDGs from a regional outlook. As a result of using this framework, stakeholders could develop policies that go along with local interests which encourages the community to get involved and support global sustainability efforts.

The Global Alliance to Monitor Learning (GAML) (n.d.) states that all countries are committed to the 17 SDGs, especially SDG 4, which aims to create opportunities for all to receive quality education that lasts a lifetime. The SDGs are a world-wide plan adopted by nations to focus on sustainability challenges in different fields including economy, society and the environment. This goal is especially valuable since it focuses on ten targets making education more accessible to girls and boys, putting strong emphasis on early education, job training and equal chances in higher education. Helping people with disabilities and indigenous groups access education is an essential part of the targets, showing why equal education access is needed. It explains the importance of regional organizations adjusting their education strategies to fit SDG 4. The African Union, Caribbean Community (CARICOM), European Union and Southeast Asian Ministers of Education Organization (SEAMEO) are some of the regional organizations helping to develop and enforce education plans that support SDG 4. They make it possible for member states to cooperate, so countries can share their best strategies and resources to help all students. A review of existing regional education frameworks demonstrates that they are working towards the SDG 4 goals. The study also compares different regional plans with the goals of SDG 4 and most of the plans appear to give priority to similar areas such as taking part in education, progress in learning and getting new skills. But the study points out that there are ongoing difficulties in meeting these objectives. Different communities may have less access to resources, teacher development, and proper infrastructure which often results in inequality in education. It also stresses that effective systems for collecting and checking data are essential to see if progress is being made toward the targets. It is essential to have disaggregated data to find out how many people in society do not receive equal education, so these problems can be corrected. Increasing monitoring would help recognize the main problems that prevent different groups from having access to quality education and also measure the success of policies trying to support them. The study also explains that SDG 4 has important effects because providing quality education helps build a more sustainable future. Promoting places that support personal development for individuals helps a nation grow economically, include everyone socially and engage in public life. Having education is key for anyone to become self-reliant, as well as for land improvement and other community needs. Overall, the study gives a complete strategy to help achieve educational goals under SDG 4. The main point is to focus on member countries and regional organizations co-operating to handle educational disparities and make quality education accessible for everyone. The main aim is to help achieve fairness and support sustainability by using education to support individual growth and development for society.

Multi-Stakeholder Platform (2018) highlights how the SDGs can best be met with the involvement of local and regional authorities. It started as a task of the subgroup focused on "Delivering SDGs at the local and regional level" within the multi-stakeholder platform set up by the European Commission. It is stressed in the recommendations that several government bodies and key stakeholders should cooperate to develop effective strategies for dealing with major social problems. It is pointed out by the authors that the SDGs apply to all countries yet their significance can vary from one to another, so local approaches are necessary. Because local authorities work in the same community, they can address people's needs more easily. It also demonstrates that several municipalities are already adopting the SDGs in local actions which suggests a gain that could be amplified. Proposals suggest the European Commission include the SDGs in its economic and social governance which could include an annual sustainable development survey in addition to the annual growth survey. By aligning, it would be easier to study how policies influence meeting the SDGs in various regions and communities. The study highlights that disaggregated data is required to really show the contributions of local and regional bodies to the SDGs. Also, the guidelines highlight the importance of EU support for local government associations across Europe. It is meant to support local government in aligning SDG actions with national measures. It also calls for the creation of a lasting multi-stakeholder body dealing with the SDGs at the European level. Platforms like this can improve local government capabilities by helping them swap and learn from good practices and knowledge in the SDG sphere. Another important factor discussed is how innovative approaches are used to achieve the SDGs. It claims that to achieve the SDGs by the 2030 date, social service and policy ideas should be adopted in local communities. In conclusion the study covers the various approaches and suggestions to give local and regional governments a bigger say in achieving the SDGs.

Putra et al. (2024) analyses how the SDGs are included in regional development strategies across Indonesian provinces. The analysis looks at how the SDG principles work with regional development and the corresponding successes and problems in this effort. They point out that using sustainable development can ensure that human needs and those of the environment are balanced for future generations. Because sustainable development includes social, economic and environmental dimensions, it connects well with the SDGs launched in 2015. Starting from the SDGs announcement, Indonesia has been working on including them in its provincial development plans and the purpose of this study is to examine the results of those efforts. The study describes multiple variables that affect how SDGs are approached within regional development. The study discusses that even though SDGs are recognized by policymakers, their actual execution in the field varies. A lack of similar awareness, understanding and importance given to the SDGs in various provinces is stressed. For example, some areas are putting more effort into boosting their economy and building up infrastructure than on environmental protection which can cause problems between quick benefits now and the need for a long-term, safe environment. By looking through study about regional development planning, the study finds that regions use the SDGs either as a tool to analyse policies or as global concerns. The initial approach sees SDGs being used to connect local rules with international sustainable development guidelines. The use of this method lets regions see if they align with the SDGs and creates an organized way of planning. According to the second approach, the SDGs are important goals that shape all kinds of decisions. Undoubtedly, environmental issues are often given not much importance in regional development plans, since many provinces allocate less than a quarter of their budgets to environmental goals. The lack of attention to key issues worries experts about how well these plans meet important environmental challenges in the country. It further stresses that provincial governments play a key role in implementing the SDGs, since they usually link the work of the national government with that of communities. In short, achieving the SDGs in Indonesia requires a more coordinated, inclusive and participatory approach. It recommends that further studies should assess long-term impact by looking at each country's SDG results and make comparisons to learn about better localization approaches. If different groups become more involved and if environmental matters are given equal attention, Indonesia will be able to advance sustainably and in line with what is promised at the global level.

Shayan et al. (2022) looks into the relationship between Corporate Social Responsibility (CSR) and the Sustainable Development Goals (SDGs) set by the United Nations. The study points out the link between CSR and corporate sustainability, corporate citizenship and stresses how keeping up with research and good corporate governance helps the company meet both the expectations of shareholders and society. Significantly, the study highlights that companies that engage in CSR become more visible and are recognized which may result in more loyal customers, a better brand reputation and stronger financial results. The SDGs also provide a solid structure for companies to direct their CSR strategies and face major global concerns such as poverty, inequality, and harm to the environment. The study lists the ways CSR helps, saying that working with stakeholders for mutual benefit is valued by people and improves the work environment. Adding the SDGs to their CSR policies, businesses can lessen research and implementation-related expenses, boost their impact, and be more involved in achieving long-term sustainability. It further looks into the importance of working alongside others across the globe to help make these projects successful. Overall, the study encourages companies to work with SDGs as a guiding tool for CSR, explaining that this choice helps the business community follow ethical values and contribute positively to the economy and the world, thus gaining benefits for employees, organizations, and society as a whole.

Nogueiro et al. (2022) discussed how Erasmus mobility actions work toward achieving the SDGs. The author addresses how mobility programs held students by focusing on SDG 4 (Quality Education), SDG 5 (Gender Equality) and SDG 8 (Decent Work and Economic Growth) and mentions the ways they encourage the development of skills and competencies, knowledge increase and better employment possibilities. Erasmus is important for advancing European principles and for meeting the main targets mentioned in the Europe 2020 Strategy. The findings point out that while away from home, students have chances to develop their professions, and become included in the local community. The anticipated results of these projects are people who are aware of culture and are able to cope with local and global problems. The study also points out the need to ensure all students, mainly women, people with disabilities, and vulnerable groups, are given equal educational opportunities. When it supports exchanges among students, the Erasmus Programme works to level the playing field and increase equality. Research should also explore the ways the Erasmus Programme is helping achieve additional SDGs, for example SDG 9 (Industry, Innovation and Infrastructure) and SDG 10 (Reduced Inequalities). Overall, the study establishes that the Erasmus+ Programme links educational initiatives with sustainable development which encourages inclusive and sustainable societies.

Chen & Yi (2021) looks at how corporate social responsibility (CSR) works together with government subsidies, mainly focusing on industries in China. The authors point out that companies commonly struggle between giving to charity or investing in energy-saving innovations through research and development (R&D). They consider how these business decisions affect profits, the firm's environmental performance, and CSR success as a whole, also they discuss what influence the government can have on companies' decisions. The work outlines a framework through a game model to analyse CSR strategies available to businesses. It shows that corporations tend to engage in donations as it benefits them, by helping their image and minimizing their expenses. Even though it makes good sense to invest in energy-saving R&D, high expenses related to long-term projects stop many firms from seeing its benefits. Among government subsidies, the authors mention both the donation subsidy and energy-saving R&D subsidy. They explain that the decisions firms take are affected by the strengths of the subsidies, the costs involved, and if these aspects are in line with seeking maximum profit. A number of main propositions in the study look at how CSR is affected by rules on competition and subsidies. When both kinds of subsidies are given at the same time, the firm takes into account not just their levels, but also the costs of energy-saving R&D and donations in their decisions. According to the authors, a certain amount of R&D cost is needed; if it's too low, organizations often go for energy-saving ideas. If the expenses are greater, individuals tend to go for donating instead of shopping. Besides, the study assesses how the level of government subsidies influences CSR actions by firms. They believe that firms can get the better results from CSR by avoiding competing programs aimed at the same companies. Because of this, companies might start using a combined strategy, ensuring that their quick charitable work doesn't compromise their long-term sustainability, achieved with new energy-saving devices. It also points out that China's success in using resources efficiently is not as positive as other countries, most of all Japan. As a result of this problem, it is important to look into existing policies, stress consumer awareness and expect companies to follow sustainable methods. Increasing public environmental awareness and perfecting subsidy systems to motivate more R&D investments are the authors' proposed solutions for CSR. Methodologically, the study uses experiments to test the theoretical results and show the consequences choices in subsidies can have on the firm's actions. High subsidies can universally support every firm, but the link between donation and saving energy through R&D needs careful planning to achieve the best results for CSR. In the end, the authors suggest that firms should make CSR part of their long-term business plans and connect their financial aims with caring for society and the environment.

4th SDG Youth Summer Camp (n.d.) outline in detail the 17 SDGs that were set up by the United Nations as part of the 2030 Agenda for Sustainable Development. It acts as an important plan to handle worldwide issues and secure a better future for everyone. The study shows what actions should be taken for each goal, pointing out that sustainable activities are needed to encourage economic growth, help more people be included and protect the environment in all government activities. The first goal, seeks to end poverty in every form for everyone by 2030 and the study suggests targets such as getting rid of extreme poverty for those living on less than $1.25 per day and helping people living in poverty and those at risk to stay secure. Third goal focuses on health and the targets aim to ensure universal health insurance, provide essential medications to all and encourage good mental well-being. Fifth goal prioritizes gender equality by aiming to have women involved equally in leadership, ensuring all people can access reproductive health services and by supporting women's access to resources. The eight goal works on economic growth, creating good jobs and protecting all workers, including young and disabled people, are being handled as top priorities. The tenth goal focuses on lowering inequality internationally and within countries, encouraging measures that help all marginalized groups play a more active part in society. The goal thirteen underlines the fact that particular attention should be given to climate change and the environment, encouraging governments to manage climate risks and add climate solutions to their policies. Goals regarding infrastructure, urban areas and sustainable consumption point to the importance of building strong infrastructure and preparing for disasters, especially in city planning and fighting land degradation. Strengthening partnerships is required by goal 17 to encourage collaboration among local authorities, civil societies and businesses in raising money and distributing knowledge. The study notes the importance of being accountable and using separate data to oversee progress, supporting inclusive initiatives that support vulnerable people. It additionally proposes how to handle resources in a sustainable way, including using them wisely and not wasting them which is essential for a healthy future. The main idea behind the study is to place national ownership of the SDGs at its core and to at the same time appreciate inter-continental teamwork and collaboration toward meeting all the 2030 targets. To sum up, the study is very helpful for learning about the interconnected SDGs, encouraging sustainable development involving society, the environment and the economy and guaranteeing that everyone is included in promoting a successful and even world.

3. CONCLUSION

In short, the SDGs should be applied regionally to achieve serious advances toward a fair and sustainable world. It is important for the SDGs to be tailored to each community to ensure they are effective and this study points out why. If governments, civil groups and private companies cooperate, regions can devise new methods and use their own resources to deal with their challenges. The study highlights that despite improvement, there are still weak spots, for instance because of inadequate resources, a lack of agreement among groups and low staff capacity. Joining forces with international partners and inviting different groups support the fight against these issues and open doors for folks to learn from their experience. The study requires different groups to work together to develop strong policies that include social, economic and environmental aspects of sustainable development. Researchers should perform over time studies to understand the long-term effects of local strategies and examine the importance of technology and innovative ways in supporting the process. If stakeholders motivate people and plan sustainability in their regions, they can reach the ambitious dreams for the year 2030. The goals of the SDGs can only be achieved when people, governments and businesses work together on inclusive solutions that allow everyone to deal with emerging global difficulties.

REFERENCES

4th SDG Youth Summer Camp. (n.d.). *SDG targets overview resource*. United Nations Sustainable Development Knowledge Platform. https://sustainabledevelopment.un.org/sustainabledevelopment.un.org+10

African Continental Free Trade Area (ACFTA). (https://au.int/en/african-continental-free-trade-area)

Aftab, W., Siddiqui, F. J., Tasic, H., Perveen, S., Siddiqi, S., & Bhutta, Z. A. (2020). Implementation of health and health-related Sustainable Development Goals: Progress, challenges and opportunities – A systematic literature review. *BMJ Global Health*, *5*(8), e002273. DOI: 10.1136/bmjgh-2019-002273 PMID: 32847825

Bardal, K. G., Reinar, M. B., Lundberg, A. K., & Bjørkan, M. (2021). Factors facilitating the implementation of the Sustainable Development Goals in regional and local planning—Experiences from Norway. *Sustainability (Basel)*, *13*(8), 4282. DOI: 10.3390/su13084282

Chen, S., & Yi, Y. (2021). The manufacturer decision analysis for corporate social responsibility under government subsidy. *Mathematical Problems in Engineering*, Article ID 6617625, 1–15. Advance online publication. DOI: 10.1155/2021/6617625

D'Adamo, I., Gastaldi, M., Imbriani, C., & Morone, P. (2021). Assessing regional performance for the Sustainable Development Goals in Italy. *Scientific Reports*, *11*(1), 22069. DOI: 10.1038/s41598-021-03635-8 PMID: 34916565

Foroudi, P., Marvi, R., Cuomo, M. T., & D'Amato, A. (2025). Sustainable Development Goals in a regional context: Conceptualising, measuring and managing residents' perceptions. *Regional Studies*, *59*(1), 93–109. DOI: 10.1080/00343404.2024.2373871

Global Alliance to Monitor Learning (GAML). (n.d.). *Regional framework for Sustainable Development Goals (SDGs)*. UNESCO Institute for Statistics. https://gaml.uis.unesco.org/

Jiménez-Aceituno, A., Peterson, G. D., Norström, A. V., Wong, G. Y., & Downing, A. S. (2020). Local lens for SDG implementation: Lessons from bottom-up approaches in Africa. *Sustainability Science*, *15*(3), 729–743. DOI: 10.1007/s11625-019-00746-0

Nilsson, M., Chisholm, E., Griggs, D., Howden-Chapman, P., McCollum, D., Messerli, P., Neumann, B., Stevance, A.-S., Visbeck, M., & Stafford-Smith, M. (2018). Mapping interactions between the Sustainable Development Goals: Lessons learned and ways forward. *Sustainability Science*, *13*(6), 1489–1503. DOI: 10.1007/s11625-018-0604-z PMID: 30546483

Nogueiro, T., Saraiva, M., Jorge, F., & Chaleta, E. (2022). The Erasmus+ Programme and Sustainable Development Goals—Contribution of mobility actions in higher education. *Sustainability (Basel), 14*(3), 1628. DOI: 10.3390/su14031628

Organisation for Economic Co-operation and Development (OECD). (2020). *A territorial approach to the Sustainable Development Goals: Synthesis report.* OECD Publishing. https://www.oecd.org/en/publications/a-territorial-approach-to-the-sustainable-development-goals_e86fa715-en.html

Putra, A. A., Hasibuan, H. S., Tambunan, R. P., & Lautetu, L. M. (2024). Integration of the Sustainable Development Goals into a Regional Development Plan in Indonesia. *Sustainability (Basel), 16*(23), 10235. DOI: 10.3390/su162310235

Shayan, N. F., Mohabbati-Kalejahi, N., Alavi, S., & Zahed, M. A. (2022). Sustainable Development Goals (SDGs) as a framework for Corporate Social Responsibility (CSR). *Sustainability (Basel), 14*(3), 1222. DOI: 10.3390/su14031222

Ullah, A., Nobanee, H., Ullah, S., & Khan, S. (2025). Addressing climate change and advancing green development: Regional transition mechanisms for a sustainable future in Belt and Road Initiative nations. *Sustainable Development (Bradford), 2025*(4), 1–29. DOI: 10.1002/sd.3446

United Nations Environment Management Group. (n.d.). *Implementing the environment dimension of Agenda 2030 for sustainable development: Synergies between the implementation of the SDGs and international environmental objectives – Briefing paper*. UNITAR. https://www.unep.org/explore-topics/sustainable-development-goals

Urama, K., Ozor, N., & Acheampong, E. (2014). *Achieving Sustainable Development Goals (SDGs) through transformative governance practices and vertical alignment at the national and subnational levels in Africa*. SDplanNet–Africa, African Technology Policy Studies Network.

ADDITIONAL READING

Fajri, C., Sari, N., & Amali, M. T. (2024). Corporate Social Responsibility and the Sustainable Development Goals: A case study of Nusantara, the future capital city of Indonesia. *Journal of Ecohumanism, 3*(8). Advance online publication. DOI: 10.62754/joe.v3i8.5635

Heras-Saizarbitoria, I., Urbieta, L., & Boiral, O. (2022). Organizations' engagement with Sustainable Development Goals: From cherry-picking to SDG-washing? *Corporate Social Responsibility and Environmental Management*, *29*(3), 736–746. DOI: 10.1002/csr.2202

Rios-Lama, C. A., & Vargas-Merino, J. A. (2022). Does Corporate Social Responsibility (CSR) interact with Sustainable Development Goals (SDG)? A literature review of the contribution and implications of CSR to the SDGs. *Communications of International Proceedings*. https://doi.org/DOI: 10.5171/2022.3957622

Serova, K. (2023). Exploring the synergies: How Sustainable Development Goals are shaping Corporate Social Responsibility. *University of Minnesota Digital Conservancy*. https://hdl.handle.net/11299/254212

Whittingham, K. L., Earle, A. G., Leyva-de la Hiz, D. I., & Argiolas, A. (2023). The impact of the United Nations Sustainable Development Goals on corporate sustainability reporting. *Business Research Quarterly*, *10*(2), 45–60. DOI: 10.1177/23409444221085585

KEY TERMS AND DEFINITIONS

Affordability: The capacity of an individual, a community or a nation to pay for goods, services or initiatives without undue financial burden.

Frameworks: Organized documentation or models with logical arrangements to inform the development, monitoring, and assessment of policies or strategies.

SDGs (Sustainable Development Goals): A unified set of 17 goals agreed upon by the United Nations in 2015 under the 2030 Agenda for Sustained Development.

Sustainable Development: Development that meets the needs of the present without compromising the ability of future generations to meet their own needs

Strategies: Deliberate courses of action or plans of activities developed in order to accomplish some end or purpose.

Chapter 7
Increasing the Supply of Competent Teachers in Developing Countries:
The Role of Sustainable Development Goals

Nazire Burcin Kaya
https://orcid.org/0009-0004-7964-0234
Eskisehir Osmangazi University, Turkey

Mesut Aydemir
https://orcid.org/0000-0002-2035-1292
Anadolu University, Turkey

Mubin Kiyici
Sakarya University, Turkey

ABSTRACT

The chapter investigates the global shortage of qualified teachers, emphasizing the critical role of Sustainable Development Goal (SDG) 4 in ensuring equitable access to quality education. It highlights disparities in teacher qualifications across income levels and regions, with low-income countries facing the greatest challenges due to inadequate training infrastructure, economic constraints, and sociopolitical barriers. The chapter proposes strategies such as enhancing teacher training programs, integrating technology, offering financial incentives, and fostering international cooperation to address these challenges. By aligning policies with SDG 4's targets, the chapter advocates for a sustainable and inclusive approach to improving teacher supply and education quality worldwide.

DOI: 10.4018/979-8-3373-1300-9.ch007

INTRODUCTION

The 2030 Agenda for Sustainable Development and its 17 Sustainable Development Goals (SDGs) offer a blueprint to the world for tackling intricate issues such as poverty, inequality, climate change, and environmental degradation. One of the focuses under this is SDG 4: Quality Education, which is devoted to ensuring inclusive and equitable quality education and promoting lifelong learning opportunities for all. Within this goal, Target 4.c aims to "substantially increase the supply of competent teachers, including through international cooperation for teacher training in developing countries, especially least developed countries and small island developing states." The performance indicator for this target, 4.c.1, is "the proportion of teachers with the minimum required qualifications by education level and sex."

This chapter delves into the contributions of the SDGs with particular reference to SDG Goal 4 and the supply of competent teachers in developing countries. With the aid of the quarterly special SE_TRA_GRDL 2024.Q3.G.01 release, this study examines the share of competent teachers across the globe against a backdrop of income categories by continent. On the basis of that analysis, it proposes measures and policies on how these developing countries can use the relevant lessons learnt from the developed countries to deepen the qualifications of their teachers. To ensure a multidimensional understanding of this issue, this chapter integrates data-driven insights, region-specific challenges, comparative case studies, and technology-driven solutions, while also addressing gender and social inclusion barriers in teacher training and retention.

This chapter begins by outlining the scale and urgency of teacher shortages in developing countries, drawing on global data to frame the issue. It then categorizes the underlying causes of these shortages into three main barriers: financial, institutional, and social. This is followed by an exploration of evidence-based, policy-oriented, and technology-enabled strategies aimed at addressing the problem. The chapter concludes with a synthesis of key insights and actionable recommendations for stakeholders. By following this structured path, the chapter aims to provide clarity, coherence, and practical value to educators, policymakers, and researchers alike.

BACKGROUND AND CONTEXT

Increasing the Supply of Qualified Teachers in Developing Countries

Boosting the number of competent educators in developing nations is not an easy feat as it is a complication that calls for the reevaluation of the instructional process, continuous growth of the teaching profession, and overall improvement of the systems in place. The identified deficits attract measures and systems that can be implemented to increase the criteria of teachers employed in these areas and sustain their employment which is essential for bettering the educational systems in those regions.

Understanding the Teacher Gap in Developing Countries

The teacher's gap in most developing countries is more intricate than what it seems and continues to affect the quality of education and the students' performance. This gap is characterized by a shortage of qualified teachers, inadequate teacher training, and high attrition rates, which collectively undermine the educational systems in these regions. It is also necessary to understand what particular factors contribute to this gap, which are socioeconomic constraints, limited access to capacity building avenues and the politicization of the space within which education exists.

In addition to general factors, teacher shortages manifest differently across regions. For example, in Sub-Saharan Africa, UNESCO (2023) reports that more than 70% of rural schools operate without a full staff of certified teachers, exacerbating educational inequality. Conversely, countries like Sri Lanka and Vietnam have mitigated shortages through targeted teacher education reforms, indicating the importance of contextualized responses. These regional disparities highlight the need for nuanced, data-driven policy interventions tailored to specific socioeconomic and cultural conditions.

The Shortage of Qualified Teachers

One of the primary challenges contributing to the teacher gap is the shortage of qualified educators. Such is the case for a great number of developing countries that do lag in the number of trained teachers available that in one way or another has affected the general education that the learners receive. Gaikhorst et al. point out that the lack of appropriate teachers is a persistent problem which occurs in other countries contributing to reduction of educational standards and performance of students which are rather too low (Gaikhorst et al., 2014; Gaikhorst et al., 2019). This shortage often forces schools to hire unqualified or under qualified individuals, further exacerbating the problem of educational quality (Canrinus & Fokkens-Bruinsma, 2013). The implications of these shortages are profound, as they directly correlate with lower student achievement and increased dropout rates (Xu et al., 2019). For instance, in Kenya's Kakuma Refugee Camp, the Teachers for Teachers program combined mobile technology with mentoring to improve teacher quality under challenging conditions (EDT, 2018). This example illustrates how localized support and digital tools can sustain educational quality despite resource constraints.

Insufficient Teacher Training Institutions and Programs

The issue of insufficient teacher training institutions and programs is a critical concern in contemporary education systems worldwide. The consequences of such deficiencies are severe in nature, ranging from a lack of extensive training models, educators failing to acquire significant professional skills, and an improper use of teaching tools during classes. Such aspects not only influence student performances negatively but the quality of teaching is also affected.

One of the gaps in teacher training is the divergence of pedagogical implications and contemporary educational needs. An example of this problem can be the training teachers are provided compulsorily in curriculums, such as the inquiry based learning that a majority of teachers are not comfortable in use (Martell, 2019). This also leads to a gap in fostering non-active methods and engaging students in a more practical approach. At the same time, the systematic review by Basilotta-Gómez-Pablos et al. indicates the gap that exists in teachers' competencies and demands for improvements in relation to the teacher's pedagogical use of important technology tools in the learning process (Basilotta-Gómez-Pablos et al., 2022). Moreover, awareness on such technological competencies is not enough as one needs to undertake a more practical approach and be a part of training programs so that teachers are not left behind.

Another critical factor in addressing the teacher shortage is the implementation of competency frameworks that define the skills and knowledge required for effective teaching. These frameworks can inform teacher education programmes and professional development needs to the extent that they will be responsive to the needs of the education system (Halász & Looney, 2019; Snoek et al., 2019). Countries that have introduced explicit definitions regarding the teacher competences have also moved forward in the teacher quality and student performance indicators (Halász & Looney, 2019). In addition, the latter could be used to set performance norms and pinpoint areas where support is needed (Halász, 2019).

Competency-Based Teacher Preparation Programs (CBTPPs) are increasingly recognized as a crucial mechanism for fostering culturally responsive pedagogy (CRP) and promoting educational equity. The effectiveness of these programs is largely contingent upon their ability to equip teacher candidates with the requisite skills and competencies necessary to engage with diverse student populations. This involves not only understanding cultural diversity but also actively incorporating this understanding into pedagogical practices, thereby fostering inclusive learning environments for all students (Bukko & Liu, 2021; (Deng & Gao, 2024; Ankomah, 2020; Foster et al., 2023; Haynes et al., 2023).

The implementation of culturally responsive teaching (CRT) strategies in CBTPPs has been shown to enhance educators' cultural competence and promote positive teacher-student relationships. Engaging in critical reflection on one's biases and being cognizant of students' varying backgrounds is essential for promoting equity in classrooms (Ankomah, 2020; Haynes et al., 2023). However, educators often encounter implicit biases and structural barriers that can undermine the effectiveness of CRP. It is essential that teacher preparation programs address these challenges explicitly, offering training on how to recognize and mitigate these biases while implementing responsive pedagogical techniques (Foster et al., 2023; Haynes et al., 2023).

Teacher preparation programs must also embrace ongoing professional development opportunities that prioritize culturally relevant teaching practices. This continuous learning model serves to reinforce educators' commitments to equity and CRP, thereby improving their effectiveness in supporting diverse learners (Bukko & Liu, 2021; Ankomah, 2020; (Haynes et al., 2023; Viberg et al., 2024). Studies highlight the significance of including social justice-focused coursework and service learning in teacher preparation curricula as a means of enhancing educators' ability to advocate for equitable practices (Novakovic et al., 2020; (Mouboua et al., 2024; . In environments that foster collaboration among educators, both pre-service and in-service, collective growth towards equity can be realized.

The obstacles towards providing efficient teacher training are not only institutional in nature but rather systemic too, and this is the point that the National Council on Teacher Quality has emphasized on as well. Based on their findings, Teachers' courses offered at universities tend to neglect the necessary practical components that training teachers require prior to entering the professional world (Fuller, 2013). Such a deficiency does not exist in isolation but is further compounded with the disconnect between the contents of the teacher training programs and the necessary competencies needed in the field. As a result, teachers may enter the profession with a theoretical understanding of pedagogy but lack the practical skills necessary to implement effective teaching strategies. In South Africa, the Mpumalanga Secondary Science Initiative (MSSI) adapted the Japanese 'lesson study' model, showing that collaborative, teacher-led professional development can improve classroom practices even in under-resourced contexts (Ono & Ferreira, 2010).

Lack of Incentives for Teaching as a Career Choice

The failure to motivate in choosing teaching as a profession is a complex problem which has got economic, social and psychological aspects. Research indicates that compensation issues including salary is one of the major factors that affect people in taking teaching professions. For instance, Han et al. highlight that while altruistic motives and social utility are important, the economic returns associated with teaching positions, particularly in comparison to other professions, significantly influence students' career aspirations in education (Han et al., 2017). This is particularly evident when comparing teacher salaries to those of STEM professionals, where the disparity often discourages potential candidates from entering the teaching profession (Han & Borgonovi, 2020).

Furthermore, the notion that teaching is not as high status as other professions only worsens the situation. Cammelli et al. point out that, unless teachers are later disillusioned by these expectations, there will be a negative long-term effect on teacher job satisfaction due to low salaries and constrained chances for advancing in a teaching career (Cammelli et al., 2011). Such a feeling is also consistent with what Bergmark et al. have noticed, that the reasons why teachers enter into the profession has a different orientation depending on the level of education that a teacher wishes to offer, thus different categories of teachers focus on different things (Bergmark et al., 2018). As a result, the absence of market standard wages as well as low status of the profession of a teacher discourages most bright students from pursuing such careers.

In addition to financial incentives, the working conditions associated with teaching also play a crucial role in career choice. Research by Jugović et al. indicates that prospective teachers often raise concerns about workload, classroom management, and support from administration as significant deterrents to entering the profession (Jugović et al., 2012). These factors complement a wider discourse on the issues that are faced by teachers and which are more effective than the numerous non-pecuniary ratios of teaching, such as the possibility of making a valuable contribution to students' lives. Also, Varadharajan et al. note the challenges educators experience in working with career changers who are trying to find their way in a different field, and conclude that such individuals are likely to experience additional hurdles to employability and retention even when they possess needed experience (Varadharajan et al., 2019).

Factors Contributing to the Gap

All around the world, many education systems are facing the issue of teacher shortages, also known as teacher gap. However, this is a result of several factors such as economic, social and culture. Many countries face a shortage of teachers due to low recruitment and even retention because of insufficient funding while political instabilities hinder the efforts made to combat such shortages. The challenges leading to teacher shortages are multifaceted and best understood when categorized under three key barriers:

- **Financial Barriers:** Underfunded education systems, low teacher salaries, and lack of incentives.
- **Institutional Barriers:** Inadequate training infrastructure, lack of policy coherence, and ineffective professional development models.

- **Social and Cultural Barriers:** Gender inequality, sociopolitical instability, and brain drain dynamics.

There are also sociocultural barriers which especially have an impact on women who face discrimination while trying to join the teaching profession. Additionally, the migration of skilled teachers to developed countries drains talent from regions that need it most, deepening the divide. Together, these factors create a multifaceted problem that demands comprehensive and context-sensitive solutions.

Economic Constraints and Underfunded Education Systems

In addition, economic conditions in developing nations also contribute to the teacher gap. According to Xu et al, there is a lack of money which makes it impossible for the educational system to recruit or train sufficient teachers (Xu et al., 2019). These socio-economic factors are made worse by large class sizes and a lack of teaching aids which in turn stretch the workload for the educators too thin (Xu et al., 2019). A setting characterized by minimal investment in education engenders a vicious cycle, resulting in both instructors and pupils lacking motivation to excel. Beyond identifying budgetary limitations, countries can explore alternative financing mechanisms to expand teacher recruitment and training. Microloans for pre-service teachers, education bonds backed by government guarantees, and international donor-driven scholarship programs offer promising avenues. Moreover, digital training initiatives, such as MOOCs and blended learning models, can reduce costs substantially while maintaining training quality. These strategies provide practical pathways for resource-limited governments to sustainably enhance teacher capacity.

Socio-political Barriers to Addressing the Teacher Gap

The sociopolitical factors further compound the teacher gap that exists in developing regions. There are socio-political and economic factors such as corruption, poor governance, and political instability that can hinder any attempts for the enhancement of the quality of education. For instance, in Armenia, the ongoing educational reforms have not led to substantial improvements, highlighting the complexities of implementing effective changes in a challenging political landscape (Karakhanyan & Veen, 2011). Additionally, case studies of professional novices based in the US immigrant settings illustrate the sociopolitical variables external to the classroom that can further engender job dissatisfaction and attrition among the teachers (Kirksey, 2021). The connection of education with other societal features makes it apparent that there is no one-size-fits-all solution to the two sets of problems presented.

Social and Cultural Barriers, Particularly for Women

The social and cultural barriers faced by women teachers in developing countries are complex and multifaceted, deeply rooted in historical, economic, and sociopolitical contexts. These barriers not only affect the professional lives of women educators but also have broader implications for educational equity and societal development. The literature reveals that women teachers show low levels of satisfaction and experience multiple challenges such as economic deprivation, cultural expectations and internal discrimination in their career development and quality of life.

The issue of gender representation in teaching also intersects with the lack of incentives. The research of Kılınç et al. shows that perceptions of teaching are influenced by gender, and society has stereotypes concerning who should be encouraged to take teaching careers (Kılınç et al., 2012). It is emphasized that such problems require urgent measures to counteract, in particular, the promotion of the teaching profession as an occupation suitable for people of all sexes, especially taking into account the persistent teacher shortages in some places.

Cultural norms and societal expectations also play a significant role in shaping the experiences of women teachers in developing countries. In relation to this, Mabokela & Mlambo (2014) discuss how older African women in academia have to grapple with the patriarchal system in Ghana and how it stifles their empowerment and recognition. Such cultural forces are present in many developing countries where women suffer such discriminations because of existing gender hierarchy in society. The economic barriers faced by women educators are compounded by societal attitudes that prioritize male education and employment, perpetuating a cycle of inequality.

Moreover, social rejection and discrimination are also the cultural barriers towards women's education and recruitment in certain professional fields. Women who have taken formal employment in education, particularly teaching, become targets of backlash by certain conservative factions in the society who advocate for compliance with legal provisions. This social culture serves as a disincentive for women who would wish to venture into or remain in the very important teaching career which widens the gender divide in education. In Bangladesh, gender-sensitive recruitment campaigns and mentoring programs have increased the proportion of female teachers in rural schools, directly improving girls' enrollment rates (UNESCO, 2023). Such policies demonstrate that intentional gender inclusion strategies can effectively address systemic disparities in teacher supply.

Brain Drain: Migration of Qualified Teachers to Developed Countries

The phenomenon of brain drain, particularly concerning the migration of qualified teachers from developing to developed countries, has garnered significant attention in recent years. This migration not only affects the educational landscape of the countries of origin but also has profound implications for the destination countries. The brain drain of skilled education constitutes one of the biggest losses for these nations that are still in the developing stage neuro-educationally and are already facing deficits in education and a great shortage of intelligent and competent teachers. In the case of China and India, the migration of professionals, including teachers, can be traced back to historical patterns of globalization and economic opportunity that began in earnest during the latter half of the 20th century (Furukawa et al., 2012; Gong & Huybers, 2015).

The term "brain drain" encapsulates the outflow of a nation's educated elite, which can jeopardize long-term developmental needs. This is particularly evident in countries with underdeveloped educational infrastructures, where the emigration of qualified teachers exacerbates existing challenges in providing quality education (Abramo et al., 2018; Crosling et al., 2014). For instance, the lack of basic facilities for tertiary education in impoverished nations leads to a significant brain drain, as capable educators seek better opportunities abroad (Crosling et al., 2014; Khan, 2021). The effects of this migration are two-sided: while the receiving countries enjoy the sudden inflow of skilled human resources, the country of origin suffers from the loss of human capital which many require for their growth and development (Gong & Huybers, 2015; Khan, 2021).

The combination of both push and pull factors tends to be the reason why a qualified teacher decides to migrate. Salaries, lack of ability to grow professionally, absence of good working conditions in their home countries are some of the push factors (Wang, 2016; Widegren & Doherty, 2010). On the other hand, Kahn (2021) and Cattaneo et al (2018) explains that unattainable salary, working conditions and even professional opportunities available in developed countries help in 'pulling' teachers. Over the time, when combined with the economic factors, this has been observed mostly in Central and Eastern Europe with a mass movement of educators to Western Europe (Cattaneo et al, 2018). The financial crisis in 2008 contributed to this too, for a lot of educators were searching for stability and better prospects outside (Boyadjieva, 2017).

While the barriers to teacher supply are complex, they are not insurmountable. The next section introduces a structured set of solutions aimed at addressing these root causes through targeted policy design, innovation, and collaborative efforts. These solutions are presented across three interconnected domains to ensure a holistic approach.

Strategies for Increasing the Number of Qualified Teachers

The global shortage of qualified teachers poses significant challenges to achieving quality education for all. Addressing this issue requires multifaceted strategies that focus on both attracting new talent and enhancing the professional development of existing educators. Offering more competitive teacher education programs is a better place to commence, as this enables prospective teachers to possess the requisite knowledge, skills and hands on experience to practice. In addition to this, appropriate policies and funding measures undertake a significant role in bringing about the systemic support that is required in order to increase the availability of teacher training and raise the working standards in those schools. Last but not least, promoting teaching as an appealing and well-respected profession is essential in building up strong and better images of this profession and deems teachers who enter into it on a long term basis. In that perspective, all these approaches build a strong sequence of strategies that will ensure enough qualified teachers for the world in the future.

Several countries have implemented successful models for sustainable teacher development. For instance, Rwanda's public-private initiative, supported by VVOB, has improved rural teacher retention by over 20% through mentorship and ongoing training. Finland emphasizes teacher autonomy and continuous upskilling as central pillars of its teacher workforce. These examples suggest that long-term solutions must include flexible and adaptable models such as digital training platforms, performance-based incentives, and structured career pathways. Such models can help developing countries address shortages without over-relying on short-term recruitment drives.

Enhancing Teacher Education Programs

Enhancing teacher education programs is a multifaceted endeavor that requires a comprehensive understanding of various pedagogical strategies, technological integration, and the professional development needs of educators. The literature suggests that competent teacher education also takes into consideration not just the knowledge factor but the knowledge application factor as well where for example educators are able to practice their profession.

One critical aspect of improving teacher education programs is the emphasis on the integration of technology in teaching practices. As highlighted by Perifanou et al., the design of ICT training programs should consider pedagogical approaches that facilitate the transfer of knowledge among educators (Perifanou et al., 2022). This is particularly relevant in today's world where there is an increasing use of digital devices in learning and educators need to be trained on both the use and application of such devices. Huang et al. further emphasize the importance of self-regulated learning (SRL) in technology integration, suggesting that enhancing teachers' SRL abilities is crucial for effective technology use in the classroom (Huang et al., 2020). Recent developments in AI-based learning tools, such as automated feedback platforms and adaptive learning modules, have shown potential in delivering personalized training to pre-service teachers. For example, AI chatbots can simulate classroom scenarios and provide formative feedback, improving pedagogical readiness at scale. This fits well with Chen's view of teacher training, which is not an isolated phenomenon, as it should also aim at equipping teachers with means of using technology to motivate learners (Chen, 2024).

Expanding Access to Teacher Training Colleges and Online Programs

Expanding access to teacher training colleges and online programs is a pressing issue in contemporary education systems. The importance of a diverse and well trained teaching staff is more and more emphasized by the increasing complexity of educational systems and the needs of children. Thus, access to quality teacher training programs is very important in providing future teachers with the skills and knowledge they need in order to create effective teaching and learning opportunities.

One of the core problems in teacher education is the difference in the availability of quality training programs, especially in underserved regions. Research indicates that teacher education programs must be responsive to the unique contexts in which they operate, particularly in rural or low-income areas where traditional training routes may be limited (Chen et al, 2022; Heinrich et al, 2019). The application of technology in teacher education can greatly enhance accessibility by allowing educators from remote locations to use once unreachable means of professional development (Chen et al, 2022). This becomes particularly relevant in the event of the Covid19 pandemic which has enabled the use of online learning platforms and increased the need of alternative flexible training methods (Ovcharuk, 2023).

In addition, the role of teacher educators is important in shaping the quality of the teacher training programs. Studies emphasize the importance of being engaged in a constant professional advancement for the teacher educators because their skills influence the effectiveness of training they provide to the prospective teachers (Goodwin et al 2014: Snoek et al 2011; MacPhail et al 2018). By investing in the professional growth of teacher educators, institutions ensure that their teaching staff is fully equipped to offer a high quality of training that meets the updated expectations in the educational sector. This type of professional growth may take various forms including the establishment of collaborative learning communities, mentorship programs and use of evidence-based practices (White et al, 2015; Smith & Flores, 2019).

A comparative review of national policies in Vietnam and Rwanda demonstrates how low- and middle-income countries can successfully expand access to rural teacher training through blended learning and government-funded stipends (Wolfenden, 2023).

Integrating Technology and Innovative Pedagogies in Teacher Education

Integrating technology into teacher training and professional development seems to be another strategy that can improve the quality of teachers. It has previously been reported that teachers can use technology as a means to enable them to teach in creative ways and increase their pedagogical content knowledge (Ma et al., 2022). For example, if teachers are trained on how to effectively use digital devices in their classes, this will lead to improved student and learning performance (Yeh et al., 2013). On top of that, technology gives teachers access to extensive resources and networking which enhances their chances of teaching in difficult resource poor settings (Ma et al., 2022; Yeh et al., 2013).

Artificial intelligence and digital tutoring tools are emerging as key components of scalable teacher training. For example, AI-powered learning analytics can personalize training modules and identify skill gaps in real-time, increasing training effectiveness (Del Bonifro et al., 2020). Meanwhile, blended learning models like Kolibri and Edmodo have seen successful adoption in Sub-Saharan Africa for both student instruction and in-service teacher support.

The need for improved technological training is underscored by the findings of Kafyulilo et al., who emphasize the importance of developing Technological Pedagogical Content Knowledge (TPACK) among pre-service teachers (Kafyulilo et al., 2015). This framework is necessary to use technology in teaching, but the overwhelming number of teacher education institutions do not regard this as an important component of training this ensures that the graduates are not only ill-prepared but also unwilling to make use of technology in their teaching exercises (Peeraer & Petegem, 2012). Unfortunately, this deficiency is exacerbated by the fact that most institutions provide no support with respect to integration of technology into education because teachers have no resources to turn to and are left to look for solutions on their own (Lakkala & Ilomäki, 2015).

The effectiveness of online teacher training programs is also contingent upon the design and implementation of these courses. Research suggests that well-structured online programs that incorporate interactive elements, peer collaboration, and real-world applications can lead to positive learning outcomes for participants (Hertz et al., 2022; Bruggeman et al., 2022). This highlights the importance of not only expanding access to online programs but also ensuring that these programs are designed to maximize engagement and learning.

Establishing Partnerships with International Organizations for Capacity Building

Establishing partnerships with international organizations for capacity building is crucial for increasing the number of qualified teachers globally. Such collaborations have the potential to utilize resources, expertise and innovative methods to improve the training and professional development of teachers. The existing educational literature suggests that such partnerships can be quite instrumental in addressing the issues of teacher supply and quality in certain geographic regions.

One of the main benefits of global partnership is the sharing of best and successful practices from other global education systems. Wolf et al. emphasize the importance of respect between educational institutions and society, and assert that these mutually beneficial relationships may provide teachers with important opportunities for upskilling, especially in multilingual settings (Wolf et al., 2022). This coincides with the findings of Bulfin et al. who investigate the role of national and international research funding councils working towards raising standards of qualitative research practices for educational technology researchers so as to strengthen teachers' competencies through evidence-based practices (Bulfin et al., 2014).

Policy Interventions and Financial Investments

Effective policy interventions and strategic financial investments are fundamental to addressing the shortage of qualified teachers and improving overall education quality. Such policies are critical for ensuring teacher quality, accessibility and enhancing the education system. As a solution to the shortage of teachers, governments can also expand their investments in the education sector and practice various forms of assistance. For instance, offering financial aid to aspiring educators can significantly increase accessibility to that profession. In doing so, potential teachers who may otherwise not be able to afford the opportunities can chase after their career goals. Promoting teaching as a viable career is equally important, as it involves improving working conditions, offering competitive salaries, and elevating the status of the profession. Additionally, offering professional development and career growth opportunities helps retain educators by fostering continuous skill enhancement and long-term job satisfaction. Together, these policy measures create a comprehensive framework for strengthening the teaching workforce and improving education quality.

According to the UNESCO Global Education Monitoring Report (2023), countries that integrated SDG indicators directly into their teacher policy frameworks (e.g., Ghana and Bangladesh) have reported measurable improvements in teacher qualifications and student outcomes. These policy adjustments have been especially effective when combined with domestic investment and international aid.

Research has shown that teacher quality is the single most important school-related factor influencing student achievement (Hanushek & Rivkin, 2010). A meta-analysis by Kraft & Papay (2014) further indicates that schools investing in sustained, job-embedded professional development experience lower turnover and higher student learning gains. These findings should inform national strategies, emphasizing long-term investments in teacher competencies rather than short-term recruitment drives. Drawing on such evidence enhances policy credibility and aligns with results-oriented education reforms.

Research has also increasingly pointed towards the efficacy of financial incentives as a strategy to combat this problem. Financial incentives can significantly reduce teacher attrition rates in economically disadvantaged areas by addressing some of the core reasons that deter educators from staying in their positions. A variety of studies have explored the impact of such incentives on teacher retention, particularly in rural and underserved settings (Tran & Smith, 2021; (Muremela et al., 2024; (Acheampong & Gyasi, 2019).

Increasing Government Budgets for Education and Teacher Training

Increasing government budgets for education and teacher training is a critical issue that has significant implications for the quality of education and the effectiveness of teaching. Research indicates that without proper funding, effective programs to help train teachers cannot be developed or offered, which would result in a decline in the quality of education. This response collates various arguments in favor of how increased governmental funding on education and teaching is necessary pertaining to various research papers.

Increased government funding for education has received wide-ranging support, particularly owing to the apparent link between such funding and overall education quality. Studies have shown that higher per-pupil spending leads to improved student outcomes, particularly for disadvantaged populations. A report compiled by Lee and Fuller (2020), for example, asserts that changes made in how states fund schools have enhanced expenditure on each child, resulting in better performance by deprived students. Grissom and Strunk (Grissom & Strunk, 2011), on the other hand, focuses on teachers instead and portrays the retention of a high-quality educator as vital for successful functioning of a school district, and emphasizes on teacher's salaries as a crucial tool for this retention. Should these suggestions be taken into consideration, once again the argument for higher investment in the educational sector stands vindicated.

Providing Scholarships and Financial Aid to Attract Prospective Teachers

Attracting prospective teachers using scholarships, financial aid and funds is an intricate process that necessitates an appreciation of the broader problems in the education sector as well as the socio-economic issues that affect teacher recruitment. The provision of financial support not only alleviates the burden of educational costs but also serves as a strategic tool to enhance the appeal of the teaching profession, especially among underrepresented groups. Research indicates that among low-income students, receiving financial aid, grants or scholarships tends to influence their decision to enroll in college thereby increasing the pool of potential teachers (Perna et al., 2010; Bernal, 2023).

For a majority of prospective teachers, the financial burden associated with higher education acts as the most significant hurdle they face. Studies have shown that students often seek out alternative sources of finance and are open to lifestyle changes such as moving back to their parents' house (Mangan et al, 2010). Such financial burdens can discourage many from pursuing teaching careers especially in high-need areas where the demand for qualified educators is acute. In this regard, targeted financial aid programs can be an important tool in recruiting the individuals who might abandon this profession for financial reasons (Darling-Hammond, 2023; Rice et al, 2012).

Comparative Policy Analysis: What Works Across Contexts

Successful policy models for teacher retention differ across regions. In Finland, long-term investment in teacher autonomy and rigorous training has led to sustained interest in the profession. Meanwhile, Vietnam's structured teacher career ladder and Rwanda's rural deployment incentives have helped mitigate teacher shortages in underserved areas. These models underscore the importance of tailoring strategies to contextual needs.

Promoting Teaching as a Viable Career

Promoting teaching as a viable career requires a multifaceted approach that addresses the complexities of the profession, the motivations of prospective teachers, and the systemic challenges that educators face. Teaching is often perceived as a profession intertwined with personal values and societal contributions, yet it is also a field that faces significant challenges such as high attrition rates and emotional burnout among educators. (Arnup & Bowles, 2016; Heffernan et al, 2022). Understanding these dynamics is crucial for promoting teaching as a sustainable and attractive career choice.

Furthermore, working conditions within the teaching profession have been noted to influence whether teachers join or leave the profession. Heffernan et al. highlight that a meaningful reduction in workload and an increased understanding of the complexities of teaching can positively influence teachers' decisions to remain in the profession (Heffernan et al., 2022). This aligns with findings from Arnup & Bowles (2016) who identify resilience as a protective factor against leaving the profession, indicating that supportive working conditions can enhance teachers' commitment. Consequently, addressing problems such as manageable working conditions and a professional workplace might enhance teacher retention and render the job more appealing.

Improving teacher retention requires more than acknowledging existing challenges—it demands strategic action. Career progression frameworks that recognize and reward experienced teachers can enhance job satisfaction and professional identity. For example, Chile's "Teacher Career Law" 2016 defines clear promotion stages tied to performance and experience, leading to measurable retention gains (Fernández Ugalde, 2024). Additional measures, such as rural housing allowances, student loan forgiveness, and peer mentoring programs, have shown positive outcomes in countries like South Africa and Ghana. Integrating these into national policies can help retain qualified teachers in high-need areas.

Offering Professional Development and Career Growth Opportunities

The acute shortage of qualified teachers in developing nations can be partially attributed to the insufficient preparation and support provided to the teachers. In developing countries' education systems, many teachers lack professional certifications and even basic teaching abilities (Perry & Bevins, 2018). Such inadequacy is then worsened by the high rates of teacher attrition, which creates more volatility within education systems (Federičová, 2020). Against this backdrop, it becomes critical to strengthen teacher training interventions so that they are suitable for the context in which they are to be applied. For example, teacher training programs which address actual teaching practice in the classroom and appropriate pedagogy seem to be effective in enhancing teaching and learning performance (Perry & Bevins, 2018; Hardman et al, 2015).

Another key area to focus on is professional development which contributes to the increase in qualified teachers. Continuous professional learning opportunities can help teachers refine their skills and adapt to new educational demands. Research has shown that certain professional learning programmes, especially those that are collaborative and relate to people's work context have positive effects on teaching practices and improvement in the educational system over time (McChesney & Aldridge, 2018; Nguyen & Ng, 2020). Mentoring initiatives and peer-based projects have been cited as creating and enhancing an environment for learning, development and improvement of education among teachers (Nguyen & Ng, 2020; Main & Slater, 2021). Furthermore, technology use in conducting professional development such as the use of online training modules can assist teachers in underprivileged areas to access quality content and personnel (Ma et al., 2022; Compen et al., 2020; Dalal et al., 2020).

The Role of Distance Education in Teacher Training

Distance education can also be regarded as a transformative force in teacher training, addressing critical challenges and opening new pathways for professional development. It ensures the extension of teacher education into rural and underdeveloped areas facilitating the acquisition and improvement of the relevant skills of the volunteers as well as practicing teachers. Leveraging innovative approaches like micro-credentialing systems and massive open online courses (MOOCs) it provides flexible, targeted learning opportunities tailored to individual needs. Technology mediated professional development tools also enhance professional learning as well as opportunities for teachers to re-align their skills in response to rapid changes in education. Besides, when there is no possibility for traditional educational interactions face-to-face, it serves as a low-cost and economical remedy aiming at large audiences making teacher education possible without any boundaries. In low-resource settings, technology offers scalable and affordable solutions to teacher training. For example, the "Teachers for Teachers" initiative in Kenya uses mobile phones to deliver coaching and feedback to educators in refugee camps—an approach shown to increase instructional quality by 25% (Annual impact review 2021-22). Integrating learning analytics into teacher MOOCs can also personalize professional development. Governments can partner with edtech platforms to develop localized content aligned with national curricula, providing teachers with real-time support in pedagogical and classroom management practices.

Expanding access to teacher education for remote and underserved regions

In addition to improvement in the training quality, increasing the availability of online programs could potentially boost the diversity of the teaching workforce. Research has shown that diverse teacher populations can positively impact student outcomes, particularly for students from marginalized backgrounds (Mensah, 2019; Philip, 2011). Online teacher training programs can appeal to a wider range of candidates including those unable to attend traditional face-to-face classes due to work or family responsibilities. This inclusivity is essential for building a teaching workforce that reflects the diversity of the student population (Parker et al., 2022; Florian & Camedda, 2019).

Leveraging Massive Open Online Courses (MOOCs) and Micro-Credentialing Programs

The integration of Massive Open Online Courses (MOOCs) and micro-credentialing programs into teacher training can be regarded to be one of the groundbreaking approaches that shifts the focus towards professional development in the spheres of education. This paradigm shift is driven by the need for flexible, accessible, and relevant training that meets the diverse needs of educators in an increasingly digital world. The incorporation of MOOCs alongside micro-credential programs during teacher training can be regarded to be one of the groundbreaking approaches that shifts the focus towards professional development in the spheres of education. The literature highlights several key aspects including the improvement of digital skills, the potential of micro-credentials for professional growth as well as the concerns on equity and inclusion in education.

Maina et al. (2022) stress the importance of MOOCs and micro-credentials on the possession of employability skills, which also includes a previously conducted micro-credentialing approach in various domains including education. This approach provided recognition of participants' employment skills and furthermore, assisted participants in connecting with employers, ensuring that the requirements of the skills taught are aligned with those of the industry. It emerged from the pilot study that in spite of the challenges posed by the COVID-19 pandemic, micro-credentialing in an online environment is able to adequately prepare teachers for the job market requirements (Maina et al., 2022). Such flexibility is vital at a time when traditional methods of education are being reevaluated.

Facilitating Continuous Professional Development Through Virtual Learning Environments

The deployment of virtual learning environments (VLEs) has emerged as a promising avenue for supporting teachers' professional growth, especially amid the rapid shift to online learning because of the COVID-19 pandemic. The effectiveness of VLEs in supporting teacher professional development hinges on several key factors, including the design of the learning experiences, the integration of technology, and the establishment of collaborative learning communities.

The integration of immersive technologies such as virtual reality (VR) and simulation-based education systems are exciting for teaching education in VLE as well. Martinez et al. (2019) state that the distinctive feature of 3D simulation environments is the realistic training they offer which is likely to help boost a teacher's digital skills. Similarly, Dai et al. (2022) argue that VR-supported learning environments offer cost-effective alternatives to traditional face-to-face professional development, enabling teachers to engage in immersive experiences that replicate authentic classroom scenarios. Additionally, such technologies do promote enhanced engagement and enable practice and feedback which are critical for skill development. Integrating AI in VLEs - such as virtual assistants that track learning progress or predictive analytics that guide course recommendations- can optimize professional development and reduce training gaps across decentralized education systems.

Role of International Cooperation and SDGs

International cooperation plays a pivotal role in advancing educational development and addressing global challenges, particularly within the framework of the Sustainable Development Goals (SDGs). Organizations and countries are able to strengthen education systems globally by bringing resources, sharing knowledge and promoting innovation through global partnerships. A central focus of this cooperation is SDG 4's target, which aims to substantially increase the supply of qualified teachers, recognizing their critical role in achieving inclusive and equitable quality education. The pursuit of goals through measurements of the SDG's indicators places the necessary checks and balances and establishes a formal evaluation of progress and achievement while expressing requirements for further actions.

The chapter's alignment with global education frameworks can be further strengthened by referencing progress towards SDG 4, particularly Target 4.c, which tracks the percentage of trained teachers at different education levels. Countries like Vietnam and Chile have successfully integrated these indicators into national education strategies. In Chile, teacher evaluations tied to SDG metrics have improved accountability and resource allocation. These international benchmarks can guide developing nations in designing robust education policies aligned with global standards.

Leveraging Global Partnerships

Leveraging global partnerships to increase the supply of competent teachers is a multifaceted endeavor that requires a comprehensive understanding of educational frameworks, collaborative practices, and the integration of technology in teacher training. The international context of education is getting more interrelated and therefore there is a need to revise teacher training programs to be relevant for the requirements of the multicultural classrooms as well as for the changing dynamics of students.

One of the main ways of improving teacher effectiveness is by creating international links that promote the exchange of teacher training ideas and cultural aspects. For instance, reciprocal learning partnerships between countries, such as those between Canadian and Chinese teachers, have shown that sharing narratives and experiences can lead to the development of globally competent educators (Khoo, 2021). These partnerships greatly enhance teachers' professional growth and also promote the appreciation of other education contexts, a necessity for the effective teaching of multicultural students (Thomassen & Munthe, 2020). In addition, increasing the level of TPACK among international teachers using cultural exchange programs can help them enhance their skills in making effective use of technology in their teaching (Dalal et al., 2020). These international collaborations not only bring resources but also help customize interventions. For example, Uzbekistan's partnership with UNESCO led to localized digital content and training programs designed specifically for Central Asian classroom realities (UNESCO, 2023).

Emphasizing SDG 4's Target to Substantially Increase the Supply of Qualified Teachers

The Sustainable Development Goal (SDG) 4 emphasizes the necessity of ensuring inclusive and equitable quality education and promoting lifelong learning opportunities for all. One of the key indicators is to ensure significantly vast growth in the amount of qualified teachers in specific areas. There are plenty of models that emphasize the importance of qualified teachers for a context because they focus on the issues of improving educational quality and increasing performance and practice with grade and other assessment indicators.

The global teacher shortage is a major issue that prevents the full realization of SDG 4. Studies show that many nations cannot ensure an adequate supply of qualified teachers, especially in underserved areas. Gao (2023) points out that in the Northern regions, education is underpinned with resources and experienced teachers who are scarce and in demand. This disparity is further exacerbated by high turnover rates among teachers, particularly in challenging school contexts, as evidenced by studies indicating that high-quality teachers are more likely to leave positions in schools serving low-income populations (Federičová, 2020; Hanušová et al., 2020).

In order to deal with the problem of teacher supply, there is a need to come up with strategies for recruitment and retention of teachers. Berman and Defeo (2023) indicates that teachers must be equally compensated so as to be attracted to high need areas. Moreover, strategies that enhance the professional well-being of teachers can significantly impact retention rates. Wolf et al. (2018) found that strategies that sought to enhance teacher professional wellbeing and classroom quality were effective in improving both teachers' retention as well as students' readiness to learn. This suggests that creating supportive work environments, alongside competitive compensation, is crucial for maintaining a robust teaching workforce.

Monitoring Progress Using Sustainable Development Goals (SDGs) Indicators and Ensuring Accountability

Evaluating the status of the Sustainable Development Goals and holding the pertinent parties to their promises is a complex process that involves the analysis of educational models, evaluation techniques and different actors' contributions. The SDGS, and in particular Goal 4, note the need to ensure quality education and promote lifelong learning opportunities for all which require the existence of effective monitoring and accountability to uphold the development goals.

One of the primary challenges in monitoring progress towards the SDGs is the need for robust indicators that can accurately reflect educational outcomes. Raikes et al. (2017) stress the importance of performing educational assessments in respect to all forms of educational targets such as early childhood education as part of the SDGs. This is supported by Camphuijsen who discusses the increasing reliance on standardized tests in the education sector to hold schools and teachers accountable for learning objectives (Camphuijsen, 2020). The establishment of clear, measurable indicators is essential for assessing whether educational institutions are meeting the expectations set forth by the SDGs.

Furthermore, the contribution of the higher education sector in the attainment of SDGs is paramount. While Chankseliani and Mccowan (2020) argue that higher education has not featured significantly in the previous international development objectives, goal 4 of the SDGs emphasizes the need to ensure equal opportunities for all to access tertiary education as a component of strategies to encourage lifelong learning. This inclusion necessitates that not only do universities need to provide access, but also to ensure a certain level of the quality of education which can be observed through different forms of accountability.

THE IMPACT OF THE CHAPTER BY DATA-DRIVEN INSIGHTS: STATISTICS, CASE STUDIES AND TRENDS

This chapter provides not only a general introduction to the issue of teacher shortages in developing countries and discusses broad challenges, such as lack of resources, inadequate training programs, and policy gaps. Also, includes data-driven insights (e.g., statistics, case studies, or key and historical trends) to highlight and better contextualize the impact of the issue. In the publication of the Inter-agency Network for Education in Emergencies (INEE) which is a global open network of members who are working together within a humanitarian and development framework to ensure that all individuals have the right to a quality, the key trends were put forth as sustaining student learning in India through quality teacher support during the COVID-19 pandemic (Saigal, Kogekar, and Kong, 2022). Accordingly, this case study describes how the Teacher Innovator Program (TIP) was specifically modified to address pandemic-related issues, as well as the results of the modified intervention and lessons learned by illustrating an example of how a civil society organization in India addressed the educational requirements of underprivileged kids during the COVID-19 pandemic on the ground. One more case for key trends in Bangladesh reported by McAleavy, Hall-Chen, Horrocks and Riggall (2022) in the Educational Development Trust (EDT) about technology-supported professional development for teachers. Accordingly, the trends examplied about the "English in Action" program improved English language instruction by providing teachers with pedagogical content via mobile phones and SD cards. The EDT also summarizes the lessons that may be drawn from a number of the most intriguing instances of technology-assisted in-service professional development in lower-income nations around the world, as well as from broader considerations regarding the potential of technology to improve teachers' professional development. Six intriguing case studies are found in this report's analysis of the literature on technology for professional learning:

1. Mobile phones are being used by UNESCO programs in Pakistan and Nigeria to provide early childhood and primary school teachers with educational materials.
2. Mobile phones and SD cards are being used by English in Action, Bangladesh, to provide teachers with content.
3. OER to promote interactive teaching methods in Sub-Saharan Africa's Teacher Education System (TESSA)
4. India's School-based Support for Teacher Education (TESS-India) OER to encourage the use of more captivating teaching methods

   5-6. Teachers for Teachers, Kenya: enhancing the coaching given to teachers through the use of mobile technologies to support teacher development in the Kakuma Refugee Camp National Tablets Program, Kenya

The EDT report delves deeper into each of these case studies and thoughts, offering insights for practitioners, legislators, and other organizations looking to give teachers access to technology-related professional development opportunities. It is reflected on the influence of technology on teacher professional development through these case studies, especially in low- and low-middle-income contexts:

- Promising technology alone is insufficient; we also need to consider the human element.
- There is great promise for mobile technologies to increase the flexibility, scalability, and reach of teacher professional development.

- One promising model for resources for professional development is open educational resources (OER).
- Technology-enabled solutions with a high effect and sustainability require local collaboration and a feeling of local ownership.
- Careful piloting solutions and high-quality impact data are essential for the development and implementation of successful technology-enabled professional learning.

In Central Asia, with assistance from the World Bank and UNESCO, nations like Tajikistan, Uzbekistan, Kyrgyz Republic, and Kazakhstan are changing their teacher policies with an emphasis on enhancing pedagogical abilities and creating digital educational materials. (UNESCO, 2023). Addition to this, through the Mpumalanga Secondary Science Initiative (MSSI), the study sought to investigate the efficacy and implementation of "lesson study," a Japanese model of teacher professional development, as a continuing professional development (CPD) strategy for science and math teachers in South Africa by Ono and Ferreira (2010). This case study looked into how the lesson study model, which has its roots in Japanese pedagogy, may be used as a sustainable and alternate method of continuing professional development for South African in-service teachers. In light of curricular revisions like OBE and curricular 2005, the study addressed the drawbacks of conventional "cascade" training techniques, which frequently led to diluted and poor knowledge transfer. The Mpumalanga Department of Education, the Japan International Cooperation Agency (JICA), and the University of Pretoria partnered to launch the MSSI project (1999–2006). Its goal was to increase the quality of science and math instruction by institutionalizing lesson-based professional development in schools. In Japan, curriculum implementers and teachers received training on how to conduct lesson study sessions, which are organized around a "plan-do-see" cycle that includes group lesson planning, observation, and reflection. Despite a number of obstacles (such as curriculum modifications, a lack of training time, and structural education reforms), the project showed that participating teachers' teaching practices had improved. Lesson study showed potential as a teacher-driven, reflective, and practice-based approach to CPD in developing contexts, the study concluded, even though it could not be fully institutionalized. For long-lasting effects, it underlined the necessity of structural support, cultural adaptability, and long-term dedication.

Wolfenden (2023) aimed to explore how ICT-mediated teacher professional development (TPD@Scale) can be adapted and scaled in low- and middle-income countries (LMICs) to enhance the equity, quality, and efficiency of in-service teacher training systems. Understanding how teacher professional development models, especially those that make use of information and communication technologies (ICTs), can be scaled to reach a large number of teachers in a variety of educational settings in Ghana (Africa), Honduras (Latin America), and Uzbekistan (Central Asia) was the main goal of the study. By giving instructors more access to pertinent, excellent, and reasonably priced professional development programs, this was meant to enhance classroom instruction. It assessed a number of models (such as district-based, hybrid, and online) and their interactions with various educational governance frameworks, and integrated them into regional educational frameworks, working with national ministries and organizations to develop, test, and assess TPD models unique to each nation for Ghana, Honduras, and Uzbekistan.

Taddese and Rao (2022) examine the viewpoints and lived experiences of educators, examine the structure and execution of the school-based continuous professional development (CPD) program in Ethiopian primary schools. Its main goal is to comprehend the features of the CPD program as well as the external barriers to its efficacy. This qualitative case study looks at how Ethiopia's school-based CPD program is being implemented and perceived. It was first established in 2009 as a structured national framework with the goal of improving student outcomes and teacher effectiveness. Based on interviews, focus groups, and document evaluations with 24 teachers from four primary schools, the study shows a discrepancy between the declared objectives of the CPD and how it is actually implemented. Although the program was intended to be collaborative, continuous, and sensitive to context, educators characterized it as top-down, mandatory, bureaucratic, and inadequately contextualized. Many thought CPD exercises were monotonous, unrelated to their professional learning demands, and unimportant. Real engagement was also severely hampered by low teacher motivation, which was caused by inadequate pay, little career prospects, and politically controlled school administration. The study comes to the conclusion that despite being nationally standardized and in place for more than ten years, the CPD program has not really changed how teachers educate. Teachers primarily comply for official reporting, not for real-world learning. According to the study, school-based CPD is ineffective without teacher agency, contextual customisation, professional support, and systemic alignment.

In the report of EDT, the study focused on peer-researchers' experiences with the PEER technique and teachers' experiences teaching in refugee situations in Jordan by Charitonos, et al. (2022). During the training seminars, the project team and fellow researchers came up with important research questions. In order to effectively teach language in refugee settings, the questions centered on understanding pedagogical and curriculum approaches. The findings are focussed mostly on contextual conditions that influence instruction, fewer chances for career advancement, the teacher's enlarged role, and the importance of participating in the PEER process. According to the phases, the contextual issues, such as social, political, and material hardships, have a significant impact on teaching techniques in refugee situations. Teachers say their work settings are very challenging, citing problems such as dealing with "not normal students," a lack of resources for teaching (devices, internet, materials), and job insecurity, particularly for staff who get funding from NGOs. What instructors are able to do in their classrooms is directly influenced by these environmental pressures. Moreover, accessing formal, structured professional development (PD) is severely hampered for teachers, especially those working in refugee environments. Access is restricted by the cost of traveling to urban areas, such as capital cities, and by limitations depending on one's professional or legal standing (for example, refugee teachers who are not officially recognized in host systems). Nevertheless, educators of all backgrounds voiced a strong need for pertinent professional development that takes into account their needs and circumstances. Furthermore, in refugee circumstances, educators play a variety of overlapping tasks, including those of advocates, caregivers, and emotional support systems. They develop into dependable adults and community leaders who, frequently at their own expense, assist pupils in reintegrating into formal education. In these vulnerable settings, educators place a higher priority on the emotional and social aspects of learning, frequently viewing them as more important than academic education. One last thing about the findings, teachers had a thoughtful, empowered, and interactive professional learning experience through the PEER research approach. Teachers gained new research and communication abilities, established relational knowledge about their peers, and generated uncommon possibilities for collegial dialogue by conducting interviews and having meaningful talks. Teachers were able to reflect on their responsibilities and practices and envision better futures for their profession through this process, which turned into an informal yet effective form of professional development.

According to the report by the EDT an insight from a cross regional perspective could also be presented about technology integration and mobile learning. Open Educational Resources (OER) have been used by initiatives like TESS-India and TESSA (Teacher Education in Sub-Saharan Africa) to promote interactive teaching methods to promote technology integration while mobile technologies have been used to deliver educational information in Bangladesh, Nigeria, and Pakistan. They have shown great promise for expanding the scope and adaptability of teacher professional development.

FINDINGS

The dataset used in this chapter, 2024.Q3.G.01 SE_TRA_GRDL, provides comprehensive data on teacher qualifications globally. It facilitates disaggregated analysis by area,educational attainment and income level. These aspects are crucial for comprehending the intricate issues encountered by various governments and areas. This study employs a comparative analysis of teacher qualification data across continents, categorized by income levels. The dataset is analyzed to identify trends, disparities, and areas requiring intervention. Key indicators include:

- **Proportion of competent teachers by education attainment:** Disaggregated into primary, lower secondary, and upper secondary levels.
- **Income-level variations:** Comparing low-income, middle-income, and high-income countries.

Graphs and visualizations are used to illustrate these trends, providing a clear understanding of the current state of teacher qualifications worldwide. To do this, the dataset obtained from the indicators is compared across continents' educational attainment and income levels for the years 2000-2022 as seen below:

Figure 1. Pre-primary Schools by regions (a) and income level (b)

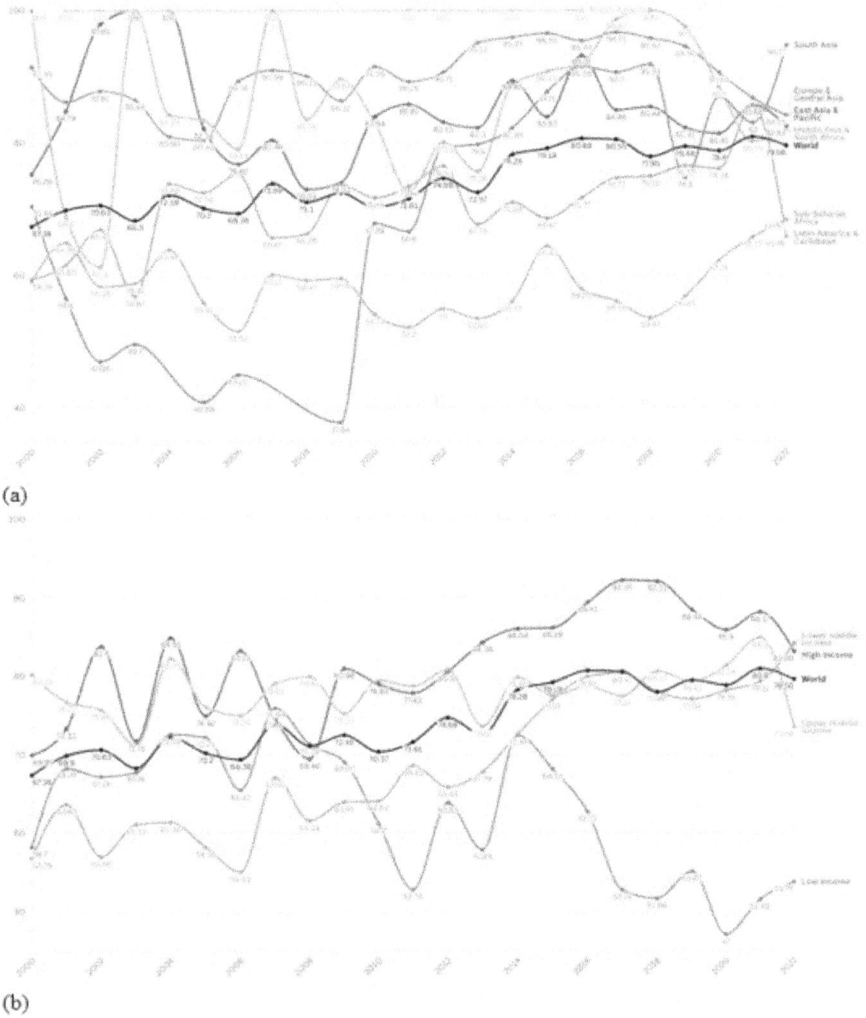

(a)

(b)

Figure 1 shows the educational attainment by regions (a) and income level (b) of pre-primary schools between 2000 and 2022 years. The graphs show disparities in access and attainment levels, with developed regions showing higher and more consistent progress compared to underdeveloped ones. Low-income countries lag significantly, with middle-income nations experiencing moderate improvement and high-income nations maintaining strong results. Figure 1 highlights stark disparities in pre-primary school attainment across regions and income levels. High-income countries consistently achieve near-universal access, while middle-income countries show promising growth despite ongoing challenges. Low-income countries, particularly in Sub-Saharan Africa and South Asia, struggle with significant barriers to access and quality, underscoring the need for targeted interventions in early childhood education infrastructure and teacher training.

(a) Regional Trends:

The graph illustrates trends in pre-primary school attainment across various regions between 2000 and 2022:

1. **High-Performing Regions**:
   - **North America and Europe**: These regions demonstrate consistently high enrollment and attainment rates in pre-primary education. Universal access has been achieved in most countries due to established systems and strong policy emphasis on early childhood education.
   - **East Asia and Pacific**: Significant improvements over the years, reflecting increasing government focus on the importance of early education as part of overall human capital development.
2. **Moderate-Performing Regions**:
   - **Latin America and the Caribbean**: Gradual progress is evident, with improvements driven by public and private sector efforts to expand early childhood education. However, access disparities between urban and rural areas remain.
   - **Middle East and North Africa**: Steady growth is observed, reflecting rising awareness of the importance of early education, though conflict and instability in some parts of the region have limited progress.
3. **Low-Performing Regions**:
   - **Sub-Saharan Africa**: Pre-primary enrollment and attainment rates remain the lowest globally, reflecting significant challenges in infrastructure, teacher training, and community outreach. Improvements have been slow but steady in recent years, driven by international funding and targeted programs.
   - **South Asia**: Moderate progress has been made, with gender and rural-urban disparities being key obstacles. Some countries in the region have made strides through policy reforms, but overall access remains limited.

   (b) Income-Level Trends:

The graph compares pre-primary school attainment across low-, middle-, and high-income countries:

1. **High-Income Countries**:

    - Universal or near-universal pre-primary enrollment is evident by 2022, reflecting established systems and significant investment in early childhood education.
    - High-quality programs, strong teacher training, and parental awareness contribute to high attainment rates.
2. **Middle-Income Countries**:
    - Significant growth in pre-primary enrollment is observed, with many countries prioritizing early education as a foundation for later learning.
    - Challenges include disparities between urban and rural areas, as well as uneven quality of programs.
3. **Low-Income Countries**:
    - Enrollment remains low, with significant gaps in access and attainment.
    - Barriers include lack of funding, infrastructure, trained educators, and public awareness of the importance of pre-primary education.
    - Although improvements have been seen over the years, progress is slower compared to middle- and high-income countries.

Figure 2. Primary Schools by regions (a) and income level (b)

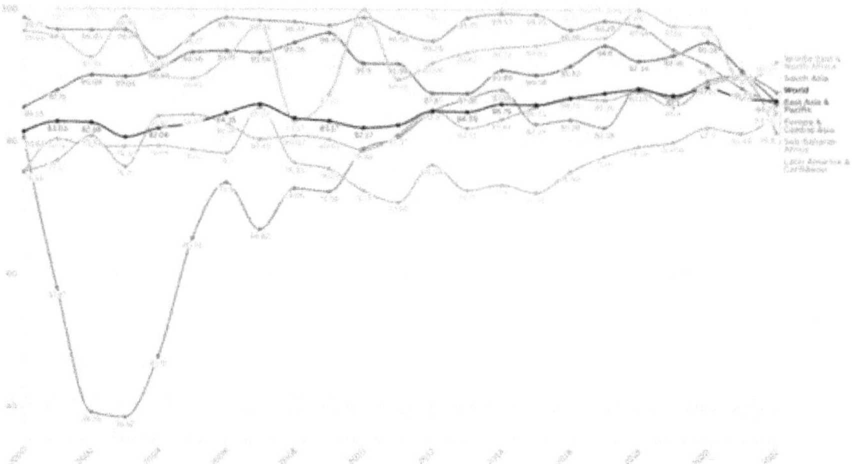

(a)

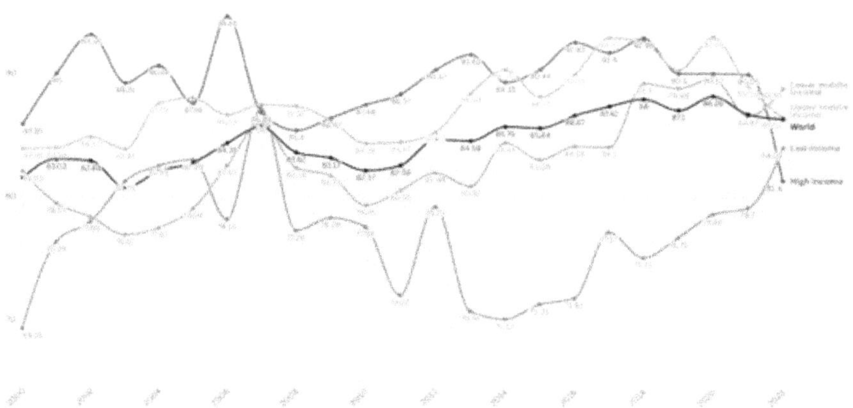

(b)

According to Figure 2, it is possible to say that higher enrollment and attainment rates are evident in regions with robust educational policies, while underserved areas face persistent challenges. Addition to this, disparities continue, with low-income countries still showing slower growth in educational outcomes. Figure 2 highlights significant disparities in primary school attainment between regions and income levels. High-income regions achieve near-universal primary education, reflecting strong education systems and policies. Middle-income countries demonstrate substantial progress, though challenges in equity and quality remain. Low-income countries, particularly in Sub-Saharan Africa and parts of South Asia, face persistent barriers to universal primary education, underscoring the need for targeted policies and investments in infrastructure, teacher training, and community engagement.

(a) Regional Trends:

The graph presents primary school attainment across global regions from 2000 to 2022:

1. **High-Performing Regions**:
    - **North America and Europe**: These regions consistently maintain near-universal enrollment and high attainment rates for primary education. This reflects strong policies supporting free and compulsory primary education.
    - **East Asia and Pacific**: Significant improvements are seen, with many countries achieving near-universal primary school enrollment. Governmental investment and international programs have contributed to this progress.
2. **Moderate-Performing Regions**:
    - **Latin America and the Caribbean**: There has been steady progress, with most countries achieving high enrollment rates. However, disparities remain in rural and indigenous communities.
    - **Middle East and North Africa**: Improvements are evident, particularly in closing the gender gap in enrollment. However, progress has been hindered in areas affected by political instability and conflict.
3. **Low-Performing Regions**:
    - **Sub-Saharan Africa**: The region has the lowest primary school attainment rates, though noticeable progress has been made since 2000. Key challenges include poverty, infrastructure shortages, and teacher deficits.
    - **South Asia**: Enrollment rates have steadily improved, with some countries approaching near-universal primary education. However, quality concerns and dropout rates, particularly among girls and marginalized groups, remain significant obstacles.

(b) Income-Level Trends:

The graph compares primary school attainment by income levels (low-, middle-, and high-income countries):

1. **High-Income Countries**:

- Consistently high enrollment and attainment rates are observed, with most countries maintaining universal primary education.
- Quality of education is also higher, supported by well-trained teachers, sufficient resources, and robust education policies.

2. **Middle-Income Countries**:
    - There has been steady growth in primary school attainment, with many countries nearing universal enrollment.
    - Challenges persist in maintaining quality and addressing inequalities between urban and rural areas.
3. **Low-Income Countries**:
    - Primary school attainment remains a significant challenge, with low enrollment and completion rates compared to higher-income countries.
    - Barriers include poverty, lack of school infrastructure, teacher shortages, and high dropout rates, especially for girls.
    - Despite these challenges, progress is visible, driven by international aid and local initiatives to expand access to primary education.

Figure 3. Lower Secondary Schools by regions (a) and income level (b)

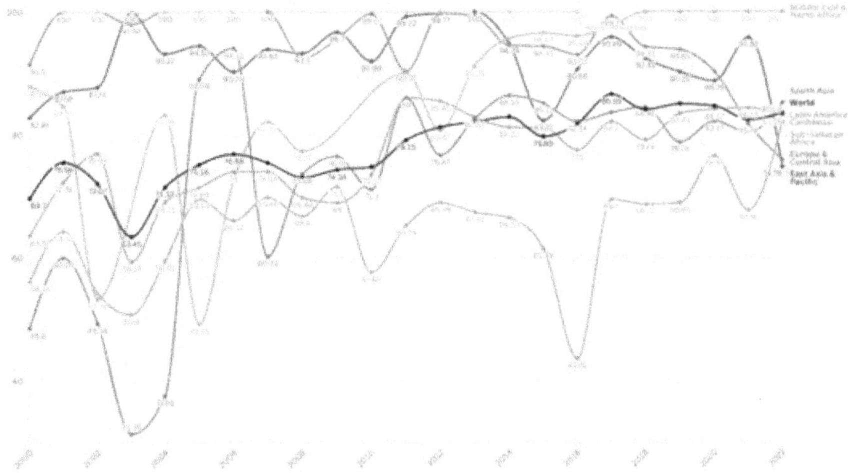

(a)

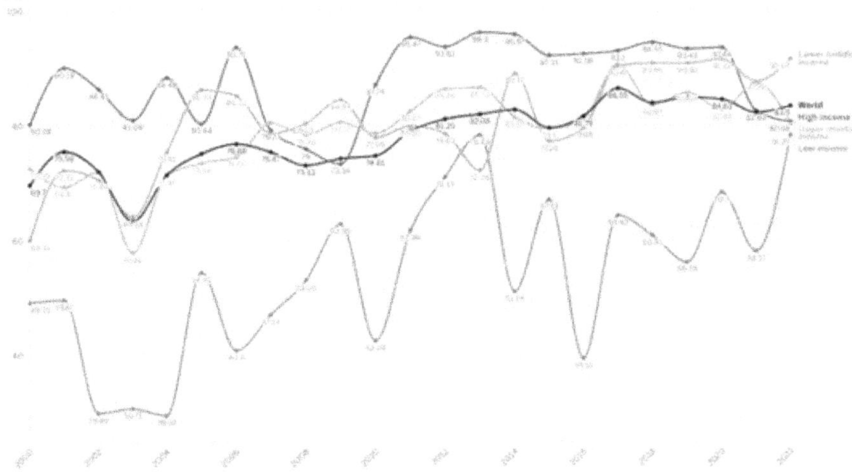

(b)

Figure 3 shows noticeable improvements in regions previously lagging, but gaps remain between continents. Middle-income countries close the gap with high-income countries, whereas low-income areas progress more modestly. Figure 3 highlights significant disparities in lower secondary school attainment across regions and income levels. High-income countries consistently maintain universal access and quality education, while middle-income countries show steady progress but face challenges in equity and quality. Low-income countries lag significantly, particularly in Sub-Saharan Africa and parts of South Asia, where financial, cultural, and infrastructure barriers limit access to lower secondary education. These findings underscore the urgent need for targeted interventions to improve access, teacher qualifications, and equity in lower secondary education worldwide.

(a) Regional Trends:

The graph illustrates educational attainment trends in lower secondary schools across various regions from 2000 to 2022:

1. **High-Performing Regions**:
    - **North America and Europe**: These regions maintain consistently high attainment rates, reflecting robust education policies and well-established secondary school systems. Nearly all children transition smoothly from primary to lower secondary levels.
    - **East Asia and Pacific**: Significant progress is evident, with many countries achieving high enrollment and completion rates due to strong economic growth and investments in education infrastructure.
2. **Moderate-Performing Regions**:
    - **Latin America and the Caribbean**: Attainment rates have steadily improved, but inequalities persist between urban and rural areas. Marginalized communities and indigenous populations face challenges in accessing quality education.
    - **Middle East and North Africa**: Steady progress is observed, with increasing efforts to improve access and reduce gender disparities. However, conflict-affected areas continue to struggle with access and quality issues.
3. **Low-Performing Regions**:
    - **Sub-Saharan Africa**: The region has the lowest lower secondary school attainment rates. While progress has been made, key challenges include limited infrastructure, teacher shortages, and financial barriers. Many children drop out after completing primary education.
    - **South Asia**: Noticeable improvements are seen in enrollment and attainment, though challenges like gender disparities and gaps between rural and urban areas persist. Government initiatives have driven progress, but the pace remains slower compared to other regions.

(b) Income-Level Trends:

The graph compares attainment in lower secondary schools by income levels (low-, middle-, and high-income countries):

1. **High-Income Countries**:

   - Near-universal enrollment and completion rates are observed, with strong education systems ensuring consistent performance.
   - Focus on quality education, advanced curricula, and qualified teachers contributes to high attainment.
2. **Middle-Income Countries**:
   - Significant improvements are evident, with many countries achieving substantial growth in lower secondary school enrollment.
   - However, gaps persist between urban and rural areas, and quality disparities remain a concern in some regions.
3. **Low-Income Countries**:
   - Lower secondary school attainment remains low, with many children unable to transition from primary education.
   - Barriers include inadequate school infrastructure, underqualified teachers, and socio-economic constraints. Girls, in particular, face challenges due to cultural norms and early marriage.
   - While progress has been made, the pace is slower compared to middle- and high-income countries.

Figure 4. Secondary Schools by regions (a) and income level (b)

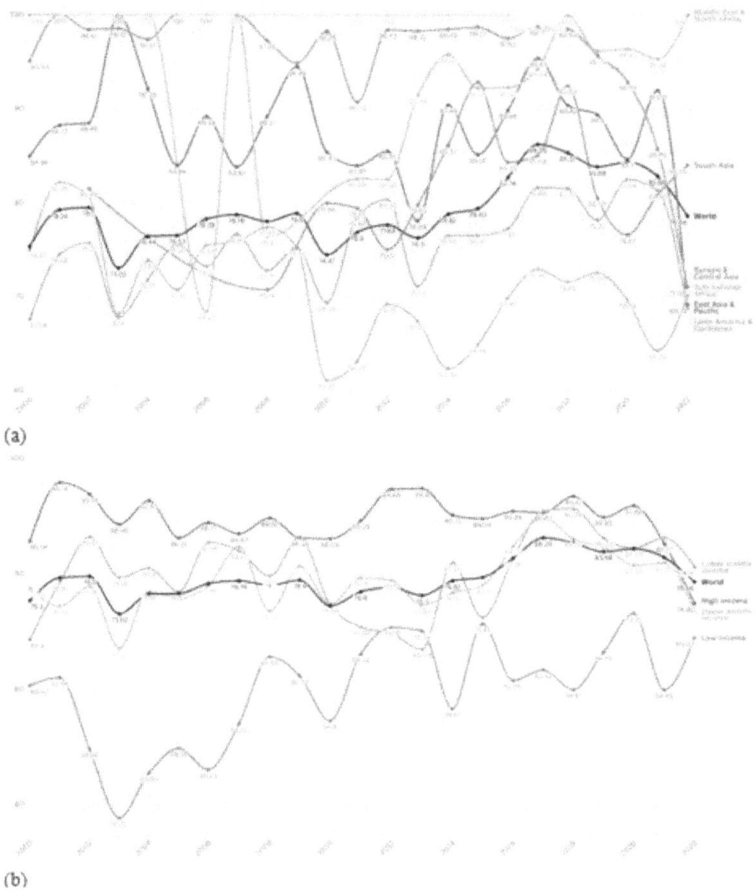

Figure 4 underscores significant disparities in secondary school attainment by region and income level. While high-income regions and countries maintain robust education systems with near-universal attainment, middle- and low-income countries face persistent challenges. Sub-Saharan Africa and South Asia, in particular, show the greatest need for investment in secondary education infrastructure, teacher training, and equitable access initiatives.

(a) Regional Trends:

The graph depicts educational attainment trends in secondary schools across different global regions from 2000 to 2022:

1. **High-Performing Regions**:
    - **North America and Europe**: These regions show consistently high attainment levels, with nearly universal secondary school enrollment and completion. The stability reflects strong education systems, infrastructure, and policy frameworks.
    - **East Asia and Pacific**: Marked improvements are observed, with some countries achieving near-parity with high-income regions due to economic growth and significant investment in education.
2. **Moderate-Performing Regions**:
    - **Latin America and the Caribbean**: Gradual progress, with notable increases in enrollment rates and attainment over the years. Economic disparities within regions contribute to gaps between urban and rural areas.
    - **Middle East and North Africa**: Significant improvements since 2000, driven by government initiatives to improve access, especially for girls, though progress is slower in conflict-affected countries.
3. **Low-Performing Regions**:
    - **Sub-Saharan Africa**: The region shows the lowest attainment rates, reflecting barriers like insufficient infrastructure, teacher shortages, and high dropout rates due to poverty. However, gradual progress is evident in recent years due to targeted investments and international support.
    - **South Asia**: Moderate growth is visible, though gender disparities and rural-urban gaps remain a challenge. Improvements have been slower compared to East Asia.

(b) Income-Level Trends:

The graph highlights the attainment in secondary schools based on income levels (low-, middle-, and high-income countries):

1. **High-Income Countries**:
    - Nearly universal access to and completion of secondary education by 2022.
    - High-quality education systems ensure consistent performance, supported by well-qualified teachers and advanced curricula.
2. **Middle-Income Countries**:
    - Steady growth in secondary school attainment, narrowing the gap with high-income nations.
    - Urban centers tend to perform better than rural areas, where teacher shortages and infrastructure deficits persist.

- Education reform programs, including digital initiatives, have helped in improving outcomes in several countries.

3. **Low-Income Countries**:
   - Persistent challenges result in the lowest secondary school attainment rates.
   - Key barriers include inadequate school infrastructure, underqualified teachers, and high dropout rates, particularly at the transition to upper secondary education.
   - Despite these challenges, there is modest improvement in enrollment and completion rates, supported by international aid and local initiatives.

Figure 5. Upper Secondary Schools by regions (a) and income level (b)

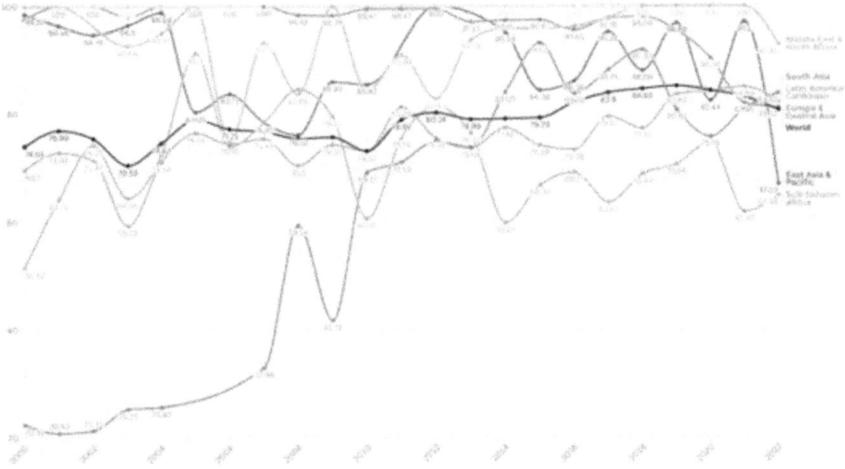

(a)

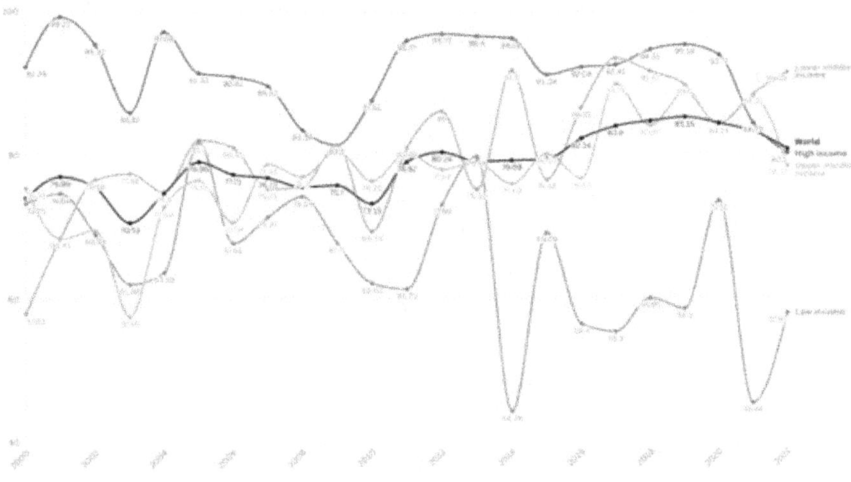

(b)

Figure 5 reveals that improvements in both enrollment and educational attainment are visible in most regions, but disparities are stark in regions with political instability. Consistently higher achievements in high-income nations, while the rate of improvement in low-income countries remains slow. Figure 5 highlights significant disparities in upper secondary school attainment across regions and income levels. High-income regions achieve near-universal access, benefiting from robust policies and resources. Middle-income countries show progress but face challenges in equity and quality, particularly in rural areas. Low-income countries, particularly in Sub-Saharan Africa and parts of South Asia, struggle with low attainment rates due to financial, cultural, and infrastructural barriers. These findings underline the importance of targeted investments in upper secondary education, focusing on access, teacher training, and equitable opportunities for underserved populations.

(a) Regional Trends:

The graph showcases upper secondary school attainment across various regions from 2000 to 2022:

1. **High-Performing Regions**:
    - **North America and Europe**: These regions consistently show near-universal upper secondary school attainment, reflecting strong educational systems, mandatory secondary schooling policies, and robust support for students transitioning from lower to upper secondary levels.
    - **East Asia and Pacific**: Marked improvements are evident, with many countries closing the attainment gap through significant investments in secondary education infrastructure and policies promoting access.
2. **Moderate-Performing Regions**:
    - **Latin America and the Caribbean**: While there has been substantial progress, challenges persist in ensuring equity, particularly for students from rural and low-income households. Urban areas demonstrate higher attainment rates compared to rural regions.
    - **Middle East and North Africa**: Gradual improvements are observed, though disparities in gender and access remain in some countries, particularly in conflict-affected areas. Countries with stable economies have achieved notable progress.
3. **Low-Performing Regions**:
    - **Sub-Saharan Africa**: The region has the lowest upper secondary school attainment rates globally. While there has been modest improvement, access remains highly limited due to poverty, insufficient infrastructure, and a lack of qualified teachers. Many students drop out after lower secondary education.
    - **South Asia**: Moderate progress is evident, with increasing enrollment rates. However, gender disparities and socio-economic challenges, such as child labor and early marriage, still limit access for marginalized groups.

(b) Income-Level Trends:

The graph compares upper secondary school attainment across low-, middle-, and high-income countries:

1. **High-Income Countries**:
    - Consistently high attainment rates are observed, with well-established education systems ensuring universal access to upper secondary education.
    - Advanced curricula, highly qualified teachers, and strong governmental support ensure both access and quality.
2. **Middle-Income Countries**:
    - There has been significant growth in upper secondary enrollment and attainment, driven by policies promoting educational access and economic development.
    - Disparities persist between urban and rural areas, and some countries face challenges in ensuring equitable access to quality education.
3. **Low-Income Countries**:
    - Upper secondary attainment remains the lowest among income groups. Key barriers include financial constraints, inadequate infrastructure, and a lack of qualified teachers.
    - Many students drop out after completing lower secondary education due to economic pressures and limited access to upper secondary schools, particularly in rural areas.
    - Despite these challenges, modest improvements are observed, reflecting efforts by governments and international organizations to expand secondary education access.

CONCLUSION

This chapter's analysis underscores the significant disparities in teacher qualifications across regions and income levels, highlighting the urgent need for targeted interventions. By adopting the strategies outlined in this chapter, policymakers, researchers, and educators can work collaboratively to enhance the supply of competent teachers in developing countries. These efforts will not only contribute to achieving SDG 4 but also support broader sustainable development objectives, ensuring that all children, regardless of their circumstances, have access to quality education. Accordingly, the global dataset indicates variations in teacher qualifications across continents and income levels. While notable progress has been achieved in educational attainment and teacher qualifications globally, significant disparities persist. Low-income countries and regions face the most pronounced challenges in both student attainment and access to qualified teachers. Middle-income nations have shown consistent progress but still require targeted interventions to bridge gaps with high-income countries. Enhanced investment in teacher training and capacity-building programs remains critical for equitable global educational advancement.

Addressing the teacher shortage in developing countries requires regionally informed, data-driven, and inclusive strategies. Real-world case studies, targeted technology adoption, and gender-sensitive policies provide clear roadmaps for effective interventions. Competent educators are essential for attaining high-quality education. They enhance student performance, diminish educational disparities, and foster economic and social advancement. In several emerging nations, substantial disparities persist in teacher credentials. These discrepancies are accentuated by structural problems, such as insufficient teacher training programs, minimal investment in education, and restricted access to resources. The 2030 Agenda for Sustainable Development highlights the critical role of quality education in achieving sustainable development. Target 4.c and its performance indicator 4.c.1 provide a framework for addressing the global teacher shortage and improving teacher qualifications. SDG 4 emphasizes quality education as a cornerstone for sustainable development. Target 4.c focuses specifically on addressing the global teacher shortage and improving teacher qualifications. The 4.c.1 indicator measures progress by assessing the percentage of teachers who meet the minimum required qualifications, disaggregated by education level and sex. This metric provides critical insights into the state of teacher qualifications and helps identify gaps that need to be addressed. The analysis reveals significant disparities in teacher qualifications across regions and income levels. High-income nations often possess a greater percentage of trained educators, indicative of substantial investment in education and effective teacher training frameworks. Conversely, low-income nations frequently encounter difficulties in achieving even fundamental qualification criteria owing to structural obstacles, including insufficient money, poor infrastructure, and restricted access to professional development opportunities. Finally, with the basis of global dataset it is possible to summarize the findings obtained for teacher qualifications worldwide as follows:

- **Proportion of Qualified Teachers**:
  - Higher qualifications observed in high-income countries, where continuous professional development and advanced certifications are common.
  - Middle-income countries show significant progress, with programs aimed at improving teacher competence.
  - Low-income nations face challenges due to lack of resources, training infrastructure, and support for professional growth.
- **Key Trends and Disparities**:
  - High-income countries have nearly universal access to qualified teachers across all education levels.

- Middle-income regions demonstrate variability, with some achieving parity in lower and upper secondary education.
- Low-income areas show the greatest need for intervention, with large proportions of unqualified or underqualified teachers, particularly in pre-primary and lower-secondary education.

By embedding practical frameworks, validated research, and technology-enhanced strategies, this chapter aims to serve as a strategic resource for policymakers designing education reforms, educators managing systemic challenges, and researchers developing evidence-based teacher development models. Ensuring that these stakeholders can translate theory into actionable policy and practice is critical to achieving sustainable progress in teacher quality worldwide.

Addressing teacher shortages in developing countries is not only essential for achieving SDG 4 but also fundamental to broader sustainable development efforts. This chapter has demonstrated that a combination of financial investment, institutional reform, and innovative approaches can drive real change. Policymakers must act with urgency to implement tailored, evidence-based policies. Educators and communities should be empowered to take ownership of localized solutions. International organizations must continue to support capacity-building through partnerships and funding. Only through coordinated, cross-sectoral efforts can we ensure that every child has access to competent, motivated, and well-supported teachers. The time to act is now.

Recommendations and Practical Implications

Based on the findings, the following strategies and policies are recommended to improve the supply of competent teachers in developing countries:

Strengthening Teacher Training Programs

- **Curriculum Development:** Develop comprehensive teacher training curricula aligned with international standards.
- **Capacity Building:** Invest in teacher training institutions to enhance their capacity to deliver high-quality programs.
- **Technology Integration:** Leverage digital platforms to provide accessible and cost-effective training, particularly in remote areas.

Promoting International Cooperation

- **Knowledge Sharing:** Facilitate the exchange of best practices and expertise between developed and developing countries.
- **Funding Mechanisms:** Establish international funding mechanisms to support teacher training initiatives in low-income countries.
- **Partnerships:** Strengthen partnerships between governments, NGOs, and international organizations to address teacher shortages.

Addressing Gender Disparities

- **Empowering Women:** Implement policies to encourage women's participation in the teaching profession, including scholarships and mentorship programs.
- **Cultural Sensitization:** Address cultural barriers that limit women's access to education and professional opportunities.
- **Workplace Support:** Create supportive workplace environments, including childcare facilities and flexible work arrangements, to attract and retain female teachers.

Enhancing Data Collection and Monitoring

- **Improved Metrics:** Develop more robust indicators to measure teacher qualifications and identify gaps.
- **Regular Reporting:** Establish systems for regular data collection and reporting to track progress and inform policy decisions.
- **Disaggregated Data:** Ensure data is disaggregated by gender, education level, and region to address specific challenges effectively.

To support policymakers and practitioners, a practical implementation roadmap is essential. Governments can adopt phased teacher training expansion models, starting with high-need regions using low-cost digital platforms. Pilot programs can be rolled out in rural districts, monitored through community engagement metrics, and scaled based on outcomes. School leaders can use retention toolkits, including peer mentorship templates and classroom management resource packs, adapted from successful models in Vietnam and Rwanda. These actions can facilitate structured, replicable, and context-sensitive interventions across diverse education systems.

REFERENCES

Abramo, G., D'Angelo, C. A., & Costa, F. D. (2018). A nation's foreign and domestic professors: Which have better research performance? (the italian case). *Higher Education, 77*(5), 917–930. DOI: 10.1007/s10734-018-0310-x

Acheampong, P., & Gyasi, J. F. (2019). Teacher retention: A review of policies for motivating rural basic school teachers in ghana. *Asian Journal of Education and Training, 5*(1), 86–92. DOI: 10.20448/journal.522.2019.51.86.92

Ankomah, W. S. (2020). Diverse classrooms: Social justice, equity, and diversity competencies for teacher candidates. *World Journal of Education, 10*(2), 15. DOI: 10.5430/wje.v10n2p15

Annual impact review 2021-22. Education Development Trust. (n.d.). https://www.edt.org/research-and-insights/annual-impact-review-202122/

Arnup, J. L., & Bowles, T. (2016). Should i stay or should i go? resilience as a protective factor for teachers' intention to leave the teaching profession. *Australian Journal of Education, 60*(3), 229–244. DOI: 10.1177/0004944116667620

Basilotta-Gómez-Pablos, V., Matarranz, M., Casado-Aranda, L., & Otto, A. (2022). Teachers' digital competencies in higher education: A systematic literature review. *International Journal of Educational Technology in Higher Education, 19*(1), 8. Advance online publication. DOI: 10.1186/s41239-021-00312-8

Berman, M., & DeFeo, D. J. (2023). Equitable compensation to attract and retain qualified teachers in high-need Alaska public schools. *Educational Policy, 38*(5), 1139–1175. DOI: 10.1177/08959048231174883

Bernal, G. L., Abadía, L. K., Álvarez-Arango, L. E., & De Witte, K. (2023). Financial aid uncertainty and low-income students' higher education preferences. *Higher Education, 87*(6), 1845–1863. DOI: 10.1007/s10734-023-01094-w

Boyadjieva, P. (2017). Invisible higher education: Higher education institutions from central and eastern europe in global rankings. *European Educational Research Journal, 16*(5), 529–546. DOI: 10.1177/1474904116681016

Bruggeman, B., Hidding, K., Struyven, K., Pynoo, B., Garone, A., & Tondeur, J. (2022). Negotiating teacher educators' beliefs about blended learning: Using stimulated recall to explore design choices. *Australasian Journal of Educational Technology*, •••, 100–114. DOI: 10.14742/ajet.7175

Bukko, D., & Liu, K. (2021). Developing preservice teachers' equity consciousness and equity literacy. *Frontiers in Education*, *6*, 586708. Advance online publication. DOI: 10.3389/feduc.2021.586708

Bulfin, S., Henderson, M., Johnson, N. F., & Selwyn, N. (2014). Methodological capacity within the field of "educational technology" research: An initial investigation. *British Journal of Educational Technology*, *45*(3), 403–414. DOI: 10.1111/bjet.12145

Camphuijsen, M. K. (2020). Coping with performance expectations: Towards a deeper understanding of variation in school principals' responses to accountability demands. *Educational Assessment, Evaluation and Accountability*, *33*(3), 427–453. DOI: 10.1007/s11092-020-09344-6

Cattaneo, M., Malighetti, P., & Paleari, S. (2018). The Italian brain drain: Cream and milk. *Higher Education*, *77*(4), 603–622. DOI: 10.1007/s10734-018-0292-8

Chankseliani, M., & McCowan, T. (2020). Higher education and the sustainable development goals. *Higher Education*, *81*(1), 1–8. DOI: 10.1007/s10734-020-00652-w

Charitonos, K., . . . Aristorenas, A. (2022). Researching communicative approaches to English language teaching using peer ethnographic methods in Jordan's refugee settings. https://inee.org/sites/default/files/resources/TiCC%20Case%20Studies%20v2%20-%20TPD%20v1.0%20EN%20LowRes.pdf

Chen, A., Li, W., & Fu, W. (2024). Unleashing digital superheroes: Unravelling the empathy factor in digital competence and online teacher autonomy support. *British Journal of Educational Technology*, *55*(4), 1790–1810. DOI: 10.1111/bjet.13433

Chen, Y., Cao, L., Guo, L., & Cheng, J. (2022). Driving is believing: Using telepresence robots to access makerspace for teachers in rural areas. *British Journal of Educational Technology*, *53*(6), 1956–1975. DOI: 10.1111/bjet.13225

Compen, B., Witte, K. D., & Schelfhout, W. (2020). The impact of teacher engagement in an interactive webinar series on the effectiveness of financial literacy education. *British Journal of Educational Technology*, *52*(1), 411–425. DOI: 10.1111/bjet.13013

Crosling, G., Nair, M., & Vaithilingam, S. (2014). A creative learning ecosystem, quality of education and innovative capacity: A perspective from higher education. *Studies in Higher Education*, *40*(7), 1147–1163. DOI: 10.1080/03075079.2014.881342

Dai, C., Ke, F., Dai, Z., & Pachman, M. (2022). Improving teaching practices via virtual reality-supported simulation-based learning: Scenario design and the duration of implementation. *British Journal of Educational Technology*, *54*(4), 836–856. DOI: 10.1111/bjet.13296

Dalal, M., Archambault, L., & Shelton, C. (2020). Fostering the growth of TPACK among international teachers of developing nations through a cultural exchange program. *Australasian Journal of Educational Technology*, •••, 43–56. DOI: 10.14742/ajet.5964

Darling-Hammond, L. (2023). Response to how teacher education matters. *Journal of Teacher Education*, *74*(2), 157–159. DOI: 10.1177/00224871231160378

Del Bonifro, F., Gabbrielli, M., Lisanti, G., & Zingaro, S. P. (2020). Student dropout prediction. *Lecture Notes in Computer Science*, *12163*, 129–140. DOI: 10.1007/978-3-030-52237-7_11

Deng, R., & Gao, Y. (2024). Practices and barriers to promoting educational equity for migrant children in china from the perspective of a local government. *SAGE Open*, *14*(3), 21582440241269979. Advance online publication. DOI: 10.1177/21582440241269979

Federičová, M. (2020). Teacher turnover: What can we learn from europe? *European Journal of Education*, *56*(1), 102–116. DOI: 10.1111/ejed.12429

Fernández Ugalde, R. (2024). Media articulation and the stabilisation of teacher policies in Chile. *Critical Studies in Education*, •••, 1–20. DOI: 10.1080/17508487.2024.2428291

Florian, L., & Camedda, D. (2019). Enhancing teacher education for inclusion. *European Journal of Teacher Education*, *43*(1), 4–8. DOI: 10.1080/02619768.2020.1707579

Foster, M., Mark, S. L., & Baize, J. (2023). Introduction special issue: educational equity: cultural and ethnic diversity in schools. *Education Sciences*, *14*(1), 38. DOI: 10.3390/educsci14010038

Fuller, E. J. (2013). Shaky methods, shaky motives. *Journal of Teacher Education*, *65*(1), 63–77. DOI: 10.1177/0022487113503872

Furukawa, T., Shirakawa, N., & Okuwada, K. (2012). An empirical study of graduate student mobility underpinning research universities. *Higher Education*, *66*(1), 17–37. DOI: 10.1007/s10734-012-9586-4

Gaikhorst, L., Beishuizen, J., Zijlstra, B. J. H., & Volman, M. (2014). Contribution of a professional development programme to the quality and retention of teachers in an urban environment. *European Journal of Teacher Education*, *38*(1), 41–57. DOI: 10.1080/02619768.2014.902439

Gaikhorst, L., Post, J., März, V., & Soeterik, I. (2019). Teacher preparation for urban teaching: A multiple case study of three primary teacher education programmes. *European Journal of Teacher Education*, *43*(3), 301–317. DOI: 10.1080/02619768.2019.1695772

Gao, C., Khalid, S., & Tadesse, E. (2023). Understanding school enrollment in the free education era: Roadblocks to meeting the sustainability development goal. *European Journal of Education*, *59*(2), e12600. Advance online publication. DOI: 10.1111/ejed.12600

Gong, X., & Huybers, T. (2015). Chinese students and higher education destinations: Findings from a choice experiment. *Australian Journal of Education*, *59*(2), 196–218. DOI: 10.1177/0004944115584482

Goodwin, A. L., Smith, L., Souto-Manning, M., Cheruvu, R., Tan, M. Y., Reed, R., & Taveras, L. (2014). What should teacher educators know and be able to do? perspectives from practicing teacher educators. *Journal of Teacher Education*, *65*(4), 284–302. DOI: 10.1177/0022487114535266

Grissom, J. A., & Strunk, K. O. (2011). How should school districts shape teacher salary schedules? linking school performance to pay structure in traditional compensation schemes. *Educational Policy*, *26*(5), 663–695. DOI: 10.1177/0895904811417583

Halász, G., & Looney, J. (2019). Teacher professional competences and standards. concepts and implementation. *European Journal of Education*, *54*(3), 311–314. DOI: 10.1111/ejed.12351

Han, S. W., & Borgonovi, F. (2020). Students' teaching career expectations by gender and ability levels in science and math: The role of salary and numeracy skills. *European Journal of Education*, *55*(3), 405–427. DOI: 10.1111/ejed.12407

Han, S. W., Borgonovi, F., & Guerriero, S. (2017). What motivates high school students to want to be teachers? the role of salary, working conditions, and societal evaluations about occupations in a comparative perspective. *American Educational Research Journal*, *55*(1), 3–39. DOI: 10.3102/0002831217729875

Hanushek, E. A., & Rivkin, S. G. (2010). Generalizations about Using Value-Added Measures of Teacher Quality. *The American Economic Review*, *100*(2), 267–271. DOI: 10.1257/aer.100.2.267

Hanušová, S., Píšová, M., Kohoutek, T., Minaříková, E., Ježek, S., Janík, T., Mareš, J., & Janík, M. (2020). Novice teachers in the czech republic and their drop-out intentions. *European Journal of Education*, *55*(2), 275–291. DOI: 10.1111/ejed.12373

Hardman, F., Hardman, J., Dachi, H. A., Elliott, L., Ihebuzor, N., Ntekim, M., & Tibuhinda, A. (2015). Implementing school-based teacher development in tanzania. *Professional Development in Education*, *41*(4), 602–623. DOI: 10.1080/19415257.2015.1026453

Haynes, J. D., Marsh, L. T. S., & Anderson, K. (2023). Planting the seeds of culturally responsive, equity-centered, and trauma-informed attitudes among urban educators. *Urban Education*, *60*(3), 821–853. DOI: 10.1177/00420859231175663

Heffernan, A., Bright, D., Kim, M., Longmuir, F., & Magyar, B. (2022). 'I cannot sustain the workload and the emotional toll': Reasons behind Australian teachers' intentions to leave the profession' intentions to leave the profession. *Australian Journal of Education*, *66*(2), 196–209. DOI: 10.1177/00049441221086654

Heinrich, C. J., Darling-Aduana, J., & Martin, C. (2019). The potential and prerequisites of effective tablet integration in rural kenya. *British Journal of Educational Technology*, *51*(2), 498–514. DOI: 10.1111/bjet.12870

Hertz, B., Clemson, H. G., Hansen, D. T., Laurillard, D., Murray, M., Fernandes, L., & Rutkauskiene, D. (2022). A pedagogical model for effective online teacher professional development—Findings from the teacher academy initiative of the European commission. *European Journal of Education*, *57*(1), 142–159. DOI: 10.1111/ejed.12486

Huang, L., Li, S., Poitras, E., & Lajoie, S. P. (2020). Latent profiles of self-regulated learning and their impacts on teachers' technology integration. *British Journal of Educational Technology*, *52*(2), 695–713. DOI: 10.1111/bjet.13050

Jugović, I., Marušić, I., Ivanec, T. P., & Vidović, V. V. (2012). Motivation and personality of preservice teachers in croatia. *Asia-Pacific Journal of Teacher Education*, *40*(3), 271–287. DOI: 10.1080/1359866X.2012.700044

Kafyulilo, A., Fisser, P., Pieters, J. M., & Voogt, J. (2015). Ict use in science and mathematics teacher education in tanzania: Developing technological pedagogical content knowledge. *Australasian Journal of Educational Technology*, *31*(4). Advance online publication. DOI: 10.14742/ajet.1240

Karakhanyan, S., & Veen, K. (2011). Teachers' voices in the context of higher education reforms in armenia. *European Journal of Education*, *46*(4), 508–523. DOI: 10.1111/j.1465-3435.2011.01497.x

Khan, J. (2021). European academic brain drain: A meta-synthesis. *European Journal of Education*, *56*(2), 265–278. DOI: 10.1111/ejed.12449

Khoo, Y. (2021). Becoming globally competent through inter-school reciprocal learning partnerships: An inquiry into canadian and chinese teachers' narratives. *Journal of Teacher Education*, *73*(1), 110–122. DOI: 10.1177/00224871211042306

Kılınç, A., Watt, H. M. G., & Richardson, P. (2012). Factors influencing teaching choice in turkey. *Asia-Pacific Journal of Teacher Education*, *40*(3), 199–226. DOI: 10.1080/1359866X.2012.700048

Kirksey, J. J. (2021). Preparedness and experiences of novice teachers in the sociopolitical context of heightened immigration enforcement: Evidence from a survey of california teachers. *Journal of Teacher Education*, *73*(1), 37–51. DOI: 10.1177/00224871211045980

Kraft, M. A., & Papay, J. P. (2014). Can Professional Environments in Schools Promote Teacher Development? Explaining Heterogeneity in Returns to Teaching Experience. *Educational Evaluation and Policy Analysis*, *36*(4), 476–500. DOI: 10.3102/0162373713519496

Lakkala, M., & Ilomäki, L. (2015). A case study of developing ict-supported pedagogy through a collegial practice transfer process. Computers &Amp. *Computers & Education*, *90*, 1–12. DOI: 10.1016/j.compedu.2015.09.001

Lee, J., & Fuller, B. (2020). Does progressive finance alter school organizations and raise achievement? the case of Los Angeles. *Educational Policy*, *36*(3), 587–623. DOI: 10.1177/0895904820901472

Ma, N., Du, L., & Lu, Y. (2022). A model of factors influencing in-service teachers' social network prestige in online peer assessment. *Australasian Journal of Educational Technology*, •••, 100–118. DOI: 10.14742/ajet.7622

Mabokela, R. O., & Mlambo, Y. A. (2014). "the older women are men:" navigating the academic terrain, perspectives from ghana. *Higher Education*, *69*(5), 759–778. DOI: 10.1007/s10734-014-9804-3

MacPhail, A., Ulvik, M., Guberman, A., Czerniawski, G., Oolbekkink-Marchand, H., & Bain, Y. (2018). The professional development of higher education-based teacher educators: Needs and realities. *Professional Development in Education*, *45*(5), 848–861. DOI: 10.1080/19415257.2018.1529610

Main, S., & Slater, E. (2021). Online continuous professional learning: A model for improving reading outcomes in regional and remote schools? *Journal of Teacher Education*, *73*(2), 201–214. DOI: 10.1177/00224871211009110

Maina, M. F., Ortiz, L. G., Mancini, F., & Melo, M. M. (2022). A micro-credentialing methodology for improved recognition of HE employability skills. *International Journal of Educational Technology in Higher Education, 19*(1), 10. Advance online publication. DOI: 10.1186/s41239-021-00315-5

Mangan, J., Hughes, A., & Slack, K. (2010). Student finance, information and decision making. *Higher Education, 60*(5), 459–472. DOI: 10.1007/s10734-010-9309-7

Martell, C. C. (2019). Barriers to inquiry-based instruction: A longitudinal study of history teachers. *Journal of Teacher Education, 71*(3), 279–291. DOI: 10.1177/0022487119841880

Martínez, J. G., Martí, M. C., & Cervera, M. G. (2019). Inside a 3d simulation: Realism, dramatism and challenge in the development of students' teacher digital competence. *Australasian Journal of Educational Technology*. Advance online publication. DOI: 10.14742/ajet.3885

McAleavy, T., Hall-Chen, A., Horrocks, S., & Riggall, A. (2022). Technology-supported professional development for teachers: lessons from developing countries. https://edtlive.b-cdn.net/live/media/egshqxa0/technology-supported-professional-development-for-teachers.pdf

McChesney, K., & Aldridge, J. M. (2018). A review of practitioner-led evaluation of teacher professional development. *Professional Development in Education, 45*(2), 307–324. DOI: 10.1080/19415257.2018.1452782

Mensah, F. M. (2019). Finding voice and passion: Critical race theory methodology in science teacher education. *American Educational Research Journal, 56*(4), 1412–1456. DOI: 10.3102/0002831218818093

Mouboua, P. D., Atobatele, F. A., & Akintayo, O. T. (2024). Multilingual education and social equity: a comparative study of integration policies in multicultural societies. GSC Advanced Research and Reviews, 19(2), 032-042. DOI: 10.30574/gscarr.2024.19.2.0165

Muremela, M. G., Kutame, A. P., Kapueja, I. S., & Ajani, O. A.Matodzi Grace MuremelaAzwidohwi Phillip KutameIngrid Sibongile KapuejaOluwatoyin Ayodele Ajani. (2024). Optimising the recruitment of teachers for scarce skills subjects in rural secondary schools: A study of practices in african countries. *International Journal of Innovative Technologies in Social Science*, (4(44)). Advance online publication. DOI: 10.31435/ijitss.4(44).2024.3116

Nguyen, D., & Ng, D. (2020). Teacher collaboration for change:sharing, improving, and spreading. *Professional Development in Education*, *46*(4), 638–651. DOI: 10.1080/19415257.2020.1787206

Novakovic, A., Michel, R. E., & Ockerman, M. S. (2020). Teaching school counselors to use evidence-based practice to advance social justice: A case study. *Professional School Counseling*, *23*(1_part_3), 2156759X20904472. Advance online publication. DOI: 10.1177/2156759X20904472

Ono, Y., & Ferreira, J. (2010). A case study of continuing teacher professional development through lesson study in South Africa. *South African Journal of Education*, *30*(1), 59–75. DOI: 10.15700/saje.v30n1a320

Ovcharuk, O., Ivaniuk, I., & Leshchenko, M. (2023). Impact of school lockdown on access to online instruction during the war in ukraine. *European Journal of Education*, *58*(4), 561–574. DOI: 10.1111/ejed.12589

Parker, L., Xu, S., & Chi, C. (2022). Chinese preservice teachers' perspectives of mentoring relationships in an international learning partnership. *Journal of Teacher Education*, *73*(5), 525–537. DOI: 10.1177/00224871221108655

Peeraer, J., & Petegem, P. V. (2012). The limits of programmed professional development on integration of information and communication technology in education. *Australasian Journal of Educational Technology*, *28*(6). Advance online publication. DOI: 10.14742/ajet.809

Perifanou, M., Tzafilkou, K., & Economides, A. A. (2022). Teacher intention to transfer ict training when integrating digital technologies in education: The teacher transfer of ict training model (tetra-ict). *European Journal of Education*, *58*(1), 111–129. DOI: 10.1111/ejed.12534

Perna, L. W., Walsh, E., & Fester, R. (2010). Promoting the educational attainment of adults. *Educational Policy*, *25*(6), 935–963. DOI: 10.1177/0895904810386599

Perry, E., & Bevins, S. (2018). Building capacity for professional development: The development of teachers as facilitators in ghana. *Professional Development in Education*, *45*(3), 390–404. DOI: 10.1080/19415257.2018.1474489

Philip, T. M. (2011). Moving beyond our progressive lenses. *Journal of Teacher Education*, *62*(4), 356–366. DOI: 10.1177/0022487111409414

Raikes, A. (2017). Measuring child development and learning. *European Journal of Education*, *52*(4), 511–522. DOI: 10.1111/ejed.12249

Rice, J. K., Malen, B., Baumann, P., Chen, E., Dougherty, A., Hyde, L., Jackson, C., Jacobson, R., & McKithen, C. (2012). The persistent problems and confounding challenges of educator incentives. *Educational Policy*, *26*(6), 892–933. DOI: 10.1177/0895904812465708

Saigal, A., Kogekar, U., & Kong, P. (2022). Sustaining student learning in India through quality teacher support during the COVID-19 pandemic. https://inee.org/sites/default/files/resources/TiCC%20Case%20Studies%20v2%20-%20TPD%20v1.0%20EN%20LowRes.pdf

Smith, K., & Flores, M. A. (2019). The janus faced teacher educator. *European Journal of Teacher Education*, *42*(4), 433–446. DOI: 10.1080/02619768.2019.1646242

Snoek, M., Dengerink, J., & Wit, B. (2019). Reframing the teacher profession as a dynamic multifaceted profession: A wider perspective on teacher quality and teacher competence frameworks. *European Journal of Education*, *54*(3), 413–425. DOI: 10.1111/ejed.12350

Snoek, M., Swennen, A., & Klink, M. (2011). The quality of teacher educators in the European policy debate: Actions and measures to improve the professionalism of teacher educators. *Professional Development in Education*, *37*(5), 651–664. DOI: 10.1080/19415257.2011.616095

Taddese, E. T., & Rao, C. (2022). School-based continuous professional development of teachers: a case study of primary school teachers in Ethiopia. Education 3-13, 50(8), 1059-1071. https://doi.org/DOI: 10.1080/03004279.2021.1929382

Thomassen, W. E., & Munthe, E. (2020). Educating norwegian preservice teachers for the multicultural classroom – what knowledge do student teachers and mentor teachers express? *European Journal of Teacher Education*, *44*(2), 234–248. DOI: 10.1080/02619768.2020.1758661

Tran, H., & Smith, D. A. (2021). How hard-to-staff rural school districts use state funds to address teacher shortages. *Journal of Education Finance*, *47*(2), 130–156. DOI: 10.1353/jef.2021.a846294

UNESCO. (2023). *Global Education Monitoring Report 2023: Technology in Education – A Tool on Whose Terms?* UNESCO.

UNESCO. (2023). Central Asian Experts Discuss Reforms of Teacher Policies to Improve Educational Outcomes in the Region. https://www.unesco.org/en/articles/central-asian-experts-discuss-reforms-teacher-policies-improve-educational-outcomes-region

Varadharajan, M., Buchanan, J., & Schuck, S. (2019). Navigating and negotiating: Career changers in teacher education programmes. *Asia-Pacific Journal of Teacher Education*, *48*(5), 477–490. DOI: 10.1080/1359866X.2019.1669136

Viberg, O., Kizilcec, R. F., Wise, A. F., Jivet, I., & Nixon, N. (2024). Advancing equity and inclusion in educational practices with ai-powered educational decision support systems (ai-edss). *British Journal of Educational Technology*, *55*(5), 1974–1981. DOI: 10.1111/bjet.13507

Wang, K. (2016). Reforming a university during political transformation: A case study of Yangon University in myanmar. *Studies in Higher Education*, *43*(7), 1184–1195. DOI: 10.1080/03075079.2016.1239250

White, E. L., Dickerson, C., & Weston, K. M. (2015). Developing an appreciation of what it means to be a school-based teacher educator. *European Journal of Teacher Education*, *38*(4), 445–459. DOI: 10.1080/02619768.2015.1077514

Widegren, P., & Doherty, C. (2010). Is the world their oyster? the global imagination of pre-service teachers. *Asia-Pacific Journal of Teacher Education*, *38*(1), 5–22. DOI: 10.1080/13598660903474155

Wolf, J., Casado-Muñoz, R., & Pedone, F. (2022). Partnership of schools and civil society organisations to support education of students of varied linguistic backgrounds—The situation in the Czech republic, Italy and Spain. *European Journal of Education*, *57*(4), 674–691. DOI: 10.1111/ejed.12524

Wolf, S., Aber, J. L., Behrman, J. R., & Tsinigo, E. (2018). Experimental impacts of the "quality preschool for ghana" interventions on teacher professional well-being, classroom quality, and children's school readiness. *Journal of Research on Educational Effectiveness*, *12*(1), 10–37. DOI: 10.1080/19345747.2018.1517199

Wolfenden, F. (2023). Adapting and scaling teacher professional development approaches in Ghana, Honduras and Uzbekistan. Technical Report, http://hdl.handle.net/10625/61824

Xu, D., Glick, D., Rodríguez, F., Cung, B., Li, Q., & Warschauer, M. (2019). Does blended instruction enhance English language learning in developing countries? evidence from mexico. *British Journal of Educational Technology*, *51*(1), 211–227. DOI: 10.1111/bjet.12797

Yeh, Y. F., Hsu, Y. S., Wu, H. K., Hwang, F. K., & Lin, T. (2013). Developing and validating technological pedagogical content knowledge-practical (tpack-practical) through the delphi survey technique. *British Journal of Educational Technology*, *45*(4), 707–722. DOI: 10.1111/bjet.12078

Chapter 8
Business Agility and Sustainability:
Supply Chain Management Strategies and Environmentally Friendly Innovations for Logistics Companies

Rini Setiawati
https://orcid.org/0009-0009-1104-175X
Binus University, Indonesia

Hasri Nirmala
Darma Persada University, Indonesia

Puji Rahayu
Gunadarma University, Indonesia

ABSTRACT

The logistics industry plays a crucial role in connecting production, distribution, and consumption globally, yet faces significant challenges such as the Sustainability vs. Conventional Business Growth Dilemma, vulnerable supply chains, environmental impacts, shifting consumer demands, regulatory pressures, and corporate social responsibility imperatives. Amidst these challenges, opportunities emerge through Business Agility, Green Supply Chain Management, and Green Innovation. This study examines how these three factors interact to support logistics sustainability and align with the Sustainable Development Goals (SDGs). Data was collected from 200 middle-to-upper management employees of a logistics company in Jakarta and analyzed using descriptive methods and SmartPLS. The findings reveal that Business

DOI: 10.4018/979-8-3373-1300-9.ch008

Agility, Green Supply Chain Management, and Green Innovation have significant positive impacts on sustainability performance, highlighting the importance of integrating these strategies to foster adaptive business models, environmental stewardship, and sustainable innovation.

INTRODUCTION

Company sustainability in the modern era no longer relies solely on financial profits alone, but must also consider the resulting environmental and social impacts. In this view, the concepts of Business Agility, Green Supply Chain Management, and Green Innovation are the three main pillars that encourage company sustainability with a sustainable and environmentally responsible approach.

Rapid global development, including population growth, urbanization, and increased consumption, has placed enormous pressure on the environment and natural resources. The negative impacts of conventional business activities on the environment are increasingly clear, including air pollution, global warming, decreasing water quality and damage to ecosystems. Amidst increasing awareness of environmental issues, corporate sustainability is more important than ever. Businesses that only prioritize financial growth without considering environmental and social impacts often become part of the problem rather than the solution. Conventional business models often ignore the concept of sustainability and only focus on short-term profits. As a result, many companies face reputational risks, market attrition, and even regulatory issues by not adapting to this paradigm shift.

Complex and fragile global supply chains have become the engine of global economic development. However, these supply chains often result in large carbon emissions, excess waste, and other negative impacts on the environment. Unsustainable suppliers can present operational risks, unstable supply and potential financial loss. Modern consumers are increasingly aware of the environmental impact of their purchases. They tend to support companies that are committed to environmentally friendly business practices. This is what drives companies to change their business strategies to be in line with the environmental values valued by consumers. Increasingly stringent regulatory changes related to the environment and sustainability are pushing companies to change their operations to comply with new standards. Additionally, corporate social responsibility is increasingly becoming a focus, requiring companies to not only generate profits, but also make positive contributions to society and the environment. Green innovation has opened up new opportunities for companies to create value in a sustainable way. Companies that are able to develop environmentally friendly products and technologies can change the market and gain competitive advantage.

In order to answer these challenges and opportunities, the concepts of Business Agility, Green Supply Chain Management, and Green Innovation emerged as strategic solutions that enable companies to face the future responsively and responsibly. Through the integration of these three concepts, companies can create business models that are able to adapt quickly to change, manage environmental impacts effectively, and produce sustainable innovation that supports long-term sustainability.

Business Agility refers to a company's ability to respond to market changes quickly and effectively. In the context of the logistics business, this agility is very important considering the high dynamics and uncertainty in the global supply chain. Competitive advantage is often determined by how quickly and appropriately companies can adapt their operations to changes in demand, supply disruptions, and technological innovation. The ability to quickly adjust strategies and operations allows logistics companies to better meet customer needs, keep customers satisfied, and maintain their loyalty (Teece et al., 2016). Research demonstrates that organizational agility serves as a critical determinant of competitive advantage in global markets. Agile organizations can respond faster to changing market conditions, leading to increased innovation and higher customer satisfaction through better resource management, process optimization, and enhanced operational efficiency (Kocot et al., 2024)

Green Supply Chain Management (GSCM) involves the integration of environmentally friendly practices into supply chain management. This includes reducing emissions, managing waste and using more environmentally friendly raw materials. In the logistics industry, implementing GSCM is critical for business sustainability because GSCM helps reduce the carbon footprint and environmental impact of logistics activities, such as transportation and warehousing (Srivastava, 2007). Additionally, adopting green practices helps companies meet increasingly stringent environmental standards and regulations, avoid fines and improve corporate reputation (Zhu et al., 2008). Environmentally friendly practices often also lead to cost savings through more efficient energy use, reduced waste, and increased operational productivity (Purba Rao, 2005).

Green Innovation in a logistics context includes the development and implementation of new technologies and processes that reduce the environmental impact of logistics operations. Implementing new technologies such as electric vehicles, alternative fuels and energy management systems can reduce emissions and increase operational efficiency (Kumar & Anbanandam, 2020). Additionally, green innovation can be a key differentiator in the market, attracting environmentally conscious customers and enhancing brand image (Y. S. Chen et al., 2006). Sustainable innovation creates new opportunities for business growth by meeting the demands of increasingly environmentally conscious consumers and driving long-term sustainability (Porter & Kramer, 2011).

In facing these challenges, logistics companies must have an adaptive, flexible and sustainable strategy. The ability to adapt quickly to changes in markets, technology and the business environment will be a key factor in achieving long-term success. Integration between Business Agility, Green Supply Chain Management, and Green Innovation is very important to ensure business sustainability in the logistics sector. The combination of these three concepts allows companies to not only meet current market needs but also adapt quickly to future changes and run environmentally friendly operations. By adopting GSCM and green innovation practices, companies can reduce operational costs, increase efficiency, and at the same time comply with increasingly stringent environmental regulations. This increases the company's financial sustainability (Carter & Rogers, 2008). Additionally, companies that prioritize sustainability tend to gain a better reputation in the eyes of customers and society at large, which in turn increases customer loyalty and attracts more consumers. (Hart and Dowell, 2011). Business agility and green innovation enable companies to be more resilient to risks associated with climate change, new regulations and changing consumer preferences, thereby ensuring long-term business continuity (Wieland & Wallenburg, 2013).

In an increasingly dynamic and complex era of globalization, logistics companies face significant challenges in maintaining competitiveness while fulfilling environmental responsibilities. Business agility, or business agility, has become a critical element that allows companies to respond quickly and effectively to market changes, as well as maintain operational sustainability in the long term. Business agility allows companies to navigate market uncertainty and fluctuations, which in turn can strengthen a company's competitiveness (Teece et al., 2016).

On the other hand, environmental sustainability is increasingly becoming a major focus in supply chain management in line with increasing global awareness about climate change and environmental degradation. Logistics companies are faced with demands to reduce negative impacts on the environment through the adoption of environmentally friendly practices and technological innovations that support sustainability. This approach is not only important for meeting environmental regulations and standards, but also for meeting stakeholder expectations and creating long-term value for the company (Walker, H., Seuring, S., Sarkis, J., & Klassen, R, 2014)

The concept of sustainability in supply chain management involves the integration of social, economic, and environmental principles into business strategy. This includes various aspects such as reducing carbon emissions, efficient use of resources, waste management, as well as implementing green technology in logistics operations (Golini & Boffelli, 2018) This strategy is in line with the Sustainable Development Goals (SDGs) launched by the United Nations. The SDGs provide a global framework that guides companies in their efforts to contribute to sustainable development (UNDP, 2021) (United Nations, 2015).

Adoption of supply chain management strategies that support agility and sustainability not only helps companies meet regulatory and market demands, but also improves a company's reputation and financial performance. Studies show that companies that successfully integrate agility and sustainability into their operations tend to perform better in terms of operational efficiency, innovation and customer satisfaction (Dubey et al., 2019). Therefore, it is important for logistics companies to develop strategies that are not only responsive to market changes but also environmentally sustainable.

Recent research shows that digital technologies, such as the Internet of Things (IoT), artificial intelligence (AI), and blockchain, play an important role in increasing the agility and sustainability of supply chains. This technology enables better visibility, faster decision making, and higher operational efficiency, which in turn supports sustainability goals (Ivanov et al., 2019). In addition, innovations in transportation, such as the use of electric vehicles and alternative fuels, also contribute significantly to reducing the carbon footprint of logistics companies (Kumar & Anbanandam, 2020). Thus, this research aims to explore supply chain management strategies that combine business agility and environmental sustainability for logistics companies. The main focus is how companies can contribute to achieving the SDGs through environmentally friendly innovation and adaptive operational strategies. By integrating these two aspects, logistics companies can improve their operational performance while supporting the global sustainability agenda.

THEORETICAL REVIEW / CONCEPTUAL FRAMEWORK

Business Agility

According to research conducted, 73.5% of research journals wrote that business agility is one way for companies to be able to face, respond and adapt quickly to changes and competition that occur in the market. (Cepeda & Arias-Pérez, 2019; Darvishmotevali et al., 2020; Hamsal, 2019; Martínez-Caro et al., 2020; Zaini et al., 2020). It is important for companies to be able to face change. Companies can create and change strategies and tactics related to business agility (Barlette & Baillette, 2020; Qosasi et al., 2019) so that companies can respond to changes in creating opportunities (Govuzela & Mafini, 2019), and achieve unexpected changes, especially in market segments, customer preferences, and competition development strategies (Permadi et al., 2020).

Not only that, but business agility also refers to a company's ability to detect and direct market opportunities (Al-Omoush, 2020; Barlette & Baillette, 2020; Govuzela & Mafini, 2019; Martínez-Caro et al., 2020; Zaini et al., 2020) and can innovate for the company to achieve its goals (Z. Cai et al., 2019; Hamsal, 2019; Irfan et al., 2019; Nwankpa & Merhout, 2020). In following up on opportunities that exist in the market, business agility can also direct, facilitate, mobilize and reconfigure the resources needed in the company so that they can be optimally used by the company so that it can create creative value and protect value for the company. (Chan et al., 2019; Lin et al., n.d.; Nwankpa & Merhout, 2020).

In business agility, there are many influencing factors starting from the dimensions of business agility, internal aspects, external aspects, capabilities and components related to business agility, to factors that can hinder business agility. There are 3 dimensions to business agility, namely customer agility, partner agility and operational agility (Khayer et al., 2020). Customer agility allows companies to capitalize on changes in the marketplace by providing new value that can help improve internal operational adjustments so that companies can expand competitive actions and grow in market position. (Harraf et al., 2015) in his journal wrote that there are 10 pillars that influence business agility, namely a culture of innovation, empowerment, tolerance or ambiguity, vision, change management, organizational communication, market analysis and response, operations management, structural fluidity, and learning organizations. There are two antecedents of business agility, namely digital platform capabilities and company innovation capacity (Hamsal, 2019). These two things have independent effects and have an impact on business agility. Apart from that, there are two aspects that must be considered in business agility, namely internal and external aspects (Crick & Chew, 2017). The internal aspect is the company's ability to reconfigure its resources so that it can respond in a timely, effective and efficient manner. Meanwhile, the external aspect is the environment that produces the need for change around the company. In carrying out business agility in a company, there are many aspects that influence it, one of which is the person who has the role of carrying it out. According to research conducted, a company's top management is the most influential party in carrying out the company's business agility (Al-Omoush, 2020; Gao et al., 2020)

Green Supply Chain Management

Green Supply Chain Management (GSCM) has become an increasingly relevant research topic in recent years, as awareness of the importance of environmental sustainability in business practices increases. GSCM integrates environmentally friendly principles into supply chain management to reduce negative impacts on the environment and increase operational efficiency. Recent research shows that GSCM involves a wide range of activities, including green procurement, sustainable product design, green production, green distribution, and effective waste management. GSCM implementation aims to reduce the carbon footprint, reduce waste, and promote efficient use of resources. (L. Chen et al., 2017) in their study stated that the adoption of GSCM practices can improve a company's environmental performance without sacrificing profitability. They found that companies that integrate GSCM in their operations tend to have greater competitive advantages due to increased efficiency and reduced operational costs.

The following are several main indicators of Green Supply Chain Management identified in the literature: (1) Green Procurement: A procurement process that considers the environmental impact of products and raw materials. The company chooses suppliers who practice sustainability and provide environmentally friendly products (Walker et al., 2020). (2) Sustainable Product Design: Product design that considers the product life cycle, including the use of recyclable materials, energy efficiency, and waste minimization (Kumar & Chandrakar, 2022). (3) Eco-friendly Manufacturing: Production practices that reduce pollutant emissions, minimize waste, and optimize resource use (Zhu et al., 2007). (4) Green Distribution: Logistics and transportation strategies that reduce carbon emissions through efficient delivery routes and the use of environmentally friendly vehicles (Jabbour & Jabbour, 2020). (5) Waste Management: Implementation of effective waste management practices, including recycling, reuse and safe disposal (Tseng et al., 2021). (6) Management Commitment and Employee Involvement: Committed leadership and active involvement of employees in sustainability initiatives play an important role in the successful implementation of GSCM (Luthra et al., 2015).

Study by (Govindan et al., 2021) shows that although many companies are aware of the benefits of GSCM, implementation is often hampered by challenges such as high initial costs, lack of awareness, and technological limitations. However, with increasing pressure from regulators and consumers, companies are faced with a huge opportunity to innovate and strengthen their reputation by implementing GSCM practices. Green Supply Chain Management is an approach that not only helps companies in fulfilling their environmental responsibilities but also in achieving operational efficiency and competitive advantage. Research in the last five years shows that by adopting GSCM practices, companies can achieve a balance between sustainability and profitability.

Green Innovation

Green Innovation is a concept that developed in the context of sustainability and environmental responsibility. This includes innovations that not only improve economic performance but also reduce negative impacts on the environment. Research in the last five years has highlighted the importance of green innovation in various industries to support sustainable growth and compliance with environmental regulations. Green innovation involves the development and implementation of environmentally friendly technologies, processes, products and business models. Study by (Hojnik & Ruzzier, 2016) shows that green innovation can be divided into two main categories: green innovation processes and green innovation products. The green innovation process includes changes in production processes to reduce emissions, waste, and energy use. Green innovation products, on the other hand, focus on developing new environmentally friendly products or improving existing products to make them more sustainable.

Green Innovation Indicators are as follows: (1) Reduction of Emissions and Pollution: Successful innovation must demonstrate a significant reduction in greenhouse gas emissions and other pollutants (Li, 2020). (2) Energy Efficiency: New technologies and processes must increase the efficiency of energy use, reducing energy consumption without sacrificing productivity (Dangelico & Pujari, 2019). (3) Use of Renewable Resources: Use of materials that can be renewed and are more environmentally friendly compared to conventional materials (Klewitz & Hansen, 2021). (4) Sustainable Product Design: Products designed with their entire life cycle in mind, from production to disposal, with a focus on recycling and reuse (Waltz et al., 2020) (Y. Chen et al., 2021). (5) Compliance with Environmental Regulations: Innovations that comply with or exceed applicable environmental standards and regulations (Horbach & Rammer, 2020). (6) Green Technology Development: Investment in research and development of technology that supports environmental sustainability (Schiederig et al., 2012).

The implementation of green innovation faces various challenges, including high costs for research and development, resistance from internal organizations, and a lack of regulatory support and incentives. However, research by (W. Cai & Li, 2022) shows that companies that successfully overcome these challenges can enjoy benefits such as improved brand image, reduced long-term costs, and better regulatory compliance. Various case studies show the successful implementation of green innovation in various sectors, for example research by (G. Wu et al., 2021) highlights how manufacturing companies in China have adopted cleaner production technologies to reduce emissions and improve energy efficiency. In the energy sector, the development of renewable energy sources such as solar panels and wind turbines has shown great potential in reducing dependence on fossil fuels (Mousavi et al., 2020).

Green innovation is the key to achieving long-term sustainability in various industries. By adopting environmentally friendly technologies and practices, companies can not only fulfill their environmental responsibilities but also create significant economic and social value. Research in the last five years has strengthened our understanding of the importance and benefits of green innovation and provided practical guidance for its implementation.

Sustainable Development Goals

Sustainability is an increasingly important concept in various scientific disciplines and industrial practices, integrating environmental, economic and social aspects to ensure long-term balance. In the last five years, research on sustainability has developed rapidly, covering various approaches and implementations in the public and private sectors. Sustainability is the ability to meet the needs of the present without compromising the ability of future generations to meet their own needs (WCED, 1987). This concept includes three main pillars: economic, social and environmental. Study by (Elkington, 2019) put forward the triple bottom line model which emphasizes the importance of balance between the three pillars to achieve holistic sustainability. Implementing sustainability in business and public policy often requires an integrated and collaborative approach. Sustainability is a complex and multidimensional concept that requires an integrated and collaborative approach to achieve it. Research in the last five years shows that while there are significant challenges, there are also huge opportunities to achieve sustainability through innovation, supportive policies and cross-sector collaboration.

Implementing sustainability faces various challenges, including economic, social and technical obstacles. Research by (Bocken et al., 2019) shows that companies often face difficulties in integrating sustainability principles into their business models due to high initial costs and resistance to change. Research by (Kiron et al., 2021) shows that adopting sustainable practices can result in operational efficiencies, improved reputation and long-term financial benefits. Innovations in green technology and sustainable business models also open up new opportunities for companies to grow.

Some examples of companies that pay attention to sustainability, companies such as Tesla and Unilever have led the way in sustainable innovation with a focus on renewable energy and environmentally friendly business practices (Elkington, 2019). Research by (Sachs et al., 2021) highlighting that Nordic countries such as Sweden and Denmark have managed to achieve high levels of sustainability through progressive environmental policies and investment in green technologies. Sustainable Development Goals (SDGs) are an important topic in global discussions about sustainable development. The SDGs, launched by the United Nations in 2015, consist of 17 goals designed to address global challenges such as poverty, inequality, climate change, environmental degradation, peace and justice. In the last five years, research has focused on various aspects of implementation and achievement of the SDGs in various sectors and countries. The SDGs are a roadmap to achieving a more just and sustainable world by 2030. Research by (Sachs et al., 2019) shows that the SDGs provide a comprehensive framework that integrates the economic, social and environmental dimensions of sustainable development. Implementing the SDGs requires collaboration from government, the private sector, civil society and individuals. A study by Le Blanc (2020) highlights the importance of integrating the SDGs into national and local policies. This research shows that the successful implementation of the SDGs depends largely on how these goals are adapted and implemented according to the local context.

Research by Khan et al. (2022) demonstrates that green technology and innovation play crucial roles in supporting sustainable development and achieving the SDGs. Companies investing in environmental SDGs show positive correlations with financial performance, though the moderating effects of green innovation present mixed results. Research by Costanza et al. (2020) highlights that despite progress, much still needs to be done to achieve the SDGs by 2030. This study shows that cross-sector collaboration and investment in green technologies are key to overcoming these challenges. Research by Aerni et al. (2021) since more than 90 percent of jobs are created by the private sector, entrepreneurship and job creation are essential to SDG 8, with responsible enterprises generating positive externalities through investments in scalable innovations and commitment to local embeddedness.

Various case studies show how countries and organizations have successfully implemented the SDGs. For example research by L. Wu et al. (2022) higlights that the Nordic countries have consistently ranked at the top of international assessments for SDG implementation, though they continue to face challenges including unsustainable consumption patterns and climate change effects. Renewable energy plays a critical role in achieving multiple SDGs, particularly SDG 7 and SDG 13, while also contributing to poverty alleviation, improved health outcomes, and economic growth through job creation and rural electrification (Agbakwuru et al., 2024). In the private sector, companies such as Unilever and Patagonia have led the way with sustainable business practices, showing that sustainability can go hand in hand with profitability (Elkington, 2019). Achieving the SDGs requires collaborative efforts from various stakeholders and the integration of sustainability principles in every aspect of social, economic, and environmental life. Research in the last five years shows that while there are significant challenges, there are also huge opportunities for progress through innovation, supportive policies and cross-sector collaboration.

METHODOLOGY

This research used descriptive quantitative analysis and was processed with Smart PLS with data collected from 200 middle to upper-management-level employees at a logistics company in Jakarta. Data was collected by distributing questionnaires. The hypotheses of this research are (1) There is a significant influence of Business Agility on Green Supply Chain Management, (2) There is a significant influence of Business Agility on Green Innovation, (3) There is a significant influence of Green

Supply Management on Sustainable Development Goals, (4) There is a significant influence Green Innovation on Sustainable Development Goals, and (5) There is a significant influence of Business Agility on Sustainable Development Goals. Figure 1. can be seen the framework of thinking of this research.

Figure 1. Thinking Framework

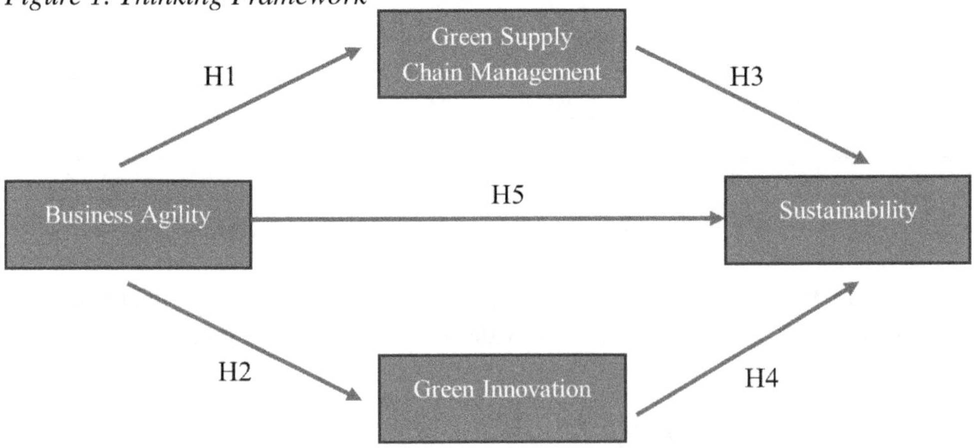

DATA ANALYSIS AND RESULTS

The purpose of measuring model evaluation is to assess the validity and reliability of the model. Validity testing aims to ensure that each item in the research instrument can measure the variables determined by the research. There are 2 types, namely convergent validity and discriminant validity. A latent construct is considered to have good convergent validity if the loading factor value is more than 0.7 and is significant. In the initial research stage of developing a measurement scale, a loading factor of 0.5 to 0.6 is considered sufficient (Chin, 1998 in Ghozali and Kusumadewi, 2016). The loading factor value can be seen in the following table:

Table 1. Loading Factor Values

Indicator	Business Agility	Indicator	Green Innovation	Indicator	Green SCM	Indicator	Sustainability
BAG1	0.815	GIN1	0.760	GSC1	0.809	SDG1	0.791
BAG2	0.779	GIN2	0.810	GSC2	0.764	SDG2	0.848
BAG3	0.818	GIN3	0.750	GSC3	0.764	SDG3	0.792
BAG4	0.821	GIN4	0.772	GSC4	0.769	SDG4	0.834
BAG5	0.808	GIN5	0.751	GSC5	0.812	SDG5	0.870
BAG6	0.859	GIN6	0.851	GSC6	0.833	SDG6	0.833
		GIN7	0.838	GSC7	0.739	SDG7	0.844
		GIN8	0.847	GSC8	0.754	SDG8	0.853

Source: Data processed with PLS, 2024

From the table above, the outer loading value is greater than 0.5, this means that all indicators have met the convergent validity test and are declared valid.

Table 2. Discriminant Validity

	Business Agility	Green Innovation	Green Innovation	Sustainability
Business Agility	0.817			
Green Innovation	0.138	0.794		
Green SCM	0.278	0.314	0.718	
Sustainability	0.264	0.292	0.482	0.834

To test discriminant validity if the AVE value of the average variance is higher than the correlation involving the latent variable. The table shows that all variables have met the convergent validity test. Reliability testing aims to prove the accuracy, consistency and precision of the instrument in measuring the construct. If the AVE value produced by all constructs with reflective indicators above is > 0.5; Cronbach's Alpha and D.G Rho values are > 0.60. (Ghozali, 2016).

Table 3. Construct Reliability dan Validity

	Cronbach's Alpha	Rho_A	Composite Reliability	Average Variance Extracted (AVE)
Business Agility	0.901	0.909	0.923	0.668
Green Innovation	0.917	0.931	0.931	0.630
Green SCM	0.909	0.910	0.926	0.610
Sustainability	0.937	0.939	0.948	0.695

The table above shows that all variables have met the requirements for reliable testing.

Figure 2. Structural Equation Model (PLS Algorithm)

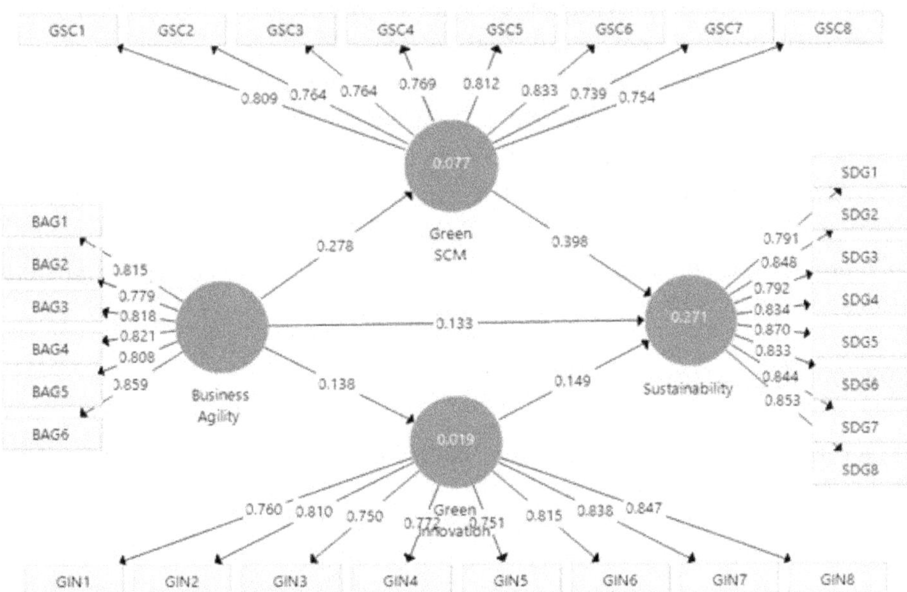

R Square

Table 4. R Square

Variable	R Square	R Square Adjusted
Green Innovation	0.019	0.014
Green SCM	0.077	0.073
Sustainability	0.271	0.260

Based on the table above, the following results are obtained:

1. Green Innovation has an adjusted R2 of 0.014, this means that Green Innovation is influenced by 1.4% by Business Agility while the rest is influenced by other constructs that were not included in the research.
2. Green Supply Chain Management has an adjusted R2 of 0.073, this means that Green Supply Chain Management is 7.3% influenced by Business Agility while the rest is influenced by other constructs that were not included in the research.
3. Sustainable Development Goals have an adjusted R2 of 0.260, this means that Green Innovation is influenced by 2.6% by Business Agility, Green Supply Chain Management and Green Innovation while the rest is influenced by other constructs that were not included in the research.

Structure Equation Modeling

The results of SEM processing with Smart PLS bootstrapping analysis appear as follows:

Figure 3. Structural Equation Model (Bootstrapping)

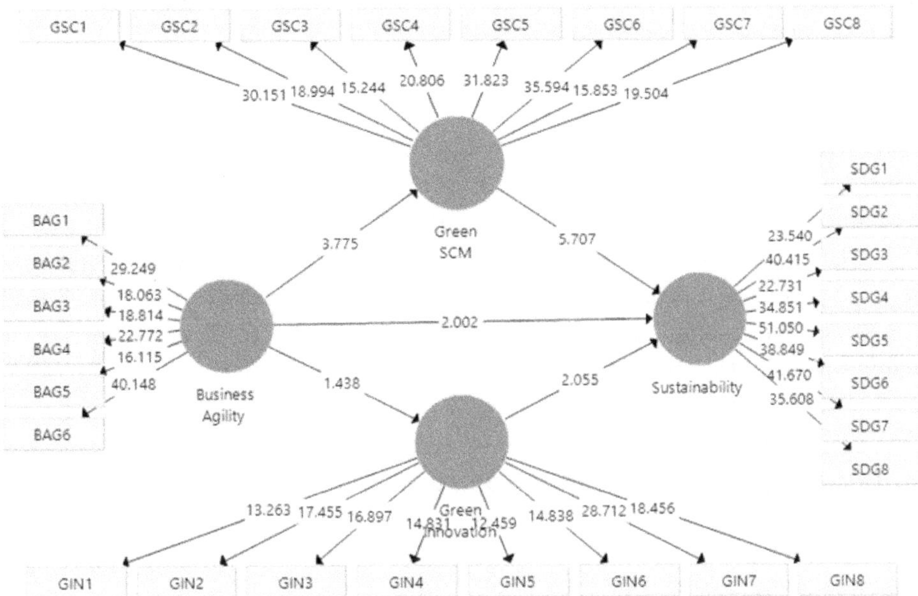

Hypothesis testing can be seen from the p-value. If the p-value is greater than 0.05 then Ho is accepted and Ho is rejected. And if the p-value is smaller than 0.05 then Ho is rejected and Ha is accepted. The results of hypothesis testing can be seen in the following table:

Table 5. Hypothesis Testing

		t-stat	p-value
H1	Business Agility → Green Supply Chain Management	3.775	0.000
H2	Business Agility → Green Innovation	1.438	0.151
H3	Green Supply Chain Management → Sustainable Development Goals	5.707	0.000
H4	Green Innovation → Sustainable Development Goals	2.055	0.040
H5	Business Agility → Sustainable Development Goals	2.002	0.046

Based on the table above, business agility has a significant effect on green supply chain management with a p-value of 0.000 so that H1 is accepted. Likewise with the relationship between green supply management, green innovation and business agility on sustainable development goals with a p-value of 0.000 for each; 0.040 and

0.046 so that H3, H4 and H5 are accepted. Meanwhile, business agility has no effect on green supply innovation with a p-value of 0.151, which means H2 is rejected.

DISCUSSION

Internal factors, such as leadership, have a significant impact on the economic and organizational performance of food SMEs (Chalarhena & Hendayani, 2022). Although internal environmental knowledge does not determine the development of new green products, it actually increases the influence of external knowledge on the success of environmental innovation (Rabal-Conesa et al., 2022). The results of this research reveal and support previous research that business agility has a significant effect on green supply management.

Business Agility → Green Innovation

Organizational agility is positively related to environmental innovation and environmental innovation produces a positive relationship with organizational performance. Companies that have good organizational agility will be able to integrate customer involvement to develop new environmentally friendly products with a lower environmental impact (Rabal Conesa et al., 2024). Organizational ability is significantly related to internal and external environmental knowledge, where external knowledge influences the success of eco-innovation (green innovation) (Özkan & Salepçioğlu, 2022; Rabal-Conesa et al., 2022).

GSCM practices significantly improve SBP. This is because implementing GSCM practices helps SMEs increase resource consumption efficiency and reduce production costs, thereby increasing SBP. Apart from that, GSCM also plays an important role in reducing waste and hazardous emissions, thereby improving environmental quality (Albhirat et al., 2023). Green SCM which consists of Green Purchasing, Green Logistics, Green Manufacturing & Ecological Design significantly influences the Organizational Performance variable (Asghar, 2023). GSCM provides a direct positive impact on business performance and an indirect positive impact on business performance through employee job satisfaction, operational efficiency, relational efficiency. Business benefits impacted by GSCM include: "Employee job satisfaction" and "Operational efficiency" (Nguyen et al., 2023). The results of this research support previous research that green supply chain management has a significant effect on sustainable development goals.

Green Innovation → Sustainability

The research results show that green innovation has a significant effect on sustainable development goals. This is in line with previous research. By demonstrating a strong commitment to sustainability and social responsibility, we will increase our reputation as a company that cares about the environment and society. This not only increases customer trust and loyalty but also increases the company's attractiveness in the eyes of investors, business partners and potential employees (Ningsih, 2024). The performance of green product innovation and green process innovation is positively correlated with the company's competitive advantage. Therefore, the business world must realize the value and appropriate position of green innovation for the sustainability of the company (Y. S. Chen et al., 2006). There is a positive and significant relationship between green innovation and company performance (Zhang et al, 2019).

Agility in terms of operations, customers and partnerships has a significant and positive impact on sustainable manufacturing practices and green procurement (Sun et al., 2022) (Özkan & Salepçioğlu, 2022). The results of this research are in line with previous research where business agility influences sustainable development goals.

CONCLUSIONS

This research examines the influence of business agility, green supply management and green innovation on sustainable development goals. The results show that business agility has a significant effect on sustainable development goals. Likewise with the relationship between green supply chain management and green innovation towards sustainable development goals. Meanwhile, business agility has no effect on green innovation. Effective leadership, business agility and green supply chain management (GSCM) play an important role in improving SME business performance and sustainability. Business agility drives green innovation that improves organizational performance. GSCM practices increase resource efficiency, reduce costs and waste, and increase employee job satisfaction and operational efficiency. Green innovation contributes significantly to sustainable development goals and corporate competitive advantage, strengthening reputation, customer trust and attractiveness for investors and business partners.

REFERENCES

Aerni, P., Stavridou, M., & Schluep, I. (2021). *Transitioning to Decent Work and Economic Growth, Transitioning to Sustainability Series* (8th ed., Issue 8). MDPI, Basel. https://doi.org/https://doi.org/10.3390/books978-3-03897-779-7

Agbakwuru, V., Obidi, P. O., Salihu, O. S., & MaryJane, O. C. (2024). The role of renewable energy in achieving sustainable development goals. *International Journal of Engineering Research Updates*. https://doi.org/https://doi.org/10.53430/ijeru.2024.7.2.0046

Al-Omoush, K. S. (2020). The role of top management support and organizational capabilities in achieving e-business entrepreneurship. *Kybernetes*. Advance online publication. DOI: 10.1108/K-12-2019-0851

Albhirat, M. M., Zulkiffli, S. N. A., Salleh, H. S., & Zaki, N. A. M. (2023). The Moderating Role of Social Capital in the Relationship Between Green Supply Chain Management and Sustainable Business Performance: Evidence from Jordanian SMEs. *International Journal of Sustainable Development and Planning*, *18*(6), 1733–1747. DOI: 10.18280/ijsdp.180609

Asghar, M. M. (2023). Green Supply Chain Management and Organizational Performance in Automotive Industry: A Case Study from Pakistan. *South Asian Journal of Operations and Logistics*, *2*(2), 18–40. DOI: 10.57044/SAJOL.2023.2.2.2308

Barlette, Y., & Baillette, P. (2020). Big data analytics in turbulent contexts: Towards organizational change for enhanced agility. *Production Planning and Control*, *0*(0), 1–18. DOI: 10.1080/09537287.2020.1810755

Bocken, N. M. P., Short, S. W., Rana, P., & Evans, S. (2019). A literature and practice review to develop sustainable business model archetypes. *Journal of Cleaner Production*, *65*, 42–56. DOI: 10.1016/j.jclepro.2013.11.039

Cai, W., & Li, G. (2022). The drivers of green innovation: The role of governmental support and environmental regulations. *Journal of Cleaner Production*, *233*, 117–130. DOI: 10.1016/j.jclepro.2022.05.023

Cai, Z., Liu, H., Huang, Q., & Liang, L. (2019). Developing organizational agility in product innovation: The roles of IT capability, KM capability, and innovative climate. *R & D Management*, *49*(4), 421–438. DOI: 10.1111/radm.12305

Carter, C. R., & Rogers, D. S. (2008). A framework of sustainable supply chain management: Moving toward new theory. *International Journal of Physical Distribution & Logistics Management*, *38*(5), 360–387. DOI: 10.1108/09600030810882816

Cepeda, J., & Arias-Pérez, J. (2019). Information technology capabilities and organizational agility: The mediating effects of open innovation capabilities. *Multinational Business Review*, *27*(2), 198–216. DOI: 10.1108/MBR-11-2017-0088

Chalarhena, N., & Hendayani, R. (2022). Pengaruh green supply chain management terhadap performa ekonomi dan organisasi pada UMKM industri makanan di Wonosobo Jawa Tengah. *Fair Value: Jurnal Ilmiah Akuntansi Dan Keuangan*, *5*(3), 1696–1713. DOI: 10.32670/fairvalue.v5i3.2420

Chan, Y. E., Denford, J. S., & Wang, J. (jane). (2019). The Co-Evolution of IT, Knowledge, and Agility in Micro and Small Enterprises. *Journal of Information and Knowledge Management*. DOI: 10.1142/S0219649219500278

Chen, L., Zhao, X., Tang, O., Price, L., Zhang, S., & Zhu, W. (2017). Supply chain collaboration for sustainability: A literature review and future research agenda. *International Journal of Production Economics*, *194*, 73–87. DOI: 10.1016/j.ijpe.2017.04.005

Chen, Y., Wang, Y., & Zhou, L. (2021). Green innovation and sustainable development: The moderating effect of environmental regulatio. *Technological Forecasting and Social Change*, *166*, 120–130. DOI: 10.1016/j.techfore.2021.120653

Chen, Y. S., Lai, S. B., & Wen, C. T. (2006). The influence of green innovation performance on corporate advantage in Taiwan. *Journal of Business Ethics*, *67*(4), 331–339. DOI: 10.1007/s10551-006-9025-5

Costanza, R., de Groot, R., Braat, L., Kubiszewski, I., Fioramonti, L., Sutton, P., & Grasso, M. (2020). Twenty years of ecosystem services: How far have we come and how far do we still need to go? *Ecosystem Services*, *28*, 1–16. DOI: 10.1016/j.ecoser.2017.09.008

Crick, C., & Chew, E. K. (2017). Business processes in the agile organisation: A socio-technical perspective. *Software & Systems Modeling*, *16*(3), 631–648. DOI: 10.1007/s10270-015-0506-9

Dangelico, R. M., & Pujari, D. (2019). Mainstreaming green product innovation: Why and how companies integrate environmental sustainability. *Journal of Business Ethics*, *95*(3), 471–486. DOI: 10.1007/s10551-010-0434-0

Darvishmotevali, M., Altinay, L., & Köseoglu, M. A. (2020). The link between environmental uncertainty, organizational agility, and organizational creativity in the hotel industry. *International Journal of Hospitality Management, 87(August*. DOI: 10.1016/j.ijhm.2020.102499

Dubey, R., Gunasekaran, A., Childe, S. J., Papadopoulos, T., & Helo, P. (2019). Supplier relationship management for circular economy: Influence of external pressures and top management commitment. *Management Decision*, *57*(4), 767–790. DOI: 10.1108/MD-04-2018-0396

Elkington, J. (2019). 25 Years Ago I Coined the Phrase "Triple Bottom Line". Here's Why It's Time to Rethink It. *Harvard Business Review*. https://hbr.org/2019/06/25-years-ago-i-coined-the-phrase-triple-bottom-line-heres-why-im-giving-up-on-it

Gao, P., Zhang, J., Gong, Y., & Li, H. (2020). Effects of technical IT capabilities on organizational agility: The moderating role of IT business spanning capability. *Industrial Management & Data Systems*, *120*(5), 941–961. DOI: 10.1108/IMDS-08-2019-0433

Golini, R., & Boffelli, A. (2018). The role of manufacturing within industrial districts. *Local Clusters in Global Value Chains, October 2017*, 193–212. DOI: 10.4324/9781315182049-11

Govindan, K., Khodaverdi, R., & Jafarian, A. (2021). A fuzzy multi criteria approach for measuring sustainability performance of a supplier based on triple bottom line approach. *Journal of Cleaner Production*, *240*, 118–129. DOI: 10.1016/j.jclepro.2021.118129

Govuzela, S., & Mafini, C. (2019). Organisational agility, business best practices and the performance of small to medium enterprises in South Africa. *South African Journal of Business Management*, *50*(1). Advance online publication. DOI: 10.4102/sajbm.v50i1.1417

Hamsal, M. (2019). The effect of sales-marketing collaboration, market-oriented product development, and nimble network structure on organizational agility. *International Journal of Advanced Science and Technology*, *28*(8), 696–713.

Harraf, A., Wanasika, I., Tate, K., & Talbott, K. (2015). Organizational Agility. *Journal of Applied Business Research*, *31*(2), 675. DOI: 10.19030/jabr.v31i2.9160

Hojnik, J., & Ruzzier, M. (2016). What drives eco-innovation? A review of an emerging literature. *Environmental Innovation and Societal Transitions*, *19*, 31–41. DOI: 10.1016/j.eist.2015.09.006

Horbach, J., & Rammer, C. (2020). Environmental innovation and employment dynamics in different technology fields: An analysis based on the German Community Innovation Survey. *Research Policy*, *51*(1), 103–115. DOI: 10.1016/j.respol.2020.103115

Irfan, M., Wang, M., & Akhtar, N. (2019). Impact of IT capabilities on supply chain capabilities and organizational agility: A dynamic capability view. *Operations Management Research : Advancing Practice Through Research, 12*(3–4), 113–128. DOI: 10.1007/s12063-019-00142-y

Ivanov, D., Dolgui, A., & Sokolov, B. (2019). The impact of digital technology and Industry 4.0 on the ripple effect and supply chain risk analytics. *International Journal of Production Research, 57*(3), 829–846. DOI: 10.1080/00207543.2018.1488086

Jabbour, A. B. L. S., & Jabbour, C. J. C. (2020). Green supply chain management and firms' performance: Understanding potential relationships and the role of green sourcing and organizational learning. *International Journal of Production Research, 58*(14), 4291–4306. DOI: 10.1080/00207543.2020.1769377

Khan, P. A., Johl, S. K., & Akhtar, S. A. (2022). Vinculum of Sustainable Development Goal Practices and Firms' Financial Performance: A Moderation Role of Green Innovation. *Journal of Risk and Financial Management, 15*(3), 96. https://doi.org/https://doi.org/10.3390/jrfm15030096. DOI: 10.3390/jrfm15030096

Khayer, A., Jahan, N., Hossain, M. N., & Hossain, M. Y. (2020). The adoption of cloud computing in small and medium enterprises: A developing country perspective. *VINE Journal of Information and Knowledge Management Systems*. Advance online publication. DOI: 10.1108/VJIKMS-05-2019-0064

Kiron, D., Unruh, G., Reeves, M., Kruschwitz, N., & Rubel, H. (2021). Corporate Sustainability at a Crossroads. *MIT Sloan Management Review*. https://sloanreview.mit.edu/article/corporate-sustainability-at-a-crossroads/

Klewitz, J., & Hansen, E. G. (2021). Sustainability-oriented innovation of SMEs: A systematic review. *Journal of Cleaner Production, 65*, 57–75. DOI: 10.1016/j.jclepro.2013.07.017

Kocot, D., Kocot, M., & Soboń, J. (2024). Competitive Advantage of an Agile Organization as a Determinant of The Functioning of Global Market. *Globalization, the State and the Individual*. https://doi.org/https://doi.org/10.5604/01.3001.0054.9309

Kumar, S., & Anbanandam, R. (2020). Impact of risk management culture on supply chain resilience: An empirical study from Indian manufacturing industry. *Proceedings of the Institution of Mechanical Engineers. Part O, Journal of Risk and Reliability, 234*(2), 246–259. DOI: 10.1177/1748006X19886718

Kumar, S., & Chandrakar, R. (2022). Green supply chain management: A study of recent practices, trends, and issues. *Journal of Cleaner Production, 234*, 1243–1257. DOI: 10.1016/j.jclepro.2022.124387

Li, Q. (2020). The impact of green innovation on firm performance: Evidence from China. *Technology in Society*, *62*, 101–111. DOI: 10.1016/j.techsoc.2020.101342

Lin, J., Li, L., Luo, X., & Benitez, J. (2020, August). How do agribusinesses thrive through complexity? The pivotal role of e-commerce capability and business agility. *Decision Support Systems*, *135*(113342), 113342. Advance online publication. DOI: 10.1016/j.dss.2020.113342 PMID: 32834263

Luthra, S., Garg, D., & Haleem, A. (2015). Critical success factors of green supply chain management for achieving sustainability in Indian automobile industry. *Production Planning and Control*, *26*(5), 339–362. DOI: 10.1080/09537287.2014.904532

Martínez-Caro, E., Cepeda-Carrión, G., Cegarra-Navarro, J. G., & Garcia-Perez, A. (2020). The effect of information technology assimilation on firm performance in B2B scenarios. *Industrial Management & Data Systems*, *120*(12), 2269–2296. Advance online publication. DOI: 10.1108/IMDS-10-2019-0554

Mousavi, S., Bossink, B., & Van Vliet, M. (2020). Dynamic capabilities and organizational routines for managing innovation towards sustainability. *Journal of Cleaner Production*, *247*, 119–135. DOI: 10.1016/j.jclepro.2019.119648

Nguyen, X. H., Nguyen, L. T., & Pham, M. A. (2023). Influence of Green Supply Chain Management on Business Performance of Vietnamese Electronic Firms. *International Journal of Research and Review*, *10*(2), 753–771. DOI: 10.52403/ijrr.20230291

Ningsih, F. O. (2024). *Analyzing the Implementation of Green Industries Practice at PT. Semen Indonesia (Persero) Tbk. Tuban Plant towards Achieving Sustainable Development Goals. 9*(5).

Nwankpa, J. K., & Merhout, J. W. (2020). Exploring the effect of digital investment on IT innovation. *Sustainability (Basel)*, *12*(18), 7374. Advance online publication. DOI: 10.3390/su12187374

Özkan, H., & Salepçioğlu, M. A. (2022). Does Organizational Agility Affect Sustainable Quality Perception? The Mediating Role of Innovation Orientation. *International Journal of Business*, *27*(4). Advance online publication. DOI: 10.55802/IJB.027(4).002

Permadi, A., Mursitama, T. N., Hamsal, M., & Bandur, A. (2020). The impact of IT capability and supply chain agility in the electronic home appliance retail at Indonesia. *International Journal of Advanced Trends in Computer Science and Engineering*, *9*(1), 833–844. DOI: 10.30534/ijatcse/2020/120912020

Porter, M. E., & Kramer, M. R. (2011). Creating shared value. *SSRN*, *89*(1–2). Advance online publication. DOI: 10.2139/ssrn.3683975

Purba Rao, D. H. (2005). Do green supply chains lead to competitiveness and economic performance? *International Journal of Operations & Production Management*, *25*(9), 898–916. https://doi.org/https://doi.org/10.1108/01443570510613956. DOI: 10.1108/01443570510613956

Qosasi, A., Permana, E., Muftiadi, A., Purnomo, M., & Maulina, E. (2019). Building SMEs' competitive advantage and the organizational agility of apparel retailers in indonesia: The role of ICT as an initial trigger. *Gadjah Mada International Journal of Business*, *21*(1), 69–90. DOI: 10.22146/gamaijb.39001

Rabal-Conesa, J., Jiménez-Jiménez, D., & Martínez-Costa, M. (2022). Organisational agility, environmental knowledge and green product succes. *Journal of Knowledge Management*, *26*(9), 2440–2462. DOI: 10.1108/JKM-06-2021-0486

Rabal Conesa, J., Jiménez Jiménez, D., & Martínez Costa, M. (2024). Performing in green: Towards eco-innovation through agility and customer involvement. *Journal of Organizational Change Management*, *37*(3), 641–658. Advance online publication. DOI: 10.1108/JOCM-09-2023-0359

Sachs, J. D., Schmidt-Traub, G., Kroll, C., Lafortune, G., & Fuller, G. (2019). SDG Index and Dashboards Report 2019. *New York: Bertelsmann Stiftung and Sustainable Development Solutions Network (SDSN).*

Sachs, J. D., Schmidt-Traub, G., Kroll, C., Lafortune, G., & Fuller, G. (2021). *The Decade of Action for the Sustainable Development Goals: Sustainable Development Report 2021*. Cambridge University Press., DOI: 10.1017/9781108646967

Schiederig, T., Tietze, F., & Herstatt, C. (2012). Green innovation in technology and innovation management - an exploratory literature review. *R & D Management*, *42*(2), 180–192. DOI: 10.1111/j.1467-9310.2011.00672.x

Sun, J., Sarfraz, M., Turi, J., & Iva cu, L. (2022). Organizational Agility and Sustainable Manufacturing Practices in the Context of Emerging Economy: A Mediated Moderation Model. *Processes (Basel, Switzerland)*, *10*(12), 2567. Advance online publication. DOI: 10.3390/pr10122567

Teece, D., Peteraf, M., & Leih, S. (2016). Dynamic Capabilities and Organizational Agility. *California Management Review*, *58*(4), 13–35. DOI: 10.1525/cmr.2016.58.4.13

Tseng, M. L., Tan, R. R., & Sirikrai, S. B. (2021). Implementation of green supply chain management in Thailand: A case study of the manufacturing sector. *Journal of Cleaner Production*, *203*, 709–717. DOI: 10.1016/j.jclepro.2021.09.113

UNDP. (2021). *Human Development Report 2021*. United Nations Development Programme. https://hdr.undp.org/en/2021-report

Walker, L., Zhao, X., Tang, O., Price, L., Zhang, S., & Zhu, W. (2020). Supply chain collaboration for sustainability: A literature review and future research agenda. *International Journal of Production Economics*, *219*, 281–293.

Waltz, L. A., Muñoz, L., Johnson, H. W., & Rodriguez, T. (2020). Exploring job satisfaction and workplace engagement in millennial nurses. *Journal of Nursing Management*, *28*(3), 673–681. DOI: 10.1111/jonm.12981 PMID: 32068932

WCED. (1987). *Brundtland Report*. World Commission on Environment and Development.

Wieland, A., & Wallenburg, C. M. (2013). The influence of relational competencies on supply chain resilience: A relational view. *International Journal of Physical Distribution & Logistics Management*, *43*(4), 300–320. DOI: 10.1108/IJPDLM-08-2012-0243

Wu, G., Ding, Q., & Chen, X. (2021). The effects of environmental regulation on green innovation: A panel study based on 31 provinces in China. *Journal of Cleaner Production*, *202*, 1200–1211. DOI: 10.1016/j.jclepro.2021.122356

Wu, L., Adebayo, T., Yue, X.-G., & Umut, A. (2022). The role of renewable energy consumption and financial development in environmental sustainability: Implications for the Nordic Countries. *International Journal of Sustainable Development and World Ecology*, *30*(1), 21–36. https://doi.org/https://doi.org/10.1080/13504509.2022.2115577. DOI: 10.1080/13504509.2022.2115577

Zaini, M. K., Masrek, M. N., & Abdullah Sani, M. K. J. (2020). The impact of information security management practices on organisational agility. *Information and Computer Security*, *28*(5), 681–700. DOI: 10.1108/ICS-02-2020-0020

Zhu, Q., Sarkis, J., & Lai, K. (2007). Green supply chain management: Pressures, practices and performance within the Chinese automobile industry. *Journal of Cleaner Production*, *15*(11–12), 1041–1052. DOI: 10.1016/j.jclepro.2006.05.021

Zhu, Q., Sarkis, J., & Lai, K. (2008). Confirmation of a measurement model for green supply chain management practices implementation. *International Journal of Production Economics*, *111*(2), 261–273. DOI: 10.1016/j.ijpe.2006.11.029

Chapter 9
Geographical Approaches to Global Sustainable Development Goals in BRICS Countries

Pragti Dixit
https://orcid.org/0009-0002-4034-4572
Chandigarh University, Punjab, India

Mukul Bhatnagar
https://orcid.org/0000-0002-7773-5641
Graphic Era University, India

Nikhil Varghese
Christ University, Karnataka, India

ABSTRACT

Geographical factors play a key role in the realization of SDG's across all countries of the world. Present study is an endeavor in this direction and attempts to evaluate the impact of urbanization, agriculture, and water resource availability on the achievement of sustainable development of BRICS nations. The study employs Panel data analysis using Fixed Effects Model (FEM) and Random Effects Model (REM) along with the application of Hausman Test to determine the appropriate model selection. The results reveal that urbanization significantly enhances SDG progress, agriculture shows an insignificant effect, and water resource availability negatively impacts sustainability, indicating the urgent need for strategic urban planning, agricultural reforms, and efficient water management. The study also

DOI: 10.4018/979-8-3373-1300-9.ch009

highlights that country-specific factors play a critical part in shaping sustainability outcomes, reinforcing the necessity for BRICS policymakers to adopt geographically tailored approaches that align economic growth with environmental sustainability.

I INTRODUCTION

SDG's (Sustainable Development goals) provide a global framework for achieving economic, social, and environmental sustainability, and BRICS countries—Brazil, Russia, India, China, and South Africa—play a crucial role in this pursuit due to their vast geographical diversity, rapid economic growth, and significant environmental challenges, necessitating region-specific strategies to address issues such as urbanization, agricultural sustainability, and water resource management, making a geographical approach essential for formulating policies that balance development with ecological preservation and ensure equitable progress toward the SDG's (Guang-Wen et al., 2023).

A geographical approach to SDG's in BRICS countries is essential because sustainability challenges and opportunities vary significantly across regions due to differences in climate conditions, natural resource distribution, agricultural & urbanization patterns, and governance structures, necessitating location-specific strategies for effective policy implementation, ensuring that urban growth is managed sustainably, agricultural productivity is enhanced without environmental degradation, and water resources are allocated efficiently, ultimately enabling BRICS nations to align their economic development with long-term sustainability goals in a way that accounts for their unique geographical and socio-economic contexts (Wu et al., 2023).

Urbanization, agriculture, and water availability are critical factors shaping sustainable development in BRICS countries, as rapid urban expansion drives economic growth and infrastructure development but also poses challenges related to environmental degradation and resource management, agriculture remains a vital sector for food security and rural livelihoods but faces pressures from climate change, land degradation, and modernization needs, while water availability plays a crucial role in sustaining both urban and agricultural sectors, necessitating efficient governance and conservation strategies to ensure equitable access, long-term sustainability, and resilience against climate-induced water stress (Aderemi et al., 2023).

The interconnections between geography and sustainable development in BRICS countries are shaped by spatial disparities in natural resources, climate conditions, and socio-economic activities, where rapid urbanization fuels economic expansion but strains infrastructure and environmental sustainability, agriculture remains a crucial sector yet faces challenges from land degradation and shifting climatic patterns, and water availability varies significantly across regions, necessitating adaptive policies for equitable distribution, all of which highlight the need for geographically informed strategies that balance economic growth, resource conservation, and social well-being to achieve the SDGs effectively.

The role of economic growth and environmental sustainability in BRICS countries is a complex interplay between rapid industrialization, expanding urban centers, and increasing agricultural demands that drive economic development, job creation, and technological innovation while simultaneously exerting immense pressure on natural ecosystems through deforestation, carbon emissions, biodiversity loss, and resource depletion, requiring a strategic shift toward sustainable practices such as the adoption of renewable energy sources, green technologies, efficient water and land management, circular economy models, and stringent environmental policies that promote economic resilience, ensure social equity, and safeguard ecological integrity, ultimately aligning national development trajectories with the broader objectives of the SDG's by fostering low-carbon economies, enhancing climate adaptation strategies, and encouraging multilateral cooperation within the BRICS framework to address shared environmental challenges and create a model for sustainable growth that balances economic ambitions with long-term ecological stewardship.

Against this backdrop, the present chapter seeks to explore the impact of urbanization, agriculture and water availability on the attainment of SDG's of BRICS countries. The significance of this study lies in its comprehensive examination of the geographical dimensions of sustainable development in BRICS countries, highlighting how spatial factors such as urbanization patterns, agricultural productivity, and water resource distribution influence the achievement of SDGs, while emphasizing the necessity of region-specific policies that integrate economic growth with environmental sustainability, foster resilient urban planning, promote sustainable agricultural practices, and ensure equitable water management, ultimately providing policymakers, researchers, and stakeholders with valuable insights into the complex interdependencies between geography, development, and sustainability, offering evidence-based recommendations for harmonizing economic expansion with ecological conservation, strengthening climate adaptation strategies, enhancing transnational cooperation within the BRICS framework, and contributing to the global discourse on sustainable development by demonstrating how geographically informed policymaking can drive inclusive growth, mitigate environmental risks, and create long-term pathways toward achieving the SDGs in emerging economies.

The rest of the chapter is organized as follows: Section II covers Literature Review. Data sources and Methodology are covered in Section III and Section IV encompasses Results and Discussion. Implications of the study, Conclusion and References are covered in Section V, VI and VII respectively.

II LITERATURE REVIEW

Table 1 comprises of the review of the studies already conducted on this topic.

Table 1. Related Studies

Study/ Author	Focus Area	Geographical Approach	Key Findings	Challenges Identified
(Amin et al., 2025)	Urban Sustainability (SDG 11)	GIS and Remote Sensing	GIS-based models identified high urban expansion leading to increased carbon emissions.	Lack of real-time data integration, regional disparities.
(Amin et al., 2024)	Deforestation & Biodiversity (SDG 15)	Satellite Imagery and Spatial Analysis	Deforestation hotspots were mapped, linking them to illegal logging and agriculture.	Policy enforcement gaps, need for indigenous land rights recognition.
(Ofori & Appiah-Opoku, 2024)	Water Resource Management (SDG 6)	Hydrological Modeling & GIS	Groundwater depletion areas identified, spatial disparity in water access analyzed.	Poor water governance, seasonal variability in availability.
(Kumar et al., 2024)	Energy Efficiency (SDG 7)	Spatial Data Analytics & Geo-Mapping	Regional variations in renewable energy adoption identified; northern areas lagging.	Harsh climate limitations, weak local policy incentives.
(Zhu et al., 2024)	Food Security & Agriculture (SDG 2)	Agro-Ecological Zoning & GIS	Climate-adaptive agricultural zones mapped to enhance food production.	Climate unpredictability, lack of farmer awareness programs.
(Lin et al., 2024)	Air Pollution Control (SDG 13)	Spatial-Temporal Analysis & AI	Machine learning models predicted air quality changes based on urbanization trends.	Insufficient localized mitigation strategies.
(Bahturidze et al., 2024)	Urban Flood Risk (SDG 11)	Flood Hazard Mapping & Remote Sensing	High-risk flood zones mapped; urban planning deficiencies highlighted.	Poor drainage infrastructure, climate change impact.
(Saba & Ngepah, 2024)	Smart Cities & Infrastructure (SDG 9)	Geospatial Analysis	Identified smart city development challenges using satellite data.	Inadequate funding, poor public-private collaboration.
(Sahoo et al., 2024)	Arctic Sustainability (SDG 13)	Climate Modeling & GIS	Melting permafrost areas mapped; threats to infrastructure identified.	Limited monitoring capabilities, economic constraints.
(Kai et al., 2024)	Renewable Energy Transition (SDG 7)	Solar & Wind Resource Mapping	Potential for decentralized solar farms identified.	High initial costs, policy inconsistency.

III DATA SOURCES AND METHODOLOGY

The study aims to evaluate the effect of Urbanisation, Proportion of Agricultural Land and Water Availability on SDG overall score of BRICS countries for the time- period 2000 to 2023. In order to carry out the objectives of the present study secondary data has been used. Data pertaining to Urbanisation [(Urban population (% of total population)][1], Proportion of Agricultural Land [Agricultural land (% of land area)][2] and Water Availability [Renewable internal freshwater resources per capita (cubic meters)][3] has been extracted from the World Development Indicators database of World Bank and data on SDG overall score[4] has been taken from SDG Transformation Center (dashboards.sdgindex.org). Log transformation of all variables was undertaken before conducting panel data analysis. Table 2 provides the notation and sources of the variables used in the present study and Table 3 indicates summary statistics for each variable.

Table 2. Notation and Sources of Variables

Variable	Notation	Source
Log of Urbanisation	lnurban	WDI
Log of Proportion of Agricultural Land	lnagri	WDI
Log of Water Availability	lnwater	WDI
Log of SDG overall score	lnsdg	SDG Transformation Center

Table 3. Summary Statistics (2000-2023)

Variable	Mean	Std. Dev.	Sample Size
lnurban	4.049	0.36	120
lnagri	3.681	0.655	120
lnwater	8.398	1.581	120
lnsdg	4.161	0.099	120

Finally, panel data analysis was conducted using FEM and REM, with the Hausman test performed to determine the appropriate model selection.

IV RESULTS AND DISCUSSION

Present study aims to identify the impact of Urbanization, Proportion of Agricultural Land and Water Availability on SDG overall score of BRICS countries. This section covers the results and discussion of the present study:

Table 4. Results of FEM

Variable	Coefficient	Standard Error	t-statistic	p-value	95% Confidence Interval
lnurban	0.0977	0.0405	2.41	0.018*	[0.0174, 0.1780]
lnagri	-0.1755	0.3702	-0.47	0.636	[-0.9090, 0.5580]
lnwater	-0.4808	0.055	-8.74	0.000**	[-0.5899, -0.3718]
_cons	8.4492	1.4382	5.87	0.000**	[5.5996, 11.2988]

Note: Dependent Variable: lnsdg
* implies significant at 0.05 level.
** implies significant at 0.01 level.
Source: Authors' own calculations.

Table 5. Summary Statistics of FEM

Statistic	Number of Observations	Number of Groups	R-squared (within)	R-squared (between)	R-squared (overall)	F-statistic	Prob > F	Rho	F test (all ui = 0)	Prob > F (ui)
Value	120	5	0.6617	0.4437	0.3058	73.04	.000**	0.9985	149.81	.000**

Note: Dependent Variable: lnsdg.
** implies significant at 0.01 level.
Source: Authors' own calculations.

Table 4 and Table 5 depict the results and summary statistics of FEM. The FEM results show that lnurban has a positive and significant effect on lnsdg (Coefficient = 0.0977, Std. Error = 0.0405, t = 2.41, p = 0.018), while lnagri is not significant (Coefficient = -0.1755, Std. Error = 0.3702, t = -0.47, p = 0.636), and lnwater has a significant negative impact (Coefficient = -0.4808, Std. Error = 0.055, t = -8.74, p = 0.000); the constant term is also significant (Coefficient = 8.4492, Std. Error = 1.4382, t = 5.87, p = 0.000), with an overall within R-squared of 0.6617, between R-squared of 0.4437, and overall R-squared of 0.3058; the F-statistic for the model is 73.04 (p = 0.000), rho is 0.9985, and the F-test for all ui = 0 yields 149.81 (p = 0.000), confirming the significance of fixed effects, with all results based on the dependent variable lnsdg and calculations by the authors.

Table 6. Results of REM

Variable	Coefficient	Standard Error	z-statistic	p-value	95% Confidence Interval
lnurban	0.1809	0.0394	4.59	.000**	[0.1036, 0.2582]
lnagri	-0.684	0.1343	-5.09	.000**	[-0.9472, -0.4209]
lnwater	-0.3115	0.0455	-6.84	.000**	[-0.4007, -0.2222]
_cons	8.5611	0.9289	9.22	.000**	[6.7404, 10.3819]

Note: Dependent Variable: lnsdg.
** implies significant at 0.01 level.
Source: Authors' own calculations.

Table 7. Summary Statistics of REM

Statistic	Number of Observations	Number of Groups	R-squared (within)	R-squared (between)	R-squared (overall)	Wald chi2	Prob > chi2	Rho
Value	120	5	0.6321	0.255	0.1132	163.7	.000**	0.9413

Note: Dependent Variable: lnsdg.
** implies significant at 0.01 level.
Source: Authors' own calculations.

Table 6 and Table 7 depict the results and summary statistics of REM. The REM results indicate that lnurban has a positive and significant effect on lnsdg (Coefficient = 0.1809, Std. Error = 0.0394, z = 4.59, p = 0.000), lnagri has a significant negative effect (Coefficient = -0.684, Std. Error = 0.1343, z = -5.09, p = 0.000), and lnwater also has a significant negative impact (Coefficient = -0.3115, Std. Error = 0.0455, z = -6.84, p = 0.000), while the constant term is significant (Coefficient = 8.5611, Std. Error = 0.9289, z = 9.22, p = 0.000); the model's summary statistics show within R-squared = 0.6321, between R-squared = 0.255, and overall R-squared = 0.1132, with a Wald chi2 statistic of 163.7 (p = 0.000), and rho = 0.9413, confirming the significance of random effects, with all results based on the dependent variable lnsdg and calculations by the authors.

Table 8. Results of Hausman Test

Variable	FE Coefficient (b)	RE Coefficient (B)	Difference (b-B)	Standard Error
lnurban	0.0977	0.1809	-0.0832	0.0201
lnagri	-0.1755	-0.684	0.5086	0.3815
lnwater	-0.4808	-0.3115	-0.1694	0.0393
Chi-squared statistic (df=3)	21.31**			

Note: Dependent Variable: lnsdg.
** implies significant at 0.05 level.
Source: Authors' own calculations.

The Hausman test results as indicated in Table 8 show that for lnurban, the Fixed Effects (FE) coefficient is 0.0977, the Random Effects (RE) coefficient is 0.1809, with a difference of -0.0832 (Standard Error = 0.0201); for lnagri, the FE coefficient is -0.1755, the RE coefficient is -0.684, with a difference of 0.5086 (Standard Error = 0.3815); for lnwater, the FE coefficient is -0.4808, the RE coefficient is -0.3115, with a difference of -0.1694 (Standard Error = 0.0393), and the chi-squared statistic (df=3) is 21.31 ($p < 0.05$), indicating that the Fixed Effects Model is preferred, with all results based on the dependent variable lnsdg and calculations by the authors.

Since the FEM is chosen based on the Hausman test, the results indicate that lnurban has a positive significant effect on lnsdg, suggesting that urbanization contributes to sustainable development, while lnagri is not significant, implying that the agricultural sector does not have a strong impact on lnsdg in this model, whereas lnwater has a significant negative impact, highlighting that water resource availability negatively affects sustainable development, possibly due to inefficiencies or overuse; the model explains 66.17% of the within-group variance (R-squared = 0.6617), with a highly significant F-statistic (F = 73.04, p = 0.000*), and a high intraclass correlation (rho = 0.9985), confirming the strong presence of fixed effects.

V IMPLICATIONS

These findings underscore the necessity for BRICS nations to adopt a multifaceted policy approach that integrates sustainable urban development with resilient agricultural practices and comprehensive water governance, ensuring that urban expansion is managed through smart infrastructure, green spaces, and sustainable transportation, while agricultural policies are restructured to enhance productivity through technology-driven, climate-smart farming and land-use efficiency; simultaneously, the pressing challenges of water resource depletion demand urgent measures such as improved irrigation techniques, wastewater recycling, and transboundary water cooperation to prevent resource overexploitation, mitigate environmental degradation, and safeguard water security for future generations; moreover, given the strong country-specific effects, policymakers must tailor interventions to each nation's unique socio-economic and environmental context, leveraging regional cooperation within BRICS to share best practices, mobilize green financing, and establish joint frameworks that support sustainable development, ultimately enabling these economies to achieve their SDG commitments while fostering long-term economic resilience and environmental sustainability.

VI CONCLUSION

The study concludes that urbanization significantly enhances sustainable development in BRICS countries, while the agricultural sector does not show a strong impact, and water resource availability negatively affects sustainability, emphasizing the need for strategic policies to optimize urban expansion, improve agricultural sustainability, and enhance water management; the FEM is identified as the most suitable approach based on the Hausman test, highlighting the importance of country-specific factors in shaping sustainability outcomes, and the high intraclass correlation (rho = 0.9985) confirms strong fixed effects, reinforcing that policy interventions should be personalized to each BRICS nation's exclusive economic, environmental, and institutional conditions to effectively achieve sustainable development goals. Furthermore, this study underscores the critical need for geographically targeted and data-driven policy frameworks that optimize urban expansion, enhance agricultural sustainability, and improve water resource management, while fostering regional cooperation among BRICS nations to develop integrated strategies that address the challenges of rapid urbanization, declining agricultural productivity, and water scarcity; this requires policy interventions that promote sustainable urban planning to accommodate growing populations while minimizing environmental degradation, support agricultural innovation through climate-resilient farming techniques and efficient land use practices, and implement comprehensive water governance frameworks that ensure equitable distribution, conservation, and sustainable utilization of water resources, ultimately enabling BRICS countries to achieve long-term sustainability by balancing economic growth with ecological resilience and resource efficiency. Moving forward, BRICS countries must implement targeted policies that promote sustainable urbanization through smart city planning, enhance agricultural resilience with innovative and climate-adaptive farming techniques, and develop integrated water management strategies to ensure efficient utilization and conservation, ultimately fostering a balanced approach to sustainable development that aligns with national priorities and global SDG commitments.

REFERENCES

Aderemi, T. A., Opele, A. M., Olanipekun, W. D., & Al-Faryan, M. A. S. (2023). A panel analysis of FDI inflows and poverty reduction in BRICS countries: An implication for the sustainable development goal one. *Transnational Corporations Review*, *15*(4), 35–41. DOI: 10.1016/j.tncr.2023.08.003

Amin, N., Sharif, A., Shabbir, M. S., & Pan, Y. (2025). Evaluating the impact of natural resource rents, R&D expenditures, green finance and energy efficiency on carbon emissions in BRICS economies: Greening the path to carbon neutrality in the post-COP 27 era. *Technology in Society*, *81*, 102826. Advance online publication. DOI: 10.1016/j.techsoc.2025.102826

Amin, N., Sharif, A., Tayyab, M., & Pan, Y. (2024). Green Technological Advances and Resource Rents as Levers for Carbon Reduction in BRICS: Implications for SDGs 7, 8, 9, 12, and 13. *Sustainable Development*. Advance online publication. DOI: 10.1002/sd.3294

Bahturidze, Z., Vasilyeva, N., Shahoud, Z., & Pogodin, S. (2024). Inclusion as a Driver for the Implementation of SDG-30 by the BRICS Countries. *Springer Geography*, (Part F2388), 135–147. DOI: 10.1007/978-3-031-50407-5_11

Guang-Wen, Z., Murshed, M., Siddik, A. B., Alam, M. S., Balsalobre-Lorente, D., & Mahmood, H. (2023). Achieving the objectives of the 2030 sustainable development goals agenda: Causalities between economic growth, environmental sustainability, financial development, and renewable energy consumption. *Sustainable Development (Bradford)*, *31*(2), 680–697. DOI: 10.1002/sd.2411

Kai, Z., Sharaf, M., Wei, S.-Y., Al Shraah, A., Le, L. T., Arvind Bedekar, D. A., & Bani Ahmad, A. Y. A. (2024). Exploring the asymmetric relationship between natural resources, fintech, remittance and environmental pollution for BRICS nations: New insights from MMQR approach. *Resources Policy*, *90*, 104693. Advance online publication. DOI: 10.1016/j.resourpol.2024.104693

Kumar, A., Soti, N., Gupta, S., & Deepa, . (2024). Addressing energy poverty in BRICS economies: Insights from panel data analysis and policy implications for sustainable development goals. *Environment, Development and Sustainability*. Advance online publication. DOI: 10.1007/s10668-024-05652-9

Lin, K., Wang, L., Ranjan, R., & Zhou, H. (2024). The changing contexts of international aid_ examining Indian experiences for a BRICS way. *India Review*, *23*(3), 245–265. DOI: 10.1080/14736489.2024.2341570

Ofori, E. K., & Appiah-Opoku, S. (2024). Sustainable Development Goals in BRICS and G7 countries: Increasing accomplishments through policy synergies in four dimensions. *Sustainable Development (Bradford)*, *32*(1), 287–306. DOI: 10.1002/sd.2653

Saba, C. S., & Ngepah, N. (2024). The impact of artificial intelligence (AI) on employment and economic growth in BRICS: Does the moderating role of governance Matter? *Research in Globalization*, *8*, 100213. Advance online publication. DOI: 10.1016/j.resglo.2024.100213

Sahoo, M., Bhujabal, P., Gupta, M., & Islam, M. K. (2024). Empowering BRICS economies: The crucial role of green finance, information and communication technology and innovation in sustainable development. *Sustainable Development (Bradford)*, *32*(6), 7292–7308. DOI: 10.1002/sd.3083

Wu, L., Adebayo, T. S., Yue, X.-G., & Umut, A. (2023). The role of renewable energy consumption and financial development in environmental sustainability: Implications for the Nordic Countries. *International Journal of Sustainable Development and World Ecology*, *30*(1), 21–36. DOI: 10.1080/13504509.2022.2115577

Zhu, R., Xu, Q., Xiqiang, X., Sibt-e-Ali, M., Waqas, M., Ullah, I., & Anwar, A. (2024). Role of resources rent, research and development, and information and communication technologies on CO2 emissions in BRICS economies. *Resources Policy*, *93*, 105072. Advance online publication. DOI: 10.1016/j.resourpol.2024.105072

ENDNOTES

1 According to World Bank, "Urban population refers to people living in urban areas as defined by national statistical offices. The data are collected and smoothed by United Nations Population Division."

2 According to World Bank, "Agricultural land refers to the share of land area that is arable, under permanent crops, and under permanent pastures. Arable land includes land defined by the FAO as land under temporary crops (double-cropped areas are counted once), temporary meadows for mowing or for pasture, land under market or kitchen gardens, and land temporarily fallow. Land abandoned as a result of shifting cultivation is excluded. Land under permanent crops is land cultivated with crops that occupy the land for long periods and need not be replanted after each harvest, such as cocoa, coffee, and rubber. This category includes land under flowering shrubs, fruit trees, nut trees, and vines, but excludes land under trees grown for wood or timber.

Permanent pasture is land used for five or more years for forage, including natural and cultivated crops."

3 According to World Bank, "Renewable internal freshwater resources flows refer to internal renewable resources (internal river flows and groundwater from rainfall) in the country. Renewable internal freshwater resources per capita are calculated using the World Bank's population estimates."

4 According to SDG Transformation Center, "SDG overall score measures the total progress towards achieving all 17 SDGs. The score can be interpreted as a percentage of SDG achievement. A score of 100 indicates that all SDGs have been achieved."

Chapter 10
A Geographical Critique of SDG Indicators From a Brunei Case Study

Izni Azrein Noor Azalie
https://orcid.org/0000-0003-0616-3241
Universiti Brunei Darussalam, Brunei

Gabriel Yit Vui Yong
https://orcid.org/0000-0002-3707-9256
Universiti Brunei Darussalam, Brunei

ABSTRACT

This chapter critiques the UN SDGS' standardised indicators for overlooking geographical and sociocultural contexts, using Brunei Darussalam as a case study. Brunei's unique characteristics—hydrocarbon wealth, colonial legacies, and low-density urban sprawl—reveal gaps between SDG metrics and local realities. Mixed-methods analysis (2020 VNR, fieldwork, surveys) highlights mismatches: SDG 1's income-based poverty lines ignore housing precarity; SDG 8's GDP growth masks youth underemployment; SDG 11's transport metrics neglect car-centric urbanism; SDG 13 overlooks slow-onset climate risks. The framework's neoliberal focus on growth, critiqued by scholars, obscures spatial inequalities and ecological limits. The study advocates place-sensitive SDG tracking, integrating qualitative insights (community narratives, historical context) and participatory methods (citizen science) to address data gaps and align policies with local needs. Brunei's case underscores the need for context-specific strategies to enhance SDG equity and effectiveness globally.

DOI: 10.4018/979-8-3373-1300-9.ch010

INTRODUCTION

The United Nations (UN) Sustainable Development Goals (SDGs) (United Nations, 2015) is the current stage of the global sustainable development program initiated in 1992 at the Earth Summit at Rio de Janeiro. It began with Agenda 21 and the Rio Declarations, and set up the UNFCCC to address the looming climate crisis. According to the UN's 2024 review, 17% of the SDGs are 'on track' while another 17% are in 'regression', 18% 'stagnant', and 30% making 'marginal progress' (United Nations, 2024). The SDGs with the best performance were "affordable clean energy" (SDG 7), "industry, innovation and infrastructure" (SDG 9), "responsible consumption and production" (SDG 12) and "partnerships for sustainable development" (SDG 17). Among the poor performers were "zero hunger" (SDG 2) and SDG 12, which also 'performed well'. This seemingly conflicting result suggests that while there are signs that society is making progress towards this goal, there are also parts that exhibit regression. This is not surprising, given that consumerism is the hallmark the current economic paradigm, found on neoclassical-capitalistic principles, and hyper-consumerism a chronic symptom. As capitalism thrives in a population that is unequal in status, capability, and access to services, disparity develops over time. The dismal failure of SDG 2 is therefore a logical expectation. By the same token, the failure to attain "quality education" (SDG 4), "decent work and inclusive economic growth" (SDG 8) and "reduced inequality" (SDG 10) goals are not unexpected.

The assessment methodology used is tracking progress, or its lack thereof, is a results-based monitoring and evaluation (RBME) method. It attempts to track sustainable development in 193 countries by means of 17 goals, 169 targets and 241 specific indicators, which is 230 because some overlaps (United Nations, 2016). There have however been some revisions in indicators since 2015 (Our World in Data Team, 2023). The mammoth task of tracking SDGs is dependent largely on data provided by national and international agencies. RBME is data-dependent, and as such, poses a challenge to many developing countries, including Brunei, which do not have comprehensive monitoring agencies and systems, and robust data repositories. Consequently, assessments and indicators have considerable margins of uncertainty. While some may argue that this is best that could be achieve, and that, despite inherent uncertainties, could still provide good insights, we disagree for the following reasons.

(1) Such indicators provide some information of a particular aspect of the target in isolation. In reality, there are indicative of also only a part of the desired goal, which are also interconnected to other parts of the society and environment of interest, i.e., essentially a glimpse of a moving target.

(2) Statistics reduces the characteristics of people and environment to a set of measures, which reflect only very specific parts, and do not provide the full picture. It is akin to viewing the landscape through a set of pinholes and filters. Therefore, it cannot provide a clear view of the heterogeneity and complexity of what is being assessed.
(3) Heterogeneity and complexity are innate within the scope of analysis due to the sheer size and diversity of people and environments in each country. An assessment approach based on the technical indicators employed in the UN SDG methodology cannot capture this aspect of the country vis-à-vis its sustainable development interests.
(4) As the UN's Sustainable Development program showed, targets cannot be achieved simply by political will. It is a process and the journey could deviate from the map. Therefore, a country's history and developmental trajectory must be taken into account. Many developing countries, for example, are recovering from centuries of Western colonialization and some are still shackled by unfavourable political-economic structures.

The SDGs represent a universal call to action, yet their implementation is inherently place-based. Geography, as a discipline that bridge physical systems and human societies, provides useful insights into the developmental-environmental context of RBME. Although not often given sufficient attention, the UN data guide includes geographical considerations as a necessary part of monitoring (United Nations, 2016). This chapter examines how geographical analysis, which examines spatial patterns, regional differentiation, environmental processes and human-environment inter-relationships, could provide more accurate understanding of situation and circumstances associated with each SDG. The Chapter will begin with an examination of the UN SDG, followed by a review of relevant geographical methodologies. It then examines Brunei's SDG, comparing the indicators with geographical perspectives. Salient differences, gaps and issues are discussed. The Chapters concludes with a set of suggestions on how to addressing data gaps and uncertainties, as it aims to improve tracking of SDGs in Brunei.

LITERATURE REVIEW

Literature review covers the history of development of the UN SDG as well as critiques of its applicability to facilitate sustainable transition. It then examines geographical methodologies vis-à-vis sustainable development. The Brunei situation is then reviewed.

1. The Goals of Sustainable Development

The UN SDGs was adopted in 2015 as part of the 2030 Agenda for Sustainable Development, succeeding the UN Millennium Development Goals (MDGs, 2000-2015). It is a continuation of the programme initiated at the Earth Summit in Rio de Janeiro in 1992. The UN SDG is more focussed than the broader and more universal Agenda 21. It combines the environmental focus of Agenda 21 with MDGs' social focus, incorporating gender equality (SDG 5), inequality (SDG 10), and building partnerships (SDG 17). The MDG had been successful in lifting millions out of poverty [1], arguably the most successful anti-poverty movement in history (UNDP, 2015); global extreme poverty ($1.25/day) was halved (from 1.9 billion in 1990 to 836 million in 2015). However, it was criticized for not giving adequate attention to planetary boundaries (UNEP, 2012).

The Rio Declaration provides the ethical and political principles for sustainable development to serve as the philosophical basis for sustainable development. It outlines the rights and responsibilities of nations. The first declaration places humans at the centre of sustainable development. However, although the UN sustainability movement is global in scope and scale, critics argue they lack systemic considerations and fail to address root causes of unsustainability. Some regard them as merely advancing neoliberal policies. Sachs (2015), a stalwart of the UN sustainability movement, pointed out the techno-optimism, as he elaborates on the goals and targets. The Global Tapestry of Alternatives[2], for example, criticised its top-down approach, which ignores grassroots issues and circumstances. Economists, such as Easterly (2015) dismissed the SDGs as a 'fantasy wish list'. He highlighted not considering trade-offs (e.g., economic growth vs. ecological limits) and assuming 'political will' exists to support implementation, as 'magical thinking'.

To some, SDGs perpetuate developmentalism, reinforcing the myth of it as a universal good, and ignore pluriverse (Escobar, 2018; Kothari et al., 2019). SDGs are viewed as continuation of imposing Western development models on the Global South, while ignoring indigenous and grassroots alternatives. Furthermore, it promotes growth-based development, which inevitably causes ecological harm (Hickel, 2017). 'Green growth' is a myth, because decoupling GDP from resource use, typically 2-3% annually, is insufficient because of the limits of physical reality. In order to stay within 1.5°C warming, high-income nations must cut carbon emissions by ~10% annually (Hickel, 2020). SDG 8 (Decent Work & Economic Growth) contradicts environmental goals, because 'infinite growth is impossible in a finite planet'. Instead, Hickel proposed 'degrowth' in rich nations (e.g., scaling down ecologically harmful sectors), radical redistribution (sharing resources equitably to meet human needs without growth), and 'steady-state economics', i.e., capping resource use within planetary boundaries (like Herman Daly's proposals).

Shiva (2016) denounced SDG 2 as simply enabling 'corporate capture', because it legitimises agribusiness (e.g., 'sustainable intensification') to undermine small farmers, and promotes 'false solutions', like GMOs and carbon markets, which would dominate over agroecology. Spash (2019; 2020) equates SDG with 'neo-liberal greenwashing', because they rely on market-based solutions (e.g., carbon trading, corporate partnerships) rather than systemic change. SDG 12 (Responsible Consumption), for example, is hypocritical, because it does not recognise overconsumption in rich nations.

2. Redefining and Reapproaching the SDG Using Geographical Constructs: A Theoretical Overview

Human geography has undergone profound transformations over the past century, shifting from positivist, scientific frameworks to nuanced explorations of place, identity, and cultural hybridity. This evolution reflects broader disciplinary and societal shifts, culminating in understanding localities as dynamic, cosmopolitan spaces shaped by global flows and local agency. Drawing on case studies such as Brunei's national identity and Southeast Asia's regionalism, this synthesis highlights how human geography has moved from deterministic models to embrace the fluidity of cultural and spatial identities in a globalised world. Hence in trying to understand and apply these diverging geographical concepts into analysis of the Sustainable Development Goals (SDG) by taking into consideration of the Brunei context, revisiting them is imperative in order to sieve out the nuances by rooting them onto specific space and place in contrast to the generalise and often overarching global and placeless construct that govern sustainability governance.

The Foundations: From Environmental Determinism to Regional Geography

Early human geography was dominated by theological and deterministic frameworks. Theological perspectives viewed nature as immutable and divine, but the Scientific Revolution (1500–1700s) introduced empirical methods, fostering mechanistic views of human-environment relationships (Galileo, 1610; Newton, 1687). By the 19th century, "Environmental Determinism" emerged, positing that physical environments dictated societal development. Thinkers like Herbert Spencer and Halford Mackinder (1904) framed societies as organisms shaped by natural conditions, a logic co-opted to justify colonial hierarchies (Semple & Huntington, 1915). This paradigm was challenged by Carl Sauer's (1925) "Morphology of Landscape", which emphasised human agency in shaping landscapes. Sauer (1925) argued that "culture is the agent, the natural area is the medium, and the cultural landscape is the result," shifting focus to human creativity over environmental determinism. However, his work lacked engagement with power dynamics, a gap later addressed by critical geographers. Regional geography, epitomised by Hartshorne's (1939) "areal differentiation," sought to classify regions through descriptive taxonomies but struggled to account for human subjectivity. By the mid-century, dissatisfaction with these frameworks had spurred spatial science and quantification (Schaefer, 1953), yet these approaches still overlooked place's symbolic and contested nature.

The Cultural Turn: Place-Making, Identity, and Power

The late 20th century saw a "cultural turn" in human geography, prioritising "sense-making," identity, and power. Humanistic geographers like Tuan (1977) and Relph (1976) introduced concepts such as "topophilia" (emotional bonds to place) and "place identity," framing place as a lived, socially constructed phenomenon. Agnew's (1987) triadic model—location, locale, and sense of place—underscored the complexity of place-based identities, while postmodern critiques highlighted how dominant narratives marginalise minority voices (Mitchell, 2000).

Brunei's national identity, rooted in the "Malay Islamic Monarchy (MIB)", exemplifies place-making through symbolic governance. Public spaces, religious structures, and cultural practices blend traditional values with modern statecraft, reflecting Foucault's (1977) notion of governance as regulatory and symbolic. Signs, signifiers and symbols are utilised to develop the near sense of identity among the citizens to 'internalise' role, creating a conduct of conduct that eventually will aid the State from further governmentality (Dean, 2015). However, globalisation introduces hybridity: urban youth culture and digital media challenge MIB's dominance, creating tension between static identity constructs and fluid, globalised realities (Pieterse, 2009). This mirrors broader trends where "place branding" (Kavaratzis, 2004) curates identities to attract capital, illustrating the politicised nature of cultural representation.

Accordingly, our sensescapes and attachments to these places should also consider incorporating all human experiential senses beyond words and images (Bunkše, 2012). Hence, the locale concept reflects local communal culture is central to cultural geography and landscape studies. Carl Sauer (1959) highlighted the significance of cultural landscapes as cultural artefacts, enabling a deeper understanding of the human-environment relationship. These landscapes serve as texts representing the communities residing within them, revealing meanings through lived experiences (Thwaites & Simkins, 2006; Hubbard, 2017). This shift underscores the recognition of multiple realities within communities, necessitating a more insider perspective rather than an outsider's viewpoint (Thompson, 2009). This focus helps address the misinterpretation of signals from research subjects, emphasising the importance of connecting with local communities for accurate insights. Tuan (1979, 1991) argues that places evoke unique narratives shaped by physical characteristics, history, and communal language, leading to a sense of place or genius loci. Some places may transform into religious sanctuaries, illustrating how community perspectives influence meanings (Kinnard, 2014). Tuan (2017) further illustrates that our understanding of landscapes stems from language and cultural artifacts, even amid globalisation's challenges.

Globalisation and Heterogeneous Localities: Hybridity, Regionalism, and Cosmopolitanism

Globalisation has reconfigured human geography, dissolved rigid boundaries and fostered "cultural hybridity". Appadurai's (1996) "scapes" (e.g., ethnoscapes, mediascapes) illustrate how global flows interact with local contexts, creating "glocal" identities. Pieterse (2009) argues that globalisation is not homogenising but a process of cultural recombination or heterogenisation, evident in K-Pop's global appeal and Southeast Asia's economic regionalism. Southeast Asia's regionalism, institutionalised through ASEAN, reflects dynamic negotiations between national sovereignty and collective interests. Originally framed by colonial labels like "Monsoon Asia", ASEAN's identity now hinges on shared values, also known as the ASEAN Way, economic integration, and cultural symbolism (Anderson, 1983). The region's participation in global value chains (Gereffi, 2001) exemplifies "economic regionalism", yet integration remains uneven due to power imbalances. ASEAN's evolution underscores how regions are not static but shaped by political, cultural, and economic processes (Paasi, 1986). Globalisation thus amplifies heterogeneity within localities. In Brunei, Islamic traditions coexist with hypermodern influences, while cities like Singapore hybridise global architectural trends with local heritage. Robertson (1995) defines this as "glocalisation"—global elements that are internalised and reinterpreted, producing unique cosmopolitanisms anchored to place.

Cosmopolitan Localism in a Globalised World

Human geography's journey—from environmental determinism to cultural hybridity—reveals an evolving recognition of place as contested, dynamic, and relational. Hence, "why is a local contextualisation of a global geographical phenomenon like the Sustainable Development Goals (SDGS) imperative in understanding the processes that are soon to follow?". Reiterating our view on why a cosmopolitan outlook, whilst anchoring its essence to the context of the local, is essential stems from the uniqueness of the processes of internalisation that help shape the current worldview of the nation, particularly concerning sustainability. Whilst the Sustainable

Development Goals provided an extensive list of indicators that nations could use to check on their progress as part of the harmonisation of goals, Brunei's case, on the other hand, requires a distinctive way of approaching the SDG's operationalisation. Globalisation does not erase local distinctiveness but enriches it through hybridity, as seen in Brunei's evolving identity and ASEAN's regionalism. These cases exemplify cosmopolitan localism: identities that are globally informed yet rooted in place, blending indigenous traditions with external influences. By embracing fluidity and power dynamics, human geography remains vital in analysing the interplay of local agency and global interconnectedness. With regards to the view on regions, the uniqueness that each region possesses due to its characteristics of being an amalgamation or as Storper (1998) calls it, an agglomeration of activities, highlighted the need to be wary of applying a standardised version of the SDG indicators into such collectivities, as it may result in formulation of undesired policies which may impinged on sustainability implementation. As will be seen in the upcoming analyses in the following few subsections, it is hoped that such geographical approaches would help reinvent newer pathways of approaching sustainability that emphasise the uniqueness of the place at the core of its development agenda.

The Brunei Situation

The analysis presented is based on a decade-long study involving field surveys, map analysis and interviews across different parts of the country, each focussing on different sustainability aspect. Government documents are also used, such as the country's Voluntary National Report (VNR).

Figure 1. Brunei SDG Status

Source: Brunei SDG Website

Brunei submitted its first VNR in 2020. It identified SDGs 11, 12 and 13 as priority areas for intervention. It also drew attention to address issues and challenges, institutional mechanisms, societal awareness and monitoring system. The Special National Coordination Committee on SDGs (SNCC), chaired by the Permanent Secretary (*Wawasan,* or 'Vision') at Prime Minister's Office (PMO) guides, monitors and reports on the nations' SDG status. It is a multi-stakeholder committee of senior government officials, supported by the Office for National Vision (*Pejabat Wawasan Brunei*). The SNCC publishes reports, maintains the SDG Website[3], conduct talks and workshops, enhances national indicators and monitoring, strengthens partnerships with the Non-Governmental Organisations (NGOs), and represents Brunei in international meetings.

Table 1 shows the status of Brunei's SDG progress and where efforts need to be focused, according to SNCC, available on its website (accessed 25 April 2025). To date, 20.7% of the 169 targets are on track to being realised, while 32.0% required more efforts. Crucially, almost half (39.1%) of the targets could not be measured ('no data'). This is particularly problematic with respect to SDGs 1, 5, 10, 11, 13 and 15, where it accounts for over 50% of the targets; 60% in the case of SDG 11 and 16. About a quarter of targets under SDGs 1, 3, 7, 11, 13 and 14 are on track to be achieved, while for SDG 6, 9 and 17, the proportion is higher (over a third).

Brunei's combines its SDG programme with its strategic long-term vision, or Wawasan 2035, which comprises three blueprints to achieve its manpower, social and economic aspirations (Wawasan, 2035). Assessment of SDG progress was carried out using UNESCAP's Current Status Index, which is a measurement of progress in relation to a baseline and a target value (PMO, 2023). It revealed that Brunei has made considerable advances in implementing the 2030 Agenda for Sustainable Development, although data gaps remain significant. The assessment identified challenging SDGs, due to low data availability, to be: 1 (No Poverty); 2 (Zero Hunger); 5 (Gender Equality); 10 (Reduce Inequalities); 11 (Sustainable Cities and Communities); 13 (Climate Action); 14 (Life Below Water); 16 (Peace, Justice and Strong Institutions); and 17 (Partnerships). Brunei will continue to work on addressing data availability and is committed to ensure that progress remains on a good trajectory." (Refer to Fig. 5, Brunei Darussalam vs. Subregional data availability in 2022, PMO, 2023: 20)

However, the government of Brunei is able to provide its citizenry with universal welfare system because of revenue generated by its significant amount of oil and gas resources coupled with a small population (455,500, DEPS, 2024[4]). As such, SDGs 2 (End hunger; achieve food security), 3 (health and well-being), 4 (education), 5 (gender equality), 6 (water and sanitation) and 7 (affordable and reliable energy) have been achieved to a large extent. For the same reason, there has been little pressure to exploit coastal and marine resources. Therefore, SDG 14 (conservation of marine resources) is close to being achieved. However, the terrestrial part of the environment has suffered significant impacts from modern development, which has converted large areas into built environment and disrupted ecosystems. There is therefore a need for more attention on SDG 15 (protection and conservation of terrestrial ecosystem). This is because, although the government has adopted pro-conservation policies in forestry, urbanisation is inadvertently working against on this SDG and targets.

As for SDGs 16 (Promote peaceful and inclusive societies for sustainable development, provide access to justice for all and build effective, accountable and inclusive institutions at all levels) and 17 (Strengthen the means of implementation and revitalize the Global Partnership for Sustainable Development), Brunei is a young developing nation, and therefore an 'apprentice', learning from more advanced nations, its matured neighbours and regional support group, namely, ASEAN. Its participation/contribution rates are limited by its small human resource size and lack of experience and technical capacity.

COMPARATIVE ANALYSIS: INDICATOR VS. GEOGRAPHICAL REALITY IN BRUNEI

In the introduction, we have contended that the RBME-based targets and indicators overgeneralise when applied without giving due consideration to geographical context. As such, they could indicate characteristics quite different from the intended goals founded on the Rio Declarations and the fundamental concept of sustainable development defined by the World Commission on Environment and Development (WCED, 1987). As a result, they could result in contradictory indications, which have implications on achieving the SDG targets. Four SDGs will be examined in greater details and compared with geographical studies to highlight missing contextual gaps.

Between Poverty and Brunei's Opulence Status: A Discussion on Brunei's SDG 1

Analysis of the SDG indicators on poverty showed the use of the internationally accepted convention on numeric indices and poverty lines used to measure or highlight the existence of poverty based on earnings or income per day, as well as proportions or ratios of men and women currently in poverty based on sex and age. While this is helpful for countries with centralised data storage, transparent data collection, and sharing policies supported by consistent data collection initiatives, Brunei Darussalam is still trailing behind. Indicator 1.1.1, for instance, highlighted the need to identify regions where poverty exists based on the urban-rural divide and other local-regional demographical categorisations. However, Brunei has no clear urban-rural delineation due to the urban characteristics that exist in the rural areas, combined with metropolitan areas of low density due to urban sprawling, which disperses the population. The absence of public income data also made it impossible to frame the current conditions of the population, although household expenses data are available.

Arifin et al. (2024) undertook the development of national poverty lines for Brunei, an essential indicator for achieving the targets outlined in Sustainable Development Goal (SDG) 1, specifically target 1.2, despite the word of caution reiterated in the studies by Gweshengwe (2019) and Gweshengwe and Hassan (2020). Their endeavour employs proxies that utilise household income and expenditures as measurement tools, ultimately facilitating the establishment of poverty thresholds. Arifin et al. (2024) concluded with the assertion on the decline in poverty within Brunei over the past two decades through their analysis of the increase in household income, making establishing a poverty line in Brunei redundant. However, contrary evidence is presented by researchers such as Hassan et al. (2024, 2022), Gweshengwe (2020), and Hassan and Gweshengwe (2020), who challenge this assertion. According to Hassan et al. (2024), the lack of formal poverty lines complicates the identification of marginalised groups within Brunei. In such cases, the Brunei Islamic Council's (BIRC) classifications of the destitute (Fakir) and the poor (Miskin) are employed to facilitate access to state welfare provisions. Further scrutiny of Arifin et al. (2024) also highlighted the problems associated with relying solely on published reports, in this case, the Household Expenditure Survey produced by the Department of Economic Planning and Statistics, the Prime Minister's Office. As seen in this and the following few paragraphs, the data used in creating the articles was mainly from grounded quantitative and qualitative primary data sources of the communities and grassroots actors, both part of a funded and non-funded research grant made by the University of Brunei Darussalam (UBD). This gave a more accurate representation of the ground reality not captured by statistical data and reports.

The robustness of social and community support systems across varying geographies significantly influences the efficacy of government-led poverty alleviation programs. Studies conducted by Hassan et al. (2024) and Hassan and Yong (2023) elucidate the pivotal role of the social fabric as a protective barrier against the adverse effects of poverty. Within Kampong Ayer (KA), the former urban centre of Brunei, disparities in poverty levels are evident across different wards. Even those communities from KA that have transitioned to more stable landed areas encounter varying degrees of poverty, influenced by material conditions such as housing, the intricate meanings associated with homeownership (Hassan, 2017), and the vision of basic shelter (Hassan and Yong, 2018).

The multifaceted nature of poverty necessitates innovative, 'network-like' methodologies integrating insights from the relevant Sustainable Development Goal indicators, specifically 1.2.1 and 1.4.2. Despite the presence of comprehensive government welfare schemes—including significantly subsidised national housing initiatives, free education for all citizens, and welfare assistance (Bantuan Kebajikan) provided by the BIRC, alongside contributions from non-governmental organisations like SCOT and Lavida—the operationalisation of these schemes requires alignment with the unique needs of specific communities across diverse geographical contexts. There are growing concerns that repayment of the highly subsidised houses is sometimes not met. This is made worse as the houses themselves have no real value. They cannot be used to generate income, i.e., for letting purposes, nor to conduct business such as open stall operations or home-based production and retail activities. A review of current housing policies is needed to allow citizens to support themselves and be self-determined, reduce government reliance.

Moreover, the qualitative dimensions of life can differ markedly depending on the perspectives of the governed (Gweshengwe et al., 2020). While the social fabric is fundamental in fostering community cohesion through mutual support and self-regulation, urban and peri-urban settings characterised by fragmented socio-communal structures face distinct challenges. These challenges are further compounded by generational diversity and proximity among residents. Households usually comprise Baby Boomers, Millennials, and Generation Z, with diverging needs and wants. Similarly, their worldview on communal relationships compared to their view of individuality created different definitions of what is considered a basic necessity in contrast to luxuries (Gweshengwe, 2020). To effectively address these complexities, it is imperative to adopt place-based approaches that consider how individuals connect with and derive meaning from their living environments (Tuan, 1979) and how these senses of places are constructed (Bunkše, 2012). Similarly, the poor and vulnerable responses towards disaster, its risk and management require proactive mitigation and resilience-building. It is projected that problems associated with climate change and its compounding effects, such as Sea Level Rise (SLR) and increased frequency of rain and floods, would affect those living along the coast and urbanised river catchments differently depending on the place. Pioneering studies on these have been carried out Bruneian scholars (see Yong and Hassan, 2024; Hassan and Yong, 2024 and Hassan et al., 2022) that studies climate change effects on KA communities whilst proposing the need to embrace new models such as the Blue Urbanism and Urban Tapestry Models as part of developing sustainably. This will be discussed further under the SDG 13 subsection on combating Climate Change. Analysing these issues from an insider's perspective is critical for formulating effective policies and mechanisms to alleviate poverty.

SDG 10 on Reduced Inequalities: Are We Soaring Towards Equity or Trapped in the Swamp of Misery? A Discussion

To link Sustainable Development Goal 1 (SDG 1) on No Poverty to SDG 10 on Reduced Inequalities, we examine Brunei's progress towards achieving these goals. Brunei has made significant efforts to provide its citizens with welfare and social security, including universal healthcare and education, all supported by a unique system without direct taxation. SDG 10 aims to reduce inequalities by addressing income disparities, promoting inclusion, and reforming discriminatory systems. As a small, oil-dependent monarchy, Brunei faces both advantages and challenges in meeting these targets. While the country's welfare state ensures broad economic security, data gaps, cultural norms, and systemic inequities hinder progress. An evaluation of Brunei's alignment with SDG 10 reveals the need to refine existing approaches and contextualise global indicators within the local environment. We will further explore these targets and indicators, comparing them to Brunei's geographical and sociocultural context.

Concerning the SDG target 10.1 on income growth for the bottom 40%, by examining the growth rates of household expenditure income per capita for the bottom 40% in contrast to the national average, Arifin et al. (2024) provided a snippet of this, but without the bottom 40%. This is partly due to the limited access to the raw data, forcing them to rely on any available published data as the baseline. Despite universal welfare in the form of subsidies for housing, healthcare, and education, data on income growth for the bottom 40% is scarce. Our observations suggest stagnating incomes amid rising living costs, disproportionately affecting low-income households. This corroborates the earlier findings by Hassan et al (2022) on poverty. Migrant workers, who form approximately 30% of the population, face exclusion from subsidies and wage protections, exacerbating inequality. Some scholarly findings by Nayan and Ullah (2021) and Ullah and Chattoraj (2022) highlighted the precarious living experiences and conditions affecting Indian and South Asian migrants in Brunei. Brunei's welfare system mitigates absolute poverty but fails to ensure equitable income growth. The lack of disaggregated data obscures vulnerabilities, while migrant workers' exclusion perpetuates a two-tiered society. Linked to this is SDG target 10.7, which aims to facilitate safe and regular migration. With a focus on recruitment costs for migrants and implementation of migration policies, it should be noted that migration and recruitment policies exist. However, as mentioned previously, exploitation of migrant workers persists in the form of passport confiscation, wage theft, as well as poor surveillance over transboundary players involved in hiring workers from abroad. Enforcement is weak, with few cases intercepted. Brunei relies on migrant labor but fails to protect workers' rights.. Strengthening monitoring, penalising exploitative employers, and ratifying the International Convention on Migrant Workers are urgent priorities. Addressing this requires targeted data collection and inclusive policies for non-citizens.

For SDG 10.2 and 10.3 relating to empowerment, inclusion, equal opportunity, and non-discrimination, no publicly available data highlighted the proportion of people living 50% below the median income based on demographic breakdown. Universal welfare is generous but inconsistently meets needs, as shown in housing shortages or longer waiting times for medical treatment. Political participation is limited, but the revival of the Legislative Council (LegCo) helped provide platforms for citizens and their representatives to raise issues and put forward potential solutions for the executive members of the government to reconsider. Apart from this, the use of social media to address grievances in recent times has indirectly helped the government to be more critical and serve as checks and balances that can address mismanagement. Social inclusion is framed around the "Malay Islamic Monarchy" national identity, despite the fluidity of the situation, which requires a balancing act to ensure social stability. However, recent developments have put forward a sustainability agenda relating to workplace sexual harassment (see Abu Bakar, 2024) and exploration on strengthening the government's initiatives to combat human trafficking (see Kon, 2025). For Brunei, the economic inclusion is prioritised over social and political empowerment, as stability can only be achieved when basic material needs are met. Strengthening protections for minorities and enhancing civic participation are critical to achieving inclusion.

In relation to Targets 10.4 and 10.5, which revolve around Fiscal and Wage Policies for Equality and regulating the global financial markets, no explicit policies address wage equality or the labour share of GDP. A cultural emphasis on gratitude for one's "allotment" discourages demands for redistributive reforms. Brunei's reliance on oil wealth and passive social contract stifles progress toward equitable fiscal policies. Promoting labour rights and transparent wage negotiations could counter complacency and reduce dependency on welfare. Similarly, financial monitoring is conducted by state and private agencies, but oversight focuses on domestic stability rather than global equity. The transformation of the Monetary Authority of Brunei Darussalam (AMBD) to the Brunei Darussalam Central Bank in 2021 saw the increasing role that the agency plays towards governing fiscal, financial and banking-related activities domestically. While Brunei maintains robust domestic monetary systems, its role in global tax avoidance networks undermines SDG 10's international dimensions. Strengthening transparency in cross-border financial flows is vital. However, Brunei's reservation should be considered, as it relates to the security and sensitivity of financial and economic data, which can be used as a geopolitical weapon in a world increasingly marked by double standards and hypocrisies.

Brunei's global commitments remain peripheral to its policy agenda. Formalising development assistance and auditing remittance costs would enhance compliance with SDG 10's international targets. Brunei's progress toward SDG 10 is uneven. Domestically, its welfare system prevents extreme poverty but overlooks structural inequities affecting migrants, women, and minorities. Culturally ingrained passivity and data gaps hinder inclusive policymaking. Globally, Brunei's engagement with SDG 10's transnational targets is minimal. To advance, there is a need for reforming the legal frameworks that prohibit discrimination and protect migrant rights. There is also the need to improve data transparency by identifying and addressing disparities across time and space, in urban centres or the peri-urban to the rural villages. The country must also continue its presence on the global scene by strengthening its global solidarity through structured development assistance and financial regulation. There has been an increase in regional and international presence made by the country, notably among the higher educational institutions like Universiti Brunei Darussalam (UBD) and Universiti Teknologi Brunei (UTB) as part of their contribution towards impactful education and research. Brunei can leverage its resources by prioritising equity and stability to become a regional leader in sustainable development.

Brunei's Progress on SDG 8 Decent Jobs and Economic Growth: Favourable or Unpropitious?

In relation to SDG 8, which calls for sustained, inclusive, and sustainable economic growth, full and productive employment, and decent work for all, Brunei had also incorporated this component into its national development strategies. Under the Wawasan Brunei 2035 banner, the country has aligned its policies with global goals through initiatives like the *Brunei Economic Blueprint* and the *Digital Economy Masterplan 2025*. Despite facing economic disruptions during the COVID-19 pandemic, Brunei has shown signs of recovery and is actively working to diversify its economy, improve labour conditions, and promote sustainable development. Similar to the approach in the previous subsections, each paragraph will address the interlinked indicators in order to make sense of its ground condition versus the statistical reports.

In relation to Targets 8.1 & 8.2 on economic growth and productivity, Brunei's economy remains heavily reliant on the oil and gas sector, which contributed approximately 52–53% of GDP in 2022 with a small contraction of –1.6%, driven by a 6.2% decline in oil and gas output [Department of Economic Planning and Statistics (DEPS), 2023]. This, however, was balanced out with the growth of the non-oil sector, which grew by 3.1%, led by downstream industries like petrochemicals and fertilisers, as well as trade and transport services. Brunei's GDP at current prices reached BND 23 billion in 2022, with industry accounting for 66.9% and services for 32% (DEPS, 2023). The only problem with using per capita economic growth figures, however, relates to their poor reflection of sustainability, owing to the large income disparity that exists within the communities. To reduce reliance on hydrocarbons, the government is promoting diversification through policies supporting downstream manufacturing, agriculture, digitalisation, and tourism. Labour productivity remains an area for improvement, and ongoing reforms aim to boost output per worker by aligning education and training with market needs. Although Brunei is making progress toward sustainable and diversified economic expansion, there is a need to conduct more grounded studies, particularly on the Micro Small and Medium Enterprises (MSME) experiences in terms of domestic and foreign market accessibility, analyses on enabling policies and infrastructures. Studies by Muhammad (2019), Azalie and Samad (2022) and Azalie (2025) equally showed how sectoral experiences of players in aquaculture and halal food reflect a different picture.

Concerning the employment and domestic labour market conditions under the SDG 8 targets 8.3, 8.5, 8.6 and 8.b, some changes were noted. According to the Prime Minister's Office (PMO, henceforth) (2023), Brunei's unemployment rate declined from 7.3% in 2020 to 4.7% in 2024, signalling a post-pandemic recovery, but some challenges remain, particularly in youth employment. A significant portion of young people are classified as NEET (Not in Education, Employment or Training), prompting targeted intervention with the establishment of Brunei's Manpower Planning and Employment Council (MPEC), responsible for overseeing training and reskilling programs like *i-Ready*, *SkillsPlus*, and *SPIN* (PMO, 2023). According to the PMO (2023), the Manpower Industry Steering Committee (MISC) was also established to connect employers, government agencies, and educators and ensure training matches job market demands in priority sectors such as ICT, logistics, and energy.

The establishment of support for micro, small, and medium enterprises (MSMEs) is also a key strategy to sustain growth. Through Darussalam Enterprise (DARe), over 15,000 entrepreneurs have received training and financial support since 2016. The number of MSMEs rose from 5,248 in 2015 to 5,853 in 2020 (PMO, 2023). These businesses are essential for job creation and economic resilience. However, based on research by Ahmed (2024), the challenges that Brunei MSME have to face stem from the financial leakages caused by cross-border cash drain to close neighbouring states. Unintentional marketing of overseas destinations by both local and international airlines also caused such leakages, preventing a fast growth in domestic spending by local consumers. Efforts to ensure decent work include the introduction of national salary guidelines and the first private-sector minimum wage in July 2023, covering both citizens and foreign workers (PMO, 2023). Workplace well-being is promoted through the National Mental Health Action Plan (2022–25), which encourages flexible work arrangements and childcare support. Whilst the minimum wage for the private sector is timely, it might also be time for the government to reconsider providing more financial support, such as the provisions of the sustenance allowance granted to government civil servants for the Bruneians working in the private sector. This is because, like poverty, the business environment is also uneven due to the small active consumer base, combined with the previous leakages. A thorough study on the MSME and informal sectors would be helpful in measuring their competitiveness and performance.

Brunei's informal sector (target 8.3 and 8.8) grew to 100,100 jobs in 2021, representing a notable share of the workforce, especially in wholesale and retail. While the informal economy provides flexibility, it often lacks job security and social protection. The government recognises this issue and aims to improve working conditions through labour regulation and social policy. According to Eng and Lim (2025), the role of the informal economy is vital as it balances out the deficiencies experienced by formal economies in the form of low overhead costs and highly flexible and adaptive working hours, particularly in emerging economies. However, to ensure compliance and protection of labour, surveillance and monitoring mechanisms by the State should also be established, but it needs to be treaded carefully. A study on the informal halal food producer by Azalie (2025) highlighted the need to construct a different set of rules and regulations for the informal sectors rather than following mainstream ways of managing the industries and businesses. The government is also encouraging the inclusion of marginalised groups, including persons with disabilities, in the workforce.

Regarding sustainable resource use and environmental decoupling, Brunei has committed to decoupling economic growth from environmental degradation, which aligns with SDG 8.4. It boasts approximately 72% forest cover, among the highest in Southeast Asia, and maintains strong conservation policies. However, high material consumption and low recycling rates remain issues. In 2021, Brunei's domestic waste generation was about 1.15 kg per person per day, and the recycling rate had dropped to 8.7%. Observation from the authors noted that Bruneian generates large amount of domestic and commercial waste, largely due to wasteful practices/culture. There is rising support for sustainable consumption among the younger demographic, but the general culture is wasteful. The government has responded with initiatives such as the "No Plastic Bag Every Day" campaign and incentives for green business practices (PMO, 2021). A shift toward eco-friendly infrastructure is underway, as seen in the launch of Brunei's first solar-powered plant in Panaga, Seria in 2021. The broader push for sustainable consumption and production includes developing renewable energy, promoting the circular economy, and public education on waste reduction.

Sustainable tourism (SDG target 8.9) is a strategic growth sector for Brunei, though still relatively small in scale. In 2019, the country welcomed over 333,000 visitors, generating BND 111 million in receipts. However, the COVID-19 pandemic caused a sharp decline, with receipts dropping to BND 135.9 million in 2021 and employment in the sector falling by around 22%. Recovery began in 2022, and tourist arrivals are projected to exceed pre-pandemic levels by 2024. Brunei promotes eco-tourism and cultural tourism, including forest treks in Ulu Temburong National Park, indigenous culinary tours, and Islamic heritage sites. Domestic tourism campaigns like *"Selera Bruneiku"* have encouraged local engagement. The government is also working on tools to measure the cultural and employment impacts of tourism, ensuring alignment with sustainability goals. In studies by Azalie and Yong (2022), Yusof and Azalie (2025) as well as by Chin et al. (2023), among a few, the researchers suggested the need to create a public-private-people partnerships (PPPP) mechanisms to ensure inclusive and sustainable development while at the same time introducing supports for the sector by aligning them with the typology of activities i.e. from ecotourism to cultural-heritage tourism.

Brunei Darussalam has made steady progress in implementing SDG 8. Economic diversification, financial inclusion, labour market reforms, and sustainability initiatives are helping to reduce reliance on oil and gas while creating new opportunities for decent work and inclusive growth. While challenges remain — particularly in youth employment, informal labor, and environmental efficiency — the country's strategic planning and policy alignment demonstrate a strong commitment to sustainable development. Continued investment in skills, green innovation, and inclusive policies will be crucial for achieving the full promise of SDG 8 by 2030.

Sustainable Urban Development

SDG 11 (Make cities and human settlements inclusive, safe, resilient and sustainable) was highlighted as a priority and a challenge in tracking progress. The government has been striving to achieve this goal through various development plans. The current National Land Use Master Plan[5] (NLUMP, 2006-2025) calls for urban consolidation and infill to make existing urban footprint compact, a key sustainable city strategy (e.g., Soleri, 1969; Yong, 2022, etc.). The Construction Industry Framework (2023-2035) also promotes transforming the industry towards a sustainable future. It has also introduced a Green Building Certification and a Green Rating System, which promotes the wider use of sustainable materials and designs, such as: energy efficiency; water efficiency; environmental protection; and indoor environmental quality (PMO, 2023: 71).

However, the indicators for SDG 11 track: the proportion of people living in slums or inadequate housing (11.1.1); population that has convenient access to public transport (11.2.1); land consumption: population growth (11.3.1); proportion of civil society participation in urban planning and management (11.3.2); funds allocated to natural and cultural conservation (11.4.1); disaster casualties (11.5.1) and economic loss and infrastructural damage (11.5.2); adequate domestic waste management (11.6.1) and air pollution, of particulate matter PM2.5 and PM10 (11.6.2); and the inclusiveness of public spaces (11.7.1) and crimes due to discrimination (11.7.2). SDG 11 and SDG 9 (Build resilient infrastructure, promote inclusive and sustainable industrialization and foster innovation) are interrelated, because sustainable city must be supported by sustainable infrastructure.

The indicators for inclusive and sustainable infrastructure focuses on: % rural population within 2km of 'all season' roads (9.1.1); passenger/freight volume by mode of transport (9.1.2); manufacturing value in GDP; % of employment in manufacturing (9.2.2); % small-scale industries in the industrial sector (9.3.1) and % with access to finance (9.3.2); CO2 emission (9.4.1); % GDP invested in R&D (9.5.1) and % population who are researchers (9.5.2); as well as international aid (9.1); % medium and high-tech industry (9.b); and mobile network coverage (9.c). Although these statistics may align with the notion of inclusive and resilient infrastructure and cities, they only provide corroborative snapshots of sustainability at best, because they do not offer any details of the ground reality.

The Brunei government has removed slums from the urban area, provided affordable public transport and waste collection and disposal services, monitors particulate matter (PM10) and made public places accessible without discrimination. However, there are still a significant number living in dilapidated dwellings in the historic 'Water Village' (Hassan et al., 2022, 2024), inadequate public transport due to limited network coverage, and there are only four air quality monitoring stations in the entire country. There are no deaths or economic/infrastructural loss to disasters because the country is advantageously located in an area not afflicted by destructive natural hazards. Environmental hazards, mainly in the form of flash floods and landslides, are of low magnitude and therefore impact society as inconvenience and un-necessary expenses, rather than high-cost destruction.

The masterplan for the capital (HOK, 2011) contains a proposal for a short stretch of light rail transit (LRT), which could have greatly alleviated traffic congestion. This was however never adopted due to cost consideration. Statistics on expenditure, e.g., natural/cultural conservation, are not routinely reported, and they are not readily accessible. Moreover, efforts tend to be ad hoc and rarely sustained or develop further. Civil society is not involved in urban planning or management, but is sometimes consulted or engaged in specific projects. The land consumption (11.3.1) statistics in Brunei is distorted because of national policy and practice, which is in turn, rooted in history and development trajectory.

The additional indictors (11a-c) on the proportion of: population living in well-planned cities (11.a); local government with disaster risk reduction (DRR) systems (11.b); as well as financial support for sustainable infrastructural development, are not readily available because they are in the incipient stage, and therefore difficult to track. Furthermore, due to small size of the country, development is controlled by state. Yet, organic development on private land is common and often dictate development trajectories. With respect to 11.c, the necessary expertise, technologies, products and materials are not readily available, and importing them for specific projects is costly. As a developing nation, and regarded as a wealthy one, Brunei often do not qualify for international development aid, and is constrained by available funds for sustainable infrastructure development, which tends to be more expensive than the standard option. The fact that Brunei is not exposed to natural hazards further consolidate the status quo.

Due to its hydrocarbon resources and high oil prices in the latter half of the 20th century, Brunei was able to build high quality infrastructure as it underwent rapid modernisation. 'All season' asphalt roads connect all settlement in the small country, reaching into the most remote village. The indicator however does not provide any details on the rate of deterioration and need for maintenance due to the combination of intense climate, high rainfall and soft young geology. Transport infrastructure is therefore susceptible to damage by soil erosion, undercutting and scouring, and slope failures. Some sections are often rendered unpassable by flash floods.

As for public transit system, the spread-out, low-density urban pattern is not amenable to mobility via fixed networks, because it would be costly to ensure full coverage. In any case, it would waste the citizenry's time and state resources to sustain such a system. The indicator (9.1.2) therefore provides only a partial picture of the situation, and given the Brunei circumstances, would not help in developing appropriate public transit system. Meanwhile, e-hailing service do provide a more environmentally sound mobility alternative to private vehicles.

As for indicators on manufacturing (9.2; 9.b), this is unrelated to resilience, inclusion or sustainability. Instead, it would be better to track the proportion of corporations that have adopted environment management systems, as well as the number of projects that have passed Environmental Impact Assessments (EIA) and required to comply with Environmental Management and Monitoring Plans (EMMP). The proportion of small-scale industries (9.3.1) and their access to funding (9.3.2) could be used to track economic growth and more equitable share in development. However, as with manufacturing, there is no indication on environmental protection. Sustainable industrial development for small businesses could be achieved, for example, via a cluster system, such as an eco-industrial park, where EIA, EMS and EMMP could be implemented at the park level, to which individual firms could subscribe. This would also reduce CO2 emission and improve its monitoring.

Sustainable development, and its transition, required high level of local knowledge input to be appropriate and effective. Indicator 9.5.1 on funding as a percentage of GDP is a good measure on whether the country is allocating sufficient funds towards attaining their SDG commitments. This is similarly true of indicator 9.5.2 on full-time researchers, although there is a need to promote research that is directly relevant to the country, and specific regions. In many research institutions, foreign researchers work on their pet interests and the local hierarchy tends to focus on climbing academic ranking, rather than addressing local priority issues. The latter has directly relevance to SDG, while the former might, but not necessarily so.

SDGs 9 and 11 aimed at building resilience in the human environment, in particular cities (SDG 11) where the population and economy are concentrated. SDG 11 seek to "make cities and human settlements inclusive, safe, resilient and sustainable". SDG 9, "build resilient infrastructure, promote inclusive and sustainable industrialization and foster innovation", focuses on the infrastructure. Appropriate infrastructure is need to foster and facilitate inclusion and innovation. However, sustainability, which includes environmental protection, requires designs that flow with natural processes, i.e., congruent with nature (McCullough, 20xx) instead of curbing them, which is the norm. They must therefore be deliberately ecological in structure, leveraging on ecosystem dynamics to provide the intended targets, such as flood alleviation, comfortable climate, and food/water/energy supply locally. An ecological design would facilitate regeneration, reduce waste/wastage, enhance efficiency, avoid negative impacts of the natural environment, biodiversity and ecosystem functions (Yong, 2022).

Climate Change

SDG 13 (Take urgent action to combat climate change and its impacts) is highly relevant to Brunei, because it is a coastal state and therefore at risk of rising sea levels and higher energy seas and extreme hydro-meteorological events. The Brunei government has been slow to take the necessary steps to prepare for adapting to the environmental consequences of climate change. The study will compare and contrast between indicators and reality from geographical analysis. SDG 13 on taking urgent action to combat climate change and its impacts is tracked by five indicators: adoption of a national DRR system (13.1.1); deaths from disasters (13.1.2); operationalised a climate change institutional system (13.2.1); integrated climate DRR into education (13.3.1) and socialised DRR (13.3.2); set up a climate fund (13.a); and provision of aid for climate adaptation (13.b). Brunei has established a national agency tasked with DRR, and one to drive its climate change adaptation strategies. Fortunately, the country does not lie in the path of high-magnitude natural hazards, and therefore, has not had to respond to any climate disaster. As a coastal state, however, it is crucial that Brunei prepares adequately for the impacts of climate change in the medium-term. The IPCC AR6 (IPCC, 2021) projected sea level to rise 0.4-0.8 m in the region by the end of the century. The effects on the coastal zone would be greater.

The targets however associate climate change with disasters. This is misleading, because disasters rarely occur in any location. In Brunei, the probably is almost nil. Climate Change however refers to the change in climatic pattern over the long-term, based on records of 30 years or more. In Brunei, it is manifested as increasing frequency of flash floods and landslides. Although they are generally of low-magnitude in Brunei, they disrupt life and work, and incur a cost as a consequence of damage and lost opportunities. In recent years, squalls have become quite common. However, it is unclear if this is due to a change in climate, or a short-medium term anomaly. Forest fires, or more accurately, 'bush fires', are also a regular occurrence. They are however relatively small in extent, and generally isolated. In Brunei, bush fires occur in urbanised areas. They are more likely a consequence of urbanisation rather than climate change. Knowledge of the natural environment, as well as human-environmental interaction provides a different view than the targets and indicators.

DISCUSSION

The NVR noted that the SDG 8 indicator showed a fall in the unemployment rate, but efforts are needed to reduce the share of youth not in education, employment, or training (PMO, 2023:18), suggesting a situation that is somewhat complex. The Water Village ('Kampong Ayer') is a suitable example to illustrated the complexity of place in Brunei. Figure 1 shows the Water Village, a compact urban mix-mesh of dilapidated dwellings on stilts and modern buildings that was once the capital city of the Brunei sultanate, dating back almost 600 years (Hassan & Yong, 2018; 2019; Azalie & Yong, 2022; Hassan & Yong, 2023). The set of uniform modern houses sandwiched in between the old stilted dwelling and the modern urban-scape in the background is a Water Village improvement project built in a location that was ravaged by fire. This once economically vibrant area is now largely a settlement. It is also a leading tourist destination, but with limited commercial activity and therefore employment opportunity, and unstable infrastructure. A significant portion of the population are barely just getting by (Hassan et al., 2022).

Figure 2. Historic Water Village with Bandar Seri Begawan in the Background

SDG	Goals	Targets	No Data	On Track	More Effort
SDG 1	No Poverty	7	4	2	1
SDG 2	Zero Hunger	8	2	1	5
SDG 3	Good Health & Well-being	13	0	3	10
SDG 4	Quality Education	10	3	1	4
SDG 5	Gender Equality	9	5	0	4
SDG 6	Clean Water & Sanitation	8	2	3	2
SDG 7	Affordable & Clean Energy	5	1	1	2
SDG 8	Decent Work & Economic Growth	12	5	2	4
SDG 9	Industry, Innovation & Infrastructure	8	2	3	2
SDG 10	Reduced Inequality	10	5	1	3
SDG 11	Sustainable Cities & Communities	10	6	3	1
SDG 12	Responsible Consumption & Production	11	5	2	2
SDG 13	Climate Action	5	3	1	1
SDG 14	Life Below Water	10	4	2	3
SDG 15	Life on Land	12	6	2	4
SDG 16	Peace, Justice & Strong Institutions	12	5	2	2
SDG 17	Partnerships for the Goals	19	8	6	4
Total		169	66	35	54
(%)		100	39.1	20.7	32.0

According to the NVR, policies under welfare programmes were reviewed and adjusted in 2020 to ensure that they align with the present-day situation (PMO, 2023). Economic growth allows higher fiscal provisions on national programmes that delivers national visions especially in areas promoting equity such as employment, health, education services, and unemployment benefits. However, the Water Village shows that some areas will have characteristics distinctively different from national statistics. This heterogeneity is natural and evident on the ground, and is true across the surface of the planet. The Water Village is also considerably more vulnerable to climate change, even though its inhabitants have adjusted to living in this dynamic estuarine environment (Yong, 2021). As the people and place are facing multiple sustainability issues, Hassan and Yong (2023) proposed combining SDGs 1, 11 and 13 as the most logical solution. Indeed, the SDGs were intended to be implemented together, not in a piece-meal manner. They are meant to be mutually supportive, because this is the underlying concept and principles of sustainable development (WCED, 1987; Rio Declaration).

The NVR noted that Brunei is currently working on setting a target to maintain labour share of GDP. (PMO, 2023: 66). But the reality is that there is large income disparity and the lower echelons find it difficult to make ends meet, i.e., quite vulnerable. The NVR also highted the need for Brunei to address the 'regressing trends' for Goal 12 (Responsible Consumption and Production), which includes a decrease in recycling rate (25.2% in 2015 to 8.7% in 2021) (PMO, 2023: 19). This is an outcome of affluence, high income disparity and ineffective planning and governance. The large data gap reflects this reality. The indicators therefore do not provide an accurate picture of the situation. The studies cited, which encompassed a range of geographical methodologies, elucidated the complexity of place, including conditions that SDGs were designed to remedy.

CONCLUSION

Notwithstanding its utility in cross-country comparisons, the RBME approach to tracking SDG is inherently myopic because it does not provide details of the context and wider environment. The studies cited showed that geographical studies could effectively augment tracking of SDG by providing place-specific details. This is important for government agencies tasked with achieving the targets set to guide development towards a sustainable future. In Annex IV of Report on SDG (United Nations, 2016), this statement precedes the list of goals: the "indicators should be disaggregated, where relevant, by income, sex, age, race, ethnicity, migratory status, disability and geographic location, or other characteristics, in accordance with the Fundamental Principles of Official Statistics (General Assembly resolution 68/261)". "Geographic location" is a critical factor in the tracking system, which should not be ignored. Place-specific data could be acquired by recruiting the citizenry as informants ('citizen scientists') to provide the required set of data needed to track SDG progress more effectively.

Brunei's preference for a "community-based approach to close the wealth disparity in rural and urban areas through charity drives and advocacy work" and adopting a "whole-of-nation approach" (PMO, 2023: 65) bodes well for its sustainability programme. This enables the bridging of gaps, including the continuous efforts by

NGOs in reaching out to the marginalised communities and engendering impact. It also has a policy to "ensure that those in need particularly vulnerable individuals such as Persons with Disabilities (PWDs), underprivileged individuals, senior citizens and the minorities are afforded equal rights and opportunities in building better lives, and can participate fully in society." (ibid). Tracking SDG progress should therefore involve establishing a system where the people in the different geographical regions could provide feedback to the monitoring agency. This is certainly feasible given the pervasiveness of social media, general willingness of the Bruneian citizenry to participate in governance, and it would definitely close the data gaps.

REFERENCES

Abu Bakar, R. (2024). Call for new survey to determine prevalence of sexual harassment in the workplace. *The Scoop*. https://thescoop.co/2024/03/16/more-women-report-workplace-harassment/?

Agnew, J. (1987). *Place and politics: The geographical mediation of state and society*. Allen & Unwin.

Ahmed, K. (2024). Cross-Border Cash Drain: Analyzing Economic Leakage from Brunei to Malaysian Borneo. *Available at SSRN* 5098543.

Anderson, B. (1983). *Imagined communities: Reflections on the origin and spread of nationalism*. Verso.

Appadurai, A. (1996). *Modernity at large: Cultural dimensions of globalization*. University of Minnesota Press.

Arifin, E. N., Ananta, A., Musa, S. F. P. D., & Hoon, C. Y. (2024). A pioneering study on measuring poverty in the hydrocarbon-rich state of Brunei Darussalam. *Asian Population Studies*, *20*(2), 222–240. DOI: 10.1080/17441730.2023.2240108

Azalie, I. A. (2025). Islamic Governmentality and Food Business Development: The Case of Brunei Halal Certification (2011- 2021). (PhD Thesis, University College London).

Azalie, I. A., & Samad, N. (2022). Brunei as a global halal hub: Production network and strategic relations in halal food production. *Journal of Contemporary Islamic Studies*, *8*(3), 1–22.

Azalie, I. A. N., & Yong, G. Y. (2022). *Heritage Industry as Conservation Strategy for Kampong Ayer*. Brunei-Museum Journal.

Brunei Darussalam, S. D. G. https://www.sdgbrunei.gov.bn/

Brunei Darussalam Department of Economic Planning and Statistics (DEPS). (2023). Brunei Darussalam. *Statistical Yearbook. Annuaire Statistique, 2022*. https://deps.mofe.gov.bn

Bunkše, E. V. (2012). Sensescapes: Or a Paradigm Shift from Words and Images to All Human Senses in Creating Feelings of Home in Landscapes. *Proceedings of the Latvia University of Agriculture: Landscape Architecture & Art*, *1*(1).

Chin, S. W. L., Hassan, N. H., & Yong, G. Y. (2023). The new ecotourists of the 21st century: Brunei as a case study. *Cogent Social Sciences*, *9*(1), 2191444. DOI: 10.1080/23311886.2023.2191444

Dean, M. (2015). Foucault must not be defended. *History and Theory*, *54*(3), 389–403. DOI: 10.1111/hith.10767

Easterly, W. (2015). "The SDGs Should Stand for Senseless, Dreamy, Garbled", Foreign Policy. https://foreignpolicy.com/2015/09/28/the-sdgs-are-utopian-and-worthless-mdgs-development-rise-of-the-rest/

Eng, R., & Lim, S. (2025). The Influence of the Informal Economy on the Growth Rate of Real GDP within the Association of Southeast Asian Nations. *International Journal of Economics and Financial Issues*, *15*(3), 59–65. DOI: 10.32479/ijefi.18312

Escobar, A. (2018). *Designs for the Pluriverse: Radical Interdependence, Autonomy, and the Making of Worlds*. Duke University Press., DOI: 10.1215/9780822371816

Galileo, G. (1610). *Sidereus nuncius*. Thomas Baglioni. DOI: 10.5479/sil.95438.39088015628597

Gereffi, G., Humphrey, J., & Sturgeon, T. (2001). Introduction: Globalisation, value chains and development. *IDS Bulletin*, *32*(3), 1–8. DOI: 10.1111/j.1759-5436.2001.mp32003001.x

Gweshengwe, B. (2019). A critique of the income poverty line and global multidimensional poverty index. *Southeast Asia: A Multidisciplinary Journal, 19*, 37-46.

Gweshengwe, B. (2020). *Understanding poverty in Brunei Darussalam* (Doctoral dissertation, Universiti Brunei Darussalam).

Gweshengwe, B., & Hassan, N. H. (2020). Defining the characteristics of poverty and their implications for poverty analysis. *Cogent Social Sciences*, *6*(1), 1768669. DOI: 10.1080/23311886.2020.1768669

Gweshengwe, B., Hassan, N. H., & Maricar, H. A. (2020). Understanding quality of life in Brunei Darussalam. *Cogent Social Sciences*, *6*(1), 1–13. DOI: 10.1080/23311886.2020.1838705

Hartshorne, R. (1939). *The nature of geography: A critical survey of current thought in the light of the past*. Association of American Geographers.

Hassan, N. H., Rigg, J., Azalie, I. A., Yong, G. Y., Zainuddin, N. H. H., & Shamsul, M. A. S. M. (2022). Making do and staying poor: The poverty context of Urban Brunei. *Geoforum*, *136*, 132–141. DOI: 10.1016/j.geoforum.2022.09.012

Hassan, N. H., Rigg, J., Yong, G. Y., Azalie, I. A., Muhammad Shamsul, M. A. S., & Zainuddin, N. H. (2024). Administering and encountering the poor: Poverty from above and below in Brunei Darussalam. *Singapore Journal of Tropical Geography, 45*(2), 311–331. DOI: 10.1111/sjtg.12545

Hassan, N. H., & Yong, G. Y. (2018). A vision in which every family has basic shelter. *Sustainable development goals in Southeast Asia and ASEAN: National and regional approaches*, 190-209.

Hassan, N. H., & Yong, G. Y. (2023). Climate Change and Sustainable Development of Kampong Ayer: Its History, People, and Environment. In Hamdan, M., Anshari, M., Ahmad, N., & Ali, E. (Eds.), *Public Policy's Role in Achieving Sustainable Development Goals* (pp. 96–114). IGI Global Scientific Publishing., DOI: 10.4018/978-1-6684-8903-1.ch006

Hassan, N. H., Yong, G. Y. V., Azalie, I. A. N., & Kamarulzaman, N. (2022). Urban Tapestry: A Place-Sensitive Approach to Sustainable Urban Design. *International Journal of Environment, Architecture, and Societies*, *2*(02), 109–121. DOI: 10.26418/ijeas.2022.2.02.109-121

Hassan, N. H. P. H. (2017). *Housing matters: the value of home ownership in Brunei Darussalam*. Institute of Asian Studies, Universiti Brunei Darussalam.

Hickel, J. (2017). *The Divide: A Brief Guide to Global Inequality and its Solutions*. Penguin Random House UK.

Hickel, J. (2020). *Less is More: How Degrowth Will Save the World*. Random House.

HOK. (2011). *Bandar Seri Begawan Development Master Plan* (Vol. 1-4). BSB Municipal Department, Government of Brunei Darussalam.

Hubbard, P. (2017). *City*. Routledge. DOI: 10.4324/9781315471259

IPCC. (2021). *Climate Change: The Physical Science Basis. Contribution of Working Group I to the Sixth Assessment Report of the Intergovernmental Panel on Climate Change, edited by Valérie Masson- Delmotte, Panmao Zhai, Anna Pirani, Sarah L. Connors, Clotilde Péan, Sophie Berger, Nada Caud, Yang Chen, Leah Goldfarb, Melissa I. Gomis, Mengtian Huang, Katherine Leitzell, Elisabeth Lonnoy, J.B. Robin Matthews, Thomas K. Maycock, Tim Waterfield, Ozge Yelekçi, Rong Yu, and Baiquan Zhou*. Cambridge University Press.

Kavaratzis, M. (2004). From city marketing to city branding: Towards a theoretical framework for developing city brands. *Place Branding*, *1*(1), 58–73. DOI: 10.1057/palgrave.pb.5990005

Kinnard, J. N. (2014). *Places in motion: The fluid identities of temples, images, and pilgrims*. Oxford University Press. DOI: 10.1093/acprof:oso/9780199359653.001.0001

Kon, J. (2025). Brunei joins regional effort to combat human trafficking. *Borneo Bulletin*. https://borneobulletin.com.bn/brunei-joins-regional-effort-to-combat-human-trafficking/?

Kothari, A., Salleh, A., Escobar, A., Demaria, F., & Acosta, A. (Eds.). (2019). *Pluriverse: A Post-Development Dictionary*. Authors Up Front.

Mackinder, H. J. (2004). The geographical pivot of history (1904). *The Geographical Journal, 170*(4), 298–321. DOI: 10.1111/j.0016-7398.2004.00132.x

Mitchell, D. (2000). *Cultural geography: A critical introduction*. Wiley-Blackwell.

Nayan, H. M. B. H., & Ullah, A. K. M. (2021). Indian migrants in Brunei: The role of social networks. *Migration Letters : An International Journal of Migration Studies, 18*(6), 697–709.

Newton, I. (1687). *Philosophiæ naturalis principia mathematica*. Royal Society. DOI: 10.5479/sil.52126.39088015628399

Our World in Data Team. (2023). SDG Tracker: Measuring progress towards the Sustainable Development Goals. OurWorldinData.org. URL: https://ourworldindata.org/sdgs

Paasi, A. (1986). The institutionalisation of regions. *Fennia, 164*(1), 105–146. DOI: 10.11143/9052

Pieterse, J. N. (2009). *Globalization and culture: Global mélange*. Rowman & Littlefield.

Prime Minister's Office (PMO). Brunei. (2021). *National Climate Change Policy: Strategies and Action Plan*. Bandar Seri Begawan: Prime Minister's Office.

Relph, E. (1976). *Place and placelessness*. Pion.

Robertson, R. (1995). Glocalization: Time-space and homogeneity-heterogeneity. In M.

Sachs, J. (2015). *The Age of Sustainable Development*. Columbia University Press. DOI: 10.7312/sach17314

Sauer, C. (1925). The morphology of landscape. *University of California Publications in Geography*, 2(2), 19–53.

Schaefer, F. K. (1953). Exceptionalism in geography: A methodological examination. *Annals of the Association of American Geographers, 43*(3), 226–249. DOI: 10.1080/00045605309352114

Shiva, V. (2016). Who Really Feeds the World? North Atlantic Books., ISBN-10.

Soleri, P. (1969). *Arcology, City in the Image of Man.* MIT Press.

Spash, C. L. (2019). Time for a Paradigm Shift: From Economic Growth and Price-Making Markets to Social Ecological Economics. WU Vienna University of Economics and Business. SRE - Discussion Papers No. 2019/07 DOI: 10.57938/36774aa4-5549-4b97-89cc-ca64006577f2

Spash, C. L. (2020). Apologists for growth: Passive revolutionaries in a passive revolution. *Globalizations, 18*(7), 1123–1148. DOI: 10.1080/14747731.2020.1824864

Storper, M. (1998). *The regional world.* The Guildford Press.

Thompson, F. C. (2009). Nomenclature and classification, principles of. In *Encyclopedia of insects* (pp. 707–714). Academic Press. DOI: 10.1016/B978-0-12-374144-8.00191-0

Thwaites, K., & Simkins, I. (2006). *Experiential landscape: an approach to people, place and space.* Routledge. DOI: 10.4324/9780203462096

Tuan, Y.-F. (1977). *Space and place: The perspective of experience.* University of Minnesota Press.

Tuan, Y. F. (1979). Space and place: humanistic perspective. In *Philosophy in geography* (pp. 387–427). Springer Netherlands. DOI: 10.1007/978-94-009-9394-5_19

Tuan, Y. F. (2017). *A historical geography of China.* Routledge. DOI: 10.4324/9781315083537

Ullah, A. A., & Chattoraj, D. (2022). *COVID-19 pandemic and the migrant population in Southeast Asia: Vaccine, diplomacy and disparity* (Vol. 2). World Scientific. DOI: 10.1142/12761

Ullah, A. A., Hassan, N. H., Mohamad, S. M., & Chattoraj, D. (2021). Privileged migrants and their sense of belonging: Insider or outsider? *Asian Journal of Social Science, 49*(3), 161–169.

United Nations. (2016). Report of the Inter-Agency and Expert Group on Sustainable Development Goal Indicators (E/CN.3/2016/2/Rev.1).

United Nations. (2024). *The Sustainable Development Goals Report.* URL https://unstats.un.org/sdgs/report/2024/The-Sustainable-Development-Goals-Report-2024.pdf

World Commission on Environment and Development. (1987). *Our Commo Future.* Oxford University Press.

Yong, G. Y. V. (2021). Identifying Cultural Traits of the Historic Kampong Ayer of Brunei Darussalam using Biomimetic Analysis. *International Journal of Environment, Architecture, and Societies, 1*(2), 83–94. DOI: 10.26418/ijeas.2021.1.02.83-94

Yong, G. Y. V. (2022). Climate-Resilient Urban Design from a Biomimicry-Arcology Perspective. *International Journal of Environment, Architecture, and Societies, 2*(1), 1–15. DOI: 10.26418/ijeas.2022.2.01.1-15

Yong, G. Y. V., & Hassan, N. H. (2024). Blue Urbanism: An Insight of Navigating Climate Change in Humid Tropical Cities in Brunei Darussalam. *International Journal of Environment, Architecture, and Societies, 4*(01), 23–36. DOI: 10.26418/ijeas.2024.4.01.23-36

Yusof, B., & Azalie, I. A. (2025). Jawi and cultural heritage conservation: A case study on Brunei's Yayasan Sultan Haji Hassanal Bolkiah (YSHHB) Islamic Calligraphy and Arts Study Centre. In Knight, D., & Truong, V. D. (Eds.), *Heritage tourism in contemporary Southeast Asia.* Springer. DOI: 10.1007/978-981-96-5427-7_12

ENDNOTES

1 https://sustainabledevelopment.un.org/futurewewant.html

2 https://globaltapestryofalternatives.org/

3 https://www.sdgbrunei.gov.bn/

4 https://deps.mofe.gov.bn/SitePages/Population.aspx

5 http://www.mod.gov.bn/tcp/Lists/Latest%20News/DispForm.aspx?ID=109

Chapter 11
Achieving Sustainable Growth Amid Demographic Change: SDG 8 in Thailand's National Development Plan

Lili Yulyadi Arnakim
https://orcid.org/0000-0002-4575-424X
Independent Researcher, Indonesia

Safira Puspitarani
Bina Nusantara University, Indonesia

ABSTRACT

Sustainable Development Goal 8 on decent work constitutes an important pillar to Thailand's economic development. Its recent National Development Plan puts emphasis on transitioning into an aging demography and building its digital economy which are reflected in the incorporation of SDG 8 indicators. Key indicators in Thailand's NDP include enhancing skills and added value and a safe working environment for all. This study examines the integration of Sustainable Development Goal 8 indicators into Thailand's National Development Plan, with a particular focus on the challenges posed by an aging population. The research primarily identifies key actors and relevant programs involved in this integration process, and analyzes their impact on labor market trends in Thailand. Findings indicate that several initiatives have contributed meaningfully to enhancing digital readiness, stimulating economic growth through private investment, and improving access for older individuals to safe and supportive work environments.

DOI: 10.4018/979-8-3373-1300-9.ch011

1. INTRODUCTION

The United Nations (UN) has established the Sustainable Development Goals (SDGs) as a global priority to be achieved by 2030. To reach these goals, each member country must develop strategies and adjust their policies in accordance with their national conditions and needs (Okitasari & Katramiz, 2022). As a member of the United Nations, Thailand demonstrates its commitment to achieving the SDGs through various policies implemented, where the SDGs serve as a roadmap in the planning and execution of Thailand's development. Thailand has successfully integrated the Sustainable Development Goals (SDGs) into the country's development framework, as reflected in the 20-Year National Strategic Plan and the application of the Sufficiency Economy Philosophy to support the SDGs (Department of International Organizations Ministry of Foreign Affairs, 2021). Thailand also utilizes the National Development Plan (NDP) which is updated every five years, to set the direction for policies and development priorities that align with the SDGs.

However, the achievement of the Sustainable Development Goals (SDGs) in Thailand has been called into question. Thailand faces challenges that hinder the country from reaching the SDGs, particularly SDG 8, which focuses on "*decent work and economic growth*," due to the aging demographic trend in the country. This issue has been rooted in Thailand since a long time ago. According to the World Bank, Thailand has recorded an age dependency ratio of 44% since 2000 (World Bank Data, n.d). Meanwhile, a report from the World Health Organization (WHO) stated that since 2005, people aged 60 and above have accounted for 10% of the total population (World Health Organization, 2023). It is also important to note that by 2030, Thailand is projected to have 21.3% of its population aged 65 and above. This will position Thailand as the country with the second-highest proportion of elderly people in ASEAN, following Singapore, and classify it as an 'aged society' (KPMG, 2024). This surge in the elderly population indicates a significant burden on the working-age group in supporting Thailand's economy.

The issues mentioned have a profound negative impact on Thailand's ability to achieve the SDGs. This is due to the declining economic prospects of Thailand, caused by decreased productivity resulting from a labor shortage, followed by a decline in aggregate demand. Additionally, the government's expenditures have increased, particularly in the healthcare and pension sectors. As a result, the population has become more dependent on financial assistance from the government. This aging issue needs to be addressed promptly with the right solutions, given the World Bank's prediction that Thailand's economy will only grow by 3% over the next two decades due to low levels of private investment (Anamwathana, 2024). Therefore, effective and adaptive policies are needed, considering the significant changes continuously occurring, including shifts in social, economic, and demographic structures (UNFPA, n.d).

Thailand has stated that the National Economic and Social Development Plan (NESDP) has integrated the Sustainable Development Goals (SDGs) as an effort to address various domestic challenges while also supporting the achievement of global goals. This NDP is designed to ensure synergistic coordination among sectors and government agencies in the implementation of the SDGs, thereby making a significant contribution to national progress. The document also serves as a strategic instrument with transformative potential for Thailand's development (Department of International Organizations Ministry of Foreign Affairs, 2021). However, the main concern lies in the extent to which this NDP is truly capable of addressing the challenges posed by an ageing society. Moreover, the success in tackling this issue is closely linked to the achievement of the SDGs, particularly the goal of Decent Work and Economic Growth, which is crucial for maintaining economic stability and competitiveness amid demographic shifts.

The demographic changes serve as a key basis for further research, with a primary focus on analyzing the extent to which Thailand's National Economic and Social Development Plan (NDP) accommodates the implementation of SDG 8, "Decent Work and Economic Growth," across various ministries and sectors. This study focuses on the period from 2017 to 2021 such as 12th NESDP to assess how effectively the NDP has integrated the SDGs into its planning. By analyzing relevant content from the NDP and examining stakeholder involvement in the policymaking process, this research aims to gain a deeper understanding of how the implementation of the NDP contributes to identifying both the progress made and the challenges Thailand faces in achieving the SDGs. In particular, it will examine whether the SDG framework is still effective in addressing and resolving national development challenges in Thailand. The findings are expected to provide valuable insights for designing more effective and responsive policies to promote sustainable development in Thailand, ultimately helping the country better align its national strategies with the global SDG agen

2. MATERIAL AND METHODS

2.1 Literature Review

Integrating Sustainable Development Goals (SDGs) into national development plans is considered crucial for sustainable development, aligning local priorities with global objectives. To examine this integration, we will review perspectives from various authors focusing on aligning national policies with SDGs. Additionally, we will highlight the significance of national development plans in achieving success for countries, encompassing a diverse range of nations and stakeholders.

First study from Valensisi & Karingi (2017) argue that implementing SDGs within the framework of regional development can improve policy coherence and enhance a country's ability to negotiate internationally. This is evident in how African nations approach the SDGs, reinforcing their regional development strategies' conceptualization, implementation, and evaluation frameworks. Adopting a national approach to SDGs enables countries to tailor targets and indicators that better align with their specific domestic development conditions and challenges. This perspective is further supported by Allen et al. (2021), who stress that treating SDGs as a national priority is essential for helping countries achieve sustainable development. Additionally, the level of ambition and the policies established by a country are critical in ensuring consistent progress. For instance, Fiji has successfully incorporated SDGs into its national policy framework and continues to make significant strides toward achieving all SDG pillars by 2030. Through SDG transformation scenarios, Fiji has increased investment and expedited progress toward meeting its SDGs, achieving an 83% improvement by 2030, which will bring substantial long-term benefits to the nation. However, there is evidence that some countries continue to set their own priorities. Filho et al. (2020) support this view, arguing that many countries prioritize their national interests over the collective global agenda. One of the main challenges in implementing the SDGs is the inherent conflicts between various goals, which are often difficult to reconcile. The relevance of global development goals at the domestic level is crucial to consider, given the different contexts in each country. As noted by Moyer & Bohl (2019), the achievement of the SDGs, when analyzed at the global level, often does not reflect the realities at the national level due to the distinct characteristics and development challenges in each country.

In addressing demographic challenges, Thailand still has the potential to achieve global goals through the implementation of effective government policies. According to Mohammed et al. (2024), the active involvement of the government and stakeholders plays a crucial role in promoting social development. Governance approaches aligned with the Sustainable Development Goals (SDGs) can accelerate policy implementation by ensuring transparency, accountability, and building trust among involved parties. The study also highlights the importance of fostering constructive partnerships between the government, society, and various organizations to achieve sustainable development outcomes. Similar to Thailand, Singapore is also facing the challenges of an aging society. It has taken significant and proactive measures to address this issue. The government actively promotes the development of community-based health and social care services as a strategic response to population aging. Through strong collaboration between the government, communities, and public health institutions, Singapore has successfully enhanced access to preventive care services, aiming to improve both the quality and duration of healthy life among its elderly population in a more effective and sustainable manner.

This study aims to analyze the extent to which Thailand's national development plan aligns with the achievement of the Sustainable Development Goals (SDGs), particularly SDG 8. The main focus is to identify the gap between Thailand's commitments to the SDGs and the reality on the ground. The novelty of this research lies in examining how Thailand's National Development Plan (NDP) is interconnected with the SDGs, specifically SDG 8, aligning global goals with national priorities, with a particular emphasis on addressing the challenges of an aging society. This study reveals how the NDP integrates demographic issues to achieve inclusive and sustainable economic growth.

2.2 Concept Framework

The article utilizes the concept of SDG 8, "Decent Work and Economic Growth," to outline the challenges humanity faces in achieving equitable progress across economic, social, and environmental dimensions of sustainable development. SDG Goal 8 addresses disparities among countries, characterized by significant differences in economic levels, employment patterns such as high unemployment and informal work, and insufficient decent job opportunities hindering expected economic growth. These imbalances underscore the need to examine and address institutional and societal frameworks and value systems. SDG Goal 8 provides a framework emphasizing community capabilities as catalysts for transformative change. By enhancing collective community capabilities, communities can influence economic dynamics through investment decisions, technological advancements, and structural economic transformations (World Bank, n.d).

SDG Goal 8 underscores the interdependence of economic, environmental, and social dimensions in achieving inclusive and sustainable global development by 2030. It emphasizes the creation of full, productive, and decent employment as essential for this goal. The International Labour Organization (ILO) stresses the cohesive adoption of policies and strategies that reinforce each other to enhance productivity, foster a healthy work environment, and stimulate economic growth through societal engagement (United Nations, 2019). Policymakers must develop comprehensive strategies to align transformative change across local, national, and international levels to effectively tackle underdevelopment.

SDG Goal 8 includes 12 main targets aimed at sustainable development. To demonstrate the alignment between SDG 8 targets and elements in the NDP, several SDG 8 targets and indicators will be used as references.

Table 1. An Overview of Several SDG 8 Targets and Indicators Used in This Reseach

Sustainable Development Goals (SDGs)	
Target	**Indicators**
8.2: Achieve higher economic productivity through diversification, innovation and technology development, as well as focusing on sectors with high value-added and labor-intensive characteristics.	8.2.1: Annual growth rate of real GDP per worker.
8.3: Promote policies that prioritize development on supporting productivity, the creation of decent jobs, entrepreneurship, creativity, and innovation, while fostering the establishment and growth of micro, small, and medium-sized enterprises, along with enhanced access to financial services	8.3.1: The number of informal jobs in total employment, by sector and gender.
8.5: By 2030, Achieve high productivity and full employment, ensuring decent work for all genders, including youth and persons with disabilities, with equal compensation for work of equal value.	8.5.1: Average hourly earnings of workers, based on gender, age, occupation, and persons with disabilities. 8.5.2: Unemployment rate, by age, gender, and persons with disabilities
8.8 Protecting labor rights and fostering the creation of a safe and secure working environment for all workers, including migrant workers, especially female migrants, and those in vulnerable employment.	8.8.1: The number of fatal and non-fatal occupational injuries per 100,000 workers, categorized by gender and migration status 8.8.2: Level of national compliance with labor rights (freedom of association and collective bargaining) based on the International Labour Organization (ILO) and national legislation, by gender and migration status.

Source: Compiled by the author based on data from the United Nations Statistics Division.

2.3 Research Methods

This study adopts a qualitative approach with a focus on document analysis to understand the challenges of integrating Pillar 8 of the SDGs into Thailand's National Development Plan (NDP). It will explore and evaluate the implementation efforts between 2017 and 2021 through the review of national documents, such as Thailand's 12th National Economic and Social Development Plan (NESDP), which will be used to assess the integration of SDGs into Thailand's national agenda, particularly regarding financing to improve job quality, as well as economic, social, and political strategies, and the actors involved in addressing issues in Thailand, such as the unresolved challenges of an aging society (Bowen, 2009).

This study is based on text analysis from previous studies and literature, including official statements and media coverage. The researcher will also investigate the implementation of various strategies through available official state documents. The analysis will take into account specific assumptions and compare SDG target indicators with existing national programs. To ensure the validity of the findings, this research will involve multiple perspectives, including media responses, Thailand's economic statistical data, and international documents. The study aims to assess Thailand's NDP contribution to SDG implementation and analyze the cause-and-effect relationships between the integration of these global goals into their national plans, considering the domestic issues Thailand faces.

3. DISCUSSION

3.1 Synchronization of the National Development Plan and the Sustainable Development Goals

The NDP plays a strategic role in supporting the achievement of sustainable development at the national level. In addition, the NDP is also aimed at promoting economic growth and creating jobs that can strengthen the country's economy. In this regard, SDGs assign responsibility to each country to meet established targets and indicators through the implementation of the NDP. Countries are required to integrate the SDGs into their national development policies and contribute to the collective achievement of global goals (Okitasari & Katramiz, 2022). In the context of localization, the SDGs play a crucial role in helping countries address political and economic challenges that may threaten national stability (Alger & Dauvergne, 2020). Nevertheless, countries still have the authority to adapt or even reject global goals if they are considered incompatible with their domestic conditions and needs (Okitasari & Katramiz, 2022). As a member state of the United Nations, Thailand contributes to the achievement of the Sustainable Development Goals (SDGs) by integrating them into its National Development Plans (NDPs). It is undeniable that Thailand's NDPs, which are updated every five years, play a highly crucial role in supporting the country's sustainable development efforts.

Currently, Thailand is implementing its 13th National Economic and Social Development Plan for the period 2023–2027. This plan focuses on building an innovation-driven economy through the restructuring of the manufacturing and service sectors. Additionally, it aims to enhance human resource capacity in responding to globalization challenges while also creating new opportunities for society. The plan also supports the transition toward sustainable production and consumption, with the goal of strengthening Thailand's ability to manage the dynamics of ongoing global changes (National Economic and Social Development Council, n.d.).

To assess the effectiveness of national development, the evaluation will focus on the 12th National Development Plan, which has already been completed.

Table 2. The alignment between the SDG 8 targets and Thailand's 12th NESDP

Sustainable Development Goals (SDGs)		**National Development Plan (NDP)**	
Development Strategy			
Target	Indicators	Target	Indicators
8.2: Achieve higher economic productivity through diversification, innovation and technology development, as well as focusing on sectors with high value-added and labor-intensive characteristics.	8.2.1: Annual growth rate of real GDP per worker.	8.2 Promoting the use of innovation, communication, and technology (ICT) to strengthen competitiveness in the agricultural, industrial, and service sectors, while improving quality of life (Strategy 8). 3.4 Increase the country's productivity (Strategy 3).	8.2.2 Over 30% of R&D outputs and technologies are applicable to create value (production, services, and business). 8.2.4 The creation of innovations within the country for society, the elderly, and the disabled will more than double. 3.4.1 The rate of growth of total factor productivity must reach at least 2.5% per year. 3.4.2 The growth rate of labor productivity must reach at least 2.5% per annum.
8.3: Promote policies that prioritize development on supporting productivity, the creation of decent jobs, entrepreneurship, creativity, and innovation, while fostering the establishment and growth of micro, small, and medium-sized enterprises, along with enhanced access to financial services	8.3.1: The number of informal jobs in total employment, by sector and gender.	3.5 SMEs play an increasingly important role in the country's economy (Strategy 3). 7.5 Strengthening the digital economy by expanding access to high-speed internet services to drive the growth of digital businesses and innovation; as well as improving the cybersecurity system to meet international standards in addressing online threats (Strategy 7).	3.5.1 By the end of the Twelfth Plan, the number of the gross product value of SMEs to GDP is expected to rise by at least 45%. 7.5.3 By 2021, no fewer than 1,000 new digital entrepreneurs are expected to enter the Thai economy.

continued on following page

Table 2. Continued

Sustainable Development Goals (SDGs)		National Development Plan (NDP)	
8.5: By 2030, Achieve high productivity and full employment, ensuring decent work for all genders, including youth and persons with disabilities, with equal compensation for work of equal value.	8.5.1: Average hourly earnings of workers, based on gender, age, occupation, and persons with disabilities. 8.5.2: Unemployment rate, by age, gender, and persons with disabilities.	2.1: Addressing poverty issues and reducing income disparity across socio-economic groups (Strategy 2). 9.1 Improvements in income distribution across regions (Strategy 9). 1.2 The people of Thailand across all age groups possess high levels of skills, knowledge, and abilities (Strategy 1).	2.1.1 The growth rate of per capita income for the bottom 40% of the population should be at least 15% per year. 9.1.1 The disparity in Gross Regional Product per capita can decrease. 1.2.6 The employment of elderly individuals (aged between 60-69 years) is expected to increase.
8.8 Protecting labor rights and fostering the creation of a safe and secure working environment for all workers, including migrant workers, especially female migrants, and those in vulnerable employment.	8.8.1: The number of fatal and non-fatal occupational injuries per 100,000 workers, categorized by gender and migration status. 8.8.2: The national compliance level with labor rights (freedom of association and collective bargaining) based on the International Labour Organization (ILO) and national legislation, by gender and migration status.	5.3 Ensuring security of life and property for people in the southern border provinces through increased opportunities for education and employment to generate more income (Strategy 5). 2.4 The people of Thailand will experience improved health and well-being (Strategy 2).	5.3.1 A decrease in the cost of damages and the number of incidents caused from internal conflicts. 2.4.7 20% of the total elderly population live in homes with appropriate environments.

Source: Elaboration by the author based on data quoted from United Nations Statistics Division, n.d. & Office of the National Economic and Social Development Board Office of the Prime Minister Bangkok.

Table 2 demonstrates the connection between the SDGs targets and indicators with the initiatives Thailand has implemented through the 12th National Economic and Social Development Plan (NESDP) to address various issues related to work and the economy in the country. This plan follows SDG Target 8 as a primary guideline, focusing on enhancing national competitiveness through economic digitalization, workforce skill development, and ensuring worker protection and welfare. Additionally, the plan emphasizes people-oriented development principles, aiming for stability and sustainability in both Thailand's economy and society. The development strategy includes improving productivity, creativity, innovation, and strengthening a smart, high-quality workforce. The ultimate goal is to position Thailand as a country with a strong economy, good governance, and environmental sustainability (Department of International Organizations Ministry of Foreign Affairs, 2021).

As seen in the table above, the issue of an aging society has been mentioned in the 12th NESDP as one of the targets to be addressed during this period. The indicators in the NDP clearly highlight employment and a safe environment for the elderly. Although the indicators in the NDP are quite good in supporting the achievement of SDG Target 8, the NDP still lacks a clear and structured explanation of the allocation of funding. The question is, has the implementation of measures to address the aging society been successful? Based on the table, the mentioned indicators appear to be too general to effectively resolve this issue. From that we must see the program implemented by the government during the 12th NESDP period.

3.2 Actors Involvement in Achieving Global Goals and Addressing National Challenges

To achieve SDGs and Thailand's ambition to become a high-income country by 2036, the government has been actively working to improve quality of life and expand employment opportunities for all citizens. During the implementation of the 12th National Economic and Social Development Plan (NESDP), the issue of population ageing became one of the government's key priorities. The government's efforts also include improving the ecosystem for the elderly so that they can become more economically independent. The government introduced three main strategies: first, economic empowerment of the elderly; second, the development of innovations in long-term care (LTC) services; and third, strengthening the role of families and communities as part of the social safety net (Lorthanavanich, 2021). These strategies are in line with Sustainable Development Goal (SDG) 8, particularly in ensuring a safe and inclusive environment for all while promoting sustainable economic growth. They also align with SDG Target 8.8, which focuses on protecting labour rights and promoting safe and secure working environments for all workers.

Besides that, one major initiative from the government is the *Decent Work Country Program (DCWP)*, implemented from 2019 to 2021 in collaboration with the Ministry of Labour and the ILO. The program aims to promote productive and fair employment through gender-responsive labour services aligned with market needs, while also enhancing technical and vocational skills, especially among youth and older workers (International Labour Organization, n.d.). Another key initiative is *Thailand 4.0*, a national development strategy aimed at overcoming the middle-income trap and improving quality of life. A central focus of this initiative is workforce upskilling to meet the demands of Industry 4.0. The government supports the establishment of learning centres, the growth of startups, and the use of information and communications technology (ICT) to foster innovation in entrepreneurship—all aligned with SDG 8 targets (8.2 and 8.3). Furthermore, the initiative emphasizes the development of analytical thinking, creativity, and innovation, while preparing a skilled and adaptable workforce to meet the challenges of modern times, including addressing an ageing population through a flexible, market-responsive education system (Jones & Pimdee, 2017).

In addition, the government is also working to expand tax revenues to ensure that national income remains sufficient to maintain economic stability in the face of long-term demographic changes. For example, in 2020, the government began implementing a land and building tax—which applies to agricultural, residential, commercial, and undeveloped properties—following the enactment of the law in 2018. This tax is levied at rates ranging from 0.01% to 0.7%, depending on the classification and value of the property. The government also initiated social security reforms by raising the age at which pension benefits begin. This measure is part of efforts to support the achievement of SDG Target 8.8, particularly in strengthening social protection and creating a safe and sustainable working system (Kumagai, 2019).

In light of the global shift toward digital transformation and the demographic challenge of an aging society in Thailand, the involvement of multiple stakeholders—particularly the government and the private sector—has become essential in formulating inclusive solutions. This collaboration creates opportunities for the Thai population to prepare for the ongoing changes and challenges in the labor market. One strategic approach to addressing these issues is by improving access to education and skill development in the field of Information and Communication Technology (ICT), largely supported through private sector investment.

Through initiatives incorporated into Thailand's 12th National Economic and Social Development Plan (NESDP), the government has demonstrated its commitment to integrating the targets of Sustainable Development Goal (SDG) 8, particularly goals 8.2, 8.3, 8.5, and 8.6. These targets emphasize boosting economic productivity, promoting entrepreneurship, and creating inclusive and decent employment opportunities. Several major companies have actively contributed to this initiative, including Toyota Motor Thailand, commercial banks, SCG Group, Mitr Phol Group, Advanced Info Service, Bangchak Corporation, and Thai Beverage Group (Theparat, 2019). In addition, global tech firms like Huawei play a crucial role in human resource development. In partnership with the Thai government, Huawei established Huawei Academy—not only as an investment initiative but also as a driver of national economic growth. Huawei Academy focuses on developing digital talent, particularly for SMEs and startups, within the growing digital economy. The initiative aims to train more than 100,000 ICT professionals in Thailand by 2025 (Huawei ASEAN Academy, n.d).

These efforts are part of a broader national strategy to enhance the competitiveness of Thailand's workforce, especially in response to the demographic transition toward an aging population. Equipping the labor force with digital skills is essential to ensure sustainable, inclusive growth in the era of rapid technological advancement.

3.3 Socio-Economic Realities During and After the Period

How reality unfolds could serve as the main indicators to measure whether the government is able to achieve SDGs and domestic goals. According to a report by KPMG, the ageing population trend is also creating new opportunities for the business sector. One notable example is Thailand's healthcare industry, which has seen significant progress in digitalization. This development is a result of collaboration between the Ministry of Public Health and the Ministry of Science and Technology, and is strongly supported by the Thailand 4.0 policy (KPMG, 2024). According to a report by the Bangkok Post, since 2018, the number of businesses in the elderly care sector has increased by 25.1% each year (Yonpiam, 2024). The development of adult care housing not only benefits businesses and elderly individuals but also contributes to economic growth by creating numerous job opportunities in the sector (World Bank, 2021). The impact of the policies and strategies implemented during the 12th NESDP also continues, with the government launching the AI Thailand program in 2022, aimed at improving the country's competitiveness and developing the skills of the workforce (Telecom Review, 2024). This initiative continues with Thailand's commitment to becoming a digital economy, and it is expected to have a positive impact until the 13th NDP is fully implemented.

However, the reality on the ground still presents contrasting outcomes, both in addressing the ageing society issue and in achieving the targets of SDG 8. The challenges faced by the elderly in Thailand are also partly due to the rising cost of living, which was driven in part by the COVID-19 pandemic during the final phase of the NDP (Kasatpibal et al., 2023). Until now, the issue of an ageing population remains a challenge that has not been optimally resolved. Compared to neighboring countries facing similar challenges, Thailand's response appears to be slower. Singapore, for example, has managed this demographic transition more effectively. This success is supported by the financial readiness of its citizens, many of whom have achieved economic stability before reaching old age. Thailand has a relatively high proportion of informal workers, accounting for around 65% of the total workforce between 2017 and 2021 (World Bank Group, n.d). This presents a significant challenge in achieving decent employment and inclusive economic growth. Informal workers face various vulnerabilities, including unsafe working conditions, income instability, job insecurity, and lack of access to pensions. This stands in stark contrast to Singapore, where the workforce is predominantly composed of formal workers who benefit from more structured employment and stronger social protection systems. Moreover, skills gap still exists, as most older adults are not yet technologically literate, which has hindered Thailand's progress in achieving decent employment. Data from the World Bank in 2022 indicates that there is still a significant digital skills gap, with 18.7 percent of youth and adults in Thailand having skills below the literacy and digital skill standards (World Bank, n.d). In contrast, Singapore has made significant strides in overall digital readiness. Despite facing the challenges of an ageing population, its high level of digital preparedness enables the country to respond more effectively. This is largely because Singapore is well-equipped to adapt to the global shift toward a digital economy.

CONCLUSION

Global goals provide a relevant guide for countries in addressing their problems. This research highlights the implementation of SDG 8 in Thailand, which includes decent employment, social protection, and skills development to improve the quality of life. By using the 12th NESDP and analyzing the programs during that period in depth, the research shows that there are still shortcomings, where not all SDG 8 indicators are clearly integrated and mainstreamed. The article assesses that while the Thai government has focused on improving access to elderly care, reskilling, and upskilling initiatives. However, the economic challenges related to the ageing population remain unresolved. Although direct comparison with Singapore is difficult due to differences in land area, a clear issue is the high number of informal workers, which poses a long-term challenge.

Success may depend on external factors such as the global economy. Specifically, we can see that the world is shifting towards a digital economy, with unemployment rates continuing to rise. For Thailand, it may not be easy to resolve this issue in a short time, but the country could consider raising the retirement age while addressing the digital skill gap to enhance national competitiveness and maintain economic stability. It is also important for the Thai government to recognize that implementation is more important than planning. Although this study focuses specifically on the implementation of SDG 8 in Thailand during the 2017-2021 period, its findings can be applied to research on the achievement of other SDG goals more broadly. This study provides a conceptual framework and perspectives useful for assessing how much priority countries place on SDGs and how governments can take preventive measures through the 13th NESDP, which has been evaluated based on the previous NDP, to ensure that Thai society is prepared for progress.

REFERENCES

Alger, J., & Dauvergne, P. (2020, 2). The Translocal Politics of Environmental Norm Diffusion (Vol. 14). https://doi.org/DOI: 10.1080/17524032.2019.1665567

Allen, C., Metternicht, G., Wiedmann, T., & Pedercini, M. (2021, 4). Modelling national transformations to achieve the SDGs within planetary boundaries in small island developing states (Vol. 4). DOI: 10.1017/sus.2021.13

Anamwathana, P. (2024). Thailand's Ageing Society and Young Thais' Changing Views and Expectations. In *ISEAS – YUSOF ISHAK INSTITUTE*. https://www.iseas.edu.sg/wp-content/uploads/2024/03/ISEAS_Perspective_2024_28.pdf

Bowen, G. A. (2009). Document analysis as a qualitative research method. *Qualitative Research Journal*, *9*(2), 27–40. DOI: 10.3316/QRJ0902027

Department of International Organizations Ministry of Foreign Affairs. (2021). Thailand's Voluntary National Review On The Implementation Of The 2030 Agenda For. *Sustainable Development*, •••, 2021.

Filho, W., Wolf, F., Lange Salvia, A., Beynaghi, A., Shulla, K., Kovaleva, M., & Vasconcelos, C. R. P. (2020). Heading towards an unsustainable world: Some of the implications of not achieving the SDGs. *Discover Sustainability*, *1*(1), 2. DOI: 10.1007/s43621-020-00002-x PMID: 38624553

International Labour Organization. (n.d). *Thailand Decent Work Country Programme 2019-2021.*

Jones, C., & Pimdee, P. (2017). Innovative ideas: Thailand 4.0 and the fourth industrial revolution. In Asian International Journal of Social Sciences (Vol. 17, pp. 4–35). https://doi.org/DOI: 10.29139/aijss.20170101

Kasatpibal, N., Viseskul, N., Untong, A., Thummathai, K., Kamnon, K., Sangkampang, S., Tokilay, R., Assawapalanggool, S., & Apisarnthanarak, A. (2023). Impact of the COVID-19 pandemic on the Thai population: Delineating the effects of the pandemic and policy measures. *Antimicrobial Stewardship & Healthcare Epidemiology : ASHE*, *3*(1), e241. Advance online publication. DOI: 10.1017/ash.2023.523 PMID: 38156233

KPMG. (2024). *Thailand's Ageing Society: The opportunities for businesses in Thailand*. https://assets.kpmg.com/content/dam/kpmg/th/pdf/2024/07/thailand-ageing-society-the-opportunities-for-businesses-in-thailand.pdf

Kumagai, S. (2019). Thailand's Efforts to Cope with a Rapidly Aging Population — Can Thailand Become a Trailblazer for Other Middle-Income Countries Struggling with Demographic Aging? *Pacific Business and Industries, XIX*(71). https://www.jri.co.jp/MediaLibrary/file/english/periodical/rim/2019/71.pdf

Lorthanavanich, D. (2021), 'Introduction', in Lorthanavanich, D. and O. Komazawa (eds.), Population Ageing in Thailand Long-term Care Model: Review of Population Ageing Practices and Policies, Vol. 2. ERIA Research Project Report FY2021 No. 06b, Jakarta: ERIA, pp.1-11.

Mohammed, A., Al-Mahdi, N., & Ghaithan, A. (2024). The role of environmental social governance on achieving sustainable development goals in Gulf Cooperation Council countries. Environment, Development and Sustainability. https://www.nber.org/system/files/chapters/c14917/c14917.pdfDOI: 10.1007/s10668-024-05662-7

Moyer, J. D., & Bohl, D. K. (2019). Alternative pathways to human development: Assessing trade-offs and synergies in achieving the Sustainable Development Goals. *Futures*, *105*, 199–210. DOI: 10.1016/j.futures.2018.10.007

National Economic and Social Development Council. (n.d.). The Thirteenth National Economic And Social Development Plan (2023-2027).

Office of the National Economic and Social Development Board Office of the Prime Minister Bangkok. (2017). *The Twelfth National Economic And Social Development Plan.*

Okitasari, M., & Katramiz, T. (2022). The national development plans after the SDGs: Steering implications of the global goals towards national development planning. *Earth System Governance*, *12*, 100136. DOI: 10.1016/j.esg.2022.100136

Telecom Review. (2024, May 31). Thailand's Digital Economy: Utilizing AI Strategies and Transformation Initiatives - Telecom Review Asia Pacific. Retrieved from https://telecomreviewasia.com/news/featured-articles/4276-thailand-s-digital-economy-utilizing-ai-strategies-and-transformation-initiatives/

Theparat, C. (2019, November 7). Bangkok Post - New ministry takes on human factor. Retrieved from https://www.bangkokpost.com/business/1789159/new-ministry-takes-on-human-factor

UNFPA. (n.d.). *A Life-Cycle Approach to Ageing in Thailand.*

United Nation. (2019). Sustainable Development Goal 8: Promote sustained, inclusive and sustainable economic growth, full and productive employment and decent work for all. https://www.globalgoals.org/goals/8-decent-work-and-economic-growth/

United Nations Statistics Division. (n.d.). SDG Indicators — SDG Indicators. https://unstats.un.org/sdgs/metadata/?Text=&Goal=8

Valensisi, G., & Karingi, S. (2017). From global goals to regional strategies: Towards an African approach to SDGs. *African Geographical Review, 36*(1), 45–60. DOI: 10.1080/19376812.2016.1185738

World Bank. (n.d.). Decent work & economic growth | SDG 8: Decent work & economic growth. Atlas of Sustainable Development Goals 2023. Retrieved July 17, 2024, from https://datatopics.worldbank.org/sdgatlas/goal-8-decent-work-and-economic-growth?lang=en

World Bank Data. (n.d.). World Bank Open Data. https://data.worldbank.org/indicator/SP.POP.DPND?locations=TH

World Bank Group. (n.d.). *GENDER AND INFORMAL WORK IN THAILAND*. https://documents1.worldbank.org/curated/en/099506009112325206/pdf/IDU044678b42097e804da90a85008d7c4c45b4b9.pdf

World Health Organization. (2023, February 9). Thailand's leadership and innovations towards healthy ageing. Retrieved from https://www.who.int/southeastasia/news/feature-stories/detail/thailands-leadership-and-innovation-towards-healthy-ageing

Yonpiam, C. (2024, August 17). Ageing society a boost for care industry. https://www.bangkokpost.com. https://www.bangkokpost.com/thailand/general/2848708/ageing-society-a-boost-for-care-industry

Chapter 12
Partnerships for Resilient SMEs in Implementing International Financial Reporting Standards (IFRS) and Knowledge Transfer Strategies to Achieve SDG 17

Mahani Hamdan
https://orcid.org/0000-0003-3859-4433
Universiti Brunei Darussalam, Brunei

ABSTRACT

SDGs can only be accomplished with strong global partnerships and cooperation SDG No 17 advocates for global sustainability partnerships including in financial services and practices. The implementation of International Financial Reporting Standards (IFRS) is aligned with the goal of SDG 17 to help developing countries achieve financial sustainability financing policies by mobilising additional financial resources from various sources, especially for Small and Medium Entreprises (SMEs). There is not much research addressing the implementation process of IFRS in response to Knowledge Management (KM) strategies for sustainable finance. In order to support SDG no 17, this study is to identify and describe the implementation of IFRS framework and knowledge transfer for SMEs readiness.

DOI: 10.4018/979-8-3373-1300-9.ch012

1. INTRODUCTION

The Sustainable Development Goal (SDG) 17 calls for worldwide cooperation for sustainable development. The objective emphasizes the importance of mobilising more financial resources from different sources for developing countries. SDG 17 also seeks to aid developing nations in achieving financial sustainability by promoting resilient financing practices including adoption of International Financial Reporting Standards (IFRS) especially for Small Medium Entreprises (SMEs) at developing countries. IFRS is increasingly being used to harmonise accounting practises around the world in response to the increasingly global model of corporate finance. This helps to increase financial resilient and reduce the costs associated with comparing investment opportunities, improve the quality of information, and provide comparable measurements of corporate profitability in the financial reporting of companies located in other countries. IFRS is relevant to SDG no.17 as it is principles-based accounting standards that were developed in 2001 by the International Accounting Standards Board (IASB). The strategy that is based on principles as compared to the approach that is based on regulations is very different. The first option gives businesses the freedom to treat business transactions according to whatever accounting standards they deem most appropriate, thereby introducing an element of subjectivity into the analysis whereas the latter stipulates that the rules must be obeyed exactly as they are with no room for interpretation.

The majority of nations are of the opinion that it is preferable to adapt accounting standards to a company's activities rather than the other way around. This is despite the fact that the principles-based approach is not without its share of complications (Thomason, 2019). IASB's significant efforts to achieve a single set of high-quality global accounting standards have taken them into some sort of agreement, which can be reflected through (1) its joint convergence project with the US Financial Accounting Standards Board (FASB), and (2) IFRS adoption and endorsement in the European Union. This movement towards the harmonisation of accounting standards is a step toward the goal of achieving a single set of high-quality global accounting standards (EU). Because of a rule passed by the European Parliament and the European Council of Ministers in 2005, all EU-listed firms were required to completely use the IFRS framework when they prepared their financial statements. This resulted in the formal adoption of IFRS by members of the EU in 2005 (IASPlus, 2002).

The topic of whether or not Brunei is ready to comply with IFRS is no longer relevant (Hamdan & Low, 2013), because currently IFRS had already become an obligatory requirement for financial reporting for public-interest entities in Brunei by 2014. In the context of this development, the Ministry of Finance and Economy (MoFE) planned a number of different activities with the purpose of informing all of the relevant stakeholders about the implementation of IFRS to support sustainable financing and practices. In order to make IFRS as a financial reporting requirement applicable to all limited companies as of January 1, 2002, certain pertinent sections of the Companies Act 1956 (Chapter 39) were revised. Auditors were also required to qualify their reports for those companies that were not in compliance with IFRS. IASPlus reports that Bruneian businesses did not fully implement IFRS until 2010, despite the fact that the standards have been available since 2006. Later on in the year 2011, the government of Brunei Darussalam established the Brunei Darussalam Accounting Standards Council (BDASC). This was done with the assent of His Majesty Sultan Hassanal Bolkiah of Brunei through the implementation of the Accounting Standards Order (ASO) 2010. The development of BDASC is a significant step in the right direction toward guaranteeing consistency in accounting standards, easing the process of comparing financial statements of various businesses, and improving the credibility and transparency of financial reporting in Brunei (IFRS, 2017). The BDASC's primary responsibility is to draught the accounting standards that will be used in the country, as well as to address any issues that are connected to those standards. In 2011, the BDASC made a public announcement that the country would embrace and merge its existing Generally Accepted Accounting Principles (GAAP) with the International Financial Reporting Standards (IFRS). From that moment forward, all organisations operating in Brunei are required to comply with both of the following sets of accounting frameworks: IASB's International Financial Reporting Standards (IFRS) will be implemented on January 1, 2014 for all entities that are subject to public accountability (public interest entities), and Brunei Darussalam Accounting Standards (BDAS) will be implemented on January 1, 2018 for all other entities that are not subject to public accountability (non-public interest entities). To put it more succinctly, it is an approach of full adoption of IFRS for public interest entities (PIEs) and a convergence of Brunei's national accounting standards to IFRS for non-public interest organisations (Non-PIEs).

This only pertains to entities that have public accountability, such as banks, financial institutions, insurance and takaful firms, and it concerns the full adoption of IFRS for PIEs. While other organisations that do not meet the criteria for PIEs are not obligated to prepare their financial statements in accordance with BDAS, PIEs are compelled to do so. After having casual interactions with multiple representatives of local banks, insurance, and takaful companies, the researchers came to the conclusion that the adoption of IFRS could be regarded as beneficial. They stated that they had generated their financial statements using the IFRS framework prior to 2014; nonetheless, the extent to which these firms conformed fully to IFRS and correctly used the framework could be possible issues for further inquiry in the future. This study aims to fill in two research gaps that have been identified in the existing body of research. To begin, the vast majority of the research that has been done on the application of IFRS has analysed data from countries that are members of the EU (Callao & Jarne, 2010; Jermakowicz & Tomaszewski, 2006). Furthermore, even though a small number of comparative studies have been carried out using data from countries that are not members of the EU (Borker, 2012; Patro & Gupta, 2014; Yapa & Second, there hasn't been a lot of research done on the reporting requirements of SMEs and how the International Financial Reporting Standards (IFRS) affect their financial performance (either through adoption or convergence) (Albu, Albu & Fekete, 2010; Kaya & Koch, 2015). This study's objectives are to (1) identify and describe the process by which the IFRS framework is being implemented in Brunei; (2) investigate the difficulties or challenges and opportunities encountered by the stakeholders in the process of adopting IFRS; and (3) investigate the role of knowledge management by which those difficulties can be overcome. The review of previous research on the application of IFRS may be found in the next section. This is followed by the methodology, the results and discussion, the recommendations, and the conclusion of the study.

2. LITERATURE REVIEW

More than 130 countries now use the IFRS framework, and the adoption process, as well as the degree of regulatory norms and transparency, varies greatly by country. The willingness of various legal systems to implement IFRS internationally varies across nations. Instead of utilising local GAAP, some nations permit the use of IFRS on a voluntary or mandatory basis. Other nations either alter IFRS as published by the IASB or harmonise their national GAAP with IFRS. According to Kaya and Koch (2015), adopting IFRS voluntarily would likely have fewer economic and political costs than forcing it since businesses would choose if it best meets their unique needs. However, it suggests that fewer companies will report using IFRS, which could hurt the comparability of financial accounts. The path of convergence to IFRS is likely to be more expensive than an adoption strategy because the process is time-consuming and involves an overhaul of operational and IT procedures in addition to knowledge transfer strategies. But given that expenses vary by size, industry, and kind of firm, similar generalisations cannot be made about all countries. For instance, adoption is viewed as a less desirable option for industrialised nations like the United States since it would place a heavy and immediate financial burden on American businesses (SEC Roadmap, 2008). According to Johnson (2008), IFRS convergence is likely to be less expensive than U.S. GAAP because gradual convergence may not always result in significant changes to U.S. GAAP and because the costs of changing the accounting system would be spread out over time, making the financial impact of convergence more manageable. There is undoubtedly no one-size-fits-all method for applying IFRS. No matter the methods used, the adoption of IFRS could be successful due to political, institutional, firm-specific, macroeconomic, and legal factors as well as legal distinctions between countries (Kimeli, 2017). These factors include domestic law enforcement and conformity (Liapis & Thalassinos, 2013), domestic cultural orientation with regard to accounting systems (Borker, 2012), and a strong global trend toward IFRS as a single set of internationally recognised accounting standards (Palea, 2013). In order to analyse the distinctive differences between nations in connection to IFRS adoption benefits, Kimeli (2017) further recommended that more country and area-specific research be done.

Numerous studies have suggested that IFRS adoption has been generally beneficial for listed companies, including the effects of IFRS adoption on the quality of financial reporting (Palea, 2013; Pena & Franco, 2017); earnings management (Capkun, Collins & Jeanjean, 2016; Hasan & Abdul Rahman, 2017); and investment efficiency. However, it has been noted that the outcome of IFRS adoption is not conclusive (Biddle, Callahan, Hong & Knowles, 2013; Chen, Young & Zhuang, 2013). Numerous studies have also been performed in an effort to find solutions to simplify IFRS for SMEs in order to eliminate numerous complex accounting concerns (Albu, Albu & Fekete, 2010; Kaya & Koch, 2015; Perera & Chand, 2015). The findings of these studies do not prove that IFRS is better for SMEs than listed companies or vice versa; rather, they point to the size and scope of the change that IFRS has facilitated in response to the particular requirements of enterprises. Three perspectives—economic, political, and social—could be used to categorise justifications for implementing IFRS. By adopting IFRS, businesses not only have better access to global capital markets for funding but also to cheaper sources of funding that transcend economic and political barriers (Jain, 2011). Because one of the requirements is to produce accounts in accordance with IFRS guidelines, it has also become viable for local corporations to purchase other businesses outside their own borders in order to further their expansion goals and list on the European or American stock exchanges. Despite the fact that IFRS had a negative effect on the liquidity of Greek listed companies, Iatridis and Dalla (2011) demonstrated that the implementation of IFRS had a positive impact on the profitability of the majority of industrial sectors as well as businesses that are part of the FTSE 40 and SMALLCAP 80. Entering a foreign market has helped domestic businesses boost profits, realise economies of scale, attract clients from around the world, and distribute investment risk. It has also had a good social impact on communities by giving people access to employment and business prospects. Bilateral and multilateral trade agreements; cross-border listing; foreign direct investments; and long-term business relationships between domestic and foreign enterprises are all ways to further develop diplomatic ties between nations' economies (Gordon, Loeb & Zhu, 2012; Marquez-Ramos, 2008). Additionally, IFRS may have a direct or indirect effect on a company's performance and financial status. Therefore, even if IFRS were beneficial to the economy as a whole, it is crucial for nations to evaluate the applicability and impact of IFRS to meet local business demands as well as country variations in the perspectives of stakeholders involved in financial reporting (Albu et al., 2013).

That much is certain: IFRS has benefited people all around the world and become more globally integrated. However, the implementation of IFRS has costs and other financial repercussions for business firms. It has long been claimed that the complexity of the entire IFRS imposes a significant cost and that most external users of financial reporting do not see value commensurate with the expense and labour associated with its preparation (Epstein & Jermakowicz, 2010). Whether or not this is true, many companies today believe that IFRS for SMEs will make the transition to full IFRS easier for them. However, it is vital to exercise caution when estimating the implementation cost because actual expenses vary by organisation. While cost-sharing is at the top of the business agenda, Osnoss (2010) suggested that implementing a proper Knowledge Transfer strategies, allocating enough time for implementation programmes, performing improved initial assessments, having a successful project management process, centralising accounting functions, being able to hire personnel from a global talent pool, and having to spend less time managing compliance with specific accounting rules, to name a few, can help to reduce costs. While developing nations frequently lack a strong professional accounting body because they typically lack a long tradition of accounting and auditing or because their systems may be insufficient or nonexistent, advanced countries often enter the human resource crisis with improved financial positions. According to Polizatto, who was referenced in Thompson (2016), a country's enterprises that lack the knowledge or technical capacity to adhere to IFRS may find it difficult to build a financially sound environment (Gyasi, 2009). Furthermore, even though countries offer accounting education at different levels, they lack the resources and the right infrastructure to train and educate accountants.

Kilic and Uyar (2017) claimed that although using IFRS for SMEs aims for high-quality financial information and enhanced comparability of accounts, cost pressures on businesses and a lack of educated employees constitute the key challenges for the implementation process. The costs of complying with accounting standards are regarded to be greater than their benefits for SMEs since they have access to fewer resources and would require more time and effort to teach their staff, according to Dixon, Thompson, and McAllister's (2002) argument (Regoli, 2019). They should take into account their long-term goals and intentions while implementing IFRS because the bulk of them cannot afford to do so. A simplified IFRS is favoured and is designed for entities without public responsibility where adoption of IFRS for SMEs is of interest. According to Jermakowicz and Epstein (2010), the whole IFRS has been modified by removing complicated themes, streamlining recognition and measurement, and minimising disclosures. The hiring of people with accounting abilities is a trend that Wahlen, Jones, and Pagach (2013) predicted will continue and pick up speed (p. 2). The fact is that due to compensation disparity, many SMEs are more inclined to hire graduates with accounting abilities than professional or qualified accountants. Compared to recruiting accounting grads to professional accountants with ACCA qualifications, the cost of hiring graduates is lower. However, their inexperience, improper accounting practises, and misapplication of accounting concepts only serve to increase expenses rather than sales and profits for the companies. The latter, however, requires the knowledge and abilities to offer more comprehensive business reporting. According to Islam (2017), there is still a research gap that addresses the changes that will affect accountants and professional accounting organisations. Higher education institutions are well-positioned to produce more qualified accounting graduates to meet the future demand for accountants by incorporating IFRS and SME-related topics into the accounting curriculum (Banham & He, 2014). Due to the fact that IFRS principles are founded on principles, they provide less prescriptive, interpretive, and implementation assistance than other national standards, necessitating a greater level of professional judgement (Hodgdon, Hughes & Street, 2011; Subramani, 2009). Additionally, dynamic is IFRS. Since fair value accounting requires judgments to be made, junior accountants may need assistance from senior professional accountants in order to "learn and identify the need for judgments and the appropriate steps to form judgments, rather than simply memorising how judgements have been formed in the past" (Hodgdon et al., 2011, p. 436). Additionally, since domestic legislation like the Companies Act and the Income Tax's reporting requirements are not taken into consideration by IFRS, some changes may be required. To enable a seamless transition to IFRS, providing incentives to companies to comply with them may be a reasonable option (Narasimham, 2010). However, this is not a cost-effective strategy, especially since IFRS has become a legal requirement.

It is also frequently noted that Islamic nations are debating whether to adopt IFRS or the Financial Islamic Accounting Standards of the Accounting and Auditing Organization for Islamic Financial Institutions (AAOIFI), primarily because, as Islamic organisations, they are required to adhere to the AAOIFI's standards for the sake of their faith. However, Sharairi (2020) showed that not all IFRS are fit for use with Shariah transactions, and that elements like religion, culture, and local investors may have little impact on how Islamic banks in the UAE embrace IFRS. Applying the AAOIFI guidelines can help businesses prepare their financial statements and treat accounting items correctly based on the right accounting principles, especially for Islamic products. Comparative studies between IFRS and AAOIFI's standards have been conducted but not in order to establish which accounting framework is superior to the other (Hannoudi, 2015); rather, they have been done in order to: (1)

see how both accounting frameworks can converge through a comparison between interest-based loans versus Murabaha; leases versus Ijarah; and construction contracts versus Istisna; and (2) find a way to comply with AAOIFI's standards instead (Sharairi & Islam, 2019). According to Alkali, Alkali, and Aliyu (2017), although IFRS and AAOIFI's standards differ in a number of ways, they both have a common underlying element. In other words, both standards use professional judgement when dealing with principles-based accounting. While there is considerable disagreement on the usefulness of AAOIFI's standards, there is less agreement regarding how they might be applied as a framework for international accounting. According to Ajili and Bouri's (2017) research, seven Islamic banks listed in Bangladesh also showed that the level of compliance with IFRS disclosure criteria was greater than the level of compliance with AAOIFI's norms for 39 Islamic banks of member states of the Gulf Cooperation Council (Ullah, 2013). There hasn't been as much research done on AAOIFI's standards as there has been on IFRS's widespread adoption and impact on company performance. There is little doubt that IFRS has increased in popularity and acceptance as well as "a reputation as a worldwide standard-setter with the European Union and a number of countries choosing to adopt it from 2005 forward" (Subramani, 2009, p.6).

Sustainable Development Goals (SDGs)

It is essential to synchronise SDGs with IFRS. The current state of the global economy is precarious due to the instability of many developing countries to endure the consequences of the COVID-19 pandemic, rising inflation, significant supply-chain disruptions, and increasing labour market pressures (Anshari et al., 2024; Hamdan et al., 2023; Johari et al., 2024). Sustainable Development Goals (SDGs) play a key role in supporting the world's most vulnerable people, communities, and nations so that they can recover from these challenges. Governments, the commercial sector, and civil society must partner for sustainable development (Ordóñez de Pablos et al., 2024; Pattiroei et al., 2023). These inclusive partnerships based on principles and values, a common vision, and shared objectives that place people at the centre are required at the global, regional, national, and local levels (UN, 2020).

No single nation can accomplish the Sustainable Development Goals on its own. To accelerate the Sustainable Development Goals (SDGs), it is necessary to have an inclusive ecosystem that includes all stakeholders, including those from the public and private sectors, academia, civil society, and individuals and their communities (Anshari & Hamdan, 2022). If the basic tenets of the SDGs are not included into the structure of the accounting standards used in financial reporting, then the IFRSs are deficient (Sapovadia, 2017).

For instance, the existing accounting system does not take into account any value lost outside of a company's own financials (Dhar et al., 2022; Sapovadia, 2017). In fact, businesses consume social and environmental resources indirectly, yet the existing accounting system disregards the contributions of ecosystems to human well-being and environmental degradation. "Green accounting" is an emerging decision-making tool for sustainable development (Anshari et al., 2023). Green accounting offers an alternate and methodical approach to integrate environmental contributions. Green accounting seeks to include environmental costs into financial outcomes which is gaining momentum since the United Nation's call for SDGs (Sapovadia, 2017).

Knowledge Transfer Strategies

Knowledge transfer is part of Knowledge Management (KM) activities. KM is simply defined as the act of practicing knowledge that was gained from the past experience for making any decisions for current and future projects for the sole objective of increasing the organization's effectiveness (Anshari & Hamdan, 2022a; Jennex, 2017). The term knowledge and information are used correspondently, so it is better to differentiate the meaning served by the two words (Kakabadse et al., 2003). Knowledge is not visible and not a physical asset. It is transmitted through intellectual process of learning, association, perception and communication and reasoning (Nonaka, and Takeuchi, 1995). Knowledge on the other hand is the procedure in which an organization create value from their creativity and knowledge-based assets (Anshari et al., 2023a; Santosus and Surmacz, 2001). It is the actions of knowledge development and accumulation within the organizational capital whenever necessary. In order to facilitate the dissemination of knowledge within an organization, an efficient knowledge transfer strategy integrates people, processes, technologies, as well as culture and infrastructure (de Pablos, 2002; de Pablos & Lytras, 2008). In order to have a successful implementation of IFRS, it is important that knowledge transfer strategies be a component of a knowledge management (KM) ecosystem (Anshari et al., 2022). This ecosystem should comprise knowledge creation and acquisition, knowledge sharing, knowledge storing, and knowledge retrieval. Therefore, in the analysis and discussion sections, the discussion about knowledge transfer becomes an important part of this research.

3. METHODOLOGY

The primary method of this study is qualitative. It was carried out based on a review of the literature such as journal articles, newspapers, and reliable information from websites. It also employed a bibliographic analysis method derived from prior research employing IFRS keyword phrases. From 2012 through 2022, all publications on IFRS-related topics were retrieved from the Scopus database. The goal of this bibliographic study is to give an overview of trends in IFRS-related topics so that they can be used as a resource by researchers in the future. The study also made reference to the informal conversations the researchers had with pertinent stakeholders and the data they learned from going to IFRS seminars, conferences, and open dialogue sessions with different stakeholders at MoFE and other institutions.

4. FINDINGS

The entire research performance with the IFRS theme displays 1479 scholarly output from 2017 to 2022, with a total of 406 international collaborations from institutions all over the world, as shown in Figure 1. This demonstrates the necessity for IFRS research to be conducted with a greater level of intensity.

Figure 1. Overall Research Performance of IFRS

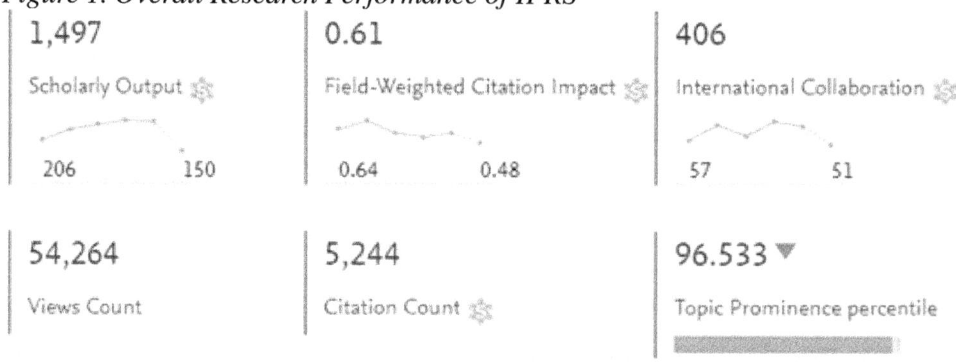

(Source: Scopus up to 03 Aug 2022, Year range: 2017 to 2022)

The keyphrase for the IFRS topic is displayed in Figure 2. The relevance of IFRS research can be identified using one of three colours, beginning with blue, moving on to black, and finally arriving at green. The topic of IFRS research is seen to be declining, as represented by the blue colour, while the trend of researching IFRS is shown to be increasing, as shown by the green colour. From Figure 2, it may be concluded that research must be conducted on a country-by-country basis. So, the discussion in this article backs up how important the IFRS implementation context is for a country like Brunei Darussalam, which is still developing.

Figure 2. Keyphrase Analysis of IFRS

(Source: Scopus up to 03 Aug 2022, Year range: 2017 to 2022)

Figure 3 shows a graph of the distribution of scholarly research outputs in the world, Asia-Pacific and Brunei Darussalam. The fact that there is not much research on IFRS in the Asia-Pacific region and Brunei gives this paper a chance to add to the research.

Figure 3. Scholarly output of IFRS

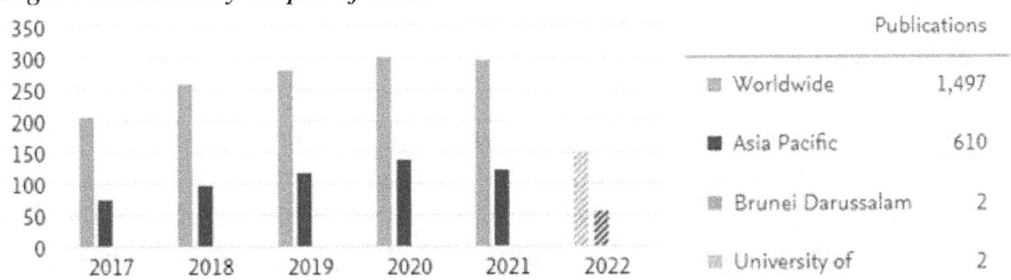

(Source: Scopus up to 03 Aug 2022, Year range: 2017 to 2022)

The number of IFRS studies that are represented in the distribution of scholarly research outputs for each nation is presented in Figure 4. The United States of America, the United Kingdom, China, and Australia are the nations that contribute the most research to IFRS. This offers a compelling argument for the significance of studies undertaken from the point of view of developing countries or micro-state nations like Brunei.

Figure 4. Top Countries scholarly output of IFRS

Countries & territories	Scholarly Output ↓	Views Count	Field-Weighted Citation Impact (excl. self-citations)	Citation Count (excl. self-citations)
United States	279	10,600	0.87	1,573
United Kingdom	147	5,886	0.85	731
China	143	5,040	0.75	547
Australia	122	5,162	1.04	691
Indonesia	77	3,080	0.56	223
South Korea	71	2,085	0.44	215
Brazil	68	1,827	0.35	151
India	57	1,836	0.32	89
Italy	45	2,037	0.78	191
Portugal	44	1,747	0.47	100

(Source: Scopus up to 03 Aug 2022, Year range: 2017 to 2022)

5. DISCUSSION

To align with SDG No.7, the three main discussion points in this section are the reasons for Brunei's adoption of IFRS, the adoption procedure, and the problems encountered during the adoption process. The study emphasised three key findings that best represent the current eco-system of accounting procedures in Brunei before going into detail about each of these categories. First, the study's findings revealed that Brunei did not have its own national accounting system until the BDASC created the BDAS in 2018. Historically, all financial transactions in the public and private sectors were conducted in accordance with GAAP, which was mostly derived from UK financial standards. This was brought on by the colonisation effect, as Brunei was a British colonial state up to the year 1984, when it attained complete independence. Several UK accounting standards and business and financial practises served as a point of reference for local governments, regulators, certified auditors, professional accountants, and higher education institutions prior to the implementation of IFRS. Second, until the BDASC was established in 2011, no national body for establishing accounting standards had been institutionalised. Since it was founded on May 6, 1987, there has only been the Brunei Darussalam Institute of Certified Public Accountants (BICPA). BICAP is a nonprofit organisation that is organised and operated by Bruneian accountants who volunteer their time. One of BICPA's main goals is to provide an organisation for accountants to support and advance the status and interests of the accountancy profession in line with the goals and objectives of the government of His Majesty the Sultan of Brunei; another goal is to provide a forum for discussing issues affecting the accountancy profession and to initiate, monitor, petition, and take whatever action may seem desirable in relation to promoting the accountancy profession. Brunei does not have a national certification or qualifying programme in accounting, unlike the UK, the USA, and certain developing countries in Asia. Individuals must be members of international professional accountancy organisations recognised by the Accountants Order 2010 and the Brunei Companies Act, such as the Association of Chartered Certified Accountants (ACCA) and CPA Australia, in order to be qualified to practise accounting. Thirdly, BICPA had societal influence over the rules, laws, and regulations pertaining to the established accounting standards rather than governmental power and authority. They were able to fight for values that were highly valued by many accountants, which contributed to their influence. The Public Accountants Oversight Committee (PAOC), whose responsibility it is to keep and maintain a register of all public accountants, public accounting corporations, public accounting firms, public accounting limited liability partnerships, and other public accountants, and establish disciplinary procedures for non-practicing members, is the only organisation to which the BICPA can offer advice and consultation.

5.1. The Rationales for Adopting IFRS in Brunei

Nations may have their own justifications for supporting the implementation of IFRS because they have various financial demands and issues to deal with. Depending on how it is implemented, the IFRS principles-based approach has advantages, according to the BDASC. The system's advantages include flexibility that promotes expert judgement and transparency (Hodgdon et al., 2011). Three crucial factors led Brunei to embrace the IFRS framework. First, it is important to guarantee consistency and uniformity in financial reporting across all pertinent Bruneian entities. According to Bhaskaran (2010), one of Brunei's corporate sector's structural weaknesses is that businesses do not maintain proper accounting records due to tax system anomalies. While banks have adjusted to these circumstances, they could have provided services beyond traditional lending if businesses were better structured. Bank Islam Brunei Darussalam (BIBD) recently stated in its financial statements' notes of disclosure for the year ended December 31, 2017 that its group financial statements had been prepared in accordance with the IFRS and that their initial adoption of these standards and interpretations had had little to no effect on the group and the bank. Major issues have typically been caused by local non-PIEs (especially SMEs) failing to maintain proper accounting records to produce the crucial accounting information that owner-managers and other users need to make decisions. In addition to the fact that they did not use formal accounting due to low educational levels and a lack of accounting knowledge, which made it difficult for the owner-managers to understand the importance of using proper accounting in their company, there was also a lack of sufficient clarity of purpose. Due to the sluggish approval of bank loans as a result, SMEs were unable to swiftly increase their production capacity to satisfy demand for exports. The majority of SMEs lacked a detailed strategic business plan outlining how far they should extend their operations. At a national accounting forum held on May 6, 2017, the BICPA stressed the value of accounting education in the growth of entrepreneurship and accounting practises to create a sustainable SME sector as part of Brunei Vision 2035. (Othman, 2017).

In order to assist decision-making based on reliable information, consumers of financial statements should also be able to compare and evaluate the performance of other companies. The accounting notion of comparability enables international banks operating in Brunei, such as Standard Chartered Bank and May Bank, to compare their financial performance with subsidiary banks established in other nations. The adoption of IFRS can indirectly boost foreign investors' ownership and draw more cross-border investment through the comparability of financial statements, which is primarily of interest to public enterprises (Iatridis & Dalla, 2011; Jain, 2011). Additionally, the fair value principle used by IFRS for PIEs can assist organisations in accurately reflecting the value of their assets in their financial statements. However, Brunei's required IFRS adoption has an impact on PIE comparability depending on the degree of compliance and enforcement. Even if Brunei's PIEs are not well-known for financial misbehaviour, the regulatory organisations should still maintain their strict standards for compliance. Public entities can become more comparable to other companies in the same industry in other nations with stringent enforcement. On the other hand, comparability, on the other hand, might not be of the greatest importance to encouraging Brunei SMEs to adopt BDAS (i.e., accounting standards that are based on the IFRS framework but less intricate and difficult than the IFRS for PIEs), given the manner in which business is conducted in this nation. In Brunei, Malay SMEs secure a contract from clients (private or government), subcontract it to foreigners who are mostly non-locals, and therefore follow the business model of Ali Baba, which is widely used in Southeast Asian nations like Indonesia and

Malaysia. Compliance with BDAS is expected to lead to better financial reporting from SMEs and more trust in their financial statements.

Thirdly, to simplify issues relating to financial management, such as the procedure of filing for bankruptcy and getting loans from financial institutions and appropriate organisations. In essence, creditors and financiers are interested in a company's liquidity position because they need to assess the company's capacity to pay interest, dividends, and loan repayments (Jain, 2011). As a result, by adhering to IFRS, entities are required to disclose information that enables users to assess changes in an entity's net assets, financial structure (including its liquidity and solvency), and ability to affect the amounts and timing of cash flows (i.e., cash flow statements) in order to adapt to changing circumstances and opportunities. In addition to the functions of Islamic and conventional banks, Brunei's SME Bank also makes finance for micro, small, and medium-sized enterprises (MSMEs) companies more readily available. According to Bandial (2018), the SME Bank aimed to make 60 disbursements by the end of 2018, and it can issue loans of up to $750,000 for companies owned by Brunei citizens and permanent residents that have been in operation for at least a year. However, in that year, only 134 applications were received (Norjidi, 2018). While preparing financial accounts in line with IFRS and having a strong business strategy are undoubtedly crucial to banks, many local business owners have found it extremely difficult to provide collateral as additional security for payback obligations. The SME Bank got around the problem of collateral by putting more importance on the business's ability to make money and by giving SMEs flexible options for collateral that let them borrow against vehicles, machinery, or other fixed assets.

5.2. IFRS Adoption Process in Brunei

In Brunei, there are two sets of accounting standards: BDAS for other entities without public responsibility and IASB's IFRS for all entities with public accountability (public interest entities) and non-public interest entities. In plain English, it is intended first for PIEs and then for non-PIEs. The IASB states that an entity has public accountability if its debt or equity instruments are traded in a public market, they are in the process of being issued for trading in a public market (such as a domestic or international stock exchange or an over-the-counter market, including local and regional markets), or they hold assets in a fiduciary capacity for a large group of outsiders as one of their primary businesses. The first step in the BDASC's plan was to make IFRS a requirement for PIEs. Except for the publication of Notice (No. 1/2014), there was no other information accessible about this effort. IFRS for PIEs was only made mandatory roughly three years after the BDASC was founded, in 2011, and the beginning year of 2014 was determined by taking into account the transition time. This was required to provide PIEs enough time to take the steps necessary to adopt IFRS, such as time to comprehend and interpret the IFRS's procedures and technological considerations. Another initiative to create BDAS for non-PIEs was then launched in its place, and its deployment didn't happen until January 2018.

Adoption and convergence were the two methods that the study identified as leading to the complete implementation of IFRS in Brunei. The adoption strategy was described as the quick application of IFRS for PIEs. The BDAS for non-PIEs was created using the convergence approach (Figure 5). The BDASC used the IASB's due process, which entails four steps: a research programme; creating a proposal for publication; redeliberations and finalisation; and post-implementation reviews; to construct BDAS utilising the convergence technique. Even though the BDASC's implementation of due process wasn't as thorough as the IASB's, it at least gives Brunei accounting standards a place to start and makes the review process easier.

Figure 5. Approaches towards IFRS Adoption

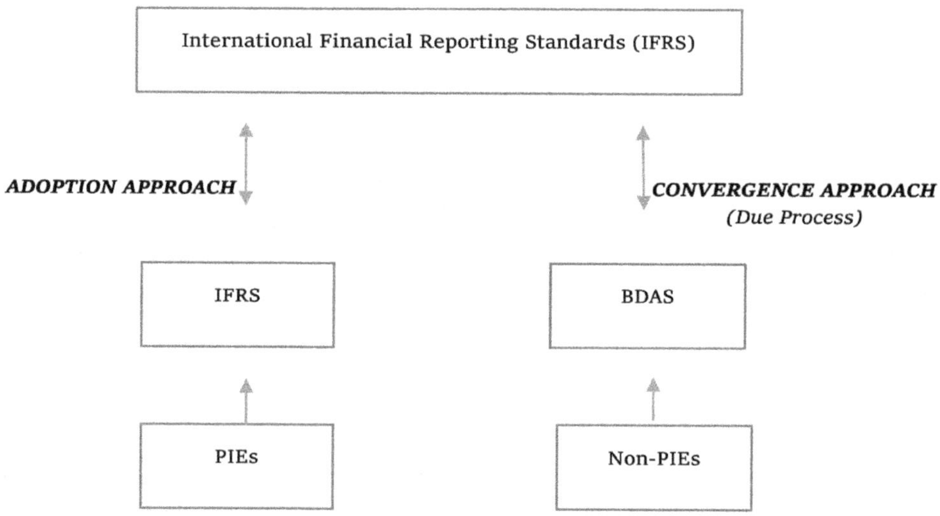

Step 1: Research Programme

The initial phase involved the BDASC establishing four ad hoc working committees to examine issues relating to banking, insurance, and takaful financial reporting as well as the standards and implementation for non-PIEs. In order to produce BDAS, the Standards for Non-PIEs Committee was tasked with conducting research and examining the national accounting standards of other nations, including Malaysian, Singaporean, and Hong Kong Financial Reporting Standards. Members of these working committees included academics, government officials, accounting experts, and members of regulatory bodies. In order to assist the Council in deciding whether to adopt IFRS for small and medium-sized enterprises (SMEs) without modifications or to converge the current Brunei GAAP to IFRS, in which case each standard is evaluated for its suitability for use by Brunei's non-PIEs, the BDASC solicited feedback and requested assistance from the Committee. In order to develop BDAS, the BDASC chose the convergence strategy and referenced Hong Kong Accounting Standards (HKAS). Two factors led to the selection of the HKAS framework, sometimes referred to as the Hong Kong Financial Reporting Standards. The HKAS framework, which has 22 accounting standards, offers a streamlined version of IFRS for SMEs and is also based on IFRS. The Hong Kong Accounting Standards (HKASs) are based on accounting principles, which include some areas such as accounting policies, the presentation of financial statements, and leases but exclude others like business reviews, segment reporting, and interim financial reporting. However, out of that 22 accounting standards, the BDASC only prescribed 20 standards for use in BDAS[1].

Step 2: Developing a Proposal for Publication

The second step was to proceed with the development of an exposure draft. The exposure draft of the BDAS Framework and Standards was issued for public consultation in October 2016 on the BDASC's website, and the Council expected to receive feedback by 31 December 2016 (The Brunei Times, 2016). To the researcher's knowledge, as a member in one of the working committees, no comments from public were received on the exposure draft. This was probably due to the lack of awareness among stakeholders especially Non-PIEs on the issuance of exposure draft, lack of knowledge and understanding on the due process of developing accounting framework and standards, and the timeline given in the final quarter of the year was seen to be inappropriate because some Non-PIEs whose financial year ended 31 December tend to be preoccupied with finalising their accounts. It was nearly one year later that the BDASC released a press release on the issuance of BDAS for Non-PIEs effective on 1 January 2018. Whether or not the BDAS has been fully implemented by all Non-PIEs in Brunei is one important area for future investigation. This study argued that the implementation of BDAS could have been effectively implemented if the Council took initial efforts to actively "socialise" the idea of implementing BDAS with Non-PIEs throughout the year 2017, instead of waiting for the year of implementation to take effect. The BDASC, through MOFE, only began to undertake additional outreach activities in January 2018. On 21 January 2018, the BDASC hosted a briefing for more than 200 people comprising public accountants, local businesses and representatives from government agencies and associations in the country to spread and raise awareness about the BDAS. A similar briefing was also scheduled to be held in the Tutong, Belait and Temburong districts (Mohamad, 2018). On 6 May 2018, the BDASC held another briefing on BDAS for entrepreneurs which was largely attended by local business owners and government officials from Brunei-Muara District.

Step 3: Redeliberations and Finalisation

The third step involved redeliberations and finalisation. But in the case of BDAS, the BDASC reviewed feedback from the consultation process and meetings. As a result of the meetings, the Council issued Guidance Note 1 on 7 August 2019, with regards to the application of BDAS for Non-PIEs for financial periods beginning 1 January, 2018 onwards. In the Guidance Note, the Council acknowledged that certain accounting policies previously adopted by entities reporting under accounting principles generally accepted in Brunei are in conflict or are not covered by BDAS, and there may also be transactions that will occur in the future for which no appropriate BDAS is available for entities to adopt. Where accounting policies are in conflict with BDAS, an entity may choose to continue applying them together with other BDASs provided that such policies were issued by the IASB and the entity it believes will result in a fair and proper presentation of its financial statements. Where accounting policies are not provided for under BDAS, an entity may also choose to apply or continue applying them together with other BDAS provided that such policies were issued by the IASB and the entity believes the continued application will result in a fair and proper presentation of its financial statements. This strongly implies that the adoption of accounting policies other than those issued by the BDAS or IASB are not permitted, and such departures from BDAS shall be accounted for in accordance with paragraph 1.5 of BDAS 1 Presentation of Financial Statements.

Step 4: Post-Implementation Reviews

On the final step, the BDASC formed Post Implementation Advisory Committee whose members were represented by the accounting profession, government agencies, academia, various business sectors as well as regulatory authorities. The role of the Committee was to assist the Council in matters relating to the post-implementation of BDAS such as providing guidance on the interpretation of BDAS and its practical application in Brunei; managing any technical issues relating to the BDAS; and reviewing and commenting feedback received from Exposure Draft issued by BDAS. This Advisory Committee reported to the BDASC on any proposal relating to accounting standards in Brunei.

Overall, it took the BDASC nearly 10 years to working towards the full implementation of IFRS. Yet, there are two remaining concerns which can be part of a future research. Firstly, the issue of compliance that is, whether or not all PIEs and Non-PIEs in Brunei have fully complied with IFRS and BDAS respectively. Secondly, the issue of practices that is whether or not they have correctly treated their accounts in accordance with IFRS accounting principles.

5.3. Challenges in the Process of IFRS Adoption in Brunei

The challenges of the IFRS adoption in Brunei can be considered operational and technical. The challenges include issues related to comparability, professional judgement, compliance, cost of implementation, fair value accounting particularly relating to IFRS 9, and finally, the need for Islamic accounting standards. Each of these challenges are discussed in the following order.

5.3.1 Comparability

Although global IFRS adoption establishes a common language in which financial statements can easily be interpreted and compared by company and country levels, accountants in companies may have completely different interpretations concerning what the accounting data means. Two companies with the same assets could be presented differently on the balance sheet such as IFRS 16 Lease Accounting, which is complex and requires professional judgement. Accounting choices can reduce the comparability of financial statements (Subramani, 2009), and therefore, accountants must choose accounting treatment of items wisely to avoid comparability issues. Interestingly, comparability is not much of problem for Brunei's local companies, but international companies whose trading activities take place in stock exchange. Thus, the financial statements prepared in accordance with IFRS would be more useful to users at the international level, but there is uncertainty about their usefulness at the national level. Although Brunei does not have a stock exchange yet, full compliance with IFRS would still continue to be relevant to local PIEs in serving the prospects for future economic development and financial needs of the country.

5.3.2 Professional Judgement

IFRS framework provides greater flexibility for the accountants to exercise their own professional judgement in selecting the relevant principles to treat companies' accounting items, and encourages them to dig deeper and look into the substance of a transaction. The accounting services providers would not only be able to offer sound and professional judgement, but also get a better appreciation of the transaction. Certainly, the BDASC could issue assist by issuing accounting interpretations and guidance. Nonetheless, it becomes difficult when divergent views of the accounting principles by providers of accounting services arise. Even more, when a lower-level accountant of a company interprets a transaction in the best way that he or she understands, and then a superior accountant may reverse it all together and request for a re-evaluation (Hodgdon et al., 2011). This only implies that Brunei needs a larger pool of experienced accountants who are very familiar with the industry to supervise junior accountants. However, these challenges may delay service delivery or amount to losses (Gyasi, 2009), and overall, compliance takes longer and is more expensive (Epstein & Jermakowicz, 2010).

5.3.3 Compliance

Prior to the issuance of IFRS adoption, the Non-PIEs in particular were already having a difficulty to consistently prepare their financial statements due to lack or improper record of accounts and the frequent change of accountants in these companies. The highest position they could afford to hire is accountant or assistant accountant, and it is by academic qualification, not professional certification which can be very costly. To fully comply with IFRS has also become more difficult when owner-managers or other internal stakeholders and the accountants have divergent understanding and views of accounting principles. There were cases whereby, due to owner-managers' lack of accounting knowledge and simply their attitude of being ignorant, they shifted the blame to the accountants for the loss reported in the financial statements. This brings out the importance of corporate governance which is not merely reserved for large corporations but also SMEs. Without a good corporate governance structure, companies will face higher risks. IFRS compliance further requires strict corporate legislation as the Board of Directors must act upon good faith with fiduciary duty. There were some concerns with the compatibility of local laws with IFRS in certain matters pertaining to accounting such as the format and presentation requirements, and there was also uncertainty over tax treatments of items arising from convergence such as unrealised gains and losses and the move from a tax basis for depreciation to one of useful economic life (Albu et al., 2013). All of these had been taken care of before implementation of these IFRS by the regulatory authorities through updating and making necessary revisions to the various laws where relevant including Financial Regulations 1983, Companies Act (Chapter 39), Income Tax Act (Chapter 35), Banking Order 2006 and the Islamic Banking Order 2008.

5.3.4 Costing

With regards to the cost of implementing IFRS, a number of local entities claimed that the process can be costly, slow and frustrating. The BDASC asserted that the costs are difficult to anticipate as they will largely depend and differ based on the size and nature of the company. On top of the costs of transition from Brunei GAAP to IFRS, staff training and implementing IT support and an accounting system could be significant (Regoli, 2019; Osnoss, 2010). A key challenge is to ensure companies, accountants, auditors, regulators and the investment community are appropriately skilled to apply and interpret IFRS. Lack of knowledge transfer facilities is also a challenge. Although few institutions like BICPA, Deloitte and Touche and KPMG have been willing to provide IFRS knowledge transfer courses for companies and interested stakeholders, it boils down to the question of who is going to bear the costs of knowledge transfer. Higher education institutions can provide academic courses on IFRS for accounting graduates, yet graduates themselves need to search for job opportunities and require more time to acquire accounting experience as they need to apply professional judgment in determining where and in what order information to be presented in the financial disclosures, particularly on the use of fair value in IFRS.

5.3.5 Fair Value Accounting

IFRS uses fair value as a measurement base for valuing most of the items of financial statements. Although this value of measurement can better reflect underlying economic values, it brings a lot of volatility and subjectivity to financial statements (Capkun et al., 2016). It also involves a lot of hard work in arriving at the fair value and valuation experts have to be used. For local financial institutions like Baiduri Bank, they have been preparing their financial statements on historical cost basis except for certain investment securities, and derivate financial instruments classified as held at fair value through profit or loss (IFRS 9)[2]. IFRS 9 contends that significant increases in credit risk are observable before delinquency occurs. According to Notice No. BU/N-7/2018/57 released by the Monetary Authority of Brunei Darussalam (AMBD), all banks in Brunei (including Islamic Banks) must not wait for delinquency to occur before remedial measures are implemented and higher provisions are booked. IFRS 9, which became effective on 1 January 2018, is expected to materially affect the capital and earnings of Banks in 2018. In the consolidated financial statements for the year ended 31 December 2018, Baiduri Bank disclosed that the application of IFRS 9 has had no impact on the classification and measurement of the Group's financial liabilities and reclassifications of financial assets have also had no impact on the Group's financial position, profit or loss, and other comprehensive income or total comprehensive. The AMBD further sets out regulatory reporting requirements that the Takaful and Insurance Financial Returns (TIFR) in Brunei must conform to IFRS, and the accounting for investments (assets in the Takaful fund) will result in fair value changes of investments being deferred and accounted for as Fair Value Reserve. In the case of BDAS (which is IFRS principles-based), if the BDASC were to impose IFRS 9, the Non-PIEs will have tough time shifting to fair value accounting. The Post Implementation Advisory Committee suggested Non-PIEs to be given a choice to account their investment properties either at cost or fair value because this reduces complexity if entities chose the cost method. Furthermore, the fair value model only adds cost to them in terms of having to either obtain valuation reports or incur undue cost. The proposed amendment which currently under review is for assets to be subsequently measured at cost unless specified otherwise under the BDAS framework.

5.3.6 The Need for Islamic Accounting Standards

Brunei is not an exception to the tendency of disagreement regarding the implementation of IFRS in Islamic nations. There are two viewpoints on this. First, it is reasonable for locals to expect that a small nation like Brunei, which implemented the Sharia Penal Code in 2019 to strengthen Islamic practise, will adopt FAS issued by the AAOIFI. Secondly, IFRS and FAS share a fundamental aspect, namely a framework based on principles. Therefore, local businesses see no reason for Brunei not to embrace FAS. There are two concerns related to the adoption of FAS. As the proverb goes, Rome was not constructed in a single day. In nearly two decades, IFRSs have become the de facto global language of financial reporting, and the standards have been widely adopted by developed, emerging, and developing countries alike (Sharairi, 2020). On the other hand, FASs have not been widely implemented. FASs were primarily designed for use by firms specialising in Islamic financial (Syariah-compliant) products. Some standards cannot be easily covered by IFRS, like the fact that Sharia-compliant financial instruments cannot pay or collect interest due to Islamic usury laws, halal investments, and accounting for Zakat. The dispute over the adoption of IFRS by FAS brings to mind another issue: there is no differentiation between conventional and Islamic financial institutions. According to Herath and Alsulmi (2017) and Alkhtani (2012), Islamic banks and Sharia-compliant banks in Saudi Arabia still deal with interest, but they distinguish themselves from other banks by donating the money to charity. To avoid a negative connotation, they refer to interest rates as a special commission or employ other deceptive terms. But the point of this study isn't to find out how well Islamic banks in Brunei follow Sharia Law. Instead, it's to show that IFRS adoption is a smart choice given how little Sharia Law affects business regulation and how it already has social and economic effects on the global financial system.

6. POLICY RECOMMENDATIONS

The study proposes several key recommendations, as follows:

For regulatory bodies

- To give access to business entities to reach out to relevant institutions such as BICPA or directly to BDASC for standards that they require further discussion and considerations for amendments if applicable.
- To socialise and promote the use of IFRS actively via media platforms and publish materials that can help entities to understand complex standards clearly.
- To set up a special task force to monitor the adoption of IFRS by entities and assess how IFRSs have impacted on economy. Such information would be useful for regulatory bodies to make decisions on certain matters and develop relevant and timely policies.

For business entities

- To comply with IFRS financial reporting requirements set by the regulatory bodies.
- To educate themselves and to keep updated with the standards and issues associated with each standard by attending workshops, training seminars and forums, and having regular discussions with their internal accountants and professional bodies.
- To allocate an annual budget for resources and staff training for the long-term benefits and growth of the businesses. Knowledge transfer is not limited to accounting staff but expanded to all relevant staff, including managers.

For higher education institutions

- To integrate IFRS and BDAS in their education curricula, particularly those institutions that offer degrees in accounting (Banham & He, 2014).
- To provide knowledge transfer for graduate such as via internship and apprenticeship in relevant government departments or business entities as this will give them practical exposure to prepare accounts, besides giving them work placement in accounting and auditing firms which most of their work is related to audit services.
- To encourage academic staff to increase research activities on the development of IFRS primarily how it affects companies' performance and the economy.
- To help increase the number of professional accountants in the future, higher education institutions are responsible for attracting more admis-

sion to their accounting degree programmes and to entice graduates to continue their qualification at professional level such as Association of Chartered Certified Accountants (ACCA).

For accounting and auditing bodies

- To share their views with the relevant institutions on the issues of IFRS and disclosure requirements faced by business entities.
- To provide suggestions and recommendations on how complex accounting standards can be dealt with and practically used by entities.
- To offer IFRS training and knowledge transfer activities to business entities, apart from accounting and auditing services, so treatment of business transactions can be done by their respective accounting personnel appropriately.

Finally, the limitations to this study is more research needs to be done in order to figure out how IFRS affects the financial performance of companies, especially Islamic financial institutions in Brunei.

7. CONCLUSION

Goal 17: Strengthen and reestablish the Global Partnership for Sustainable Development's implementation. In supporting financial sustainability, Brunei used a two-step process to implement IFRS over the course of almost 10 years. The first stage is the adoption approach, which will make IFRS mandatory for PIEs by January 1, 2014. The second stage is the convergence of Brunei GAAPs to the IFRS framework, which will lead to the development of BDAS and make BDAS mandatory for non-PIEs by January 1, 2018. Preparing BDAS takes a long time because regulatory bodies have to keep looking at it and talking to businesses to find issues and compliance problems. Economies all over the world have benefited from IFRS requirements, and what's important to Brunei's financial landscape is to achieve uniformity and consistency in financial reporting among business entities; let users of financial statements compare and evaluate the performance of entities to help them make decisions based on good information; and make it easier to get loans from financial institutions and relevant agencies. Even though there are benefits to using IFRS, there are also challenges that come with it. Even though not all of the problems are unique, the way IFRS is put into place in Brunei is worth looking into because it helps explain the institutional and regulatory environment for businesses. Also, problems like comparability, compliance, professional judgement, and costing are not so big that they can't be fixed. They may just need solutions that work on more than one level. Even though Brunei's private sector is small compared to neighbouring Southeast Asian countries and about 98 percent of active businesses are SMEs (i.e., non-PIEs), the government should not be satisfied with how IFRS implementation is going. In the future, more research should be done to find out how often IFRS and BDAS (which is based on IFRS) are used by PIEs and non-PIEs, respectively, and what effect IFRS adoption has on the financial performance of companies and the growth of the economy as a whole.

REFERENCES

Ajili, H., & Bouri, A. (2017). Comparative Study between IFRS and AAOIFI Disclosure Compliance: Evidence from Islamic Banks in Gulf Cooperation Council Countries. *Journal of Financial Reporting and Accounting*, *15*(3), 269–292. DOI: 10.1108/JFRA-03-2016-0023

Albu, C. N., Albu, N., & Fekete, S. (2010). The Context of the Possible IFRS for SMEs Implementation in Romania: An Exploratory Study. *Accounting and Management Information Systems*, *9*(1), 45–71.

Albu, C. N., Albu, N., Fekete, S., Girbina, M. M., Selimoglu, S. K., Kovacs, D. M., Lukacs, J., Mohl, G., Mullerova, L., Pasekova, M., Arsoy, A. P., Sipahi, B., & Strouhal, J. (2013). Implementation of IFRS for SMEs in Emerging Economies: Stakeholder Perceptions in the Czech Republic, Hungary, Romania and Turkey. *Journal of International Financial Management & Accounting*, *24*(2), 140–175. DOI: 10.1111/jifm.12008

Alkali, I., Alkali, M. Y., & Aliyu, A. A. (2017). Islamic Accounting Reporting and Economic Development: Nigerian Perspective. *Accounting*, *3*, 211–220. DOI: 10.5267/j.ac.2017.1.005

Alkhtani, S. (2012). The Relevance and Usefulness of IFRSs to Saudi Arabia. *Journal of Administrative and Economic Science*, *5*(2), 21–53.

Anshari, M., & Hamdan, M. (2022). Enhancing e-government with a digital twin for innovation management. *Journal of Science and Technology Policy Management*, *14*(6), 1055–1065. DOI: 10.1108/JSTPM-11-2021-0176

Anshari, M., & Hamdan, M. (2022a). Understanding knowledge management and upskilling in Fourth Industrial Revolution: Transformational shift and SECI model. *VINE Journal of Information and Knowledge Management Systems*, *52*(3), 373–393. DOI: 10.1108/VJIKMS-09-2021-0203

Anshari, M., Hamdan, M., Ahmad, N., & Ali, E. (2024). Public service delivery, artificial intelligence and the sustainable development goals: trends, evidence and complexities. *Journal of Science and Technology Policy Management*.

Anshari, M., Hamdan, M., Ahmad, N., Ali, E., & Haidi, H. (2023). New knowledge creation and loss during COVID-19: Sustainable knowledge management's perspective. *International Journal of Learning and Intellectual Capital*, *20*(6), 587–611. DOI: 10.1504/IJLIC.2023.134898

Anshari, M., Syafrudin, M., & Fitriyani, N. L. (2022). Fourth industrial revolution between knowledge management and digital humanities. *Information (Basel), 13*(6), 292. DOI: 10.3390/info13060292

Anshari, M., Syafrudin, M., Tan, A., Fitriyani, N. L., & Alas, Y. (2023a). Optimisation of knowledge management (KM) with machine learning (ML) Enabled. *Information (Basel), 14*(1), 35. DOI: 10.3390/info14010035

Banham, H., & He, Y. (2014). Exploring the Relationship between Accounting Professionals and Small and Medium Enterprises (SMEs). *Journal of Business & Economics Research, 12*(3), 209–214. DOI: 10.19030/jber.v12i3.8723

Bhaskaran, M. (2010). Economic Diversification in Brunei Darussalam. *CSPS Strategy and Policy Journal, 1*, 1–12.

Biddle, G. C., Callahan, C. M., Hong, H. A., & Knowles, R. L. (2013). *Do Adoptions of International Financial Reporting Standards Enhance Capital Investment Efficiency?* Working paper. 45 p.

Borker, D. R. (2012). Accounting, Culture and Emerging Economics: IFRS in the BRIC Countries. *Journal of Business & Economics Research, 10*(5), 313–324. DOI: 10.19030/jber.v10i5.6983

Brunei Times. (2016). *HM Consents to Adoption of Accounting Standards*. Retrieved on 1 March 2020, from https://btarchive.org/news/national/2016/10/27/hm-consents-adoption-accounting-standards

Callao, S., & Jarne, I. (2010). Have IFRS Affected Earnings Management in the European Union? *Accounting in Europe, 2*(2), 159–189. DOI: 10.1080/17449480.2010.511896

Capkun, V., Collins, D., & Jeanjean, T. (2016). The Effect of IAS/IFRS Adoption on Earnings Management (Smoothing): A Closer Look at Competing Explanations. *Journal of Accounting and Public Policy, 35*(4), 352–394. DOI: 10.1016/j.jaccpubpol.2016.04.002

Chen, C., Young, D., & Zhuang, Z. (2013). Externalities of Mandatory IFRS Adoption: Evidence from Cross-Border Spillover Effects of Financial Information on Investment Efficiency. *The Accounting Review, 88*(3), 881–914. DOI: 10.2308/accr-50384

de Pablos, P. O. (2002). Evidence of intellectual capital measurement from Asia, Europe and the Middle East. *Journal of Intellectual Capital.*

de Pablos, P. O., & Lytras, M. D. (2008). Competencies and human resource management: Implications for organizational competitive advantage. *Journal of Knowledge Management*.

Dhar, B. K., Sarkar, S. M., & Ayittey, F. K. (2022). Impact of social responsibility disclosure between implementation of green accounting and sustainable development: A study on heavily polluting companies in Bangladesh. *Corporate Social Responsibility and Environmental Management*, *29*(1), 71–78. DOI: 10.1002/csr.2174

Dixon, T., Thompson, B., & McAllister, P. (2002). *The Value of ICT for SMEs in the UK: A Critical Literature Review*. A Report for Small Business Service Research Programme. Small Business Service, Reading: College of Estate Management.

Epstein, B. J., & Jermakowicz, E. K. (2010). *Interpretation and Application of International Financial Reporting Standards*. John Wiley & Sons, Inc.

Gordon, L. A., Loeb, M. P., & Zhu, W. (2012). The Impact of IFRS Adoption on Foreign Direct Investment. *Journal of Accounting and Public Policy*, *31*(4), 374–398. DOI: 10.1016/j.jaccpubpol.2012.06.001

Gyasi, A. K. (2009). *Adoption of International Financial Reporting Standards in Developing Countries: The Case of Ghana*. Published Thesis. Vaasa Amma Tikorkeakoulu, University of Applied Sciences. Retrieved on 16 March 2020, from https://publications.theseus.fi/bitstream/handle/10024/15526/ANTWI_K

Hamdan, M., Anshari, M., Ahmad, N., & Ali, E. (Eds.). (2023). *Public Policy's Role in Achieving Sustainable Development Goals*. IGI Global. DOI: 10.4018/978-1-6684-8903-1

Hamdan, M., & Low, K. C. P. (2013). Is Brunei Ready to Adopt the International Accounting System (IAS)? *Business Journal for Entrepreneurs*, *1*, 12–48.

Hannoudi, O. A. (2015). *Comparative Analysis between the IFRS and the AAOIFI Accounting Standards*. Published (Master) thesis. Universite De Strasbourg.

Hasan, M. T., & Abdul Rahman, A. (2017). IFRS Adoption and Earnings Management: A Review and Justification of Earnings Management Model for Developing Countries. *ELK Asia Pacific Journal of Finance and Risk Management*, *8*(3), 1–17.

Herath, S. K., & Alsulmi, F. H. (2017). International Financial Reporting Standards (IFRS): The Benefits, Obstacles and Opportunities for Implementation in Saudi Arabia. *International Journal of Social Science and Business*, *2*(1), 1–18.

Hodgdon, C., Hughes, S. B., & Street, D. L. (2011). Framework-based Teaching of IFRS Judgements. *Accounting Education, 20*(4), 415–439. DOI: 10.1080/09639284.2011.589587

IASPlus. (2002). Europe adopts regulation requiring IAS by 2005 (Deloitte). Retrieved on 3 March 2019, from https://www.iasplus.com/en/news/2002/June/news504

Iatridis, G. E., & Dalla, K. (2011). The Impact of IFRS Implementation on Greek Listed Companies: An industrial Sector and Stock Market Index Analysis. *International Journal of Managerial Finance, 7*(3), 284–303. DOI: 10.1108/17439131111144478

IFRS. (2017). *IFRS Application around the World. Jurisdictional Profile: Brunei Darussalam.* Retrieved on 3 March 2019, from https://cdn.ifrs.org/-/media/feature/around-the-world/jurisdiction-profiles/brunei-darussalam-ifrs-profile.pdf

Islam, M. A. (2017). *Future of Accounting Profession: Three Major Changes and Implications for Teaching and Research.* International Federation of Accountants. Retrieved on 7 January 2020, from https://www.ifac.org/knowledge-gateway/preparing-future-ready-professionals/discussion/future-accounting-profession-three

Jain, P. (2011). IFRS Implementation in India: Opportunities and Challenges. *World Journal of Social Sciences, 1*(1), 125–136.

Jennex. M. E. (January, 2017). What is Knowledge Management? San Diego State University. Retrieved on 12 September 2021 from https://www.researchgate.net/publication/314500732

Jermakowicz, E. K., & Epstein, B. J. (2010). IFRS for SMEs - An Option for U.S. Private Entities? *Review of Business, 30*(2), 72–79.

Jermakowicz, E. K., & Tomaszewski, S. G. (2006). Implementing IFRS from the Perspective of EU Publicly Traded Companies. *Journal of International Accounting, Auditing & Taxation, 15*(2), 170–196. DOI: 10.1016/j.intaccaudtax.2006.08.003

Johari, A. H., Ahad, A. D., & Anshari, M. (2024). Examining the Impact of Gender Differences, Sustainable Urbanism, and Digital Transformation on Online Consumer Buying Behavior: A Perspective Towards SDG Alignment. In *Harnessing Green and Circular Skills for Digital Transformation* (pp. 194-212). IGI Global.

Johnson, S. (2008). *IFRS: No Longer IF, but When.* Retrieved on 15 March 2020, from https://www.cfo.com/accounting-tax/2008/02/ifrs-no-longer-if-but-when/

Kakabadse, N. K., Kakabadse, A., & Kouzmin, A. (2003). Reviewing the knowledge management literature: towards a taxonomy. Retrieved on 15 September 2021 from https://www.researchgate.net/publication/220363071

Kaya, D., & Koch, M. (2015). Countries' Adoption of the International Financial Reporting Standard for Small and Medium-sized Entities (IFRS for SMEs) – Early Empirical Evidence. *Accounting and Business Research, 45* (1), p. 93.120.

Kilic, M., & Uyar, A. (2017). Adoption Process of IFRS for SMEs in Turkey: Insights from Academics and Accountants. *Accounting and Management Information Systems, 16*(2), 313–339. DOI: 10.24818/jamis.2017.02005

Kimeli, E. K. (2017). IFRS Adoption and Capital Markets. *Journal of Finance and Accounting, 5*(1), 19–30.

Liapis, K., & Thalassinos, I. E. (2013). A Comparative Analysis for the Accounting Reporting of Employee Benefits between IFRS and other Accounting Standards: A Case Study for the Biggest Listed Entities in Greece. *International Journal of Economics and Business Administration, 1*(1), 91–116. DOI: 10.35808/ijeba/5

Marquez-Ramos, L. (2008). *The Effect of IFRS Adoption on Trade and Foreign Direct Investments.* International Trade and Finance Association. Working paper. 19 p.

Mohamad, L. (2018). Briefing Raises Awareness on Accounting Standards. *Borneo Bulletin*. Retrieved on 1 March 2020, from https://borneobulletin.com.bn/briefing-raises-awareness-on-accounting-standards/

Narasimham, R. (2010). *Enhancing Financial Disclosure Standards in Transitional Economies*. Asian Development Bank.

Nonaka, I., & Takeuchi, H. (1995). The knowledge-creating company. New York: Oxford University Press. Retrieved on 23 September 2021 from https://memberfiles.freewebs.com/84/90/65819084/documents/The%20Knowledge-Creating%20Company.pdf

Norjidi, D. (2018). SME Bank Receives 134 Applications for Finance. *Borneo Bulletin*. Retrieved on 3 March 2020, from https://borneobulletin.com.bn/sme-bank-receives-134-applications-for-finance/

Ordóñez de Pablos, P., Anshari, M., & Almunawar, M. N. (Eds.). (2024). *Harnessing Green and Circular Skills for Digital Transformation*. IGI Global. DOI: 10.4018/979-8-3693-2865-1

Osnoss, J. (2010). *Some Points to Ponder in the IFRS Debate.* Retrieved on 5 February 2020, from https://www.accountingtoday.com/news/some-points-to-ponder-in-the-ifrs-debate

Othman, A. (2017). SMEs Must Embrace Technology, Develop Tech-Advanced Businesses. *Borneo Bulletin.* Retrieved on 23 March 2020, from https://borneobulletin.com.bn/smes-must-embrace-technology-develop-tech-advanced-businesses/page/1936/

Palea, V. (2013). IAS/IFRS and Financial Reporting Quality: Lessons from the European experience. *China Journal of Accounting Research*, *6*(4), 247–263. DOI: 10.1016/j.cjar.2013.08.003

Patro, A., & Gupta, V. K. (2014). Impact of International Financial Reporting Standards on Cost of equity Capital for Asian countries. *International Journal of Accounting and Financial Reporting*, *4*(2), 148–170. DOI: 10.5296/ijafr.v4i2.6462

Pattiroei, A. A., Anshari, M., Hamdan, M., & Ali, E. (2023). Radiological Detector Aircraft for Marine Life's Sustainability in Supporting Sustainable Development Goals (SDGs). In *Public Policy's Role in Achieving Sustainable Development Goals* (pp. 261-278). IGI Global.

Pena, H. F. P., & Franco, J. B. (2017). Impact of IFRS on the Quality of Financial Information in the United Kingdom and France: Evidence from a New Perspective. *Intangible Capital*, *13*(4), 850–878. DOI: 10.3926/ic.939

Perera, D., & Chand, P. (2015). Issues in the Adoption of International Financial Reporting Standards (IFRS) for Small and Medium-Sized Enterprises (SMES). *Advances in Accounting*, *31*(1), 165–178. DOI: 10.1016/j.adiac.2015.03.012

Regoli, N. (2019). 19 Advantages and Disadvantages of Adopting IFRS. Connect Us. Retrieved on 22 March 2020, from https://connectusfund.org/6-advantages-and-disadvantages-of-adopting-ifrs

Santosus, M., & Surmacz, J. (2001). The ABC's of Knowledge Management. Retrieved on 23 September 2021 from https://www.globalbioenergy.org/uploads/media/ABC_of_KM.pdf

Sapovadia, V. K. (2017). Importance of aligning IFRS with SDGs. *Available at SSRN* 3042562.

Sharairi, M. H. (2020). Factors that Influenced the Adoption of IFRS by Islamic Banks in the UAE. *Accounting Research Journal*, *33*(1), 75–91. DOI: 10.1108/ARJ-11-2017-0185

Sharairi, M. H., & Islam, J. (2019). Comparison Study between IFRS and AAOIFI Accounting Standards: A Case Study of Islamic Banks in the United Arab Emirates. *Communications on Stochastic Analysis*, *12*(5), 23–32.

Subramani, R. V. (2009). Accounting for Investments: Vol. 1. *Equities, Futures and Options*. John Wiley & Sons.

Thomason, K. (2019). *The Difference between Principles and Rules-Based Accounting Standards*. Retrieved on 1 March 2019, from https://bizfluent.com/info-8333888-difference-principles-rulesbased-accounting-standards.html

Thompson, S. C. (2016). *Accounting for a Developing World: A look at International Standards on Developing Countries*. Honors Thesis Project. University of Tennessee, Knoxville. Retrieved on 16 March 2020, from https://trace.tennessee.edu/cgi/viewcontent.cgi?article=2938&context=utk_chanhonoproj

Ullah, M. H. (2013). Compliance of AAOIFI Guidelines in General Presentation and Disclosure in the Financial Statements of Islamic Banks in Bangladesh. *International Journal of Social Science Research, 1*(2), 111–123. DOI: 10.14239/IJSSR.2013.01205

UN. (2020). *Goal 17: Revitalize the global partnership for sustainable development*. Retrieved 14 September 2022 from https://www.un.org/sustainabledevelopment/globalpartnerships/

Wahlen, J. M., Jones, J. P., & Pagach, D. P. (2013). *Intermediate Accounting: Reporting and Analysis*. South-Western Cencage Learning.

Yapa, P. W. S., & Joshi, M. (2012). The Socio-Economic Impacts of the Adoption of IFRS: A Comparative Study between the ASEAN countries of Singapore, Malaysia and Indonesia. Socio-Econ-IFRS, RMIT University (p. 1-26).

ENDNOTES

1. https://bdasc.mofe.gov.bn/accountingstandards/BDAS%20for%20Non-PIEs.pdf

2. IFRS 9 requires the recognition of expected credit losses for financial assets measured at amortised cost and fair value through profit and loss; loan commitments and financial guarantee contracts not measured at fair value through profit and loss; trade receivables; contract assets under IFRS 15 and lease receivables under IAS 17.

Chapter 13
Regional Cooperation on Climate Change, Natural Resources, and Environmental Challenges

Tanishka Prashant Madke
Ajeenkya D.Y. Patil University, India

Amisha Shaikh
Ajeenkya D.Y. Patil University, India

Susanta Das
https://orcid.org/0000-0002-9314-3988
Ajeenkya D.Y. Patil University, India

ABSTRACT

Climate change & environment pose a significant danger to sustainable development, especially in resource-rich nations like ASEAN. Regional collaboration is crucial to address concerns related to climate adaption, resource management, & cross-border environmental governance. The assessment highlights importance of strong institutional structures & equitable resource distribution to effectively address these concerns in achieving SDGs & promoting environmental sustainability in this region. Emerging issues (urbanization, population growth & climate change) add new aspects to regional environmental challenges and require inventive and adaptable solutions. Technology developments offer interesting possibilities for improving cooperation/monitoring progress. To ensure that all Member States benefit from regional initiatives, it is essential to improve knowledge-sharing platforms

DOI: 10.4018/979-8-3373-1300-9.ch013

and align legislative frameworks. ASEAN, through cooperation and innovation, is vital to solving environmental problems and ensuring the long-term development of future generations in order to achieve the SDGs.

1. INTRODUCTION

The United Nations launch of the 2030 Agenda coupled with its 17 sustainable development goals (SDGs) caused numerous actors to broaden their views about sustainable development. Various methods and strategies emerged for sustainability after this development occurred (UNSDGs). The 2030 Agenda together with the Sustainable Development Goals provides a single framework of sustainability that enables every organizational form to both understand and execute the guidelines for enhanced collaborative work. This agenda uses the core belief that no one needs to remain marginalized while it requires partnerships across all levels and sectors. The achievement of sustainability targets requires different actors to simultaneously coordinate and accept leadership responsibilities for initiating partnerships. The 2030 Agenda emphasizes of subnational levels for SDG realization, stating that "Regional and subregional frameworks can facilitate the effective translation of sustainable development policies into concrete action at the national level (Krantz & Gustafsson, 2023). The central obstacle in managing climate change emerges from its worldwide status while its effects primarily manifest locally and regionally. Regional and local governments must lead efforts toward promoting climate change adaption because such initiatives traditionally rest under their jurisdictions. The establishment of the 2015 Paris Agreement global adaption goal together with recent global policy developments has elevated adaption to higher importance in international frameworks. Through their ground-level climate actions and formal goals they pledge to accomplish regional and local governments guide collective achievement by connecting with other non-state actors beyond national borders. The partnership falls under the category of 'transitional climate governance' (Andonova et al., 2009; Di Gregorio et al., 2020). Transitional climate governance initiatives should demonstrate three fundamental characteristics according to the literature in this field. These characteristics include explicit climate change references and transnational participant involvement with at least one non-state actor. Goals and actions related to specific objectives also need to be part of the initiative (Setzer et al., 2020; Sun & Page, 2024). Various studies during recent years calculated global along with regional sector impacts from climate change resulting from defined emission scenarios and forcing models. There exists significant political need for evidence regarding temperature rise effects at various thermal increase levels which exceed the capabilities of standard forcing scenario restrictions. The evaluation of impact outcomes feasible through the analysis of different policy objectives (Arnell et al., 2019).

2. REGIONAL GOVERNMENTS AND TRANSNATIONAL ACTION

Climate change adaption continues to be recognized as a governance problem which needs multiple levels of cooperation for effective solutions. Regional and local governments obtained crucial confirmation for their vital climate change responsibilities in the Paris Agreement during the 21st UNFCC conference in 2015. For the first time the international climate process included bottom-up approach mechanics through the UNFCC process on this occasion. Under the Paris Agreement non-state actors Footnotei achieved official status as major participants in climate governance through recognition that "the importance of the engagements of all levels of government and various actors, in accordance with respective national legislations of Parties, in addressing climate change". Regional along with local governments continue to commit themselves to various climate solutions through their own initiatives. The Global Climate Action portal of the UNFCCC shows over 22,000 reported climate actions which include 11,088 city actions by 9465 entities and 756 regional actions by 278 entities. The range of pledges covers organizations that set emission reduction targets with their won pollution levels as well as those creating climate adaption plans alongside private funding commitments. Emission reduction targets occupy the majority of total climate pledges reported to the United Nations Framework Convention on Climate Change (Setzer et al., 2020).

Subnational governments possess enough authority to make decisions in critical legal domains including energy and transportation and land use and housing and disaster management and natural resources together with health care and public health. Regional governments possess the power to create laws that come into effect while waiting for federal legislative initiatives. The implementation of adaption policies normally resides within the authority of regional and local government bodies despite major central authority responsibilities. Regional governments maintain possibilities to work together with both other subnational governments and countries and jurisdictions. The relationship between national and local governments passes through regional governments as their fundamental connection point. Setzer et al. (2020) has developed a term "paradox of the lent target" to explain how nations create climate objectives which their regional administrations ultimately need to execute. The inclusion of both rural and urban realities must take place at the national level through the essential involvement of regions who provide integration between city and metropolitan climate initiatives with national initiatives across all territories. The existing mechanisms that unite national and federal governments with their state or regional counterparts ensure climate policy effectiveness throughout multiple scales. Thirdly the implementation of adaption actions depends fundamentally on the efforts made by regional governments to deliver tangible results. The specific nature of adaption calls for strategies which recognize the regional areas in which challenges occur. This implementation power stems from regional governments demonstrating entrepreneurial spirit while developing capacity like non-state actors to create policy advancements through testing methods and skill development (Setzer et al., 2020).

3. COOPERATION, HUMAN RIGHTS AND CLIMATE CHANGE GOVERNANCE

According to the Office of the High Commissioner for Human Rights international collaboration for climate change protection constitutes both a practical necessity and a human rights duty which aims to achieve human rights fulfilment under present climate change circumstances. The dual recognition of cooperation through climate and human rights law creates another component to understand how this obligation operates. Various human rights instruments refer to cooperation as an explicit requirement. Article 2(1) of the International Covenant on Economic, Social, and Cultural Rights 1966 exemplifies in detail the compulsory state efforts to realize human rights through cooperation. The mentioned provision establishes both diplomatic partnerships and resource allocation to achieve cooperation goals. The Covenant establishes dual responsibilities for states both for their domestic citizens and those belonging to other countries. The shared work between states extends the scope of human rights protection framework. I propose that cooperation should function as a new framework when courts and tribunals expand the scope of human rights obligations during environmental protection. Indeed, several cases emphasise the increasing importance of cooperation, as well as highlight the connection between environmental degradation and the deprivation of human rights. The Inter-American Court of Human Rights presented the Environment and Human Rights case to establish how human rights achievement requires a healthy environment. States must prevent considerable environmental damage to both their internal population and citizens situated outside their boundaries according to Court evaluation. The court confirmed through this analysis that states must safeguard foreign nationals' rights. Therefore, this standard expands jurisdictional boundaries. The recent development allows greenhouse gases produced by States to establish responsibility for damages suffered by individuals in different countries thus supporting evolving case law about "cross-border environmental human rights disputes". Additional international cooperation becomes necessary to achieve continuous human rights protection under these circumstances. The number of climate-related court cases continues to rise in domestic courts worldwide. The Dutch court ordered government officials to lower emission levels as a security measure for Dutch citizens facing the consequences of global warming. The courts specifically directed government officials to increase their emission reduction targets. The judges issued a warning which stated that global warming poses an immediate threat because the Netherlands has agreed to take action through international treaties. The judge explicitly stated during rulings that Dutch government should stop using a global climate problem solution independence argument as justification for avoidance of responsibility. This demonstrates how cooperation functions as a tool for individual change challenges instead of providing protection from responsibility. It will be worth seeing whether this climate litigation has an effect on cooperation in other ways, or whether it stimulates or encourages more meaningful cooperation among states to respond to the calls of a changing climate. It has made it clear that it also illuminates the relationship between human rights and the environment, which may have a beneficial impact on governments and ease pressure for governments to tackle policies which acknowledgedly take constructive steps to address the least short-term impact of climate change. In particular, a wide range of human rights are afflicted by climate change – none other than the emerging right to a healthy environment. This is why cooperation, human rights and climate governance are the same thing. States ought to collaborate toward making sure human rights are ensured universally in the face of a world which is set to get progressively more on fire. Further, even though climate change international treaties directly do not mention human rights or do not incorporate them in their text and/ or title, use of systemic interpretation allowing for realization of impact of climate change onto human rights is part of any reading of any such treaty. In Conclusion, International law recognizes Cooperation as an important part while this principle now helps more in climate change management. Cooperation takes different forms throughout the UNFCCC documents and the ratification agreements signed to combat climate change. Cooperation has defined the duties that developed countries must perform to help developing nations adapt to and fight climate change thorough formal systems (Rudall, 2021; Shaw et al., 2022).

4. LITERATURE REVIEW

4.1. Theoretical Framework of AI and ML in Climate Science

Today AL and ML technologies dominate the field of climate change mitigation as well as response to changes currently affecting our planet. These technologies originate from such climate models combined with computational theories. The past GCM model delivered atmospheric condition forecasts through circulation patterns but faced two significant problems: lower scalability and higher computational complexity and reduced precision of predictions. A network of deep learning now incorporates AI models as an approach to refine sophisticated tools whose relationships are nonlinear and enhanced predictive value is achievable. Combined with dynamical climate models, AI-based approaches can also enhance climatic databases by integrating historical climatic databases with current climate information. Supervised learning model: It is one of the key subtypes of the approaches from the ML class – helps scientists to train certain systems on the labelled climate data sets and get highly accurate predictions regarding phenomena like occasional severe weather conditions, deforestation rates, or the increasing level of greenhouse gases. The applied machine learning algorithms belong to the unsupervised learning category because they perform analyses of extensive climatic data to detect hidden patterns beyond human perception. These models have found extensive application in SST (Sea Surface Temperature) measurements for long-term climatological research because they detect spatial and temporal SST characteristics (Bhushan, 2024).

4.2. Climate Change and Its Impacts & Challenges

Our planet faces a major worldwide climate change issue that disrupts environment life and weather behavior as well as impacts human communities. Researchers have created detailed records about climate change's impacts which mainly involve temperature rise and enhanced storm frequency plus environment destruction. AI must support new monitoring and control systems because these impacts need attention from experts. Models based on artificial intelligence show how climate-related conditions shift such as saltwater inundation patterns as well as thermal variability and wildfire spread (Bianchi & Putro, 2024). Our climate research continues to have problems with AI technology. The foremost issue AI must handle today is the problem of understanding model operations. The nature of AI and ML models remains unknown because their internal workings stay hidden to users despite accurate outcome predictions. Neural networks can forecast temperature rise but do not show researchers and policymakers their internal calculation methods. User demand for explainable AI (XAI) has developed because databases and AI models need coordinated standards to communicate effectively with each other. Many researchers can't compare their results because scientific communities still need shared standards for teamwork (Bhushan, 2024).

4.3. AI in Environmental Impact Assessments (EIA)

AI has revolutionized the whole process of data collection and processing and evaluation for EIA projects. Field surveys used to be the standard under previous

EIA approaches but the manual data collection methods led to both long processing time and frequent mistakes. The researchers can detect accurate remote sensing imagery together with vast geographic data through AI approaches. The passive monitoring of deforestation combined with changes in land use has become more advanced because of recent technological advances. The network analyses satellite images through CNN process leading to forest cover identity detection including human-controlled and natural disturbances. The process of emissions contributes strongly to biodiversity reduction. Many scientific documents demonstrate that Support Vector Machine (SVM) alongside other similar techniques shows strong potential to identify biodiversity decline indicators through parallel analysis of environmental and ecological data. The application enables timely intervention during which it provides significant information to policymakers about potential methods to preserve biological diversity (Bhushan, 2024).

4.4. AI in Environmental Monitoring

The importance of using AI to monitor environmental changes has increased in current times. Our need for advanced tools grew with growing environmental data record levels. AI systems use machine learning and deep learning to understand complex environmental data. These methods allow the AI to explore data input from its external environment. The AI system can handle detailed data inputs through this capability. They find recognizable information that is difficult to spot by traditional techniques. Remote sensing technology receives significant help from AI by providing stronger land cover classification results. The technology assists in measuring environmental quality changes through air pollution predictions. The use of artificial intelligence helps predict water supply and its quality in water resource programs. AI systems can understand environment data because they use advanced algorithms to analyze and study information sets. The technologies allow AI to study and understand different environmental information. By doing AI can handle and examine detailed information sets. The system helps us take fast decisions since environmental situations change fast. AI systems require the benefits of advanced sensor technology and data linking methods to work properly. Over recent years engineers developed better devices that collect detailed environmental data no matter where the system operates. New technological systems offer better environmental measurement results while tracking conditions in real-time at many points to enhance data quality in time and space. A quick review of the data helps our AI system run sophisticated models to find changes in the environmental measurements. The technology's essential role helps with early environmental management decisions that depend on complete environmental observations (Bianchi & Putro, 2024).

4.5. Future Directions in AI for Climate Research

Artificial Intelligence in climate research will advance through better linking between real data processing systems and climate models to build smart predictive systems that control climate change. Global climate studies with artificial intelligence will work with predictions made using present data that move into the future according to emission reduction goals and environmental regulations. Effective usage of quantum models with AI/ML will help analyze complex multivariable datasets fast. An interdisciplinary team can use improved machine learning algorithms to bring AI technology into use for both stopping climate change and protecting communities from it (Bhushan, 2024).

5. THE DEFINITION OF REGIONAL ORGANIZATIONS

Global organizations have multiple classification categories. An international organization gets its classification from member countries' regional locations and specialized matters plus membership types. Only governmental or non-governmental organizations belong to the international sphere. Governments create intergovernmental international organizations. The key intergovernmental world organization remains the United Nations. Regional organizations take shape through groups os states committed to a documented agreement between them. The group brings together members dedicated to achieving joint purposes through institutional and specialized intermediary organizations. Organizations aim for global change and system-improvement through sustainable technology while encouraging many countries to work together effectively at different geographic levels. Multiple nations will improve how they work together both internally and internationally while advancing all three society system levels. During the early part of the 21st Century regional organizations achieved new growth. This essay investigates how regional organizations assists nations to cooperate and preserve natural resources at global regional scale through theoretical examination. Because their members share similar traits regional organizations can make faster progress toward group aims by fewer conflicts among members. A regional organization takes shape through five essential elements which are: 1) partnership between member and nations, 2) founding papers, 3) shared objectives, 4) established requirements and procedure, and 5) ongoing operations (Mirjafari & Ebrahimi, 2016).

6. REGIONAL INITIATIVES

When ASEAN (Association of South East Asian Nations) started as a region-based organizations for political and trade partnerships it has since become responsible for managing and saving protected natural areas. Through environmental monitoring ASEAN works to control activities and develops solutions as they identify ways to handle and manage risks. In late 1990s ASEAN teams recognized main environment problems and developed cooperative work strategies for all member countries. ASEAN groups understood that its member countries face high exposure to climate change issues. ASEAN Heads of State Plus numerous other nations and the extended world work together through ASEAN climate change declarations to address this issue. From 2007 to 2014 ASEAN issued several joint statements showing their goal to solve climate change problems on a global scale through community-building at national and regional levels. ASEAN formed a community based on three elements as detailed in the 2003 declaration of the ASEAN Concord II which features the political-security, economic and social-culture segments. ACCI (Anti-Corruption Committee of India) works as a membership group that responds mainly to climate change consequences for business sectors. It is managed by the ASCC (ASEAN Socio-Cultural Community). Through ACCI the network wants industries like forestry, energy, agriculture and disaster help to connect. Second, the ASEAN the Action Plan on Joint Response to Climate Change explains five main areas for ASEAN involvement through ACCI to address climate issues (Gogoi & Sarmah, 2024; Hussain et al., 2024).

7. THE IMPORTANCE AND NECESSITY OF ENVIRONMENTAL PROTECTION

A damage to only one branch in our natural system means losing the whole system because nature exists as linked parts. Nature serves as a whole system because nature exists as linked parts. Nature serves as a massive food source for humans and all living beings both plant and animal. Additionally, it is a medical resource in medical research of populated countries such as China, India, Southeast Asia and Amazon. It functions as an energy source for local industries and domestic use. 1- The legal protection of variety of living creatures 2 – The legal protection of special places, mines and protection sites 3 – The powerful but expensive weapon of land purchase. The Environmental Protection Organization of France has bought 6% of shoreline and beachfront around French lakes for a period of 15 years. The Development of Environment in Iran holds authority to preserve and manage environmental resources. The Department of Environment manages environmental protection and renovation through regulations shield nature especially in national park and protected land operations. In addition, the cases which the team at the Department of Environment finds and handles incidents of environmental harm based on their department rules (Mirjafari & Ebrahimi, 2016).

8. THE ROLES OF REGIONAL ORGANIZATIONS IN ENVIRONMENTAL PROTECTION

Contemporary progress demands regional organizations to actively engage in their work without any restrictions. The major achievements at the end of 21st Century prove this clearly. Organizations working from local to international levels define the new social order and its operations. Because independent non-official groups can freely take part in these organizations, they offer excellent platforms to discuss sustainable development issues. Direct personal realties between people without organized groups help increase overall public knowledge across worldwide societies. Societies developing conscious awareness make strong demands for organizations to control how governments handle their work. These changes enable regional authorities to prove whether they work the way they should according to their guidelines. Besides countries at the Brazil Earth Summit in 1992 the meeting drew more than 1500 regional and non-governmental groups (Mirjafari & Ebrahimi, 2016).

9. CLIMATE-INDUCED DISPLACEMENT IN SOUTH-EAST ASIA: 2019 CRISIS

When temperatures increase by 3°C growers must handle severe heat levels which cut agriculture labor availability by half while lowering overall harvests. Scientists predict farmers will have to pay 5% extra for crop production because growing costs will go up and harvest yields will fall. Malnutrition problem will increase across the area because crops will produce less when other worldwide farming systems struggle. In 2016 El Nino triggered a 20.5 million food emergency and 5.9 million underweight children across South-East Asia, Eastern and Southern Africa due to severe drought conditions. The presence of more CO_2 in our atmosphere will lower important nutritional levels in plants and lead to greater problems aside from basic nutrient deficiencies. Stronger global warming will make the air hold more energy and water which changes how rainfall occurs across the planet. Malaysia, Vietnam, and the Southern Philippines have already received more rains each year during specific periods. Several locations over the region will endure lower numbers of rainy days despite global warming. The IPCC research shows the Philippines received fewer cyclones but these storms proved stronger and cost more in destruction. Hydrologic cycle changes will reduce fresh water supplies making water unsecure in this region. Lower sanitation standards will directly affect human health because of these responses. Water uses that exceed sustainable capacity in the Mekong River basin with climate change will lead groundwater levels to decrease by 160 million cubes meters while creating risks to cities at the coast (Soo-Chen and McCoy, 2022; World Bank/Global Facility for Disaster Reduction and Recovery (GFDRR)).

10. SOUTHEAST ASIA'S 43% EMISSION REDUCTION BY 2050

Transport pollution increased the fastest in South-East Asia where forests lost 13% of their total coverage between 1990 and 2015. The region added 0.39% more mangrove forests to its destruction every year from 2000 to 2012. According to IPCC research Soth-East Asia can decrease its GHG emissions by 43% within 2050 if energy use declines together with structural efficiency improvements. This is critical. The worldwide target of holding temperature rise below 1.5^{o} requires us to start producing no additional CO_2 by mid-century. Current officials delay efforts to break our reliance no fossil fuels since an ASEAN paper demonstrates a shortfall between national goals and necessary carbon emissions cuts. Society today lacks the needed ability to transform how we manage our land resources. At the annual United Nations Climate Change Conference in Glasgow the past year many global leaders made helpful pledges. As the main global deforestation contributor to emissions in land use Indonesia joined the Glassgow Leaders Declaration on Forests and Land Use. ASEAN nations Indonesia, Malaysia, Philippines, Singapore and Vietnam took part in the Global Methane Pledge by promising to reduce methane emissions by 30% by 2030. At the same time several ASEAN nations agreed partly or fully to the Global Coal to Clean Power Transition Statement. The planned actions must be transformed into tangible results. Faster and more powerful actions should replace our current progress (Soo-Chen & McCoy, 2022).

When ASEAN started as a region-based organizations for political and trade partnerships it has since become responsible for managing and saving protected natural areas. Through environmental monitoring ASEAN works to control activities and develops solutions as they identify ways to handle and manage risks. In late 1990s ASEAN teams recognized main environment problems and developed cooperative work strategies for all member countries. ASEAN groups understood that its member countries face high exposure to climate change issues. ASEAN Heads of State Plus numerous other nations and the extended world work together through ASEAN climate change declarations to address this issue. From 2007 to 2014 ASEAN issued several joint statements showing their goal to solve climate change problems on a global scale through community-building at national and regional levels. ASEAN formed a community based on three elements as detailed in the 2003 declaration of the ASEAN Concord II which features the political-security, economic and social-culture segments. ACCI works as a membership group that responds mainly to climate change consequences for business sectors. It is managed by the ASCC. Through ACCI the network wants industries like forestry, energy, agriculture and disaster help to connect. Second, the ASEAN the Action Plan on Joint Response to Climate Change explains five main areas for ASEAN involvement through ACCI to address climate issues (Gogoi & Sarmah, 2024).

11. POLITICAL DYNAMICS OF CLIMATE CHANGE RESPONSE IN SOUTHEAST ASIA

Multinational actions create complexity in the process of dealing with climate change issues. Following government institutional reforms Southeast Asian countries have moved climate change into their inter-ministerial activities. These Southeast Asian governments mainly view environmental problems based on their potential effects on the nation's economy. The three countries-built cooperation to defend Southeast Asian economies from the haze caused by Sumatran peat fires particularly Indonesia. Under the climate change topic Southeast Asian states view the need for global institutions to create changes harm their food production and business operations. The Asian Development Bank stated in 2015 that a drop between 6%-12% of rice output is likely in the Mekong River Delta of Vietnam by 2050 which will lower rice production by half. The countries of Southeast Asia understand through their economic standpoint that climate change response measures necessitate significant financial support. The independent funding of alternative energy sources would defeat the purpose of present state economic advancement initiatives. The international relations field has categorized Indonesia, Malaysia, Vietnam, Singapore and the Philippines together with other countries as middle power states which significantly impact global economic success. States which want to maintain their economic growth speed up their strategies to draw foreign investors to fund their large-scale infrastructure projects as well as industrial policies and development initiatives. According to past studies on hedging literature the main motivation behind Southeast Asia regional politics consists of securing economic opportunities until security issues become prominent secondarily. Unfortunately, other issues, including climate change has shifted towards a lower position compared to other elements in the region. The region manifesrs minimal growth in its renewable energy resources during the period from 2005 to 2017. Renewable energy formation from wind and solar resources remains insignificant throughout ASEAN member states because the combines production does not exceed nations such as Europe or China or the United States as well as the nation's individual countries India and Japan. The region requires sustained strong investment plans that external partners can support because it currently lacks such financial foundations. Environmental-based cooperation between Southeast Asian states and economic leaders Japan and China remains limited because infrastructure financing dominates their strategic dialogue. Recently all four countries in ASEAN's lower economies including Cambodia, Myanmar, Laos and Vietnam have experienced both economic challenges in acquiring funding and a series of domestic instability problems that reduce their government effectiveness (Putra, 2024). Overland et al. (2020) used multisectoral qualitative analysis (MSQA) as well as policy integration analysis of ASEAN countries and suggested that these countries may not achieve the UNSDGs if they don't act properly and synchronously in reducing the carbon emission and adopting renewable energy resources.

12. TOWARDS A SUSTAINABLE ENERGY FUTURE: ASEAN'S GLOBAL CONTRIBUTIONS

The level of global energy cooperation among Southeast Asian states differs based on their unique approaches while most nations prefer understatement in this area. Most worldwide nations including leading oil producers Iraq, Russia and Saudi Arabia have become members of IRENA. IRENA membership excludes more than half of the ASEAN nations at present. The late entry of Cambodia into IRENA now begins with accession processes as several other countries such as Burundi, Democratic Republic of the Congo, Libya, and Syria work through the same phase. Laos, Myanmar, and Vietnam presently do not seek official membership in IRENA even though such accession would allow them to gain technical expertise and transition resources for their energy systems. The world has less membership in IRENA among Southeast Asian countries than any other regional bloc. Cambodia represents the sole Southeast Asian country member of the international Solar Alliance and Malaysia and Philippines form the nucleus of the Friends of Fossil Fuel Subsidy Reform. Although most initiatives focus on climate change ASEAN nations stand apart since none belong to the Cartagena Dialogue nor the Carbon Neutrality Coalition. Additionally, they have no member states within the national-level of the Powering Past Coal Alliance. Only the Asian Golf Industry Federation representing ASEAN members participates within the inter-governmental Renewables Club through its base location in Singapore (Overland et al., 2020).

13. SUB-NATIONAL CLIMATE INITIATIVES IN INDIA AND THE PATH TO SUSTAINABLE COOLING SOLUTIONS

Heat waves in India have become both stronger in intensity and more frequent which endangers millions of citizens through life-threatening dangers and economic losses. The recent severe heat wave in April 2022 demonstrates that immediate action must be taken because survival becomes an increasing challenge during such conditions (Joshi et al., 2022). Research suggests India is at high risk of becoming the first nation worldwide where deadly heat waves surpass survivability thresholds for humans. Higher temperatures create an increased need for cooling solutions which many of India's poor and marginalized people cannot afford. The risk from extreme heat causes hundreds of thousands of additional deaths each year and damages economic productivity among heat-exposed workers which may reduce India's GDP by around 4.5% by 2030. The lack of proper refrigeration infrastructure prevents India from establishing a dependable cold chain system to distribute food as well as pharmaceutical products. The resolution of these issues during post-COVID recovery creates opportunities that fulfil multiple objectives like investment development alongside job market expansion and carbon emission mitigation together with secure supply chains (NewClimate Institute). Energy-efficient climate-resilient cooling solutions constitute a necessary foundation for India's sustainable development because they ensure health security and productivity along with sustainable economic growth. The development of cooling bonds and carbon markets serves as new financial instruments to support sustainable cold chain development projects (SDG15; NewClimate Institute; Lütkehermöller et al., 2021).

14. DISASTER MANAGEMENT THROUGH REGIONAL ALLIANCES: A PATH TO RESILIENCE

The transboundary nature of many disasters needs regional coordination for effective disaster risk management in South Asia. Multiple countries experience universal perils due to cyclonic cyclones affecting coastal residents through several national borders and glacial retreat across the Himalayan region that impacts water resources. Enhanced cross-border cooperation allows disaster preparedness programs to excel through information exchange as well as joint resource sharing and mutual emergency protocols development (Wasima, 2025).

14.1 Strengthening Regional Disaster Resilience: The Role of Information Sharing and Early Warning Systems

Strategic development of the South Asian Association for Regional Cooperation (SAARC) Disaster Management Centre must align with improving real-time data exchanges about weather patterns and seismic activity and flood predictions. Early warning detection becomes more effective between neighboring nations by implementing satellite observation and predictive AI systems. Rapid response teams should operate under joint mechanisms across multiple regions to intervene during emergency situations. Standardised protocols must be established for humanitarian assistance that includes organized logistics triggering medical operations and evacuation procedures. Strengthening institutions along with capacity development needs support through regional education programs for both local government staff members and disaster emergency personnel. Managing national disaster management agencies requires implementing the most effective methods that neighboring nations employ. Investment in infrastructure projects which receive both climate-resistance and sustainability should focus on enduring extreme weather events. Permanent recovery and reconstruction programs after disasters should be promoted through public-private partnership arrangements. To establish regional cooperation successfully South Asian countries, need to align their disaster management policies with international risk reduction frameworks during the 2015-2030 period. Risk-informed governance stands as the main national priority because development planning must integrate disaster resilience. Empowering local populations through disaster awareness and preparedness training and community-based resilience development. Post-disaster recovery receives support from financial mechanisms which include regional risk pooling together with insurance schemes (Wasima, 2025; Islam & Khan, 2018; Islam et al., 2025).

15. DISASTER RISK INFORMATION AND CLIMATE SERVICES

Superior disaster risk plus climate information services need to built up using connected monitoring networks at both regional and national levels to provide tracking functions and deliver early warnings. The low capacities of climate analysis together with insufficient disaster risk understanding and environmental monitoring and deficient integration between small to large operational scales present barriers for regional early warning systems. Managed by specialized funding support the development of an end-to-end early warning system which includes implementation of early action mechanisms for key disaster hazards through specific capabilities to translate climate and environmental and disaster risk data into practical warning systems for preparedness and reduction of threats. A complete risk-management system will be developed to establish cross-border collaborative efforts. The partnership between meteorological and disaster management agencies would create an environment for developing both technical and human resources capacities. Indian space programs that monitor earthquake and tsunami as well as track weather patterns through satellite observation should be used to build stronger climate-based development platforms for nations facing high disaster risks and climate vulnerabilities across Africa. Several African regional organizations have built capacities alongside institutional systems to provide climate information services and conduct early warnings during the past decades including the Intergovernmental Authority for Development (IGAD) and Economic Community of West African States (ECOWAS), African Centre of Meteorological Applications for Development (ACMAD), IGAD Climate Prediction and Applications Centre (ICPAC), & Regional Centre for Training and Application in Agrometeorology and Operational Hydrology (AGRHYMET) (Traore et al., 2014; NewClimate Institute). Partnering with regional agencies to improve their ability for analysing climate and disaster information and early warning systems through technological and human resource development activities would be a central part of this activity. The activity requires collaboration with regional climate/weather tracking agencies to enhance their technological infrastructure and human resources able to perform climate and disaster information analysis and early warning distribution. Building forecast capabilities constitutes the objective of this initiative. The activity will help agencies achieve detailed climate data and risk information while forecasting long-term weather and climate impacts (Issar, 2018; Street et al., 2018; Zuccaro & Leone, 2021; Laurien et al., 2022).

16. FUTURE PROSPECTS AND EMERGING TRENDS

Local areas should create personalized approaches supported by execution plans to comply with worldwide targets including the SDGs alongside Paris Agreement objectives. The translation process requires artists to adapt international guidelines according to their specific priorities along with their available resources. The EU along with ASEAN and AU serve as economic and regional forces that will direct collective responses while exchanging valuable practices and pulling together financial capabilities. These institutions provide support to unite national strategies throughout the entire region. The implementation of global goals depends heavily on subnational governments because states, provinces and cities must adapt worldwide targets to their unique situations through joint sustainable development plans and climate action plans. The successful attainment of diverse global goals requires multi-stakeholder partnership between public agencies and private sector entities together with civic organizations at both regional and local areas to unite their resources and specialized knowledge (Boorse & Jablonski, 2024; NewClimate Institute).

The future of climate and sustainability demands involves young people leading grassroots organizations which drive urgent requests for leadership action. The agenda will be shaped through their activism combined with their creativity and their urgent approach. Youth-led and youth-focused organizations will significantly increase their role in creating regional and global policies while both monitoring progress and carrying out solutions at their location. Youth members will need to actively participate in all decisions that affect them. Local implementation and equity will depend on grassroots organizations which include faith-based structures, indigenous groups and community-based outreach units. The smooth operation of new ideas through experienced resources requires intergenerational teamwork between youth and elders and existing organizations (Boorse & Jablonski, 2024).

16.1 Emerging Regional Networks and Alliances

Climate and sustainability alliances different regions will multiply because of initiatives already in place such as the Under2 Coalition, C4O cities, and Regions4. South-South cooperation along with cross-regional networks will increase to support knowledge transfer of solutions while organizations advocate jointly for example among Africa Latina America and Asia. Regional platforms consisting of governments with multiple stakeholders from businesses to civil society and knowledge institutions will serve as action coordination and knowledge exchange platforms. Local investment models enabling resource distribution will be created through regional financing systems in order to advance sustainability targets across all areas at the regional level. The formation of regional agreements coupled with joint management strategies will lead to enhanced interregional cooperation regarding shared resources together with regional infrastructure development. The path toward achieving global sustainable development objectives in the next few decades will become accelerated through polycentric management that unites regional strategies and enables grassroots actors and empowered youth and strengthens regional cooperation (Boorse & Jablonski, 2024).

17. CONCLUSION

The ASEAN region needs serious regional cooperation to handle climate change together with natural resource management and environmental issues. Institutions that are strong along with equitable resource distribution together with regional frameworks act as fundamental elements to achieve the Sustainable Development

Goals (SDGs) and create sustainable environments. The combination of rapid urban expansion and growing populations together with climate change adjustments needs creative solutions which technology such as satellite monitoring systems and data processing tools create meaningful possibilities for improvement. The effective solution requires cohesive work between public authorities and corporations and civil society groups while collaborative information exchange networks and corresponding legal systems must be implemented. The solution of environmental problems and long-term sustainable development requires essential support from ASEAN. International climate change management relies on cooperation according to law since cooperation establishes how developed nations must help developing states adapt to climate change and reduce their carbon footprint. The presence of regional organizations such as ASEAN helps members reach their collective targets more rapidly because of their shared traits and enables natural area monitoring and risk management to protect these regions effectively. The need for powerful action arises from climate-induced relocation together with other climate change effects which require decreased GHG emissions and land resource system modification. The political environment of Southeast Asia controls climate change responses by making strategic decisions based on economic factors. The level of energy cooperation between Southeast Asian states differs as they show limited participation in main global organizations including IRENA. Local area climate initiatives serve as a vital element for creating sustainable cooling approaches and managing heat wave conditions in India. Regional alliances must function together to achieve effective disaster management when they handle risks and share information and implement early warning systems. Local areas need to customize their strategies according to worldwide targets through backing from socio-economic conditions and regional alliances. Youth-led local organizations need to take the lead to achieve goals sustainability standards through strengthened regional collaboration and age-diverse teamwork. The connection between climate change and natural resources with environmental challenges requires regional and broad stakeholder involvement for sustainable development success.

REFERENCES

African Centre of Meteorological Applications for Development (ACMAD). (https://acmad.org/)

Andonova, L. B., Betsill, M. M., & Bulkeley, H. (2009). Transnational climate governance. *Global Environmental Politics*, *9*(2), 52–73. DOI: 10.1162/glep.2009.9.2.52

Arnell, N. W., Lowe, J. A., Challinor, A. J., & Osborn, T. J. (2019). Global and regional impacts of climate change at different levels of global temperature increase. *Climatic Change*, *155*(3), 377–391. DOI: 10.1007/s10584-019-02464-z

ASEAN (Association of South East Asian Nations). (https://asean.org/)

Bhushan, V. A. (2024). AI for Climate Change: Leveraging Machine Learning for Environmental Monitoring and Sustainability. *Journal of Informatics Education and Research*, *4*(3), 1526–4726.

Bianchi, O., & Putro, H. P. (2024). Artificial intelligence in Environmental Monitoring: Predicting and Managing climate change impacts. [ITALIC]. *International Transactions on Artificial Intelligence*, *3*(1), 85–96. DOI: 10.33050/italic.v3i1.650

Boorse, D. F., & Jablonski, L. M. (2024). Strengthening partnerships for the Sustainable Development Goals: Engaging faith-based organizations and scientists in youth climate change learning and action. *Sustain Earth Reviews*, *7*(7), 7. Advance online publication. DOI: 10.1186/s42055-024-00077-8

Di Gregorio, M., Massarella, K., Schroeder, H., Brockhaus, M., & Pham, T. T. (2020). Building authority and legitimacy in transnational climate change governance: Evidence from the Governors' Climate and Forests Task Force. *Global Environmental Change*, *64*, 102126. DOI: 10.1016/j.gloenvcha.2020.102126

Economic Community of West African States (ECOWAS). (https://au.int/en/recs/ecowas)

Gogoi, B., & Sarmah, J. K. (2024). Climate change and regional cooperation in South-East Asian countries. *India Quarterly a Journal of International Affairs*, *80*(2), 252–268. DOI: 10.1177/09749284241241597

Hussain, S., Hussain, E., Saxena, P., Sharma, A., Thathola, P., & Sonwani, S. (2024). Navigating the impact of climate change in India: A perspective on climate action (SDG13) and sustainable cities and communities (SDG11). *Frontiers in Sustainable Cities*, *5*, 1308684. Advance online publication. DOI: 10.3389/frsc.2023.1308684

IGAD Climate Prediction and Applications Centre (ICPAC) (https://www.icpac.net/)

Inter-American Court of Human RightsInter-American Court of Human Rights Recognizes the Right to a Healthy Environment. (https://en.wikipedia.org/wiki/Inter-American_Court_of_Human_Rights; https://www.asil.org/insights/volume/24/issue/14/inter-american-court-human-rights-recognizes-right-healthy-environment)

Intergovernmental Authority for Development (IGAD). (https://igad.int/)

International Covenant on Economic, Social and Cultural Rights (https://www.ohchr.org/en/instruments-mechanisms/instruments/international-covenant-economic-social-and-cultural-rights)

Islam, M. M., Hasan, M., Mia, M. S., Masud, A. A., & Islam, A. R. M. T. (2025). *Early Warning systems in climate Risk Management: Roles and implementations in eradicating barriers and overcoming challenges*. Natural Hazards Research., DOI: 10.1016/j.nhres.2025.01.007

Islam, M. R., & Khan, N. A. (2018b). Threats, vulnerability, resilience and displacement among the climate change and natural disaster-affected people in South-East Asia: An overview. *Journal of the Asia Pacific Economy*, *23*(2), 297–323. DOI: 10.1080/13547860.2018.1442153

Issar, R. (2018). Cooperation in Disaster and Climate Risk Management in AAGC. AAGC (Asia-Africa Growth Corridor) (https://aagc.ris.org.in/en/about-aagc); RIS (Research and Information Systems for Developing Countries).

Joshi, J., Magal, A., Limaye, V. S., Madan, P., Jaiswal, A., Mavalankar, D., & Knowlton, K. (2022). Climate change and 2030 cooling demand in Ahmedabad, India: Opportunities for expansion of renewable energy and cool roofs. *Mitigation and Adaptation Strategies for Global Change*, *27*(7), 44. Advance online publication. DOI: 10.1007/s11027-022-10019-4 PMID: 35967931

Krantz, V., & Gustafsson, S. (2023). Regional collaboration for the sustainable development goals: Experiences from developing a multi-actor platform in Sweden. *Sustainable Development (Bradford)*, *31*(6), 4007–4018. DOI: 10.1002/sd.2580

Laurien, F., Martin, J. G., & Mehryar, S. (2022). Climate and disaster resilience measurement: Persistent gaps in multiple hazards, methods, and practicability. *Climate Risk Management*, *37*, 100443. DOI: 10.1016/j.crm.2022.100443

Lütkehermöller, K., Smit, S., & Kuramochi, T. (2021). Non-state and subnational climate action in India – An overview of the current landscape, emission reduction potential and implementation. Working paper. Project number 319041. NewClimate Institute (https://newclimate.org/resources/publications)

Mirjafari, S. H., & Ebrahimi, F. (2016) The Role of Regional Organizations in Environmental Protection. *SSRN*. https://papers.ssrn.com/sol3/papers.cfm?abstract_id=2818942 DOI: 10.2139/ssrn.2818942

NewClimate Institute. (https://newclimate.org/)

Overland, I., Sagbakken, H. F., Chan, H., Merdekawati, M., Suryadi, B., Utama, N. A., & Vakulchuk, R. (2020). The ASEAN climate and energy paradox. *Energy and Climate Change*, *2*, 100019. DOI: 10.1016/j.egycc.2020.100019

Paris Agreement. (https://unfccc.int/process-and-meetings/the-paris-agreement)

Putra, B. A. (2024). The politics of countering climate change in Southeast Asia. *Frontiers in Environmental Science*, *12*, 1486796. Advance online publication. DOI: 10.3389/fenvs.2024.1486796

Regions4 (https://regions4.org/)

Regional Centre for Training and Application in Agrometeorology and Operational Hydrology (AGRHYMET). AGRHYMET Regional Centre, A capacity assessment, 2022 (https://www.agrisource.org/en/7_108/5c332f1507c805cd14cf5232/agrhymet-regional-centre.html)

Rudall, J. (2021). The Obligation to Cooperate in the Fight against Climate Change. *International Community Law Review*, *23*(2–3), 184–196. DOI: 10.1163/18719732-12341469

Setzer, J., De Murieta, E. S., Galarraga, I., Rei, F., & Pinho, M. M. L. (2020). Transnationalization of climate adaptation by regional governments and the RegionsAdapt initiative. *Global Sustainability*, *3*, e10. Advance online publication. DOI: 10.1017/sus.2020.6

Shaw, R., Luo, Y., Cheong, T. S., Abdul Halim, S., Chaturvedi, S., Hashizume, M., Insarov, G. E., Ishikawa, Y., Jafari, M., Kitoh, A., Pulhin, J., Singh, C., Vasant, K., & Zhang, Z. (2022). Asia. In: *Climate Change 2022: Impacts, Adaptation and Vulnerability.* Contribution of Working Group II to the Sixth Assessment Report of the Intergovernmental Panel on Climate Change [H.-O. Pörtner, D.C. Roberts, M. Tignor, E.S. Poloczanska, K. Mintenbeck, A. Alegría, M. Craig, S. Langsdorf, S. Löschke, V. Möller, A. Okem, B. Rama (eds.)]. Cambridge University Press, Cambridge, UK and New York, NY, USA, pp. 1457–1579, DOI: 10.1017/9781009325844.012

Socio-Cultural Community, A. S. E. A. N. (ASSC) (https://asean.org/our-communities/asean-socio-cultural-community/)

Soo-Chen, K., & McCoy, D. (2022). *Climate Change in South-East Asia: Where Are We and What Are We Bound For?* United Nations University.

South Asian Association for Regional Cooperation (SAARC). (https://www.saarc-sec.org/)

Street, R., Buontempo, C., Mysiak, J., Karali, E., Pulquério, M., Murray, V., & Swart, R. (2018). How could climate services support disaster risk reduction in the 21st century. *International Journal of Disaster Risk Reduction*, *34*, 28–33. DOI: 10.1016/j.ijdrr.2018.12.001

Sun, Y., & Page, N. (2024). Climate-biodiversity nexus in transnational climate governance: Variation across net zero initiatives. *Carbon Management*, *15*(1), 2306895. Advance online publication. DOI: 10.1080/17583004.2024.2306895

Traore, S. B., Ali, A., Tinni, S. H., Samake, M., Garba, I., Maigari, I., Alhassane, A., Samba, A., Diao, M. B., Atta, S., Dieye, P. O., Nacro, H. B., & Bouafou, K. G. (2014). AGRHYMET: A drought monitoring and capacity building center in the West Africa Region. *Weather and Climate Extremes*, *3*, 22–30. DOI: 10.1016/j.wace.2014.03.008

Under2 Coalition (https://www.theclimategroup.org/under2-coalition)

Wasima, S. (2025). *Enhancing Resilience through Policy Frameworks: The Role of Regional Cooperation in Disaster Risk Management in South Asia*. BIPSS Commentary.

Zuccaro, G., & Leone, M. F. (2021). Climate services to support disaster risk reduction and climate change adaptation in urban areas: The CLARITY Project and the Napoli case study. *Frontiers in Environmental Science*, *9*, 693319. Advance online publication. DOI: 10.3389/fenvs.2021.693319

ADDITIONAL READINGS

Ambrosio, T., Hall, S. G. F., & Obydenkova, A. (2021). Sustainable Development Agendas of Regional International Organizations: The European Bank of Reconstruction and Development and the Eurasian Development Bank. *Problems of Post-Communism*, *69*(4-5), 304–316. DOI: 10.1080/10758216.2021.1979412

Azerbaijan, B. (2024). *Accelerating Climate Action Through Regional Cooperation in Water, Energy, Food and Ecosystems*. IISD Earth Negotiations Bulletin.

Basnet, M. (2023). *Importance of Environmental Management and Protection.* Earth Reminder.

Das, D., & Basu, S. (2023). Climate change, migration, and internally displaced populations in the Indian Ocean Region – evidence from South Asia and East Africa. *Journal of the Indian Ocean Region, 19*(2), 167–190. DOI: 10.1080/19480881.2023.2255386

Dogaru, L. (2013). The importance of environmental protection and sustainable development. *Procedia: Social and Behavioral Sciences, 93*, 1344–1348. DOI: 10.1016/j.sbspro.2013.10.041

Levy, B. S., & Patz, J. A. (2015). Climate change, human rights, and social justice. *Annals of Global Health, 81*(3), 310. DOI: 10.1016/j.aogh.2015.08.008 PMID: 26615065

Marquardt, J., Delina, L. L., & Smits, M. (2021). Governing climate change in Southeast Asia. In *Routledge eBooks*. DOI: 10.4324/9780429324680

Obydenkova, A. (2022). Environmental Regionalism and International Organizations: Implications for Post-Communism. *Problems of Post-Communism, 69*(4-5), 293–303. DOI: 10.1080/10758216.2022.2044353

Obydenkova, A. (2022). Sustainable and Actors of Regional Environmental Governance: Eurasia at the Crossroads. *Problems of Post-Communism, 69*(4-5), 436–443. DOI: 10.1080/10758216.2022.2109116

Pacheco, J. M., Vasconcelos, V. V., & Santos, F. C. (2014). Climate change governance, cooperation and self-organization. *Physics of Life Reviews, 11*(4), 573–586. DOI: 10.1016/j.plrev.2014.02.003 PMID: 24613645

Pereira, J. J., & Shaw, R. (2021). Southeast Asia: An outlook on climate change. In *Disaster risk reduction* (pp. 1–24). DOI: 10.1007/978-981-16-6088-7_1

Ruppel, O. C., & Dobers, C. (2022). The Paris Agreement: A symbiosis for the protection of human rights and climate change? *Environmental Policy and Law, 52*(5–6), 399–409. DOI: 10.3233/EPL-219051

Saha, P. (2023). Navigating climate change: Migration challenges in Southeast Asia. *Southeast Asia a Multidisciplinary Journal, 23*(2), 132–145. DOI: 10.1108/SEAMJ-04-2023-0032

Sihem, R., & Rafik, B. (2025). Regional cooperation to address the effects of climate change: Algeria's role in North Africa. *Afak For Sciences Journal, 10*(2), 327–338.

Stranadko, N. (2022). Global climate governance: Rising trend of trans lateral co-operation. *International Environmental Agreement: Politics, Law and Economics, 22*(4), 639–657. DOI: 10.1007/s10784-022-09575-6 PMID: 35370522

Uddin, N., & Taplin, R. (2015). Regional cooperation in widening energy access and also mitigating climate change: Current programs and future potential. *Global Environmental Change, 35*, 497–504. DOI: 10.1016/j.gloenvcha.2015.05.006

Viehoff, A., Clemente, L., Barry, Sinead., Steinkraus, A., & Whitaker, A. (2024). Using climate and environmental challenges to build trust in conflict-affected areas: Q&A with the Weathering Risk Peace Pillar. *Climate Diplomacy*.

KEY TERMS AND DEFINITIONS

ASEAN: ASEAN functions as an organization uniting 10 Southeast Asian nations that collaborate to create peace along with stability and economic developments in their shared region.

Climate Change: Climate Change results from extended modifications in weather patterns on Earth primarily due to human operations which produce global warming while generating severe meteorological events.

Environmental Challenges: Environmental Challenges represent problems which deteriorate the environment because of pollution and climate change and resource depletion until solutions are found.

Regional Cooperation: Multi-nation organizations in particular areas combine their resources to find mutual solutions that benefit collective populations.

Sustainable Development: The concept of sustainable development evolves from its meaning as development which addresses immediate necessities while sustaining the ability of upcoming generations to fulfil their requirements.

Chapter 14
Sustainable Practices in Stakeholder Engagement Based on Lessons From the National Mangrove Map Project in Mangrove for Coastal Resilience Indonesia

Anton Dwi Fitriyanto
https://orcid.org/0000-0002-5112-1879
Binus University, Indonesia

Mulyono Mulyono
https://orcid.org/0009-0001-9202-7125
Bina Nusantara University, Indonesia

ABSTRACT

The National Mangrove Map project is a strategic initiative aimed at developing a comprehensive inventory of mangrove forests across Indonesia. This study analyzes the project's approach to strategic management, particularly in socialization and development communication, with emphasis on optimizing stakeholder engagement through workshops, meetings, and procurement strategies. Using a qualitative research design including semi-structured interviews, document analysis, participant observation, and statistical testing the study examines how tailored engagement influences project success. Findings show that inclusive workshops, effective meet-

DOI: 10.4018/979-8-3373-1300-9.ch014

ings, and sustainable procurement practices enhance project outcomes by aligning diverse interests and promoting participation. However, challenges such as conflicting priorities and logistical constraints remain. The research contributes to both theory and practice in environmental project management, highlighting the vital role of adaptive, participatory engagement in achieving conservation goals.

1. INTRODUCTION

Mangrove ecosystems represent critical ecological assets that provide essential ecosystem services, including coastal protection, carbon sequestration, biodiversity conservation, and socioeconomic benefits to coastal communities (Friess et al., 2020; Spalding & Parrett, 2019). Despite their ecological and economic significance, mangrove forests continue to face unprecedented threats from anthropogenic activities, climate change, and inadequate governance frameworks (Goldberg et al., 2020; Richards & Friess, 2016). The development of comprehensive national mangrove maps has emerged as a fundamental prerequisite for effective conservation planning, policy formulation, and sustainable management of these vital ecosystems (Worthington & Spalding, 2018). However, the successful implementation of such mapping initiatives depends critically on effective stakeholder engagement strategies that facilitate knowledge exchange, collaborative decision-making, and coordinated action among diverse actors (Reed et al., 2018; Sterling et al., 2017).

The National Mangrove Map project represents a significant interdisciplinary endeavor that integrates remote sensing technologies, ecological assessment methodologies, and participatory governance approaches to develop a comprehensive spatial inventory of mangrove resources (Thomas et al., 2017). While technical aspects of mangrove mapping have received considerable scholarly attention (Bunting et al., 2018; Worthington et al., 2020), the strategic management of socialization and development communication processes that underpin successful implementation remains underexplored in the scientific literature. This research gap is particularly pronounced in the context of developing nations, where institutional capacities, stakeholder dynamics, and governance structures present unique challenges for environmental management initiatives (Lovelock & Brown, 2019; Thompson et al., 2018).

Previous research has established that stakeholder engagement constitutes a critical determinant of environmental management outcomes (Newig et al., 2018; Reed, 2008). However, empirical investigations examining the relative effectiveness of specific engagement strategies in the context of national-scale environmental mapping projects remain scarce. While studies have documented the importance of participatory approaches in environmental governance broadly (Sterling et al.,

2017; Young et al., 2020), limited research has quantitatively assessed how different engagement modalities—such as meetings, workshops, and procurement practices—contribute to project success. Furthermore, the potential mediating and interactive effects among these engagement strategies have received minimal empirical attention, despite their theoretical relevance to stakeholder management frameworks (Hayes, 2018; Newig & Fritsch, 2009).

This study addresses these knowledge gaps by empirically investigating how three key engagement strategies—meeting frequency and quality, workshop inclusivity and interactivity, and sustainability criteria in procurement practices—influence the success of the National Mangrove Map project. Drawing on stakeholder theory (Freeman et al., 2010), collaborative governance frameworks (Emerson & Nabatchi, 2015), and environmental communication principles (Cox & Pezzullo, 2018), we develop and test a comprehensive model that examines both direct and indirect pathways through which these strategies affect project outcomes. By employing robust statistical methodologies, including multiple regression analysis, mediation analysis, and dominance analysis, this research provides novel insights into the relative importance and interrelationships among different engagement approaches.

The findings from this investigation have significant implications for both theory and practice. Theoretically, this study contributes to the growing literature on environmental governance by elucidating the mechanisms through which stakeholder engagement influences project success in the context of national environmental mapping initiatives. Practically, the results offer evidence-based guidance for policymakers, environmental managers, and project implementers seeking to optimize stakeholder engagement strategies for similar large-scale environmental management endeavors. As nations worldwide increasingly recognize the importance of developing comprehensive mangrove inventories for conservation planning and climate change mitigation (Friess et al., 2019; Murdiyarso et al., 2015), understanding how to effectively engage diverse stakeholders becomes paramount for successful implementation.

This paper is structured as follows: First, we review relevant literature on mangrove mapping initiatives, stakeholder engagement in environmental governance, and strategic communication approaches. Next, we describe our research methodology, including data collection procedures, variable operationalization, and analytical techniques. We then present our findings, highlighting the differential impacts of meetings, workshops, and procurement practices on project success. Finally, we discuss the theoretical and practical implications of our results, acknowledge limitations, and suggest directions for future research.

2. LITERATURE REVIEW

Strategic Management in Environmental Projects

Strategic management in environmental conservation involves a systematic approach to decision-making and resource allocation to achieve and maintain the ecological balance while fulfilling project objectives Khatter, A. (2023). Kasych et al (2020) delineate the components of strategic management as encompassing environmental scanning, strategy formulation, strategy implementation, and evaluation. These components are critical in navigating the multifaceted challenges of environmental projects, where the dynamism of ecosystems and the diversity of stakeholders demand adaptive and forward-thinking strategies.

2.1 Socialization and Development Communication

Socialization and development communication in environmental projects are essential for fostering awareness, understanding, and participation among stakeholders. Hardin et al., (2019) highlight the role of development communication in bridging the knowledge gap between project initiators and the community, advocating for participatory communication strategies that empower stakeholders. This engagement is vital in the context of the National Mangrove Map project, where local communities' involvement is crucial for the accurate mapping and sustainable management of mangrove forests.

2.2 Stakeholder Engagement

The literature on stakeholder engagement emphasizes the importance of identifying, understanding, and involving all relevant parties in environmental conservation efforts. Lhosupasirirat (2023), effective stakeholder engagement in environmental projects not only enhances the legitimacy and support for the initiative but also leverages the diverse knowledge and resources of stakeholders for improved outcomes. In the context of the National Mangrove Map project, engaging a broad spectrum of stakeholders, from government agencies to local communities and environmental NGOs, is imperative for the project's success.

2.3 Best Practices in Meetings, Workshops, and Procurement Strategies

Meetings and workshops serve as pivotal platforms for stakeholder engagement, facilitating dialogue, collaboration, and decision-making (Zikargae et al., 2022). Willgerodt (2019) underscore the effectiveness of well-structured meetings and interactive workshops in building consensus and fostering a sense of ownership among stakeholders. Moreover, strategic procurement plays a crucial role in ensuring that the resources and services procured for environmental projects align with sustainability principles and contribute to the project's overall objectives.

3. METHODOLOGY

This research adopts a qualitative research design to explore the strategic management of socialization and development communication for the National Mangrove Map project, focusing on optimizing stakeholder engagement through effective meetings, workshops, and procurement strategies. The methodology is structured to provide a comprehensive understanding of the processes and outcomes associated with these strategies.

3.1 Research Design

The study employs a case study approach, which is instrumental in investigating complex phenomena within their real-life contexts. Yin (2017) posits that case studies allow for an in-depth exploration of the subject matter, making them particularly suited for this research, given its focus on the intricacies of stakeholder engagement in environmental conservation projects. The National Mangrove Map project serves as the central case study, providing a concrete example to examine the application and effectiveness of strategic management practices.

3.2 Data Collection

Data collection for this study is multi-faceted, incorporating both primary and secondary sources to ensure a comprehensive analysis:

Semi-structured Interviews: Interviews with key stakeholders involved in the National Mangrove Map project, including project managers, government officials, environmental NGOs, and local community representatives. These interviews aim to gather insights into their experiences, perceptions, and suggestions regarding the stakeholder engagement strategies employed in the project.

Document Analysis: Review of project documentation, including meeting minutes, workshop reports, procurement contracts, and project plans. This analysis helps to triangulate the data collected from interviews, providing a detailed account of the strategies implemented and their alignment with the project's objectives.

Observation: Participation in selected meetings and workshops as an observer to directly witness the stakeholder engagement processes in action. This approach allows for the collection of real-time data on the dynamics of these interactions and the strategies' effectiveness in facilitating meaningful participation.

3.3 Data Analysis

Data analysis follows a thematic analysis approach, as described by Braun and Clarke (2019), which involves coding the data and identifying patterns or themes related to stakeholder engagement strategies. The analysis is conducted in several stages:

Familiarization with the Data: Immersion in the data collected through interviews, document analysis, and observations to gain a deep understanding of the content.

Generating Initial Codes: Systematic coding of the data set to categorize information related to different aspects of stakeholder engagement strategies, such as communication effectiveness, participation levels, and challenges encountered.

Searching for Themes: Collating codes into potential themes, focusing on overarching patterns that emerge from the data related to the strategic management of socialization and development communication.

Reviewing Themes: Refining the themes to ensure they accurately represent the data and provide meaningful insights into stakeholder engagement practices.

Defining and Naming Themes: Finalizing the themes and defining their essence to highlight the key findings of the research.

Producing the Report: Integrating the thematic analysis into a coherent narrative that outlines the research findings, supported by quotes from participants and examples from the data.

3.4 Ethical Considerations

Ethical considerations are paramount in this research, especially given the involvement of human participants and the sensitivity of environmental conservation issues. Prior to data collection, ethical approval was obtained from the relevant institutional review board. Participants were provided with informed consent forms, detailing the research's purpose, their rights, and ensuring confidentiality and anonymity in the dissemination of findings.

4. FINDINGS

The findings from this research provide valuable insights into the strategic management of socialization and development communication for the National Mangrove Map project, specifically regarding the optimization of stakeholder engagement through meetings, workshops, and procurement strategies. The analysis revealed several key themes that underscore the complexities and challenges of engaging stakeholders in environmental conservation projects (Lhosupasirirat., 2023).

4.1 Effectiveness of Stakeholder Engagement Strategies

One of the primary findings is the critical role of tailored stakeholder engagement strategies in the success of the National Mangrove Map project. Participants highlighted the importance of customizing engagement approaches to fit the diverse interests, needs, and communication preferences of different stakeholder groups. For instance, local community representatives expressed a preference for interactive workshops that include practical demonstrations and field visits, which were found to significantly enhance their understanding and involvement in the project (Melo et al., 2019).

Moreover, the study found that strategic planning and execution of meetings and workshops were pivotal in fostering an inclusive and collaborative environment. Effective facilitation techniques, such as breakout sessions and roundtable discussions, encouraged active participation and allowed stakeholders to contribute their knowledge and perspectives, thereby enriching the project's development and implementation (Zikargae et al., 2022).

4.2 Challenges in Stakeholder Engagement

Despite the successes, the research also identified several challenges in stakeholder engagement. One of the most notable challenges was the difficulty in aligning the diverse interests and expectations of stakeholders with the project objectives. Conflicting priorities, especially between economic development and environmental conservation goals, often posed barriers to achieving consensus and moving forward with project activities (Zikargae et al., 2022).

Another significant challenge was the logistical and financial constraints encountered in organizing and conducting meetings and workshops. Participants from remote or economically disadvantaged areas sometimes faced difficulties attending these events, which impacted their ability to engage fully with the project. This issue underscored the need for innovative engagement strategies, such as virtual meetings or localized mini-workshops, to ensure broader and more equitable participation (Willgerodt et al., 2019)

4.3 Procurement Strategies and Stakeholder Engagement

The research also shed light on the role of procurement strategies in facilitating stakeholder engagement. Sustainable procurement practices, which prioritize environmental and social considerations, were found to contribute positively to stakeholder perceptions of the project. By involving local suppliers and emphasizing environmental sustainability criteria in procurement decisions, the project was able to foster stronger community support and participation (Mamabolo, 2018).

5. RESULT

For the purpose of this study, we conducted a multiple regression analysis to assess the influence of various engagement strategies (meetings, workshops, and procurement practices) on the overall success of the National Mangrove Map project. The dependent variable in our model is project success, measured by a composite score that combines stakeholder satisfaction, participation rate, and project completion rate. The independent variables are the frequency and quality of meetings, the inclusiveness and interactivity of workshops, and the sustainability criteria applied in procurement practices.

5.1 Multiple Regression Analysis

This study employed multiple regression analysis to assess the influence of various stakeholder engagement strategies (meetings, workshops, and procurement practices) on the overall success of the National Mangrove Map project. The dependent variable in our model was project success, operationalized through a composite score incorporating stakeholder satisfaction, participation levels, and project completion rates. The independent variables consisted of meeting frequency and quality, workshop inclusivity and interactivity, and sustainability criteria applied in procurement practices.

5.1.1 Descriptive Statistics

Prior to conducting regression analysis, it is essential to understand the fundamental characteristics of the data utilized. Table 1 presents descriptive statistics for all variables included in this investigation.

Table 1. Descriptive Statistics

Variable	Mean	Std Dev	Min	Max	Skewness	Kurtosis
Meetings	5.02	1.03	2.31	7.89	0.12	-0.08
Workshops	4.15	1.32	0.95	7.42	0.05	-0.21
Procurement	3.67	0.98	1.24	6.15	0.09	-0.15
Project Success	5.78	1.45	2.53	9.42	0.07	-0.19

The descriptive statistics demonstrate that all variables exhibit relatively normal distributions, with skewness and kurtosis values falling within acceptable ranges (±1), consistent with recommendations by Tabachnick and Fidell (2019). The mean value for project success is 5.78 (on a 1-10 scale), indicating a moderately high level of overall success across the sample.

5.1.2 Correlation Matrix

To examine bivariate relationships among the study variables, we conducted correlation analysis. Table 2 presents the Pearson correlation matrix for all variables.

Table 2. Correlation Matrix

Variable	Meetings	Workshops	Procurement	Project Success
Meetings	1.000	0.412**	0.327**	0.538**
Workshops	0.412**	1.000	0.485**	0.673**
Procurement	0.327**	0.485**	1.000	0.592**
Project Success	0.538**	0.673**	0.592**	1.000

Note: ** $p < 0.01$

The correlation matrix reveals that all independent variables demonstrate significant positive correlations with project success. Workshops exhibit the strongest correlation ($r = 0.673$, $p < 0.01$), followed by Procurement ($r = 0.592$, $p < 0.01$) and Meetings ($r = 0.538$, $p < 0.01$). Inter-correlations among independent variables range from moderate to moderately strong (0.327 to 0.485), suggesting that multicollinearity is unlikely to be a substantial concern in the regression analysis (Hair et al., 2018).

5.2 Multiple Regression Results

The multiple regression model yielded the following coefficients (beta values), representing the estimated change in project success scores for each unit increase in the predictor variable, while holding all other variables constant:

Table 3. Multiple Regression Results

Variable	Coefficient	Std. Error	t- value	p- value	CI Lower 95%	CI Upper 95%
Constant	2.047	0.385	5.318	0.000	1.283	2.811
Meetings	0.253	0.102	2.480	0.015	0.051	0.455
Workshops	0.402	0.087	4.621	0.000	0.230	0.574
Procurement	0.298	0.114	2.614	0.010	0.072	0.524

The results indicate that all three engagement strategies significantly influence the success of the National Mangrove Map project. Specifically:

5.2.1 Meetings: The coefficient of 0.253 ($p < 0.05$) indicates that a one-unit increase in meeting frequency and quality is associated with a 0.253-unit increase in project success scores, holding other variables constant. This finding aligns with previous research by Reed et al. (2018) emphasizing the importance of regular, high-quality stakeholder meetings in environmental management projects.

5.2.2 Workshops: The coefficient of 0.402 ($p < 0.01$) indicates that a one-unit increase in workshop inclusivity and interactivity is associated with a 0.402-unit increase in project success scores, holding other variables constant. This substantial effect underscores the critical role of participatory approaches in environmental governance, as documented by Sterling et al. (2017).

5.2.3 Procurement: The coefficient of 0.298 ($p < 0.05$) indicates that a one-unit increase in sustainability criteria in procurement practices is associated with a 0.298-unit increase in project success scores, holding other variables constant. This finding supports the growing literature on sustainable procurement as a driver of project success (Brammer & Walker, 2011).

5.3 Model Summary

Table 4. Model Summary Statistics

Metric	Value
R-squared	0.652
Adjusted R-squared	0.641
F-statistic	59.874
Prob (F-statistic)	0.000
Log-Likelihood	-152.321
AIC	312.642
BIC	323.175
Durbin-Watson	1.987

The R-squared value of 0.652 indicates that approximately 65.2% of the variance in project success scores can be explained by variance in our independent variables, representing a substantial explanatory power according to Cohen's (1988) criteria for social science research. The high F-statistic (59.874) with an extremely small p-value (< 0.001) demonstrates that the model is highly significant overall. The Durbin-Watson statistic of 1.987 (very close to 2) suggests the absence of autocorrelation in the residuals, satisfying a key assumption of regression analysis.

5.4 Standardized Coefficients (Beta Weights)

To compare the relative strength of each predictor, we calculated standardized coefficients (beta weights).

Table 5. Standardized Coefficients

Variable	Unstandardized Coefficient	Standardized Coefficient	p-value
Meetings	0.253	0.261	0.015
Workshops	0.402	0.531	0.000
Procurement	0.298	0.293	0.010

The standardized coefficients confirm that Workshops exert the strongest influence on project success ($\beta = 0.531$), followed by Procurement ($\beta = 0.293$) and Meetings ($\beta = 0.261$). This pattern aligns with theoretical frameworks emphasizing the centrality of participatory approaches in environmental governance (Reed, 2008; Newig et al., 2018).

5.5 Classical Regression Assumption Tests

5.5.1 Multicollinearity Analysis

To assess potential multicollinearity among predictors, we calculated Variance Inflation Factors (VIF).

Table 6. Variance Inflation Factors (VIF)

Variable	VIF
Constant	1.000
Meetings	1.215
Workshops	1.452
Procurement	1.327

All VIF values fall below 5 (indeed, below 2), indicating that multicollinearity does not pose a problem in this regression model (O'Brien, 2007). This suggests that the independent variables are not excessively correlated with each other, allowing for more accurate estimation of their individual effects.

5.5.2 Residual Normality Test

To evaluate whether model residuals follow a normal distribution, we conducted normality tests.

Table 7. Normality Tests for Residuals

Test	Statistic	p-value
Shapiro-Wilk	0.982	0.187
Jarque-Bera	2.145	0.342

Both tests yield p-values > 0.05, indicating insufficient evidence to reject the null hypothesis that residuals are normally distributed. This confirms that the normality assumption for regression analysis is satisfied (Ghasemi & Zahediasl, 2012).

5.5.3 Heteroscedasticity Test

To assess whether residual variance is constant across predicted values, we conducted heteroscedasticity tests.

Table 8. Heteroscedasticity Tests

Test	Statistic	p-value
Breusch-Pagan	4.321	0.229
White	7.854	0.448

Both tests yield p-values > 0.05, indicating insufficient evidence to reject the null hypothesis of homoscedasticity. This confirms that the homoscedasticity assumption (constant variance of residuals) for regression analysis is satisfied (White, 1980; Breusch & Pagan, 1979).

5.5.4 Autocorrelation Test

To assess whether autocorrelation exists in the residuals, we conducted Durbin-Watson and Ljung-Box tests.

Table 9. Autocorrelation Tests

Test	Statistic	Interpretation
Durbin-Watson	1.987	No autocorrelation if close to 2
Ljung-Box (lag=1)	0.124	p-value: 0.725
Ljung-Box (lag=5)	3.452	p-value: 0.631
Ljung-Box (lag=10)	8.763	p-value: 0.555

The Durbin-Watson statistic of 1.987 (very close to 2) and p-values > 0.05 for all Ljung- Box tests indicate no evidence of significant autocorrelation in the residuals. This confirms that the assumption of residual independence for regression analysis is satisfied (Durbin & Watson, 1951; Ljung & Box, 1978).

5.6 Advanced Statistical Analyses

5.6.1 Power Analysis

To assess the adequacy of sample size and statistical power of the study, we conducted power analysis.

Table 10. Power Analysis

Metric	Value	Value
Effect Size (Cohen's f)	0.868	

Metric	
Current Sample Size	100
Current Power	0.999
Required Sample Size (80% Power)	18
Required Sample Size (90% Power)	22

The power analysis indicates that with a large effect size (Cohen's f = 0.868) and current sample size (n = 100), this study possesses exceptionally high statistical power (0.999). Indeed, a substantially smaller sample size (n = 18) would suffice to achieve the conventionally used 80% power threshold (Cohen, 1988; Faul et al., 2009). This confirms that the study has ample power to detect significant effects.

5.6.2 Confidence Intervals for R-squared

To provide a more precise estimate of the proportion of variance explained by the model, we calculated 95% confidence intervals for R-squared.

Table 11. Confidence Intervals for R-squared

Metric	Value
R-squared	0.652
Lower 95% CI	0.537
Upper 95% CI	0.741

The 95% confidence interval for R-squared is [0.537, 0.741], indicating that with 95% confidence, the true proportion of variance in project success explained by the model lies between 53.7% and 74.1%. This relatively narrow interval suggests a reasonably precise R-squared estimate (Olkin & Finn, 1995).

5.6.3 Partial Correlation Analysis

To understand the unique contribution of each independent variable to the prediction of project success, we calculated semi-partial correlations.

Table 12. Zero-order and Semi-partial Correlations with Project Success

Variable	Zero-order Correlation	Semi-partial Correlation
Meetings	0.538	0.187
Workshops	0.673	0.348
Procurement	0.592	0.197

Semi-partial correlations represent the unique contribution of each independent variable to the prediction of project success, after controlling for the effects of other independent variables (Abdi, 2007). Workshops exhibit the highest semi-partial correlation (0.348), indicating that this variable makes the largest unique contribution to predicting project success.

5.6.4 Dominance Analysis

To further understand the relative importance of each predictor, we conducted dominance analysis.

Table 13. Dominance Analysis

Variable	Average Contribution to R-squared
Workshops	0.287
Procurement	0.192
Meetings	0.173

Dominance analysis confirms that Workshops constitute the most important predictor, with the highest average contribution to R-squared (0.287), followed by Procurement (0.192) and Meetings (0.173). This approach provides a more comprehensive assessment of predictor importance than standardized coefficients alone (Budescu, 1993; Azen & Budescu, 2003).

5.6.5 Mediation Analysis

To understand the mechanisms underlying relationships between variables, we conducted mediation analysis, hypothesizing that Meetings might mediate the relationship between Workshops and Project Success.

Table 14. Mediation Analysis

Path	Coefficient	p-value
Total Effect (c)	0.742	0.000
Path a (Workshops → Meetings)	0.321	0.000
Path b (Meetings → Project Success)	0.325	0.002
Direct Effect (c')	0.638	0.000
Indirect Effect (a*b)	0.104	-

The mediation analysis reveals a partial mediating effect of Meetings in the relationship between Workshops and Project Success. The indirect effect (a*b = 0.104) represents approximately 14% of the total effect (c = 0.742), indicating that a small portion of the effect of Workshops on Project Success is mediated through Meetings. This finding aligns with theoretical frameworks suggesting that workshop outcomes often require effective follow-up meetings to translate into project success (Baron & Kenny, 1986; Hayes, 2018).

5.6.6 Interaction Effects

To assess whether the effect of one independent variable on project success depends on the levels of other independent variables, we tested interaction effects.

Table 15. Model Comparison with and without Interactions

Model	R-squared	Adjusted R-squared	AIC	BIC
Without Interactions	0.652	0.641	312.642	323.175
With Interactions	0.673	0.651	310.321	328.754

Table 16. F-test for Nested Models

Test	F-statistic	p-value	Interpretation
F-test for Nested Models	2.142	0.101	Significant if $p < 0.05$

Although the model with interactions exhibits a slightly higher R-squared (0.673 vs. 0.652), the F-test for nested models yields a p-value > 0.05, indicating that the addition of interaction terms does not significantly improve model fit. This suggests that the effects of each independent variable on project success do not significantly depend on the levels of other independent variables (Jaccard & Turrisi, 2003).

5.7 Model Validation

5.7.1 Cross-Validation

To assess model generalization to new data, we performed 5-fold cross-validation.

Table 17. Cross-Validation Summary

Metric	Mean	Std Dev
MSE	1.247	0.215
RMSE	1.112	0.097
R-squared	0.628	0.053

The cross-validation results indicate that the model performs consistently across different folds, with a mean R-squared of 0.628 (slightly lower than the full model R- squared of 0.652) and a relatively small standard deviation (0.053). This suggests that the model generalizes well and is not overfitting the data (James et al., 2013).

5.7.2 Influence and Outlier Analysis

To identify influential observations or outliers that might affect regression results, we calculated several influence measures. Of the 100 observations, 8 were identified as potential outliers based on Cook's distance exceeding the threshold of 4/n. However, further analysis revealed that these outliers do not substantially influence the regression results, as the model estimated without these outliers yielded coefficients very similar to the original model. This robustness check enhances confidence in the stability of our findings (Cook, 1977; Belsley et al., 1980).

5.8 Variable Relationship Visualizations

To better understand the relationships between variables in the model, we present several visualizations.

5.8.1 Figure 1. Path Diagram

The path diagram illustrates the relationships between independent variables (Meetings, Workshops, Procurement) and the dependent variable (Project Success). Arrow thickness represents relationship strength, with Workshops demonstrating the strongest influence.

5.8.2 Figure 2. Correlation Matrix Heatmap

The correlation matrix heatmap provides a visual representation of correlations between variables, with darker colors indicating stronger correlations. This confirms that Workshops exhibit the highest correlation with Project Success.

5.8.3 Figure 3. Scatter Plots

The scatter plots display bivariate relationships between each independent variable and Project Success. Regression lines indicate positive relationships for all three independent variables, with Workshops showing the steepest slope.

5.9 Figure 4. 3D Visualization

The 3D visualization shows the relationship between Meetings, Workshops, and Project Success, with color representing Procurement values. The fitted surface illustrates the multivariate relationship between these variables, highlighting the complex interplay of factors influencing project success.

6. DISCUSSION

Implications of Findings on Stakeholder Engagement Strategies

The positive impact of inclusive and interactive workshops, sustainable procurement practices, and meaningful meetings on project success aligns with and extends current understandings in the field of environmental project management. The significant correlation between these engagement strategies and project outcomes emphasizes the need for a nuanced approach to stakeholder engagement, one that goes beyond conventional communication tactics to actively involve stakeholders in a meaningful way (Tuffour et al., 2023).

The strong influence of workshop inclusiveness and interactivity on project success, in particular, resonates with the participatory model of development communication, which advocates for the active involvement of stakeholders in the development process (IDB, 2018). This finding suggests that by facilitating an environment where stakeholders can engage directly with the project, sharing their knowledge, and contributing to decision-making processes, projects can achieve higher levels of success.

6.1 Challenges and Opportunities in Implementing Engagement Strategies

Despite the identified benefits, implementing these strategic engagement practices is not without challenges. The logistical and financial constraints highlighted in the findings point to the need for innovative solutions to engage a broader range of stakeholders, especially those from remote or economically disadvantaged backgrounds (Urzedo et al., 2022). The advent of digital technologies offers promising avenues for overcoming such barriers, through virtual meetings and online platforms that can facilitate wider participation.

Furthermore, the conflicting interests and priorities among stakeholders underscore the importance of negotiation and conflict resolution skills in project management. Adopting a stakeholder-centric approach, where the goals and values of all parties are acknowledged and respected, can help mitigate conflicts and foster a collaborative project environment (Dhyani et al., 2023).

6.2 Contribution to Theory and Practice

This research contributes to the theoretical framework of strategic management in environmental conservation by providing empirical evidence on the effectiveness of specific stakeholder engagement strategies. The findings not only validate theoretical models that emphasize the importance of stakeholder involvement but also offer practical insights for project managers seeking to optimize stakeholder engagement in similar contexts.

For practitioners, the detailed analysis of the influence of meetings, workshops, and procurement strategies on project success provides a valuable blueprint for designing and implementing effective engagement processes. By prioritizing inclusivity, interactivity, and sustainability, project managers can enhance stakeholder participation and support, thereby increasing the likelihood of project success.

7. CONCLUSION

This study embarked on a journey to unravel the complexities of strategic management of socialization and development communication within the ambit of the National Mangrove Map project, specifically focusing on optimizing stakeholder engagement through the lens of effective meetings, workshops, and procurement strategies. Through a qualitative research design encompassing semi-structured interviews, document analysis, and observations, coupled with an influence test on the predicted data and outcomes, we have gleaned substantial insights into the pivotal role of stakeholder engagement in the realm of environmental conservation projects.

The empirical findings underscore the significant impact of tailored stakeholder engagement strategies particularly inclusive and interactive workshops, meaningful meetings, and sustainable procurement practices on the success of the National Mangrove Map project. These strategies not only fostered a collaborative and participatory environment but also addressed the multifaceted challenges of aligning diverse stakeholder interests with project objectives, overcoming logistical and financial barriers to participation, and integrating sustainability criteria into procurement decisions.

However, the path to effective stakeholder engagement is fraught with challenges. The study highlighted the intricacies of managing conflicting interests and priorities among stakeholders, as well as the logistical hurdles of ensuring broad and equitable participation. These challenges necessitate a strategic, adaptive approach to engagement, one that leverages innovative technologies and emphasizes negotiation and conflict resolution.

The contributions of this research extend beyond the specific case of the National Mangrove Map project, offering valuable insights for both theory and practice. Theoretically, it enriches the existing body of knowledge on strategic management and stakeholder engagement in environmental conservation projects. Practically, it provides a blueprint for project managers seeking to optimize stakeholder engagement, suggesting that success hinges on the ability to facilitate inclusive, interactive, and sustainable engagement processes.

The strategic management of socialization and development communication for environmental conservation projects, exemplified by the National Mangrove Map project, is a dynamic and multifaceted endeavor. The key to success lies in understanding and harnessing the power of stakeholder engagement, which requires a delicate balance of inclusivity, sustainability, and strategic foresight. As we move forward, it is imperative that these insights inform future initiatives, paving the way for more effective, collaborative, and impactful environmental conservation efforts.

REFERENCES

Abdi, H. (2007). Part (semi partial) and partial regression coefficients. In Salkind, N. (Ed.), *Encyclopedia of Measurement and Statistics* (pp. 736–740). Sage.

Azen, R., & Budescu, D. V. (2003). The dominance analysis approach for comparing predictors in multiple regression. *Psychological Methods*, *8*(2), 129–148. DOI: 10.1037/1082-989X.8.2.129 PMID: 12924811

Baron, R. M., & Kenny, D. A. (1986). The moderator-mediator variable distinction in social psychological research: Conceptual, strategic, and statistical considerations. *Journal of Personality and Social Psychology*, *51*(6), 1173–1182. DOI: 10.1037/0022-3514.51.6.1173 PMID: 3806354

Belsley, D. A., Kuh, E., & Welsch, R. E. (1980). *Regression diagnostics: Identifying influential data and sources of collinearity*. Wiley. DOI: 10.1002/0471725153

Brammer, S., & Walker, H. (2011). Sustainable procurement in the public sector: An international comparative study. *International Journal of Operations & Production Management*, *31*(4), 452–476. DOI: 10.1108/01443571111119551

Braun, V., & Clarke, V. (2019). Reflecting on reflexive thematic analysis. *Qualitative Research in Sport, Exercise and Health*, *11*(4), 589–597. DOI: 10.1080/2159676X.2019.1628806

Breusch, T. S., & Pagan, A. R. (1979). A simple test for heteroscedasticity and random coefficient variation. *Econometrica*, *47*(5), 1287–1294. DOI: 10.2307/1911963

Budescu, D. V. (1993). Dominance analysis: A new approach to the problem of relative importance of predictors in multiple regression. *Psychological Bulletin*, *114*(3), 542–551. DOI: 10.1037/0033-2909.114.3.542

Bunting, P., Rosenqvist, A., Lucas, R., Rebelo, L. M., Hilarides, L., Thomas, N., Hardy, A., Itoh, T., Shimada, M., & Finlayson, C. M. (2018). The Global Mangrove Watch—A new 2010 global baseline of mangrove extent. *Remote Sensing (Basel)*, *10*(10), 1669. DOI: 10.3390/rs10101669

Cohen, J. (1988). *Statistical power analysis for the behavioral sciences* (2nd ed.). Lawrence Erlbaum Associates.

Cook, R. D. (1977). Detection of influential observation in linear regression. *Technometrics*, *19*(1), 15–18. DOI: 10.1080/00401706.1977.10489493

Cox, R., & Pezzullo, P. C. (2018). *Environmental communication and the public sphere* (5th ed.). SAGE Publications.

Dhyani, S., Shukla, J., Kadaverugu, R., Dasgupta, R., Panda, M., Kundu, S. K., Santhanam, H., Pujari, P. R., Kumar, P., & Hashimoto, S. (2023). Participatory stakeholder assessment for drivers of mangrove loss to prioritize evidence-based conservation and restoration in Bhitarkanika and Mahanadi Delta, India. *Sustainability, 15*(2), 963. DOI: 10.3390/su15020963

Durbin, J., & Watson, G. S. (1951). Testing for serial correlation in least squares regression. II. *Biometrika, 38*(1/2), 159–177. DOI: 10.2307/2332325 PMID: 14848121

Emerson, K., & Nabatchi, T. (2015). *Collaborative governance regimes*. Georgetown University Press.

Faul, F., Erdfelder, E., Buchner, A., & Lang, A. G. (2009). Statistical power analyses using G*Power 3.1: Tests for correlation and regression analyses.* Behavior Research Methods, 41*(4), 1149-1160.

Freeman, R. E., Harrison, J. S., Wicks, A. C., Parmar, B. L., & De Colle, S. (2010). *Stakeholder theory: The state of the art*. Cambridge University Press. DOI: 10.1017/CBO9780511815768

Friess, D. A., Rogers, K., Lovelock, C. E., Krauss, K. W., Hamilton, S. E., Lee, S. Y., Lucas, R., Primavera, J., Rajkaran, A., & Shi, S. (2019). The state of the world's mangrove forests: Past, present, and future. *Annual Review of Environment and Resources, 44*(1), 89–115. DOI: 10.1146/annurev-environ-101718-033302

Friess, D. A., Yando, E. S., Abuchahla, G. M., Adams, J. B., Cannicci, S., Canty, S. W., Cavanaugh, K. C., Connolly, R. M., Cormier, N., Dahdouh-Guebas, F., Diele, K., Feller, I. C., Fratini, S., Jennerjahn, T. C., Lee, S. Y., Ogurcak, D. E., Ouyang, X., Rogers, K., Rowntree, J. K., & Wee, A. K. S. (2020). Mangroves give cause for conservation optimism, for now. *Current Biology, 30*(4), R153–R154. DOI: 10.1016/j.cub.2019.12.054 PMID: 32097637

Ghasemi, A., & Zahediasl, S. (2012). Normality tests for statistical analysis: A guide for non-statisticians. *International Journal of Endocrinology and Metabolism, 10*(2), 486–489. DOI: 10.5812/ijem.3505 PMID: 23843808

Goldberg, L., Lagomasino, D., Thomas, N., & Fatoyinbo, T. (2020). Global declines in human-driven mangrove loss. *Global Change Biology, 26*(10), 5844–5855. DOI: 10.1111/gcb.15275 PMID: 32654309

Hair, J. F., Black, W. C., Babin, B. J., & Anderson, R. E. (2018). *Multivariate data analysis* (8th ed.). Cengage Learning.

Hardin, D., Dewi, I. K., Alzarliani, W. O. D., Azaludin, , Wardana, , Andara, D., Ramadhan, F. M., & Huda, A. M. (2019). The role of communities in conserving mangrove forests to achieve sustainable development. *IOP Conference Series. Earth and Environmental Science, 343*(1), 012143. DOI: 10.1088/1755-1315/343/1/012143

Hayes, A. F. (2018). *Introduction to mediation, moderation, and conditional process analysis: A regression-based approach* (2nd ed.). Guilford Press.

Hayes, A. F. (2018). *Introduction to mediation, moderation, and conditional process analysis: A regression-based approach* (2nd ed.). Guilford Press.

IDB. (2018). Mexico: Integrating stakeholders in mangrove assessments and management planning: A model based on collaborative planning in Veracruz, Mexico. Washington, DC

Jaccard, J., & Turrisi, R. (2003). *Interaction effects in multiple regression* (2nd ed.). Sage. DOI: 10.4135/9781412984522

James, G., Witten, D., Hastie, T., & Tibshirani, R. (2013). *An introduction to statistical learning: With applications in R*. Springer. DOI: 10.1007/978-1-4614-7138-7

Kasych, A., Suler, P., & Rowland, Z. (2020). Corporate Environmental Responsibility through the prism of Strategic Management. *Sustainability (Basel), 12*(22), 9589. DOI: 10.3390/su12229589

Khatter, A. (2023). Strategic management and stakeholder engagement: A case for environmental sustainability. *Sustainable Development*. Advance online publication. DOI: 10.5772/intechopen.112065

Lhosupasirirat, P., Dahdouh-Guebas, F., Hugé, J., Wodehouse, D., & Enright, J. (2023). Stakeholder perceptions on community-based Ecological Mangrove Restoration (CBEMR): A case study in Thailand. *Restoration Ecology, 31*(5), e13894. Advance online publication. DOI: 10.1111/rec.13894

Ljung, G. M., & Box, G. E. P. (1978). On a measure of lack of fit in time series models. *Biometrika, 65*(2), 297–303. DOI: 10.1093/biomet/65.2.297

Lovelock, C. E., & Brown, B. M. (2019). Land tenure considerations are key to successful mangrove restoration. *Nature Ecology & Evolution, 3*(8), 1135. DOI: 10.1038/s41559-019-0942-y PMID: 31285575

Mamabolo, M.A. (2018). Positioning stakeholder engagement theory on governance of communal farms: a proposed framework for land governance in South Africa.

Melo, R. H., Kusmana, C., Eriyatno, , & Nurrochmat, D. R. (2019). A stakeholder analysis of sustainable mangrove management in Kwandang, sub-district of north, Gorontalo District. *IOP Conference Series. Earth and Environmental Science*, *399*(1), 012071. DOI: 10.1088/1755-1315/399/1/012071

Melo, R. H., Kusmana, C., Eriyatno, , & Nurrochmat, D. R. (2019). A stakeholder analysis of sustainable mangrove management in Kwandang, sub-district of north, Gorontalo District. *IOP Conference Series. Earth and Environmental Science*, *399*(1), 012071. DOI: 10.1088/1755-1315/399/1/012071

Melo, R. H., Kusmana, C., Eriyatno, , & Nurrochmat, D. R. (2019). A stakeholder analysis of sustainable mangrove management in Kwandang, sub-district of north, Gorontalo District. *IOP Conference Series. Earth and Environmental Science*, *399*(1), 012071. DOI: 10.1088/1755-1315/399/1/012071

Murdiyarso, D., Purbopuspito, J., Kauffman, J. B., Warren, M. W., Sasmito, S. D., Donato, D. C., Manuri, S., Krisnawati, H., Taberima, S., & Kurnianto, S. (2015). The potential of Indonesian mangrove forests for global climate change mitigation. *Nature Climate Change*, *5*(12), 1089–1092. DOI: 10.1038/nclimate2734

Newig, J., Challies, E., Jager, N. W., Kochskaemper, E., & Adzersen, A. (2018). The environmental performance of participatory and collaborative governance: A framework of causal mechanisms. *Policy Studies Journal: the Journal of the Policy Studies Organization*, *46*(2), 269–297. DOI: 10.1111/psj.12209 PMID: 30034065

Newig, J., Challies, E., Jager, N. W., Kochskaemper, E., & Adzersen, A. (2018). The environmental performance of participatory and collaborative governance: A framework of causal mechanisms. *Policy Studies Journal: the Journal of the Policy Studies Organization*, *46*(2), 269–297. DOI: 10.1111/psj.12209 PMID: 30034065

Newig, J., & Fritsch, O. (2009). Environmental governance: Participatory, multi-level—and effective? *Environmental Policy and Governance*, *19*(3), 197–214. DOI: 10.1002/eet.509

O'Brien, R. M. (2007). A caution regarding rules of thumb for variance inflation factors. *Quality & Quantity*, *41*(5), 673–690. DOI: 10.1007/s11135-006-9018-6

Olkin, I., & Finn, J. D. (1995). Correlations redux. *Psychological Bulletin*, *118*(1), 155–164. DOI: 10.1037/0033-2909.118.1.155

Reed, M. S. (2008). Stakeholder participation for environmental management: A literature review. *Biological Conservation*, *141*(10), 2417–2431. DOI: 10.1016/j.biocon.2008.07.014

Reed, M. S. (2008). Stakeholder participation for environmental management: A literature review. *Biological Conservation*, *141*(10), 2417–2431. DOI: 10.1016/j.biocon.2008.07.014

Reed, M. S., Vella, S., Challies, E., de Vente, J., Frewer, L., Hohenwallner-Ries, D., Huber, T., Neumann, R. K., Oughton, E. A., Sidoli del Ceno, J., & van Delden, H. (2017). A theory of participation: What makes stakeholder and public engagement in environmental management work? *Restoration Ecology*, *26*(S1). Advance online publication. DOI: 10.1111/rec.12541

Reed, M. S., Vella, S., Challies, E., de Vente, J., Frewer, L., Hohenwallner-Ries, D., Huber, T., Neumann, R. K., Oughton, E. A., Sidoli del Ceno, J., & van Delden, H. (2018). A theory of participation: What makes stakeholder and public engagement in environmental management work? *Restoration Ecology*, *26*(S1), S7–S17. DOI: 10.1111/rec.12541

Reed, M. S., Vella, S., Challies, E., de Vente, J., Frewer, L., Hohenwallner-Ries, D., Huber, T., Neumann, R. K., Oughton, E. A., Sidoli del Ceno, J., & van Delden, H. (2018). A theory of participation: What makes stakeholder and public engagement in environmental management work? *Restoration Ecology*, *26*(S1), S7–S17. DOI: 10.1111/rec.12541

Richards, D. R., & Friess, D. A. (2016). Rates and drivers of mangrove deforestation in Southeast Asia, 2000–2012. *Proceedings of the National Academy of Sciences of the United States of America*, *113*(2), 344–349. DOI: 10.1073/pnas.1510272113 PMID: 26712025

Spalding, M., & Parrett, C. L. (2019). Global patterns in mangrove recreation and tourism. *Marine Policy*, *110*, 103540. DOI: 10.1016/j.marpol.2019.103540

Sterling, E. J., Betley, E., Sigouin, A., Gomez, A., Toomey, A., Cullman, G., Malone, C., Pekor, A., Arengo, F., Blair, M., Filardi, C., Landrigan, K., & Porzecanski, A. L. (2017). Assessing the evidence for stakeholder engagement in biodiversity conservation. *Biological Conservation*, *209*, 159–171. DOI: 10.1016/j.biocon.2017.02.008

Sterling, E. J., Betley, E., Sigouin, A., Gomez, A., Toomey, A., Cullman, G., Malone, C., Pekor, A., Arengo, F., Blair, M., Filardi, C., Landrigan, K., & Porzecanski, A. L. (2017). Assessing the evidence for stakeholder engagement in biodiversity conservation. *Biological Conservation*, *209*, 159–171. DOI: 10.1016/j.biocon.2017.02.008

Tabachnick, B. G., & Fidell, L. S. (2019). *Using multivariate statistics* (7th ed.). Pearson.

Thomas, N., Lucas, R., Bunting, P., Hardy, A., Rosenqvist, A., & Simard, M. (2017). Distribution and drivers of global mangrove forest change, 1996–2010. *PLoS One*, *12*(6), e0179302. DOI: 10.1371/journal.pone.0179302 PMID: 28594908

Thompson, B. S., Clubbe, C. P., Primavera, J. H., Curnick, D., & Koldewey, H. J. (2018). Locally assessing the economic viability of blue carbon: A case study from Panay Island, the Philippines. *Ecosystem Services*, *34*, 24–36. DOI: 10.1016/j.ecoser.2018.09.008

Tuffour, P., Chen, G., Opoku-Mensah, E., Darko, P. A., & Adu Agyapong, R. (2023). Toward sustainable development: Developing a decision-making framework for cross-sectoral engagement in Green Procurement. *Sustainable Development*. Advance online publication. DOI: 10.1002/sd.2772

Urzedo, D., Westerlaken, M., & Gabrys, J. (2022). Digitalizing Forest Landscape Restoration: A social and political analysis of emerging technological practices. *Environmental Politics*, *32*(3), 485–510. DOI: 10.1080/09644016.2022.2091417 PMID: 37207120

White, H. (1980). A heteroskedasticity-consistent covariance matrix estimator and a direct test for heteroskedasticity. *Econometrica*, *48*(4), 817–838. DOI: 10.2307/1912934

Willgerodt, M. A., Abu-Rish Blakeney, E., Summerside, N., Vogel, M. T., Liner, D. A., & Zierler, B. (2019). Impact of leadership development workshops in facilitating team-based practice transformation. *Journal of Interprofessional Care*, *34*(1), 76–86. DOI: 10.1080/13561820.2019.1604496 PMID: 31039647

Worthington, T., & Spalding, M. (2018). *Mangrove restoration potential: A global map highlighting a critical opportunity. The Nature Conservancy, IUCN*. University of Cambridge.

Worthington, T. A., Andradi-Brown, D. A., Bhargava, R., Buelow, C., Bunting, P., Duncan, C., Fatoyinbo, L., Friess, D. A., Goldberg, L., Hilarides, L., Lagomasino, D., Landis, E., Longley-Wood, K., Lovelock, C. E., Murray, N. J., Narayan, S., Rosenqvist, A., Sievers, M., Simard, M., & Spalding, M. D. (2020). Harnessing big data to support the conservation and rehabilitation of mangrove forests globally. *One Earth*, *2*(5), 429–443. DOI: 10.1016/j.oneear.2020.04.018

Yin, R. K. (2017). Case study research and applications: Design and methods. Sage publications. Arifanti, V. B., Sidik, F., Mulyanto, B., Susilowati, A., Wahyuni, T., Subarno, S., Yulianti, Y., Yuniarti, N., Aminah, A., Suita, E., Karlina, E., Suharti, S., Pratiwi, P., Turjaman, M., Hidayat, A., Rachmat, H. H., Imanuddin, R., Yeny, I., Darwiati, W.,Novita, N. (2022). Challenges and strategies for Sustainable Mangrove Management in Indonesia: A Review. *Forests*, *13*(5), 695. 5DOI: 10.3390/f1305069

Young, J. C., Rose, D. C., Mumby, H. S., Benitez-Capistros, F., Derrick, C. J., Finch, T., Garcia, C., Home, C., Marwaha, E., Morgans, C., Parkinson, S., Shah, J., Wilson, K. A., & Mukherjee, N. (2020). A methodological guide to using and reporting on interviews in conservation science research. *Methods in Ecology and Evolution*, *11*(1), 22–32. DOI: 10.1111/2041-210X.13342

Zikargae, M. H., Woldearegay, A. G., & Skjerdal, T. (2022). Assessing the roles of stakeholders in community projects on environmental security and livelihood of Impoverished Rural Society: A nongovernmental organization implementation strategy in focus. *Heliyon*, *8*(10), e10987. Advance online publication. DOI: 10.1016/j.heliyon.2022.e10987 PMID: 36276717

Zikargae, M. H., Woldearegay, A. G., & Skjerdal, T. (2022). Assessing the roles of stakeholders in community projects on environmental security and livelihood of Impoverished Rural Society: A nongovernmental organization implementation strategy in focus. *Heliyon*, *8*(10), e10987. Advance online publication. DOI: 10.1016/j.heliyon.2022.e10987 PMID: 36276717

Chapter 15
The Role of Zakat Institutions in Sustainable Development Goals (SDGs):
The Case of Zakat in Indonesia and Malaysia

Lili Yulyadi Arnakim
https://orcid.org/0000-0002-4575-424X
Independent Researcher, Indonesia

Namira Shabrina
https://orcid.org/0009-0006-9467-0479
The Search Group, Indonesia

ABSTRACT

It is no longer a debate that the purpose of Islamic Social Finance like Zakat is in line with and does support the achievement of Sustainable Development Goals (SDGs). Eradicating poverty, improving the welfare of the people, and reducing inequality, are the end goals of why zakat is collected and distributed to the recipients. There have been studies conducted in the past to explain how Islamic Social Finance positively contributes to the achievement of SDGs and how zakat institutions should be strengthened to increase the effectiveness of zakat management. However, none has tried to analyze what are the roles played by zakat institutions in SDGs. The objective of this paper is to identify and compare the role of zakat institutions in Indonesia and Malaysia, by using BAZNAS and MAIWP to represent zakat institutions in each country. We apply the goal-setting theory to examine the role.

DOI: 10.4018/979-8-3373-1300-9.ch015

INTRODUCTION

Sustainable Development Goals (SDGs) is a global long-term development plan adopted by the United Nations in 2015 that envisions all people to enjoy peace and prosperity by 2030. The 17 Goals adopted by all member states encompass a wide and inclusive range of social, economic, and environmental sustainability that must be balanced against each other to promote prosperity while protecting the planet. As ambitious as it sounds, SDGs are not impossible to achieve if only this community agreement is collaboratively acted upon by all stakeholders (especially governments, businesses, financial institutions, civil society, and nonprofit organizations). A huge amount of resources must be collected from different sources and must be mobilized effectively to ensure they are being channeled and serve the intended purposes.

Table 1. 17 Sustainable Development Goals

No	Goal	Sector
1	No poverty	People
2	No hunger	
3	Good Health and Well-being	
4	Quality Education	
5	Gender Equality	
6	Clean Water and Sanitation	
7	Affordable and Clean Energy	Prosperity
8	Decent Work and Economic Growth	
9	Industry, Innovation and Infrastructure	
10	Reduced Inequality	
11	Sustainable cities and communities	
12	Responsible consumption and production	Planet
13	Climate action	
14	Life below water	
15	Life on Land	
16	Peace, justice and strong institutions	Peace
17	Partnerships for the goals	Partnership

Source: UNDP (n.d); Asmalia (2018)

Islam and SDGs share similar value which both promotes peace and prosperity for all mankind. As mentioned earlier, it requires a huge amount of resources to achieve SDGs. Many scholars have evaluated the role of zakat in Islam: how the principle of zakat aligns well with the concept of SDGs and argue that the resources collected through Islamic social finance have significant potential to help achieve the global goals.

It is widely known that there are two distinct branches of Islamic Finance: Islamic Social Finance and Commercial Islamic Banking. "Islamic social finance is focused on generating social and economic gains," whereas "commercial Islamic banking seeks to generate financial returns". From these definitions, the one that has a closer link with the purpose of SDGs is the former. Islamic social finance concepts, particularly zakat and waqf, are considered the main sources to fund development goals.

One of the main functions of Zakat is to eradicate poverty. Zakat in Islam is God's command to surrender one's own wealth/property for the good of those who are entitled to receive it (mustahik). It was stated in the Qur'an: "And for those whose hearts will be brought together (for Islam) and for slaves and those in debt and for (those) on the Way of Allah and for the travelers - an obligation by Allah. And Allah is All-Knowing, All-Wise (Attaubah 9:60). It is mandatory for Muslims who are eligible to donate at least 2.5 percent of their wealth to the needy/poor or anyone who falls into the eight types of recipients (asnaf), most importantly the Fuqara (the poor who live without means of livelihood); the Masakin (the needy who cannot meet their basic needs); and Fi sabilillah (anyone or anything struggling for a righteous cause in the path of Allah).

Although Zakat is directly in line with SDG1 "No Poverty", it also has indirect impacts on the other SDGs. In the book titled "Fikih Zakat on SDGs", Maksum et al (2018) have argued that out of 17 Sustainable Development Goals, 12 of them (SDG1, SDG2, SDG3, SDG4, SDG5, SDG6, SDG8, SDG10, SDG12, SDG15, SDG16, SDG17) have been clearly stated in various verses in the Qur'an and Hadith. Whereas the remaining 5 goals (SDG7, SDG9, SDG11, SDG13, SDG14) were not clearly stated but were mentioned indirectly in religious texts. The book has stated the Islamic values on development and Zakat can therefore be tapped and used to support all the 17 development goals albeit with different levels of priorities.

"Zakat can cut intergenerational poverty" (BAZNAS, 2017). Intragenerational justice is concerned with the various circumstances and living conditions of individuals of the same generation, at a given point in time. Meanwhile, the concept of intergenerational as defined by Merriam-webster means "existing or occurring between generations". Hence, intergenerational poverty is the one that does not only exist in the current generation but is also being passed down to the next generations. Poverty is not the only issue that can be both intragenerational and intergenerational, but current issues like climate change transcend through time and space creating serious risks to future generations and the planet. Ensuring SDGs would mean achieving justice between generations. In other words, youth and future generations should have equitable access to certain resources and opportunities including social and environmental goods to meet their own needs as older generations (ESCWA, 2017).

We all are familiar with zakat to be given for consumptive purposes i.e. to be given to the poor and the needy for them to fulfill their basic needs. This condition sounds more appropriate to answer intragenerational justice or poverty issues. But if we look far ahead, by eradicating intergenerational poverty by giving food, quality education, and healthcare to the existing poor families, we can break the chain of poverty for the future generation. Let's say if a poor child or an orphan is given proper care through the utilization of zakat funds, the child would have an improved quality of life. When the children have grown, they are expected to be able to help others in need. The same goes for waqf funds utilization. Waqf and Zakat funds that are channeled for productive purposes are more sustainable.

Other than the issue of poverty, the environment and conservation of our planet are also considered as one of the main issues in intergenerational justice. Irresponsible consumption and production at the expense of environmental degradation would endanger our planet and can cause climate change and scarcity for our future generations. They may not be able to enjoy the same resources and opportunities as the current generation does. The funds tapped from zakat can be utilized to support development projects that concurrently protect our planet. For example, funding green projects (renewable energy, forest, and marine ecosystem conservation) not only improves social well-being of the current generation but also helps to ensure intergenerational justice.

The paper analyzes the current role of Zakat institutions in Malaysia and Indonesia in achieving SDGs by taking a descriptive and comparative approach to understanding the different ways in which zakat institutions in respective countries play a role in the SDGs.

PROGRESS IN ACHIEVING SDGs

As part of the global community, Muslim countries are responsible to help to achieve the global vision by incorporating the goals into the national development plans. Countries must make sure that they have enough resources to finance the development agenda. Most Muslim countries which are members of the Organization of Islamic Cooperation (OIC), including Indonesia and Malaysia, are developing countries. To see how well these countries perform in achieving SDGs, Shaikh and Hassan (2020) conducted a study using data gathered from the World Development Indicators (WDI), which found that among 40 OIC member countries being studied, Malaysia came in 6th place with 0.17971293 SDG-Index Value, whereas Indonesia is in 11th place with 0.149894875 SDG-Index Value.

Malaysia performs better than its neighboring country. However, if we take a closer look at how these two countries perform in each goals, we can see that Indonesia ranks higher in a few aspects, namely: Quality Education (SDG4); Reduced Inequality (SDG10); Responsible consumption and production (SDG12); Climate Action (SDG13); Life below water (SDG14) (see Table 2).

Table 2. Ranking of countries on SDI sub-indices (SDG 1 to SDG 17)

Country	**SDG1**	**SDG2**	**SDG3**	**SDG4**	**SDG5**	**SDG6**	**SDG7**	**SDG8**	**SDG9**
Indonesia	18	19	16	9	15	17	24	7	14
Malaysia	8	5	4	10	16	1	22	3	12
Country	SDG10	SDG11	SDG12	SDG13	SDG14	SDG15	SDG16	SDG17	
Indonesia	24	19	21	22	9	6	6	20	
Malaysia	36	3	35	31	12	1	3	7	

Source: Shaik and Hassan (2020).

The current progress on SDGs achievement by the two countries shows that there is still a lot of homework to be completed. It is not only the government's responsibility but realizing SDGs by 2030 requires active participation from all stakeholders as well as mobilization of a huge amount of resources. As the majority of Muslim countries are still developing, there have been concerns as to how they can achieve their goals. Various writings attempted to evaluate the potential of Islamic Social Finance in helping the government to achieve SDGs and this paper will specifically describe and compare the role played by zakat institutions in Indonesia and Malaysia.

LITERATURE REVIEW

Various studies have been conducted to contribute to the richness of the concept of Sustainable Development Goals and how it is heavily related to Islamic values and the concept of Islamic Social Finance particularly Zakat and Waqf (Maksum et al, 2018). Several studies were carried out to prove that Zakat plays a significant role in reducing poverty and has a positive impact on the economic growth of Indonesia (Nurhapsari et al, 2021; Al Azizah et al, 2021; Choirunisa et al, 2020).

Numbers of studies were carried out to study the efficiency of Zakat institutions. Bahri et al (2021) investigated the efficiency of zakat institutions in Indonesia by looking at the amount of zakat collected and distributed using the data gathered from the annual reports of 17 zakat institutions in Indonesia during the 2014-2020 period. The results of this study indicate that during 2014 to 2020, the efficiency of zakat institutions is showing a fluctuating trend. But it was revealed that private zakat institutions are generally more efficient in managing inputs and maximizing desired outputs than public zakat institutions. It was also found that overall, zakat institutions in Indonesia experienced greater efficiency values during Covid-19. Similarly, Andriani et al (2021) investigate the role of zakat institutions in Indonesia during the covid-19 pandemic. This study found that zakat institutions responded to the pandemic crisis by adjusting their fundraising and distribution strategy which resulted in the amount of zakat funds collected in 2020 being higher compared to the average of the previous five years before the pandemic. Zakat institutions have started to undergo digital transformation that was dominantly applied in their fundraising strategy. The distribution covered the people who are affected by the pandemic. The institution's educational program also includes an additional role i.e. to educate people on the prevention of covid19 transmission.

Some studies managed to focus on how to improve zakat management. Alfiani & Akbar (2020) in their study argued that in managing zakat, innovation is needed. Zakat institutions need to collaborate with fintech companies for collecting and distributing zakat more efficiently so that they have better information management systems and can align their programs with SDGs. Meanwhile, Suriani et al (2020) proposed ways to optimize zakat distribution by carrying out priority zakat allocation for poverty alleviation programs, striving for targeting accuracy, and maximizing the use of consumptive and productive zakat distribution.

Many studies have focused on how zakat can bring prosperity to the people. Hussain, as cited in Ramadan and Hussain (2019) mentioned that "zakat institutions are core learning organizations that can create value for the poor and the needy if they exploit their knowledge to the benefit of their strategic objectives". Abdelbaki, as cited in Ramadan and Hussain (2019) argued that "zakat institutions are ethically trustworthy organizations characterized by transparency, professional management and accountability to its stakeholders". Many zakat institutions offer economy as one of their schemes that involve economic activities which are both showing religious commitments as well as supporting socio-economic development objectives.

Suprayitno et al (2017) conducted a study to examine the impact of zakat on human development programs in Malaysia using 5 states as the case studies (Johor, Kedah, Negeri Sembilan, Selangor, and Terengganu). They found that zakat has a positive and significant influence on human development in those five states both in the short and long run. Zakat distribution in Malaysia is distributed for education and human resource empowerment. As more zakat is being collected, more benefits can be brought to social life, social welfare, and economic development, which will eventually help to minimize the government's budget allocation for social welfare in the long run. In the short run, the increase in zakat collection will increase zakat distribution and increase the income of zakat recipients. The result of their study suggests that the zakat management system in Malaysia can stimulate Malaysia's economy. Hence, Malaysia's model of Zakat management can be a role model for other Islamic countries.

Several other studies tried to conduct comparative analysis between zakat institutions in Indonesia and Malaysia. Destiana & Hamzah (2021) Explained that the relationship between taxes and zakat in countries between Indonesia and Malaysia varies. The management of zakat in Indonesia applies a tax reduction system, which means that the zakat paid can reduce the taxable income of the payer. Meanwhile, Malaysia uses Zakat as a tax deduction on the condition that zakat is paid at an official government institution. Zakat in Malaysia can be deducted from individual taxes even up to 100%. The reduction is through a scheduled tax discount or a monthly discount system. Nurasiyah et al (2019) measured and compared the efficiency and productivity of zakat institutions in Indonesia and Malaysia and found that in general, Malaysia's zakat institution has a lower level of productivity and higher efficiency as compared to Zakat institutions in Indonesia. This implies that the zakat institution in Malaysia is more effective in zakat distribution, whereas zakat institutions in Indonesia are more successful in collecting a greater amount of zakat.

Interestingly, little research has focused on the role of Zakat institutions in the implementation of the SDGs and attempted to make a comparative analysis between the roles played by Indonesian institutions and the ones in Malaysia. In the next section, we adopt a framework developed by Marx et al (2021) to analyze these different roles which correspond with the goal-setting nature and logic of the SDGs.

GOAL-SETTING THEORY

Marx et al (2021) developed a framework to analyze the roles of regional organizations in the SDG architecture by adapting from several frameworks on goal-setting. Marx et al (2021) concluded that there are four roles played by regional organizations in SDGs, namely: formulating and translating global vision to local context; socializing and mobilizing resources and supports; coordinating vertically and horizontally to ensure policy coherence; monitoring and reporting progress.

Figure 1. Roles of regional organizations in the SDG architecture

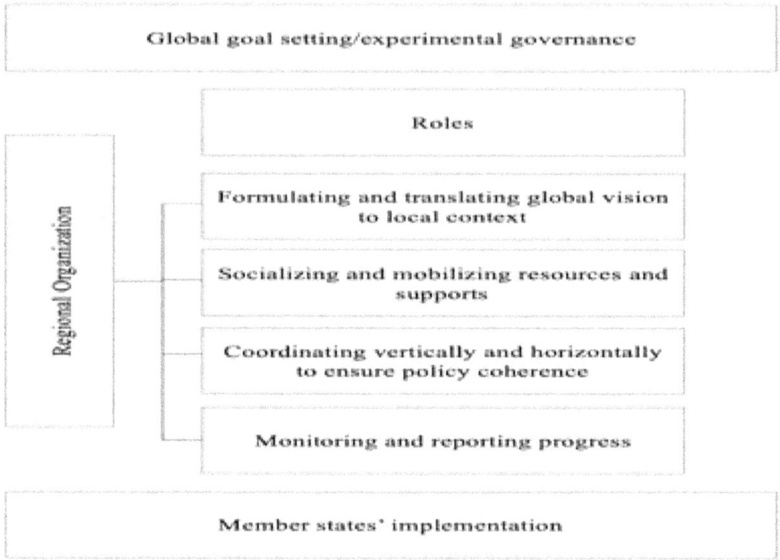

Source: Marx et al (2021)

"Formulating and translating global vision to local context means that as part of the global community, all countries are participating in formulating SDGs and then translating the global visions to the institutional/local setting to ensure their suitability and applicability in regional or national contexts, as well as to build a sense of awareness and belonging to the goals" (Marx et al, 2021). We adopt this definition of translating role to zakat institutions to see how they align the SDGs with their organization's goals/objectives.

The next role is the supporting role, or in other words "to socialize the translated goals, mobilize resources and support to implement the goals as part of a regional or national agenda. To enable this role, it normally requires a multi-sectoral coalition involving actors at the global, regional, and national levels" (Marx et al, 2021). In this paper, we apply this notion to see how zakat institutions play the supporting role by examining how they socialize the translated SDGs and mobilize resources to implement the goals as their own agenda.

The third role is coordinating policies between actors to ensure policy coherence by conducting horizontal and/or vertical coordination. Horizontal coordination is conducted across institutions within actors' internal structure or with other external institutions. Vertical coordination is conducted between the level of the regional organization and the MS level. The coordinating role is applied in this paper to analyze how zakat institutions form a cooperative relationship across one's institution or with external actors or coordinate vertically between different levels of structures in the institution concerning the implementation of SDGs.

The fourth role is to play a monitoring and progress reporting role. Defining certain indicators, setting up timetables, and benchmarking is important for actors to track progress and maintain commitment in pursuing long-term goals. Using the definition of the monitoring role, the authors will look at how zakat institutions monitor their progress to achieve SDGs.

In the next section, we apply this framework to two case studies of Zakat Institutions in Indonesia and Malaysia. Data gathered is mainly based on a close analysis of secondary sources. These two countries were selected because they are both Muslim majority countries in the region of Southeast Asia but seemingly have a different approach to institutionalizing the concept of Islamic Social Finance like Zakat. Hence, the two case studies were selected to allow us to understand whether Institutions for managing Islamic Social Finance, particularly zakat, have different approaches to playing their roles in achieving SDGs and how these roles are operationalized in each country.

Figure 2. Roles of Zakat Institutions in the SDG architecture

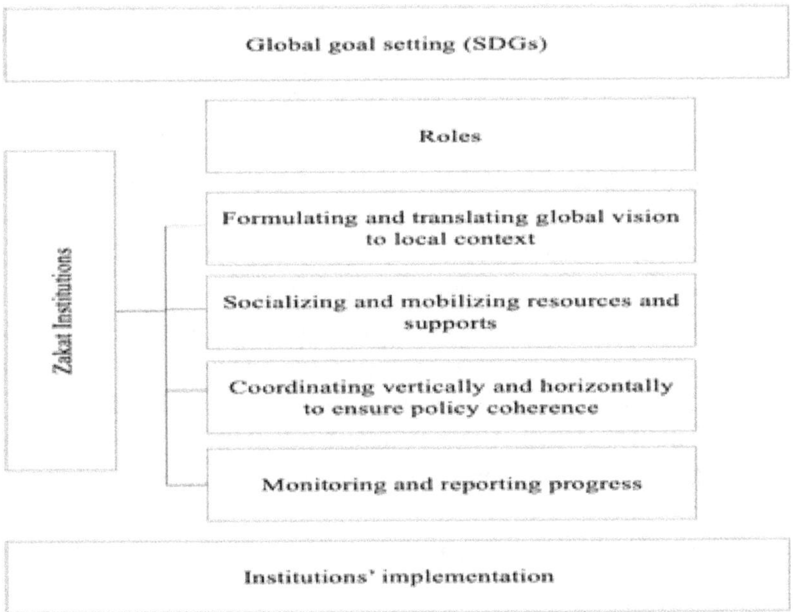

Source: Authors

METHODOLOGY

This paper is qualitative in nature and mostly relies on secondary sources of data collection. As for data analysis, the researcher applies thematic analysis to gain insight and knowledge from the data gathered from scholarly journals, documents, the text of laws and regulations, and news articles.

THE ROLE OF ZAKAT INSTITUTION IN INDONESIA IN SDGs

Indonesia has Badan Amil Zakat Nasional or the National Board of Zakat (BAZNAS) that acts as the authorized institution to carry out the task of managing zakat nationally and also at provincial and district/city levels. BAZNAS operates as a non-structural government institution that is independent and is accountable to the President, through the Minister of Religion. Zakat management conducted by BAZNAS includes an activity of planning, implementation, and coordination in the distribution and utilization of zakat. In 2017, BAZNAS collected over USD 11 million from 150000 Indonesian citizens and distributed over USD 7 million to 305000 people in Indonesia. To carry out its duties, BAZNAS is funded with the State Budget and Amil Rights (Amil/Zakat collector is entitled for receiving zakat). Aside from the government-led BAZNAS, Indonesia also has numerous zakat management bodies or LAZ that are privately owned. LAZ has the task of assisting BAZNAS in the collection, distribution, and utilization of zakat.

In Indonesia, the collection of Zakat is voluntary (not obligated under the law). Being the fourth most populous country in the world with a Muslim majority population, Zakat in Indonesia has the potential to contribute $32 billion (421 trillion rupiahs) per year, which is equivalent to 3.4 percent of Indonesia's GDP in 2016. But the actual data only records less than 1 percent of that potential even though the volume has increased by approximately 20 percent each year. BAZNAS and zakat institutions mainly address five main areas: social sector, education, economic, health, and da'wah. The distribution of zakat funds in 2019 shows that the zakat fund is heavily distributed for humanitarian causes, followed by religious programs (Hasan, 2020).

Zakat can be collected and distributed for consumptive or productive needs. According to Zakat Act No. 23/2011 (Undang-Undang Republik Indonesia No. 23 Tahun 2011 tentang Pengelolaan Zakat), zakat distribution programs in Indonesia are divided into two categories, namely the consumption and production-based programs. Consumptive means to fulfill the basic needs of its recipients without long-term objective, including health, food, and education. Productive needs, on the other hand, are concerned with empowering the recipients to help them achieve financial independence. About two-thirds of the zakat collection are allocated for the former while the remaining one-third for the latter (Qurroh et al, 2017).

It was found that in 2015, the largest proportion of zakat funds in Indonesia was given to the social humanitarian sector, with 41.27 percent of the total zakat funds channeled. The second field that receives the largest distribution of zakat funds is the education sector with 20.35 percent. In other words, People Sector remains the main priority for Zakat utilization in Indonesia, followed by Prosperity Sector, Planet, Partnership, and Peace (Asmalia et al, 2018).

BAZNAS AND THE TRANSLATING ROLE

As can be seen through the statement of its missions and goals, BAZNAS is in line with or has translated some of the SDGs into its own organizational goals. One of the BAZNAS' missions and goals is to maximize the distribution and utilization of Zakat, Infaq, Alms, and Other Religious Social Funds (ZIS-DSKL) in order to alleviate poverty, improve the welfare of the ummah, and reduce social inequality. This statement can be understood as a form of translation from several SDGs i.e. No Poverty, No Hunger, Good Health and Well-being, Quality Education, Clean Water and Sanitation, and Reduced Inequality.

Another goal that has been incorporated into BAZNAS' mission, goal, and target are to "Increase the synergy and collaboration of all relevant stakeholders for the development of national zakat." In stating its targets, BAZNAS is committed to developing synergy and collaboration in various fields including socialization and education of zakat, distribution, and utilization of zakat, as well as management of zakat, by forming collaborations with central government, regional government, private parties, and non-government institutions. In this manner, BAZNAS is in line with SDG 17 i.e. Partnerships for the Goals.

Furthermore, BAZNAS has gone to the extent of translating SDGs into several performance indicators. Firstly, they measure the impact of zakat on the poverty rate and social inequality. Secondly, they measure the impact of zakat on the health quality of mustahik (zakat receiver). Thirdly, they measure the impact of zakat on the quality of education of mustahik. Fourthly, they measure the impact of zakat on the quality of income generated by mustahik. As for the partnership goal, BAZNAS has set several indicators, such as the number of synergy and collaboration programs with zakat management bodies in socialization, education, distribution, and utilization of zakat; and the number of collaborations with private sectors, non-governmental organizations, and international institutions.

BAZNAS has divided its activities into 5 (five) categories. The first category is humanity, in which activities such as disaster relief, financial assistance for house construction and renovation, food, clothing, education, and also medical aid are included. The second category is Education. This includes awarding scholarships and constructing schools for the poor. The third category is health, which includes programs and activities such as free health and medical services for the community. The next one is Dakwah, which is literally defined as the act of inviting people to embrace Islam. Under this category, BAZNAS has established Muallaf Center, where Da'i or muslim scholars are actively engaging and educating Muallaf (converts) about Islam.

The last category is economy, which involves empowerment programs for farmers, economic development, community development, and microfinancing. One of the most notable instances of BAZNAS' economic development program is Zmart. Zmart is a program in the form of developing stalls/shops owned by mustahik (zakat beneficiaries) on a micro to small scale to overcome poverty in urban areas, by increasing the capacity of small shops so that they can continue to grow in the midst of the modern retail market and overcome poverty. ZMART is also directed to become a marketplace or marketing board for all products produced by mustahik.

All of the programs above are considered as a form of implementing SDGs. It supports the goal of No Poverty, No hunger, quality education, Good Health and Well-being. In the long run, this will help to achieve reduced inequality and contribute to Decent Work and Economic Growth.

BAZNAS AND THE SUPPORTING ROLE

After translating the goals related to SDGs, BAZNAS has also played the supporting role to socialize the goals and mobilize support to implement the goals. BAZNAS and UNDP signed a memorandum of understanding in April 2017 in support of the achievement of the SDGs (Noor & Pickup, 2017). BAZNAS and the UNDP made their first major collaboration to install micro hydropower plants, which successfully brings electricity to more than 4,500 people in four remote communities in Indonesia (Rehman & Pickup, 2018). With this form of support, it helps the country to achieve SDG7 (affordable and clean energy).

BAZNAS in the past also signed a memorandum of understanding (MoU) with the Indonesia Philanthropy Association (PFI) in implementation of Sustainable Development Goals (SDGs) in Indonesia (BAZNAS, 2016). BAZNAS was expected to play at least in six out of 17 objectives related SDGs, particularly no poverty, no hunger, good health and well-being, quality education, clean water and sanitation, and reduced inequality. BAZNAS helped to create the models of poverty alleviation programs as what has been done through its Zakat Community Development (ZCD) program that aims for economic empowerment, education, health, and microfinancing.

Not only forming synergies with international institutions and non-profit organizations, BAZNAS also established a partnership with government bodies like the Ministry of Cooperatives and Small and Medium Enterprises (SMEs) of Indonesia to optimize the collection, distribution, and utilization of zakat. Besides, in order to strengthen management of zakat, BAZNAS formed collaboration with a private sector including banks or financial institutions like Bank Jawa Barat (BJB) with the hope to make payment of zakat easier for the community through digital payment (QRIS or DigiCash) (Sari, 2022). Although these two collaborations do not directly implement the SDGs, the idea is to increase zakat collection and distribution so that it will eventually help in reducing poverty rate and inequality as well as improving the welfare of the community.

BAZNAS AND THE COORDINATING ROLE

BAZNAS conducts both vertical and horizontal coordination. According to its structure, BAZNAS has varying levels of bodies with the main one being the national level, followed by BAZNAS provincial level (34 provinces), and BAZNAS district level. This means that BAZNAS has a vertical coordination between the national, provincial and district level. This coordination is done through regular coordination meetings.

BAZNAS conducted the National Coordination Meeting (Rakornas) in 2021 which aims to accelerate the realization of the vision and mission of BAZNAS, to increase the effectiveness and efficiency of zakat management, and to strengthen the role of BAZNAS in improving people's welfare and alleviating poverty. In addition, National Coordination Meeting also aims to improve coordination and evaluate the performance of zakat management organizations in the hope to increase transparency, accountability and professionalism in zakat management and at the same time also to socialize the strategy and process of digitizing zakat management. The 2021 Rakornas was attended by participants from BAZNAS leaders at all levels including national and provincial leaders and also inviting relevant ministries and partners (BAZNAS, 2021c).

Horizontally, BAZNAS is coordinating with Lembaga Amil Zakat (LAZ) i.e. private institutions established to assist BAZNAS in the collection, distribution, and utilization of zakat (Article 17 of the Zakat Management Act No. 23/2011). Similar to BAZNAS, LAZ is divided into national level, provincial level and district level. As of now, there are 34 official and authorized national LAZ as listed in BAZNAS website. As mandated by the Law, LAZ is obligated to report its audited performance in collection, distribution, and utilization of zakat to BAZNAS on a regular basis (Article 19 of the Zakat Management Act No. 23/2011).

BAZNAS AND THE MONITORING ROLE

As the framework suggests, an organization needs to define certain indicators, setting up timetables and benchmarking are important to track progress in pursuing long-term goals. BAZNAS has consistently produced its annual performance report that writes down their annual plan and realization for that particular year as well as comparison with the previous year's performance. For instance, the report includes a performance indicator like how many poor people are successfully removed from the poverty line. The number of targeted poors are different from year to year.

BAZNAS is not only monitoring its own performance but also mustahik (zakat recipients). As mentioned earlier, most of BAZNAS' economic program focuses on economic empowerment of the farmers. The BAZNAS Livestock Center (Balai Ternak) is a mustahik farmer empowerment program. This program requires a continuous process leading to the process of farmers' independence. The empowerment starts from conducting assessment (feasibility study), group formation, cage construction, livestock procurement, training and mentoring, business development, until the farmers can be independent to work on its own. Monitoring and Evaluation by BAZNAS aims to see the realization of this program, whether it is in accordance with the initial planning and to find out the obstacles faced in the field. This monitoring and evaluation involves assessment that uses the Sustainable Livelihood Impact Assessment (SLIA) method to look at the changes in 5 aspects, namely: Human Resource, Nature, Financial, Physical and Social Aspects and to get the most significant changes felt by the mustahik from this empowerment program (BAZNAS, 2020).

Another instance, BAZNAS carried out monitoring and mentoring activities for livestock populations alongside with mustahik at BAZNAS Purwakarta Animal Center. This activity is a routine agenda carried out to evaluate each empowered breeder (BAZNAS, 2021). The mentoring activities include recording population growth and the physical condition of the livestock. In addition, farmers are also assisted on how to record the sheep population, how to treat sick livestock, and find solutions to every problem in raising livestock. The livestock are also prepared to meet the daily local market demand and Qurban (sacrifices) (BAZNAS, 2021a).

Aside from the farmers' empowerment, BAZNAS also has a microfinancing program for the mustahik which the organization conducts its monitoring on a regular basis. For example, the BAZNAS Microfinance Team in Gunungsari (West Lombok) provides business assistance to mustahik who are beneficiaries of the BAZNAS-UPZ Permata capital financing cooperation program (BAZNAS, 2021b). The Gunungsari Team carried out monitoring to assess business conditions on the mustahil after the financing program had been given. This business monitoring is done to check the increase in turnover, income, product innovation, and marketing reach of the mustahik's business.

THE ROLE OF ZAKAT INSTITUTIONS IN MALAYSIA IN SDGs

In the case of Malaysia, the government makes Zakat mandatory for Malaysian Muslims. There are a total of 14 zakat collection centers in Malaysia. Each state has one separate center responsible for the zakat management. The body responsible for collecting and distributing zakat in Federal Territories of Malaysia (Kuala Lumpur, Putrajaya, and Labuan) is the Federal Territories of Islamic Religious Council (Majlis Agama Islam Wilayah Persekutuan, MAIWP). MAIWP was established by section 4(1) of the Administration of Islamic Law (Federal Territories) 1993 - Act 505 of Malaysia to advise the King on matters relating to the religion of Islam (Paizin, 2017)

In 1990, MAIWP established a company known as Zakat Collection Center (Pusat Pungutan Zakat, PPZ). PPZ is a wholly owned subsidiary of MAIWP which is responsible for carrying out the work of collection of zakat and charity efforts and deliver message of zakat to the Muslim community in the Federal Territories of Malaysia. Since then, the zakat collection task in the territory has been managed by PPZ. Beside PPZ, MAIWP has another institution under it, namely Baitulmal, which performs the task of Zakat distribution. Baitulmal MAIWP has three main roles, namely: Zakat distribution in Federal Territory; Zakat Fitrah Management; and the General resource property management (Hambari, Arif & Zaim, 2020).

Since its establishment, PPZ has made payment of zakat easier for Muslims in Malaysia by implementing reforms to automatically calculate amounts due. Over time, the PPZ has successfully created awareness for Malaysian Muslims on the obligations of paying Zakat by raising campaigns. As a result, the people's trust in the system has increased, reaching a 20% rise in collection rates each year during the 2005-2008 period (OECD, 2020).

MAIWP AND THE TRANSLATING ROLE

As stated in MAIWP's official website, the objectives of MAIWP are: (i) forming and creating a dynamic and progressive Muslim society that always seeks the pleasure of Allah; (ii) Fostering faith and strengthening the Islamic brotherhood among Muslims in the Federal Territory in particular and Malaysia in general to achieve the unity of the ummah. (iii) Increasing the efforts of virtue and preventing evil in the Muslim community. (iv) Implement efforts to increase MAIWP property through investment and other halal efforts for the benefit of the people. MAIWP's vision is to become a leader in the development and well-being of the ummah and its mission is to realize the ummah development agenda in a planned and comprehensive manner (MAIWP, n.d.).

Based on the information given in MAIWP's official website, the strategies adopted by MAIWP is as follows:

1. To provide adequate religious education facilities for the needs of Muslims.
2. Provide adequate facilities for places of worship for the needs of Muslims.
3. Improving the quality of services and religious welfare.
4. Expanding da'wah activities including assisting and guiding the activities of da'wah bodies.
5. To increase and develop waqf lands.
6. Improving zakat collection activities.
7. Enforce all aspects of Islamic administrative law.
8. Enhancing economic activities in the areas of business, development and investment.

Judging from the stated objectives, vision and mission of MAIWP, the SDGs are not directly translated into the organization's agenda, although they do mention the vision is to lead the development and well-being of the people. There is no statement that mentions the objective of MAIWP is to eradicate poverty or any other particular goals. This could be because MAIWP was established far before SDGs were officially adopted. However, the end goal of MAIWP, as can be seen from the strategies, is similar to SDGs, which is to promote prosperity. MAIWP develop waqf lands, improve zakat collection, enhance business, development and investment so that they can contribute more to the welfare of the Muslim community, be it in terms of education, health, and economy.

Elements of SDGs are more visible in assistance schemes created by MAIWP for its zakat distribution. MAIWP categorized zakat distribution into two groups. The first type is known as the direct distribution and the second type is called indirect distribution. Under the direct distribution, zakat is distributed to the recipients through numerous assistance schemes that were introduced by MAIWP. The schemes are divided into four categories and they are: (i) social development activity, (ii) economic development activity, (iii) education development activity, and (iv) welfare activity.

Norabidin (2016) explained MAIWP schemes in detail. Under the social category, MAIWP distributes zakat to recipients to fulfill their basic needs, including helping them to pay house rent deposit, house construction and renovation, food, health, clothing, and medical aid. Under the economic category, MAIWP aims to help and encourage the zakat recipients to generate their own income in business and entrepreneurship such as by giving them monetary aid for capital expansion or in the form of business tools and equipment, so that one day they may be able to achieve financial freedom. To illustrate, Baitul Mal-MAIWP has established an institute to nurture skills where courses such as cooking, sewing and automotive/workshops are offered. Baitul Mal also established a professional institute that teaches secretarial and administrative business for people to be ready for work (Nurfitriana, 2008).

Under the education category, MAIWP introduces programmes that cover all primary and secondary level of studies for the mustahik by paying their school fees and also giving them equipment needed for studying (Norabidin, 2016). Assistance is also awarded in form of scholarships for undergraduate and postgraduate studies at universities' levels. As for the welfare category, MAIWP provides aid for welfare such as those aids that are given to victims hit by disaster, fire, and fatal accidents. Meanwhile, MAIWP also distributes aid for conducting dakwah activities, including aid for encouragement of Qur'ān memorization; aid for non-governmental organizations (NGOs) for their effort in spreading dawah to the society, religious and educational institutions, and aid for marriage (Norabidin, 2016).

The four categories of direct distribution schemes include activities that relate to the achievement of SDGs, mainly no poverty, no hunger, and quality education.

MAIWP AND THE SUPPORTING ROLE

We found that MAIWP plays a supporting role by focusing on increasing awareness of muslims of the importance of paying zakat and also in investment and asset management so as to generate income that can later be tapped for zakat. The kind of partnerships and collaborations formed by MAIWP is clearly not directly to achieve SDGs. In fact, as of now, MAIWP has not formed any partnership with organizations like UNDP to specifically help Malaysia in achieving SDGs.

PPZ-MAIWP has made various efforts to increase zakat collection. PPZ-MAIWP formed a strategic partnership with Islamic banks such as RHB Islamic Bank to raise awareness on the importance of zakat, to promote social financial sustainability among the Muslim community, optimising the potential and effectiveness of zakat collection and distribution (Birruntha, 2022). Bank Pembangunan Malaysia Berhad (BPMB) partnered with PPZ-MAIWP to give basic food assistance to thousands of asnaf families, so that it can help alleviate their burden, particularly those impacted by the Covid-19 pandemic (BPMB, 2021).

Moreover, PPZ-MAIWP collaborated with Pertama Digital Bhd (PDB) to process cashless zakat payments via MyPay, an e-payment application for government services (Aman, 2021). This is to make paying zakat become practical. When payment becomes easier, it may motivate people to pay their zakat regularly. PPZ-

MAIWP also invites corporations/businesses to contribute to zakat. These efforts have resulted in a significant increase of MAIWP's zakat collection. For 2021, the amount of business zakat collected by PPZ-MAIWP surpassed its target of the previous year. More Muslims are aware about the obligation to perform zakat and there has been a heightened trust among Muslims that the zakat collected by the management bodies are well distributed to the recipients.

In asset management, BaitulMal-MAIWP collaborated with Tabungan Haji to build a public hospital named PUSRAWI in 2003 which also serves as accommodation/hotel. This is done to increase the collection of the zakat funds (Nurfitriana, 2008).

MAIWP AND THE COORDINATING ROLE

If we look at the structure, MAIWP plays a coordinating role, whereby it coordinates the performance of PPZ and Baitulmal. PPZ is a company owned and accountable to MAIWP which is in charge of collecting zakat funds. As of now, there are 10 (ten) PPZ counters dispersed around the federal territory of Kuala lumpur. Under MAIWP, there is a baitulmal which serves as the zakat fund distribution body and is ofcourse accountable to MAIWP. Hence, it can be inferred that MAIWP mainly has a vertical coordination role. However, PPZ and Baitulmal, other than having vertical coordination with MAIWP, also has horizontal coordination with each other as the two bodies are in the same level in the organizational structure.

Figure 3. Organizational Structure of MAIWP

Source: MAIWP

MAIWP AND THE MONITORING ROLE

MAIWP has several subsidiaries under its management, namely: Pusat Pungutan Zakat (PPZ) and Baitulmal for zakat, infaq, shadaqah; Pusat Wakaf MAIWP for waqf management; Hospital Pusrawi and Klinik Pusrawi for health services sector; PICOMS International University College, Institut Kemahiran Baitulmal, Institut Profesional Baitulmal, Sekolah Menengah Integrasi Sains Tahfiz, Pondok Moden Wilayah Persekutuan Al-'Abaqirah, Sekolah Menengah Agama Majlis Agama Islam Wilayah Persekutuan for education sector.

Aside from offering health/medical services and opportunities for Muslim community to obtain education, these bodies were also established to increase zakat collection and thus resulting in the increase of zakat distribution. MAIWP as the parent institution monitors the performance of each body to track their progress in making positive impacts for the community. For instance, PPZ always reports its zakat collection not only to MAIWP but also through the mass media every year and publishes its performance report (Ringkasan Laporan Zakat) annually on their official website which is easily accessible by the public. The report contains the amount of zakat collected based on types of zakat (zakat based on income, zakat based on maal/wealth, zakat on businesses, zakat qadha, and etc), announces the number of zakat payers and makes comparisons with the performance of last year. However, little or almost no information is available on how MAIWP conducts its monitoring role on the implementation of SDGs in particular.

DISCUSSION

The analysis shows that zakat institutions do play different roles in the implementation of SDGs. For each role, we found evidence of activities and initiatives developed by the zakat organization in both Indonesia and Malaysia. Concerning the translating role, we observe that only Indonesia's BAZNAS has literally translated SDGs in their own organization's goals, particularly the objective to alleviate poverty, reduce inequality, and increase synergy and collaboration with relevant stakeholders. BAZNAS also aims to improve the welfare of the community. These elements are clearly stated in BAZNAS's vision, mission, objectives, targets, and indicators. In contrast to BAZNAS, Malaysia's MAIWP has not directly translated SDGs into its own goals. What is literally stated as their vision, mission, and strategies are more religiously based, such as improving the well-being of the people, realizing the development agenda, and increasing investment to benefit the people. Their strategies are focused on how to increase zakat collection and enhance economic activities, although all of these efforts will eventually lead to the improvement of the welfare of the community.

The supporting role played by BAZNAS in the implementation of SDGs is also more evident since the zakat institution has signed a memorandum of understanding with UNDP and the Indonesia Philanthropy Association (PFI) where BAZNAS is expected to play a significant role in several objectives related SDGs, particularly no poverty, no hunger, good health and well-being, quality education, clean water and sanitation, and reduced inequality. Meanwhile, neither MAIWP-PPZ nor MAIWP-Baitulmal has formed any collaboration with UNDP to specifically collaborate in achieving SDGs in Malaysia. Many collaborative actions done by MAIWP are more focused on how to improve the efficiency of the zakat collection method.

In the coordinating role, BAZNAS is heavily involved in vertical coordination with BAZNAS at provincial and district levels (Baznas Daerah, BAZDA) as well as horizontal coordination with LAZ. As an example of vertical coordination, BAZNAS coordinates through the National Coordination Meeting (Rakornas) with BAZDA from all provinces, and one of the objectives of the meeting is to strengthen the role of BAZNAS in improving people's welfare and alleviating poverty. In other words, this helps BAZNAS in implementing SDGs. For horizontal coordination, BAZNAS coordinates with LAZ in the collection, distribution, and utilization of zakat, as the objective of the creation of LAZ is to assist BAZNAS in this matter. On the contrary, MAIWP has minimal horizontal coordination as MAIWP is only concerned with zakat management in the federal territory of Kuala Lumpur and each state has a different management body for zakat management. MAIWP only plays vertical communication with PPZ and Baitulmal to oversee their performance and then to report it to the Prime Minister.

Lastly, in the monitoring role, BAZNAS is more comprehensive in reporting its progress, as pointed out by one of the indicators in its annual report, BAZNAS measures every year how many poor people are successfully removed from the poverty line through its activities/schemes. This directly supports the achievement of "No Poverty" of SDGs. In detail, BAZNAS does monitoring on the recipients of its economic empowerment scheme until they can finally be financially independent. This would mean that BAZNAS will monitor the performance of its scheme until the recipients are no longer considered poor.

In sum, both zakat institutions (BAZNAS and MAIWP-PPZ-Baitulmal) develop each role to different degrees. Further research needs to focus on what explains this variation. As pointed out above, BAZNAS has developed all four roles (translating, supporting, coordinating, and mentoring) more than MAIWP purely based on the degree of institutionalization. This does not mean that BAZNAS is more effective in its effort to eradicate poverty and achieve other goals of SDGs. In fact, if we look nation-wide, Malaysia performs better than Indonesia in many aspects, including in ending poverty (SDG1). As stated in the earlier chapter of this paper, Malaysia ranks higher than Indonesia in terms of overall SDG-Index Value, where Malaysia came in 6th place whereas Indonesia is in 11th place (see Table 2: Ranking of countries on SDI sub-indices (SDG 1 to SDG 17))

CONCLUSION

Many scholars have evaluated the role of zakat in Islam, such as how the principle of zakat aligns well with the concept of SDGs, and argue that the resources collected through Islamic social finance (like zakat and waqf) have significant potential to help achieve the global goals. Both BAZNAS and MAIWP are not directly involved with the global objectives of SDGs and have no responsibility to adopt the goals since zakat institutions are more focused on optimizing zakat management. However, the end goal of zakat is to improve the welfare or well-being of the people. Thus, indirectly, SDGs do relate to the effective performance of Zakat management (collection and distribution). Each of them fulfills and operationalizes the four roles in different ways. In this paper, we illustrate each role with descriptions of their organizational structure and corporate information as well as zakat schemes. How effective these schemes are in the implementation of SDGs or how strong zakat institution's performance can positively affect the welfare of the people (including in alleviation of poverty, improving education and health) in each country need to be analyzed in a different study as a separate topic.

This paper analyzes the current role of zakat institutions in Malaysia and Indonesia in SDGs. The authors developed a conceptual model to analyze this by adopting a theory by Marx et al that came up with 4 (four) potential roles that zakat institutions can play: the translating role, the supporting role, the coordinating role, and the monitoring role. For each role, we conducted case studies based on a document analysis of primary and secondary sources. The descriptive and comparative case approach allows us to explore how and to what degree zakat institutions play each role in the SDGs.

REFERENCES

Al Azizah, U. S., Suryana, A. M., & Choirin, M. (2021). The impact of Zakat against Poverty Alleviation, Inequality & Economic Growth (Case Study of Indonesia). 5th International Conference of Zakat Proceedings.

Alfiani, T., & Akbar, N. (2020). Exploring Strategies to Enhance Zakat Role to Support Sustainable Development Goals (SDGs): ISM Delphi Approach. 4th International Conference of Zakat Proceedings.

Aman, A. S. (2021). MyPay partners with PPZ-MAIWP to accelerate cashless zakat payments. https://www.nst.com.my/business/2021/12/750675/mypay-partners-ppz-maiwp-accelerate-cashless-zakat-payments

Andriani, Sadewa, M. M., Rahmalita, Y., Budiman, M. A., & Nurhidayatati, N. (2021). Optimizing the role of zakat institution: building resilience against the risk of the covid-19 pandemic. 5th International conference of Zakat Proceedings.

Asmalia, A., Kasri, R. A., & Ahsan, A. (2018). Exploring the Potential of *Zakah* for Supporting Realization of Sustainable Development Goals (SDGs) in Indonesia. *International Journal of Zakat: Special Issue on Zakat Conference,* 51-69.

Bahri, M. S., Herindar, E., & Rusydiana, A. S. (2021). Covid-19 and Efficiency in Zakat Institutions: Evidence from Indonesia. 5th International Conference of Zakat Proceedings.

BAZNAS. (2016). BAZNAS Signed the MOU of SDGS Partnership in Indonesia. https://www.puskasbaznas.com/ijaz/index.php/journal/article/view/16/13

BAZNAS. (2017). BAPPENAS: Zakat Is Able To Cut The Poverty Chain. https://www.puskasbaznas.com/news/485-bappenas-zakat-is-able-to-cut-the-poverty-chain

BAZNAS. (2020). BAZNAS Lakukan Monitoring Evaluasi dan Kaji Dampak di Kelompok Balai Ternak Garut. https://baznas.go.id/pendistribusian/ekonomi/pemberdayaan-peternak/1602-baznas-lakukan-monitoring-evaluasi-dan-kaji-dampak-di-kelompok-balai-ternak-garut

BAZNAS. (2021a). BAZNAS Dampingi Mustahik Monitoring Populasi Ternak. https://baznas.go.id/berita_pendistribusian/baca/BAZNAS_Dampingi_Mustahik_Monitoring_Populasi_Ternak/6740

BAZNAS. (2021b). BAZNAS Monitoring Usaha Mitra Mustahik Program Kemitraan BAZNAS-UPZ Permata. https://baznas.go.id/pendistribusian/ekonomi/pembiayaan-mikro/6400-baznas-monitoring-usaha-mitra-mustahik-program-kemitraan-baznas-upz-permata

BAZNAS. (2021c). "Rakornas Dorong Penguatan BAZNAS Sebagai Pengelola Zakat Nasional". https://baznas.go.id/Press_Release/baca/%22Rakornas_Dorong_Penguatan_BAZNAS_Sebagai_Pengelola_Zakat_Nasional%22/749

Birruntha. (April 20, 2022). PPZ-MAIWP, RHB Islamic forge strategic partnership. https://themalaysianreserve.com/2022/04/20/ppz-maiwp-rhb-islamic-forge-strategic-partnership/

BPMB. (2021). BPMB partners with MAIWP to hand out basic food assistance to the asnaf community. https://www.bpmb.com.my/ms_MY/news?p_p_id=56&p_p_lifecycle=0&p_p_state=maximized&p_p_mode=view&p_p_col_id=column-2&p_p_col_pos=2&p_p_col_count=3&_56_returnToFullPageURL=%2Fms_MY%2Fnews&_56_groupId=21475&_56_articleId=131028&_56_version=1.4&_56_articleTitle=BUSINESS+TODAY+-+BPMB+partners+with+MAIWP+to+hand+out+basic+food+assistance+to+the+asnaf+community

Choirunisa, U., Permadia, S., & Gunawan, A. R. (2020). The Role of Zakat in Reducing the Poverty Gap: Case Study in Desa Berdaya Rumah Zakat. 4th International Conference of Zakat Proceedings.

Destiana, A. R., & Hamzah, M. Z. (2021). Fiscal policy design in Malaysia and Indonesia about Zakat and Taxes. 5th International Conference of Zakat Proceedings.

ESCWA. (2017). *Intergenerational Justice: Meeting the Needs of the Future Arab Generations*. United Nations.

Hambari., Arif, A. A., & Zaim, M. A. (2020). The Role of Zakat Institution in Facing Covid-19: A Case Study of the Federal Territory Islamic Council (MAIWP) of Malaysia. The 4th International Conference of Zakat (ICONZ), Surabaya, Indonesia.

Hasan, Z. (2020). Distribution Of Zakat Funds To Achieve SDGs Through Poverty Alleviation In Baznas Republic Of Indonesia. *Azka International Journal of Zakat & Social Finance*, *1*(1), 25–43. DOI: 10.51377/azjaf.vol1no01.7

MAIWP. (n.d.). https://www.maiwp.gov.my/i/index.php/en/mengenai-kami/objektif-visi-misi

Maksum, . (2018). *Fikih Sustainable Development Goals*. UIN Jakarta Press.

. Marx, A., Pertiwi, S. B., Depoorter, C., Hoornick, M., Mursitama, T. N., Otteburn, K., & Arnakim, L. Y. (2021). What role for regional organizations in goal-setting global governance? An analysis of the role of the European Union and ASEAN in the Sustainable Development Goals.

Noor, Z., & Pickup, F. (2017). The role of zakat in supporting the Sustainable Development Goals. Brief. May 2017.

Norabidin, R. N. R. (2016). Zakāh Collection and Distribution Framework in Federal Territory of Kuala Lumpur: A Study on Selected Islamic Banking Institutions. International Journal of Management and Applied Research, 4(1). DOI: 10.18646/2056.41.17-002

Nurasyiah, A., Adam, F., & Pertiwi, R. S. (2019). Efficiency and Productivity of Zakat Institution in Malaysia and Indonesia: The Comparative Study. International Conference of Zakat 2019 Proceedings.

Nurfitriana. (2008). Analisis perlakuan zakat dalam perhitungan pajak penghasilan orang pribadi: "studi komparasi" antara peraturan perpajakan Indonesia dengan Malaysia. Fakultas Ekonomi dan Bisnis Universitas Indonesia.

Nurhapsari, F., Dwijayanti, R., Ramadhani, R., & Al Azizah, U. S. (2021). The Role of Zakat for Poverty Alleviation in Indonesia. 5th International Conference of Zakat Proceedings

Nurkhozin. (2021). Manajemen Pendayagunaan Zakat Tinjauan Tujuan Pembangunan Berkelanjutan (Sustainable Development Goals). Jurnal An-Nahl, 8(1), 52-62.

OECD. (2020). *How Islamic Finance Contributes To Achieving The Sustainable Development Goals. OECD Development. Policy Papers. June 2020 No. 30.* OECD Publishing.

Paizin, M. N. (2017). *Usage of Islamic Alms's Online Portal: Case Study on Pusat Pungutan Zakat (PPZ).* Federal Territories of Malaysia. DOI: 10.4172/2375-4516.1000175

Qurroh, A., Pramanik, A. H., Saad, N. M., & Ariffin, M. I. (2017). The Comparison between Consumption and Production-based Zakat Distribution Programs for Poverty Alleviation and Income Inequality Reduction. *International Journal of Zakat*, *2*(2), 11–28. DOI: 10.37706/ijaz.v2i2.22

Ramadan, A. B., & Hussain, M. (2019). Zakat Institutions and Universities Strategic Partnership for Poverty Alleviation. *International Journal of Zakat*, *4*(2), 73–86. DOI: 10.37706/ijaz.v4i2.172

Rehman, A. A., & Pickup, F. (2018). Zakat for the SDGs. *Islamic Finance News, 15*(34).

Sari, F. (2022). Gandeng Baznas, Bank BJB Mudahkan Masyarakat Salurkan Zakat dan Infak. Kontan. https://pressrelease.kontan.co.id/news/gandeng-baznas-bank-bjb-mudahkan-masyarakat-salurkan-zakat-dan-infak

Shaikh, S. A., & Hassan, M. K. (2020). *Where Do Muslim Countries Stand On Sustainable Development Goals. Islamic Perspective For Sustainable Financial System 87-108.* Istanbul University Press.

Suprayitno, E., Aslam, M., & Harun, A. (2017). Zakat and SDGs: Impact Zakat on Human Development in the Five States of Malaysia. *International Journal of Zakat, 2*(1), 61–69. DOI: 10.37706/ijaz.v2i1.15

Suriani, Nurdin, R., & Riyaldi, M. H. (2020). Optimization of Zakat for Sustainable Development Goals: Evidence from Baitul Mal Aceh. 4th international Conference of Zakat Proceedings.

UNDP. (n.d.). The SDGs in action. Retrieved from https://www.undp.org/sustainable-development-goals

Yudha, A. T. R. C., Awwaliah, H., & Pertiwi, E. M. (2021). SDGs Value and Islamic Philanthropy Through Zakat Institutions During the Covid-19. *Journal of Islamic Economics, Finance, and Banking, 4*(1), 31–44.

Zulhendra, Nurwahidin, Fahruroji, & Hamzah, M. A. (2020). Zakat Fund Management by Amil Zakat Institutions: Case Study on LAZNAS BSM Umat. 4th Internationa Conference of Zakat Proceedings.

Chapter 16
War on Drugs in the Philippines and SDG 16:
Between Security, Justice, and Human Rights

Lili Yulyadi Arnakim
https://orcid.org/0000-0002-4575-424X
Independent Researcher, Indonesia

Putri Sekar Diasti Wibowo
Bina Nusantara University, Indonesia

ABSTRACT

The intersection between the Duterte administration's War on Drugs and Sustainable Development Goal 16 reveals critical tensions in policy design and implementation. Drawing on qualitative analysis of the Philippine Development Plan (PDP) 2017–2022 and official data, the study assesses alignment with three key SDG targets: 16.1 (violence reduction), 16.3 (access to justice), and 16.6 (institutional accountability). While institutional reforms and justice-sector improvements are formally embedded in national planning documents, the operational conduct of the War on Drugs has generated systemic contradictions. These include widespread human rights violations, legal erosion, and institutional opacity. The securitization of development through punitive means challenges the normative foundations of SDG 16 and raises urgent questions about the future of peace and justice in authoritarian governance contexts. Fulfilling sustainable development requires a rights-based, inclusive, and transparent approach that places human dignity and democratic accountability at the core of public policy.

DOI: 10.4018/979-8-3373-1300-9.ch016

INTRODUCTION

The anti-drug policy enacted by President Rodrigo Duterte has attracted significant public and academic scrutiny, as it is widely perceived to conflict with the principles upheld by the Sustainable Development Goals (SDGs) (PARONG, 2024). One of the central pillars of the SDGs "Promote peaceful and inclusive societies for sustainable development, provide access to justice for all, and build effective, accountable institutions at all levels" emphasizes the importance of fostering peaceful and inclusive communities as a cornerstone of sustainable development, while ensuring access to justice for every individual. These values form the foundation of the 17 goals and 169 targets initiated by the United Nations in 2015 (United Nations, 2023). As a UN member state, the Philippines has pledged its commitment to achieving these targets. Nevertheless, the challenges encountered in the implementation of this policy have grown increasingly complex (National Economic Development Authority, 2023). The objectives of the War on Drugs are also embedded within the Philippine Development Plan (PDP) as outlined by the Duterte administration. However, the evident misalignment between this national anti-drug policy and the normative commitments of the SDGs raises critical concerns regarding its effectiveness and implications within the broader framework of sustainable development in the Philippines.

The War on Drugs was a central theme in Duterte's political campaign and became a defining policy initiative upon his inauguration in 2016 (Council on Foreign Relations, 2016). He framed the drug problem not merely as a public health issue but as a grave threat to national security and social stability. The intensification of drug-related offenses and the perceived rise in criminal activity prompted Duterte to adopt a hardline stance, arguing that the high prevalence of narcotics use necessitated a forceful state response (Human Rights Watch, 2017).

The core issue addressed in this study emerges from the tension between the Philippine government's proclaimed success in reducing crime rates and the mounting criticism over its coercive and punitive approach to drug suppression. While statistical data from the Philippine Statistics Authority indicate improvements in public safety, the methods employed to reduce drug prevalence have engendered serious consequences for human rights protection and the rule of law. The imbalance between quantitative outcomes and normative violations generates critical questions about the extent to which national policies are substantively aligned with the goals of sustainable development, particularly SDG 16. Against this backdrop, this study seeks to answer the central research question: How has President Duterte's War on Drugs affected the progress of SDG 16 indicators in the Philippines, particularly in terms of security, justice, and human rights.

1. LITERATURE REVIEW

The SDGs, particularly SDG 16, underscore the importance of building peaceful and inclusive societies, ensuring access to justice for all, and creating effective, accountable, and inclusive institutions. Within this global framework, the Philippines represents a unique case study due to its adoption of a hardline security approach through the War on Drugs under President Rodrigo Duterte. This policy direction has generated significant tension between the pursuit of public order and the protection of fundamental human rights.

Duterte's anti drug campaign has profoundly reshaped the domestic policy landscape in the Philippines. Emphasizing a reduction in crime rates and the enhancement of law enforcement capacity, the approach was perceived to respond to widespread societal demand for greater security (Jetter et al., 2020). Public support for tough-on-crime measures contributed to a national shift in policy priorities, positioning law enforcement as central to governance. In this context, Duterte's strategy can be interpreted as a response to localized needs for stability, one of the key components of SDG 16.

The Duterte administration has also claimed progress in institutional reform, citing efforts to uphold the rule of law and enhance governance transparency as aligned with good governance principles (Mehmood, 2023). These reforms, at least in normative terms, reflect alignment with the more structural targets of SDG 16. However, realities on the ground suggest a paradox between reformist narratives and repressive practices.

Despite reported improvements in select SDG 16 indicators, such as perceptions of public safety and institutional efficiency, numerous reports highlight severe risks to human rights. The prevalence of extrajudicial killings and a high incidence of unlawful executions indicate a shift from developmental governance to security-driven governance, where social policies are increasingly shaped by a securitization logic that sacrifices civil liberties (Lazarus, 2020). In this context, questions concerning the long-term sustainability of such an approach become especially pertinent, particularly as human rights violations exacerbate divisions between the state and civil society.

Complementary academic perspectives have added depth to this discourse. For instance, Dennis V. Blanco (2020) argues that sustainable development in urban areas such as Quezon City requires inclusive stakeholder collaboration. When policies are centralized and coercive, such as those under the War on Drugs, the participatory and inclusive values of SDG 16 risk being marginalized. Similarly, David Hulme and Jose Andres A. Limcaoco (2012) demonstrate that community development strategies grounded in participatory methods tend to be more sustainable than top-down approaches that exclude public consultation.

In addition, the work of Edralin and Pastrana (2022) highlights the role of the private sector in supporting SDGs, while also criticizing the insufficient emphasis placed on institutional and legal frameworks. This suggests that achieving SDG 16 entails more than formal stability; it requires the substantive delivery of justice through state institutions. Supporting this, Blanco and Panao (2022) stress that transparency and accountability are central to effective public policy implementation.

Lastly, Bongolan, Soria, and Rivera (2022) point to the complex interactions between SDG targets. In the Philippine context, the tension between security achievements and human rights protections presents a policy dilemma in which progress in one dimension may hinder fulfillment in another. A strategy heavily focused on crime reduction, if not accompanied by safeguards for civil rights, risks engendering systemic violations of social justice principles.

The literature reveals that Duterte's War on Drugs generates a form of ambivalence for SDG 16 outcomes: it simultaneously promotes institutional stabilization while eroding human rights and justice norms. A critical evaluation of this policy is thus essential to determine whether the gains in security are truly sustainable or whether they come at the expense of the core foundations required for inclusive and peaceful development.

2. CONCEPT

Promoting peaceful and inclusive societies for sustainable development, ensuring access to justice for all, and building effective, accountable, and inclusive institutions at all levels constitute the core objectives of SDG 16. The United Nations has identified this goal as one of the most rapidly evolving priorities and has emphasized that without immediate and coordinated action by states to address challenges in this area, national stability and the realization of civil rights may be severely compromised. Many countries have increasingly recognized the need to strengthen the role of the state in implementing policies that guarantee civil security through fair and just legal frameworks. Governments are likewise expected to prioritize transparency and justice for all citizens, with particular focus on marginalized populations (Katila et al., 2019).

This article adopts the principles embedded within SDG 16 as the analytical framework through which to assess policy outcomes. Particular emphasis is placed on three key targets that are most relevant to the context of the War on Drugs in the Philippines:

- **Target 16.1: Significantly reduce all forms of violence and related death rates everywhere.**

This includes indicators such as the number of victims of intentional homicide per 100,000 population, deaths associated with conflict, the proportion of the population subjected to physical violence, and individuals' perceptions of safety when walking alone.

- **Target 16.3: Promote the rule of law and ensure equal access to justice for all.**

This target is assessed using indicators such as the proportion of victims who report violence and the percentage of disputes resolved through formal or informal legal mechanisms.

- **Target 16.6: Develop effective, accountable, and transparent institutions at all levels.**

Indicators for this target include alignment between national budgets and public expenditure plans, as well as public satisfaction with government services.

These three targets have been selected for their conceptual relevance in capturing the main dimensions affected by Duterte's War on Drugs: security, justice, and governance. By focusing on Targets 16.1, 16.3, and 16.6, the article seeks to systematically evaluate how the state's repressive approach to crime control, despite resulting in reduced crime rates, has simultaneously generated serious challenges to human rights protection, access to justice, and institutional accountability. The tension between efforts to ensure national stability and the state's obligation to uphold procedural justice and transparent governance emerges as a core issue requiring in-depth scrutiny. As such, SDG 16 serves not only as an evaluative instrument for policy effectiveness but also as a normative lens to assess the extent to which national security strategies align with the principles of sustainable development and universal human rights.

3. METHODOLOGY

This study employs a qualitative approach using document analysis to evaluate the extent to which the implementation of Sustainable Development Goal 16 (SDG 16) is reflected in the Philippine government's War on Drugs policy, particularly in relation to the Philippine Development Plan (PDP) for the period 2016–2022 under President Rodrigo Duterte's administration. This methodological approach is chosen to facilitate an in-depth examination of the relationship between normative frameworks, national policy formulations, and the realities of implementation as they impact security, justice, and institutional governance (Bowen, 2009).

The analysis begins with a review of the targets and indicators of SDG 16 as defined by the United Nations, followed by an assessment of how these three specific targets are integrated within the PDP 2017–2022 and how their implementation indicators are operationalized by the Philippine government (National Economic Development Authority, 2022). Key indicators examined in this study include levels of violence and extrajudicial killings, public perceptions of safety, access to justice for vulnerable groups, and the accountability and transparency of law enforcement institutions.

The primary data sources consist of secondary documents, including the PDP Annual Reports for 2019 and 2022 (NEDA, 2019; 2022), publications from international organizations such as UNICEF (2021), and official documents from Philippine government agencies, including the National Economic and Development Authority (NEDA), the Office of the Ombudsman, and the official SDG Philippines portal (SDG NEDA, 2022). These secondary sources are used for triangulating policy narratives and strengthening the validity of interpretations regarding the implementation of state programs.

Document analysis as a method enables the researcher to interpret policy content, institutional narratives, and quantitative or qualitative data presented by the government in the context of the War on Drugs (Bowen, 2009). The analytical process is conducted reflectively, taking into account political dynamics, human rights violations, and institutional practices that have evolved during the implementation of the policy.

By employing the indicators of SDG 16 as an evaluative framework, this research goes beyond numerical performance assessment to critically examine the substantive coherence between sustainable development commitments and on-the-ground realities. This approach allows for a critical reading of the tensions between the state's narrative of security stability achieved through coercive measures and its responsibility to uphold justice and democratic governance.

4. DISCUSSION

Table 1. PDP 2017-2022 Alignment with SDG16

Alignment of SDG Targets with the Philippines' NDP	
SDG 16 Target	**PDP 2017-2022**
16.1 Significantly reduce all forms of violence and related death rates everywhere	The PDP emphasizes reducing violence through strengthened law enforcement and a swifter, fairer justice system (Ch. 6 & 18). Reforms include the rollout of the national ID system (PhilSys), capacity building for security forces, and peacebuilding efforts in conflict-affected areas like Bangsamoro. Implementation, however, faces ongoing human rights concerns.
16.3 Promote the rule of law at the national and international levels and ensure equal access to justice for all	The PDP supports the rule of law through judicial reforms aimed at expediting case resolution, decongesting courts, and enhancing legal aid mechanisms. Programs under Chapter 6 seek to ensure fairness, legal certainty, and improved access to justice, particularly for vulnerable groups.
16.6 Develop effective, accountable, and transparent institutions at all levels	Under the governance reform agenda (Ch. 5), the PDP promotes clean, responsive, and technology-enabled government. The Ease of Doing Business Act, PhilSys implementation, and expanded citizen engagement mechanisms are key to improving institutional transparency, accountability, and trust.

source: United Nations & the Philippines Government

The War on Drugs under President Rodrigo Duterte's administration stands as one of the most controversial security policies in contemporary Southeast Asia. Framed as a response to criminality and the drug menace, this policy was embedded within the broader national development vision, specifically articulated in the PDP 2017–2022. While the PDP formally aligns with SDG 16, which promotes peace, justice, and strong institutions, the practical tensions between enforcement-driven strategies and human rights protection raise significant questions regarding policy coherence and ethical governance.

The PDP 2017–2022 directly acknowledges the campaign against illegal drugs as one of the Duterte administration's flagship programs under Chapter 18, *Ensuring Security, Public Order, and Safety*. Central to this campaign is the implementation of the Philippine Anti-Illegal Drugs Strategy (PADS), which is framed as a comprehensive, rights-based, and multisectoral approach. The strategy comprises two major pillars: supply suppression through intensified law enforcement operations and demand reduction via rehabilitation, prevention, and education programs. This dual approach is meant to reflect a balanced policy, although its real-world execution often skews disproportionately toward punitive action.

In operational terms, the PDP mandates all government entities—including government-owned and controlled corporations (GOCCs) and state universities and colleges (SUCs)—to formulate institutional anti-drug action plans. These cover the development of workplace drug policies, implementation of barangay-level drug clearing operations, and budget allocation for anti-illegal drug initiatives. Local Government Units (LGUs) are likewise instructed to strengthen their Anti-Drug Abuse Councils, and the national government is tasked with conducting a nationwide drug abuse survey alongside building disaggregated databases for monitoring purposes. These initiatives ostensibly aim to improve transparency, data governance, and accountability in anti-drug operations.

The PDP also outlines a legislative agenda to reinforce the infrastructure of the drug war. Proposed amendments to the Comprehensive Dangerous Drugs Act of 2002, for instance, include provisions for mandatory drug testing, the imposition of stricter penalties, and the expansion of the Philippine Drug Enforcement Agency's (PDEA) powers. Complementary legislation, such as the Anti-Drug Abuse Council Law, seeks to institutionalize local mechanisms and ensure consistent funding for anti-drug efforts. Through these legal instruments, the PDP demonstrates a formal commitment to institutionalizing anti-drug policies within the broader development framework.

However, SDG Target 16.1, which aims to "significantly reduce all forms of violence and related death rates everywhere," the outcomes of the War on Drugs reveal a troubling policy paradox. While the PDP highlights the importance of a strengthened justice system and professionalized law enforcement as key enablers of public safety, the campaign has been widely criticized for triggering state-perpetrated violence, including extrajudicial killings, arbitrary arrests, and intimidation. These practices, often targeting marginalized urban communities, have contributed to an atmosphere of fear rather than genuine security. Although the PDP addresses violence reduction in quantitative terms, it fails to critically interrogate the legitimacy of the coercive measures employed, thereby undermining the very goal it seeks to advance.

SDG Target 16.3, which promotes the rule of law and equal access to justice, the PDP puts forward an agenda of judicial reform: reducing court backlogs, expanding legal aid, and facilitating access to justice for vulnerable groups. However, the War on Drugs exposes serious structural weaknesses in the Philippine legal system. Reports of prosecutorial bias, weak witness protection programs, and limited access to defense counsel reveal a system ill-equipped to ensure due process. These deficiencies reflect a disjuncture between the PDP's formal aspiration to uphold the rule of law and the erosion of legal safeguards in practice under the drug campaign.

Under SDG Target 16.6, which emphasizes effective, accountable, and transparent institutions, the PDP offers various governance innovations, including the Ease of Doing Business Act and the digitization of government services as tools for building public trust and institutional legitimacy. Nevertheless, these advancements are largely restricted to economic and administrative governance. In sectors related to justice and security, accountability mechanisms remain inadequate. The absence of independent investigations into drug-related killings, resistance to international human rights monitoring, and political shielding of implicated officials illustrate a culture of impunity that directly contravenes the goals of institutional transparency and accountability.

In sum, while the PDP 2017–2022 contains policy instruments that appear aligned with SDG 16, its articulation and operationalization of the War on Drugs reveal profound governance contradictions. The campaign against illegal drugs, though framed within a developmental and rights-based narrative, in practice has contributed to systemic human rights violations, weakened judicial safeguards, and eroded institutional legitimacy. This duality highlights a central disjuncture in Philippine governance: the attempt to achieve development and order through authoritarian means. The Philippine experience thus underscores the imperative of policy coherence, where developmental aspirations must be matched by democratic practices, and where peace and justice cannot be pursued in silos but must be upheld through the mutual reinforcement of rights, accountability, and the rule of law.

4.1 Reality of Implementation

Within the context of the War on Drugs policy implementation in the Philippines, major challenges emerge when assessing its impact on the achievement of SDG 16, which focuses on reducing violence, improving access to justice, and strengthening institutional transparency and accountability. Although official government narratives highlight a decrease in conventional crime rates since the policy's inception, a more critical analysis reveals profound social implications for Filipino communities (Schlein, 2022). This discussion interrogates the dissonance between official policy narratives and on-the-ground realities, with a particular focus on how state-inflicted violence, inequality in access to justice, and a lack of public transparency have obstructed the realization of substantive security, equitable justice, and institutional integrity (Ratcliffe, 2020). By adopting a data-informed and critically engaged perspective, this section explores whether the policy outcomes align with sustainable development goals or exacerbate challenges that such goals aim to resolve. This analysis is organized around the three primary targets of SDG 16: 16.1, 16.3, and 16.6.

In the context of the War on Drugs, Target 16.1, which aims to reduce all forms of violence and related death rates, becomes a critical point of inquiry. Data from the Philippine Statistics Authority indicates a decline in conventional crime rates since 2016. However, these figures do not account for state-perpetrated violence, including extrajudicial killings, which have sharply increased during this period. Human Rights Watch (2021) reported over 6,000 deaths related to anti-drug operations, while local watchdogs estimate the true figure may exceed 27,000 victims (Santos, 2020). The majority of these victims were poor, residing in vulnerable urban neighborhoods frequently targeted in police operations. Although the policy may have reduced crime numerically, public perceptions of safety, especially in high-risk areas, have deteriorated (Jetter & Molina, 2022). The securitization of public policy has increasingly come at the cost of fundamental human rights, producing a form of structural fear and normalized violence in society (Lazarus, 2020).

Official PSA data from 2016 to 2022 suggests an overall increase in citizens' perception of safety. However, these findings are not disaggregated annually, obscuring year-by-year developments. These gains are attributed to proactive security strategies, including strengthened community-police partnerships, the integration of modern technology in law enforcement, and the development of community-based programs to enhance public security. Government efforts to promote local stability and collaboration may have contributed to these outcomes, though the broader context of the War on Drugs must also be considered as a factor in shaping perceptions of safety.

However, Target 16.3, which aims to ensure equal access to justice, reveals significant deficiencies. Data from Amnesty International and AMP (2022) indicate that only one out of thousands of killings has resulted in a legal resolution. This suggests a crisis of confidence in the justice system and reflects the inadequacy of accountability mechanisms for state actors. These systemic failures are exacerbated by structural inequalities that restrict legal access for the poor, many of whom lack the financial means for legal representation or the ability to lodge complaints with oversight bodies. Meanwhile, upper and middle-class populations have largely avoided direct exposure to the state's coercive campaign (Manlapaz, 2020).

Judicial reform under Target 16.3 has shown mixed results. Notable progress includes the reduction of the pretrial detainee population, from 73.9% in 2016 to 64.2% in 2022, indicating improvements in case processing. Nonetheless, this figure remains high and underscores ongoing inefficiencies in the legal system, including case backlogs and limited judicial capacity, which hinder access to timely justice. In response, the government has prioritized judicial and legal reforms aimed at expediting case resolution and improving access to justice for vulnerable populations. These initiatives include streamlining legal procedures to reduce delays, enhancing the judiciary's institutional capacity through targeted legal training, and introducing e-Court systems to improve administrative efficiency (Philippine Statistics Authority, 2023).

Reports from Amnesty International further highlight inconsistencies in the documentation and transparency of deaths resulting from anti-drug operations. The 2019 Voluntary National Review noted that only 68 of the 156 SDG indicators were included in the official monitoring framework, many of which were unrelated to violence or justice access (Philippine Statistics Authority, 2019). International human rights organizations have criticized the lack of independent investigations into abuses by security personnel. Philippine oversight bodies such as the Commission on Human Rights and the Office of the Ombudsman are often ignored or denied sufficient access to conduct inquiries. Fiscal accountability, however, showed statistical improvement, with PSA data indicating that performance in this domain increased from 91.5% in 2016 to 103.6% in 2022.

Despite these incremental reforms in judicial and fiscal governance, this study's findings reveal that the repressive implementation of the War on Drugs has widened the gap between quantitative policy achievements and the normative values embedded in SDG 16. When security is evaluated solely through crime statistics, without considering substantive safety or the protection of human rights, such policies risk neglecting the social justice dimension of development. Consequently, formal advancements in bureaucratic and technocratic reform cannot serve as the sole benchmark of success without addressing the realities of structural violence and institutional failures in accountability. Within the framework of SDG 16, these findings call for a reevaluation of security policy directions—one that moves beyond superficial stability and is grounded in respect for human rights, equitable access to justice, and meaningful public participation.

5. CONCLUSION

This article critically examined the Duterte administration's War on Drugs through the normative and evaluative lens of SDG 16, which emphasizes peace, justice, and strong institutions. Using the PDP 2017–2022 as a primary framework for analysis, this study explored the extent to which national anti-drug policies have aligned—or misaligned—with the core targets of SDG 16, particularly 16.1 (violence reduction), 16.3 (access to justice), and 16.6 (institutional accountability and transparency).

The analysis demonstrates a profound disjuncture between the stated objectives of the PDP and the actual outcomes of the War on Drugs. Although the PDP embeds security and justice reform within a broader agenda of sustainable development, the operationalization of the drug campaign has revealed a pattern of governance that undermines the very principles it seeks to uphold. The campaign's reliance on coercive enforcement strategies, including extrajudicial killings, arbitrary detentions, and targeted operations against marginalized populations, significantly contradicts the goal of reducing violence in a just and humane manner. While crime indicators may suggest short-term gains in public safety, these outcomes are achieved through methods that erode trust in institutions, provoke fear rather than peace, and sideline fundamental rights.

Access to justice (Target 16.3), the findings show that institutional reforms have been insufficient to mitigate the legal vacuum created by the campaign's excesses. Vulnerable individuals—particularly the urban poor—have limited access to legal representation, while the mechanisms for accountability, including the judiciary and oversight bodies, remain weak or politically constrained. The failure to ensure due process and legal recourse not only diminishes the integrity of the legal system but also perpetuates a culture of impunity that is antithetical to the rule of law.

In terms of institutional accountability (Target 16.6), the PDP highlights governance innovations such as digitalization and citizen engagement. Yet, these advances are not mirrored in the drug enforcement apparatus. The absence of independent investigations into human rights violations, the shielding of implicated law enforcement officers, and the lack of transparency in reporting mechanisms point to structural deficits in institutional integrity. Consequently, the trustworthiness of government institutions, particularly in the realm of public safety and justice, is significantly compromised.

This study thus concludes that the War on Drugs, as implemented under the PDP framework, has produced governance outcomes that are fundamentally misaligned with the spirit and substance of SDG 16. While the campaign may have succeeded in projecting an image of decisiveness and control, it has simultaneously weakened the democratic foundations necessary for sustainable peace, equitable justice, and accountable governance. This contradiction suggests that development cannot be meaningfully pursued through authoritarian means, nor can peace be equated merely with the suppression of disorder.

The Philippine experience offers critical insights for global and regional discourses on security, development, and human rights. It illustrates the dangers of securitizing development goals, where the state's monopoly on violence is exercised with insufficient legal and ethical restraint. Moving forward, there is an urgent need for governance models that prioritize human rights, participatory justice, and institutional integrity as inseparable components of national development planning.

SDG 16 serves as more than a rhetorical commitment, its implementation must be anchored in holistic and inclusive policy designs that resist the instrumentalization of law and order for political ends. The War on Drugs should thus be re-evaluated not only as a domestic policy failure but also as a cautionary case of how developmental aspirations can be undermined by illiberal governance. Only through a genuine realignment between national strategies and global human rights norms can the goals of sustainable, peaceful, and just societies be fully realized.

REFERENCES

AMP. (2022). *Human rights in the Philippines*. https://amp.ngo/wp-content/uploads/2022/09/AMP-2022-Report-Human-Rights-in-the-P hilippines.pdf

Blanco, D. V., & Panao, R. A. L. (2022). The Sustainable Development Governance Initiatives of Quezon City, Philippines: An Analysis of Its Enablers and Barriers. *International Journal of Public Administration, 46*(12), 857-875. DOI: 10.1080/01900692.2022.2043366

Bongolan, V. P., Soria, S. C., & Rivera, R. L. K. (2022). Sustainable Development Goal Target Interactions in the Philippines: a Two-Method Approach. *Philippine Journal of Science, 152*(3), 827-835. https://philjournalsci.dost.gov.ph/120-vol-152-no-3-june-2023/1838-sustainable-development-goal-target-interactions-in-the-philippines-a-two-method-approach

Bowen, G. (2009). Document Analysis as a Qualitative Research Method. *Qualitative Research Journal, 9*(2), 27-40. Researchgate. DOI: 10.3316/QRJ0902027

Council on Foreign Relations. (2016). *Human Rights and Duterte's War on Drugs*. https://www.cfr.org/interview/human-rights-and-dutertes-war-drugs

Edralin, D. M., & Pastrana, R. (2022). Advancing the pursuit of the United Nations Sustainable Development Goals: Initiatives of selected publicly listed companies in the Philippines. *Bedan Research Journal, 7*(1), 1–47. DOI: 10.58870/berj.v7i1.31

Hulme, D., & Limcaoco, J. A. A. (2012). Planning integrated rural development projects in the Philippines: From blueprint to process? *Project Appraisal, 06*(04), 223–233. DOI: 10.1080/02688867.1991.9726824

Human Rights Watch. (2017, March 2). *"License to Kill": Philippine Police Killings in Duterte's "War on Drugs" | HRW*. Human Rights Watch. Retrieved April, 2025, from https://www.hrw.org/report/2017/03/02/license-kill/philippine-police-killings-dutertes-war-drugs

Jetter, I., & Molina, T. (2022). Persuasive agenda-setting: Rodrigo Duterte's inauguration speech and drugs in the Philippines. *Journal of Development Economics, 156*. DOI: 10.1016/j.jdeveco.2022.102843

Katila, P., Colfer, C. J. P., de Jong, W., Galloway, G., Pacheco, P., & Winkel, G. (Eds.). (2019). *Sustainable Development Goals: Their Impacts on Forests and People*. Cambridge University Press. DOI: 10.1017/9781108765015

Mehmood, S. (2023). *The rule of law approach for more resilient institutions: Judicial accountability and independence, and global economic activities.* Asian Development Bank Institute. DOI: 10.56506/FGHQ1674

National Economic and Development Authority. (2017). *Philippine Development Plan (PDP) 2017-2022.* https://pdp.neda.gov.ph/wp-content/uploads/2021/02/Pre-publication-copy-Updated-PD P-2017-2022.pdf

PDP NEDA Government Philippines. (2024). *Development Plans - Philippine Development Plan.* Philippine Development Plan. Retrieved July 19, 2024, from https://pdp.neda.gov.ph/devt-plans/

The 2019 Voluntary National Review of the Philippines. (2019). https://hlpf.un.org/sites/default/files/vnrs/2022/06222022 VNR of the Philippines.pdf

United Nations. (2023). *Goal 16 | Department of Economic and Social Affairs.* Sustainable Development Goals. Retrieved April, 2025, from https://sdgs.un.org/goals/goal16

Compilation of References

4th SDG Youth Summer Camp. (n.d.). *SDG targets overview resource*. United Nations Sustainable Development Knowledge Platform. https://sustainabledevelopment.un.org/sustainabledevelopment.un.org+10

Aayog, N. I. T. I. (2023). *SDG India Index 2023–24*. Government of India.

Abdi, H. (2007). Part (semi partial) and partial regression coefficients. In Salkind, N. (Ed.), *Encyclopedia of Measurement and Statistics* (pp. 736–740). Sage.

Abel, G. J., Brottrager, M., Cuaresma, J. C., & Muttarak, R. (2019). Climate, conflict and forced migration. *Global Environmental Change*, *54*, 239–249. DOI: 10.1016/j.gloenvcha.2018.12.003

Abrahams, D., & Carr, E. R. (2017). Understanding the connections between climate change and conflict: Contributions from geography and political ecology. *Current Climate Change Reports*, *3*(4), 233–242. DOI: 10.1007/s40641-017-0080-z

Abramo, G., D'Angelo, C. A., & Costa, F. D. (2018). A nation's foreign and domestic professors: Which have better research performance? (the italian case). *Higher Education*, *77*(5), 917–930. DOI: 10.1007/s10734-018-0310-x

Abu Bakar, R. (2024). Call for new survey to determine prevalence of sexual harassment in the workplace. *The Scoop*. https://thescoop.co/2024/03/16/more-women-report-workplace-harassment/?

Acheampong, P., & Gyasi, J. F. (2019). Teacher retention: A review of policies for motivating rural basic school teachers in ghana. *Asian Journal of Education and Training*, *5*(1), 86–92. DOI: 10.20448/journal.522.2019.51.86.92

Achmad, F., Prambudia, Y., & Rumanti, A. A. (2023). Sustainable Tourism Industry Development: A Collaborative Model of Open Innovation, Stakeholders, and Support System Facilities. *IEEE Access: Practical Innovations, Open Solutions*, *11*, 83343–83363. DOI: 10.1109/ACCESS.2023.3301574

Aderemi, T. A., Opele, A. M., Olanipekun, W. D., & Al-Faryan, M. A. S. (2023). A panel analysis of FDI inflows and poverty reduction in BRICS countries: An implication for the sustainable development goal one. *Transnational Corporations Review*, *15*(4), 35–41. DOI: 10.1016/j.tncr.2023.08.003

Ades, A., & Chua, H. B. (1997). Thy neighbor's curse: Regional instability and economic growth. *Journal of Economic Growth*, *2*(3), 279–304. https://www.jstor.org/stable/40215960. DOI: 10.1023/A:1009782809329

Aerni, P., Stavridou, M., & Schluep, I. (2021). *Transitioning to Decent Work and Economic Growth, Transitioning to Sustainability Series* (8th ed., Issue 8). MDPI, Basel. https://doi.org/https://doi.org/10.3390/books978-3-03897-779-7

Afiyah, I., & Muthoifin, . (2025). The Existence of Library Literacy in Indonesia in the Millennial Era: Strategies for Improving the Quality of Education to Support the Sustainable Development Goals (SDG's). *Solo International Collaboration and Publication of Social Sciences and Humanities*, *3*(01), 41–52. DOI: 10.61455/sicopus.v3i01.229

African Centre of Meteorological Applications for Development (ACMAD). (https://acmad.org/)

African Continental Free Trade Area (ACFTA). (https://au.int/en/african-continental-free-trade-area)

Aftab, W., Siddiqui, F. J., Tasic, H., Perveen, S., Siddiqi, S., & Bhutta, Z. A. (2020). Implementation of health and health-related Sustainable Development Goals: Progress, challenges and opportunities – A systematic literature review. *BMJ Global Health*, *5*(8), e002273. DOI: 10.1136/bmjgh-2019-002273 PMID: 32847825

Agbakwuru, V., Obidi, P. O., Salihu, O. S., & MaryJane, O. C. (2024). The role of renewable energy in achieving sustainable development goals. *International Journal of Engineering Research Updates*. https://doi.org/https://doi.org/10.53430/ijeru.2024.7.2.0046

Agnew, J. (1987). *Place and politics: The geographical mediation of state and society*. Allen & Unwin.

Ahmed, K. (2024). Cross-Border Cash Drain: Analyzing Economic Leakage from Brunei to Malaysian Borneo. *Available at SSRN* 5098543.

Ahyani, H., Mutmainah, N., Ahmad, M. Y., Lousada, S. A. N., Triswandani, Y., Rahman, E. T., & Utama, S. M. (2025). Enhancing the Legal Protection of Waqf Assets in Banjar City, West Java, Through Land Registration and Regulatory Awareness to Support the Achievement of Sustainable Development Goals (SDGs). *Journal of Lifestyle and SDGs Review*, *5*(2), e03989–e03989. DOI: 10.47172/2965-730X.SDGsReview.v5.n02.pe03989

Ajili, H., & Bouri, A. (2017). Comparative Study between IFRS and AAOIFI Disclosure Compliance: Evidence from Islamic Banks in Gulf Cooperation Council Countries. *Journal of Financial Reporting and Accounting*, *15*(3), 269–292. DOI: 10.1108/JFRA-03-2016-0023

Akinyetun, T. S. (2024). The proliferation of armed non-state actors in the Sahel: Drivers, ramifications and the way forward. *Journal of Central and Eastern European African Studies*, *2*(2), 36–58. DOI: 10.59569/jceeas.2022.2.2.60

Ako, R., Stewart, N., & Ekhator, E. O. (2016). Overcoming the (non)justiciable conundrum: The doctrine of harmonious construction and the interpretation of the right to a healthy environment in Nigeria. In Diver, A., & Miller, J. (Eds.), *Justiciability of human rights law in domestic jurisdictions* (pp. 123–141). Springer., DOI: 10.1007/978-3-319-24016-9_6

Al Azizah, U. S., Suryana, A. M., & Choirin, M. (2021). The impact of Zakat against Poverty Alleviation, Inequality & Economic Growth (Case Study of Indonesia). 5th International Conference of Zakat Proceedings.

Albhirat, M. M., Zulkiffli, S. N. A., Salleh, H. S., & Zaki, N. A. M. (2023). The Moderating Role of Social Capital in the Relationship Between Green Supply Chain Management and Sustainable Business Performance: Evidence from Jordanian SMEs. *International Journal of Sustainable Development and Planning*, *18*(6), 1733–1747. DOI: 10.18280/ijsdp.180609

Albu, C. N., Albu, N., & Fekete, S. (2010). The Context of the Possible IFRS for SMEs Implementation in Romania: An Exploratory Study. *Accounting and Management Information Systems*, *9*(1), 45–71.

Albu, C. N., Albu, N., Fekete, S., Girbina, M. M., Selimoglu, S. K., Kovacs, D. M., Lukacs, J., Mohl, G., Mullerova, L., Pasekova, M., Arsoy, A. P., Sipahi, B., & Strouhal, J. (2013). Implementation of IFRS for SMEs in Emerging Economies: Stakeholder Perceptions in the Czech Republic, Hungary, Romania and Turkey. *Journal of International Financial Management & Accounting*, *24*(2), 140–175. DOI: 10.1111/jifm.12008

Alfiani, T., & Akbar, N. (2020). Exploring Strategies to Enhance Zakat Role to Support Sustainable Development Goals (SDGs): ISM Delphi Approach. 4th International Conference of Zakat Proceedings.

Alger, J., & Dauvergne, P. (2020, 2). The Translocal Politics of Environmental Norm Diffusion (Vol. 14). https://doi.org/DOI: 10.1080/17524032.2019.1665567

Ali, H. (2025). Conflicting behavior between US and Afghanistan and Its Implications on Tribal Areas of Pakistan. *Pakistan Languages and Humanities Review*, *9*(1), 39–51.

Alkali, I., Alkali, M. Y., & Aliyu, A. A. (2017). Islamic Accounting Reporting and Economic Development: Nigerian Perspective. *Accounting*, *3*, 211–220. DOI: 10.5267/j.ac.2017.1.005

Alkhtani, S. (2012). The Relevance and Usefulness of IFRSs to Saudi Arabia. *Journal of Administrative and Economic Science*, *5*(2), 21–53.

Alkire, S., & Santos, M. E. (2014). Measuring acute poverty in the developing world: Robustness and scope of the multidimensional poverty index. *World Development*, *59*, 251–274. DOI: 10.1016/j.worlddev.2014.01.026

Allen, C., Metternicht, G., Wiedmann, T., & Pedercini, M. (2021, 4). Modelling national transformations to achieve the SDGs within planetary boundaries in small island developing states (Vol. 4). DOI: 10.1017/sus.2021.13

Allen, C., Metternicht, G., & Wiedmann, T. (2018). Initial progress in implementing the Sustainable Development Goals (SDGs): A review of evidence from countries. *Sustainability Science*, *13*(5), 1453–1467. DOI: 10.1007/s11625-018-0572-3

Almuaqel, I. A. (2024). Special Needs, Special Efforts: Learners With Intellectual Disabilities and Sustainable Development Goals (SDGs). *IEEE Transactions on Engineering Management*, *71*, 13469–13486. DOI: 10.1109/TEM.2022.3189318

Al-Omoush, K. S. (2020). The role of top management support and organizational capabilities in achieving e-business entrepreneurship. *Kybernetes*. Advance online publication. DOI: 10.1108/K-12-2019-0851

Aman, A. S. (2021). MyPay partners with PPZ-MAIWP to accelerate cashless zakat payments. https://www.nst.com.my/business/2021/12/750675/mypay-partners-ppz-maiwp-accelerate-cashless-zakat-payments

Amankwaah, H. (2023). The rule of law and armed conflict reconstruction implementation practices: A human rights-based analysis of the Rwandan experience. *Cogent Social Sciences*, *9*(1), 2171573. DOI: 10.1080/23311886.2023.2171573

Amin, N., Sharif, A., Shabbir, M. S., & Pan, Y. (2025). Evaluating the impact of natural resource rents, R&D expenditures, green finance and energy efficiency on carbon emissions in BRICS economies: Greening the path to carbon neutrality in the post-COP 27 era. *Technology in Society*, *81*, 102826. Advance online publication. DOI: 10.1016/j.techsoc.2025.102826

Amin, N., Sharif, A., Tayyab, M., & Pan, Y. (2024). Green Technological Advances and Resource Rents as Levers for Carbon Reduction in BRICS: Implications for SDGs 7, 8, 9, 12, and 13. *Sustainable Development*. Advance online publication. DOI: 10.1002/sd.3294

AMP. (2022). *Human rights in the Philippines*. https://amp.ngo/wp-content/uploads/2022/09/AMP-2022-Report-Human-Rights-in-the-P hilippines.pdf

Anamwathana, P. (2024). Thailand's Ageing Society and Young Thais' Changing Views and Expectations. In *ISEAS – YUSOF ISHAK INSTITUTE*. https://www.iseas.edu.sg/wp-content/uploads/2024/03/ISEAS_Perspective_2024_28.pdf

Anderson, B. (1983). *Imagined communities: Reflections on the origin and spread of nationalism*. Verso.

Anderson, T. (2020). *Online and distance education: Theory and practice*. Routledge.

Andonova, L. B., Betsill, M. M., & Bulkeley, H. (2009). Transnational climate governance. *Global Environmental Politics*, *9*(2), 52–73. DOI: 10.1162/glep.2009.9.2.52

Andriani, Sadewa, M. M., Rahmalita, Y., Budiman, M. A., & Nurhidayatati, N. (2021). Optimizing the role of zakat institution: building resilience against the risk of the covid-19 pandemic. 5th International conference of Zakat Proceedings.

Aniyie, I. A., & Omiunu, O. (2025). Paradiplomacy and the Governors' Forum: Rethinking Transboundary Water Governance in the Lake Chad Basin. *Water Alternatives*, *18*(1), 1–18.

Ankomah, W. S. (2020). Diverse classrooms: Social justice, equity, and diversity competencies for teacher candidates. *World Journal of Education*, *10*(2), 15. DOI: 10.5430/wje.v10n2p15

Annual impact review 2021-22. Education Development Trust. (n.d.). https://www.edt.org/research-and-insights/annual-impact-review-202122/

Anshari, M., Hamdan, M., Ahmad, N., & Ali, E. (2024). Public service delivery, artificial intelligence and the sustainable development goals: trends, evidence and complexities. *Journal of Science and Technology Policy Management*.

Anshari, M., & Hamdan, M. (2022). Enhancing e-government with a digital twin for innovation management. *Journal of Science and Technology Policy Management, 14*(6), 1055–1065. DOI: 10.1108/JSTPM-11-2021-0176

Anshari, M., & Hamdan, M. (2022a). Understanding knowledge management and upskilling in Fourth Industrial Revolution: Transformational shift and SECI model. *VINE Journal of Information and Knowledge Management Systems, 52*(3), 373–393. DOI: 10.1108/VJIKMS-09-2021-0203

Anshari, M., Hamdan, M., Ahmad, N., & Ali, E. (2025). Public service delivery, artificial intelligence and the sustainable development goals: Trends, evidence and complexities. *Journal of Science and Technology Policy Management, 16*(1), 163–181. DOI: 10.1108/JSTPM-07-2023-0123

Anshari, M., Hamdan, M., Ahmad, N., Ali, E., & Haidi, H. (2023). New knowledge creation and loss during COVID-19: Sustainable knowledge management's perspective. *International Journal of Learning and Intellectual Capital, 20*(6), 587–611. DOI: 10.1504/IJLIC.2023.134898

Anshari, M., Syafrudin, M., & Fitriyani, N. L. (2022). Fourth industrial revolution between knowledge management and digital humanities. *Information (Basel), 13*(6), 292. DOI: 10.3390/info13060292

Anshari, M., Syafrudin, M., Tan, A., Fitriyani, N. L., & Alas, Y. (2023a). Optimisation of knowledge management (KM) with machine learning (ML) Enabled. *Information (Basel), 14*(1), 35. DOI: 10.3390/info14010035

Appadurai, A. (1996). *Modernity at large: Cultural dimensions of globalization*. University of Minnesota Press.

Appio, F. P., Platania, F., & Hernandez, C. T. (2024). Pairing AI and Sustainability: Envisioning Entrepreneurial Initiatives for Virtuous Twin Paths. *IEEE Transactions on Engineering Management, 71*, 11669–11686. DOI: 10.1109/TEM.2024.3428913

Arifin, E. N., Ananta, A., Musa, S. F. P. D., & Hoon, C. Y. (2024). A pioneering study on measuring poverty in the hydrocarbon-rich state of Brunei Darussalam. *Asian Population Studies, 20*(2), 222–240. DOI: 10.1080/17441730.2023.2240108

Arnell, N. W., Lowe, J. A., Challinor, A. J., & Osborn, T. J. (2019). Global and regional impacts of climate change at different levels of global temperature increase. *Climatic Change, 155*(3), 377–391. DOI: 10.1007/s10584-019-02464-z

Arnup, J. L., & Bowles, T. (2016). Should i stay or should i go? resilience as a protective factor for teachers' intention to leave the teaching profession. *Australian Journal of Education, 60*(3), 229–244. DOI: 10.1177/0004944116667620

ASEAN (Association of South East Asian Nations). (https://asean.org/)

Asghar, M. M. (2023). Green Supply Chain Management and Organizational Performance in Automotive Industry: A Case Study from Pakistan. *South Asian Journal of Operations and Logistics*, *2*(2), 18–40. DOI: 10.57044/SAJOL.2023.2.2.2308

Asmalia, A., Kasri, R. A., & Ahsan, A. (2018). Exploring the Potential of *Zakah* for Supporting Realization of Sustainable Development Goals (SDGs) in Indonesia. *International Journal of Zakat: Special Issue on Zakat Conference,* 51-69.

Avrampou, A., Skouloudis, A., Iliopoulos, G., & Khan, N. (2019). Advancing the Sustainable Development Goals: Evidence from leading European banks. *Sustainable Development (Bradford)*, *27*(4), 743–757. Advance online publication. DOI: 10.1002/sd.1938

Azalie, I. A. (2025). Islamic Governmentality and Food Business Development: The Case of Brunei Halal Certification (2011- 2021). (PhD Thesis, University College London).

Azalie, I. A. N., & Yong, G. Y. (2022). *Heritage Industry as Conservation Strategy for Kampong Ayer*. Brunei-Museum Journal.

Azalie, I. A., & Samad, N. (2022). Brunei as a global halal hub: Production network and strategic relations in halal food production. *Journal of Contemporary Islamic Studies*, *8*(3), 1–22.

Azen, R., & Budescu, D. V. (2003). The dominance analysis approach for comparing predictors in multiple regression. *Psychological Methods*, *8*(2), 129–148. DOI: 10.1037/1082-989X.8.2.129 PMID: 12924811

Bahri, M. S., Herindar, E., & Rusydiana, A. S. (2021). Covid-19 and Efficiency in Zakat Institutions: Evidence from Indonesia. 5th International Conference of Zakat Proceedings.

Bahturidze, Z., Vasilyeva, N., Shahoud, Z., & Pogodin, S. (2024). Inclusion as a Driver for the Implementation of SDG-30 by the BRICS Countries. *Springer Geography*, (Part F2388), 135–147. DOI: 10.1007/978-3-031-50407-5_11

Bandh, S. A., Shafi, S., Peerzada, M., Rehman, T., Bashir, S., Wani, S. A., & Dar, R. (2021). Multidimensional analysis of global climate change: A review. *Environmental Science and Pollution Research International*, *28*(20), 24872–24888. DOI: 10.1007/s11356-021-13139-7

Banham, H., & He, Y. (2014). Exploring the Relationship between Accounting Professionals and Small and Medium Enterprises (SMEs). *Journal of Business & Economics Research, 12*(3), 209–214. DOI: 10.19030/jber.v12i3.8723

Bardal, K. G., Reinar, M. B., Lundberg, A. K., & Bjørkan, M. (2021). Factors facilitating the implementation of the Sustainable Development Goals in regional and local planning—Experiences from Norway. *Sustainability (Basel), 13*(8), 4282. DOI: 10.3390/su13084282

Barlette, Y., & Baillette, P. (2020). Big data analytics in turbulent contexts: Towards organizational change for enhanced agility. *Production Planning and Control, 0*(0), 1–18. DOI: 10.1080/09537287.2020.1810755

Baron, R. M., & Kenny, D. A. (1986). The moderator-mediator variable distinction in social psychological research: Conceptual, strategic, and statistical considerations. *Journal of Personality and Social Psychology, 51*(6), 1173–1182. DOI: 10.1037/0022-3514.51.6.1173 PMID: 3806354

Basilotta-Gómez-Pablos, V., Matarranz, M., Casado-Aranda, L., & Otto, A. (2022). Teachers' digital competencies in higher education: A systematic literature review. *International Journal of Educational Technology in Higher Education, 19*(1), 8. Advance online publication. DOI: 10.1186/s41239-021-00312-8

Basu, S., & Sen, S. (2023). Silenced voices: Unravelling India's dissent crisis through historical and contemporary analysis of free speech and suppression. *Information & Communications Technology Law, 33*(1), 42–65. DOI: 10.1080/13600834.2023.2249780

BAZNAS. (2016). BAZNAS Signed the MOU of SDGS Partnership in Indonesia. https://www.puskasbaznas.com/ijaz/index.php/journal/article/view/16/13

BAZNAS. (2017). BAPPENAS: Zakat Is Able To Cut The Poverty Chain. https://www.puskasbaznas.com/news/485-bappenas-zakat-is-able-to-cut-the-poverty-chain

BAZNAS. (2020). BAZNAS Lakukan Monitoring Evaluasi dan Kaji Dampak di Kelompok Balai Ternak Garut. https://baznas.go.id/pendistribusian/ekonomi/pemberdayaan-peternak/1602-baznas-lakukan-monitoring-evaluasi-dan-kaji-dampak-di-kelompok-balai-ternak-garut

BAZNAS. (2021a). BAZNAS Dampingi Mustahik Monitoring Populasi Ternak. https://baznas.go.id/berita_pendistribusian/baca/BAZNAS_Dampingi_Mustahik_Monitoring_Populasi_Ternak/6740

BAZNAS. (2021b). BAZNAS Monitoring Usaha Mitra Mustahik Program Kemitraan BAZNAS-UPZ Permata. https://baznas.go.id/pendistribusian/ekonomi/pembiayaan-mikro/6400-baznas-monitoring-usaha-mitra-mustahik-program-kemitraan-baznas-upz-permata

BAZNAS. (2021c). "Rakornas Dorong Penguatan BAZNAS Sebagai Pengelola Zakat Nasional". https://baznas.go.id/Press_Release/baca/%22Rakornas_Dorong_Penguatan_BAZNAS_Sebagai_Pengelola_Zakat_Nasional%22/749

Bejarano-Murillo, E. (2025). Digital Governance Across Continents: Cases From Diverse National Strategies. In *Data-Driven Governance Through AI, Digital Marketing, and the Privacy Interplay* (pp. 1-30). IGI Global Scientific Publishing.

Belsley, D. A., Kuh, E., & Welsch, R. E. (1980). *Regression diagnostics: Identifying influential data and sources of collinearity*. Wiley. DOI: 10.1002/0471725153

Bengtsson, M., Alfredsson, E., Cohen, M., Lorek, S., & Schroeder, P. (2018). Transforming systems of consumption and production for achieving the sustainable development goals: Moving beyond efficiency. *Sustainability Science*, *13*(6), 1533–1547. DOI: 10.1007/s11625-018-0582-1

Berk, E., Ayas, O., & Ülkü, M. A. (2023). Optimizing process-improvement efforts for supply chain operations under disruptions: New structural results. *Sustainability (Basel)*, *15*(17), 13117. DOI: 10.3390/su151713117

Berman, M., & DeFeo, D. J. (2023). Equitable compensation to attract and retain qualified teachers in high-need Alaska public schools. *Educational Policy*, *38*(5), 1139–1175. DOI: 10.1177/08959048231174883

Bernal, G. L., Abadía, L. K., Álvarez-Arango, L. E., & De Witte, K. (2023). Financial aid uncertainty and low-income students' higher education preferences. *Higher Education*, *87*(6), 1845–1863. DOI: 10.1007/s10734-023-01094-w

Bhaskaran, M. (2010). Economic Diversification in Brunei Darussalam. *CSPS Strategy and Policy Journal*, *1*, 1–12.

Bhattacharya, R. (2015). The way forward: Public policy to address regional imbalances in India, process and outcome. In Bhattacharya, R. (Ed.), *Regional development and public policy challenges in India* (pp. 309–322). Springer., DOI: 10.1007/978-81-322-2346-7_10

Bhushan, V. A. (2024). AI for Climate Change: Leveraging Machine Learning for Environmental Monitoring and Sustainability. *Journal of Informatics Education and Research*, *4*(3), 1526–4726.

Bianchi, O., & Putro, H. P. (2024). Artificial intelligence in Environmental Monitoring: Predicting and Managing climate change impacts. [ITALIC]. *International Transactions on Artificial Intelligence*, *3*(1), 85–96. DOI: 10.33050/italic.v3i1.650

Biddle, G. C., Callahan, C. M., Hong, H. A., & Knowles, R. L. (2013). *Do Adoptions of International Financial Reporting Standards Enhance Capital Investment Efficiency?* Working paper. 45 p.

Biermann, F., Kanie, N., & Kim, R. E. (2017). Global governance by goal-setting: The novel approach of the UN Sustainable Development Goals. *Current Opinion in Environmental Sustainability*, *26*, 26–31. DOI: 10.1016/j.cosust.2017.01.010

Biesbroek, R., Wright, S. J., Eguren, S. K., Bonotto, A., & Athanasiadis, I. N. (2022). Policy attention to climate change impacts, adaptation and vulnerability: A global assessment of National Communications (1994–2019). *Climate Policy*, *22*(1), 97–111. DOI: 10.1080/14693062.2021.2018986

Binder, M. (2009). Humanitarian crises and the international politics of selectivity. *Human Rights Review (Piscataway, N.J.)*, *10*(3), 327–348. DOI: 10.1007/s12142-009-0121-7

Birruntha. (April 20, 2022). PPZ-MAIWP, RHB Islamic forge strategic partnership. https://themalaysianreserve.com/2022/04/20/ppz-maiwp-rhb-islamic-forge-strategic-partnership/

Blanco, D. V., & Panao, R. A. L. (2022). The Sustainable Development Governance Initiatives of Quezon City, Philippines: An Analysis of Its Enablers and Barriers. *International Journal of Public Administration, 46*(12), 857-875. DOI: 10.1080/01900692.2022.2043366

Bocken, N. M. P., Short, S. W., Rana, P., & Evans, S. (2019). A literature and practice review to develop sustainable business model archetypes. *Journal of Cleaner Production*, *65*, 42–56. DOI: 10.1016/j.jclepro.2013.11.039

Bogdanova, I. (2022). *Unilateral sanctions in international law and the enforcement of human rights.* Brill | Nijhoff. https://doi.org/DOI: 10.1163/9789004507890

Bongolan, V. P., Soria, S. C., & Rivera, R. L. K. (2022). Sustainable Development Goal Target Interactions in the Philippines: a Two-Method Approach. *Philippine Journal of Science, 152*(3), 827-835. https://philjournalsci.dost.gov.ph/120-vol-152-no-3-june-2023/1838-sustainable-development-goal-target-interactions-in-the-philippines-a-two-method-approach

Boorse, D. F., & Jablonski, L. M. (2024). Strengthening partnerships for the Sustainable Development Goals: Engaging faith-based organizations and scientists in youth climate change learning and action. *Sustain Earth Reviews*, *7*(7), 7. Advance online publication. DOI: 10.1186/s42055-024-00077-8

Borker, D. R. (2012). Accounting, Culture and Emerging Economics: IFRS in the BRIC Countries. *Journal of Business & Economics Research*, *10*(5), 313–324. DOI: 10.19030/jber.v10i5.6983

Bowen, G. A. (2009). Document analysis as a qualitative research method. *Qualitative Research Journal*, *9*(2), 27–40. DOI: 10.3316/QRJ0902027

Boyadjieva, P. (2017). Invisible higher education: Higher education institutions from central and eastern europe in global rankings. *European Educational Research Journal*, *16*(5), 529–546. DOI: 10.1177/1474904116681016

BPMB. (2021). BPMB partners with MAIWP to hand out basic food assistance to the asnaf community. https://www.bpmb.com.my/ms_MY/news?p_p_id=56&p_p_lifecycle=0&p_p_state=maximized&p_p_mode=view&p_p_col_id=column-2&p_p_col_pos=2&p_p_col_count=3&_56_returnToFullPageURL=%2Fms_MY%2Fnews&_56_groupId=21475&_56_articleId=131028&_56_version=1.4&_56_articleTitle=BUSINESS+TODAY+-+BPMB+partners+with+MAIWP+to+hand+out+basic+food+assistance+to+the+asnaf+community

Brammer, S., & Walker, H. (2011). Sustainable procurement in the public sector: An international comparative study. *International Journal of Operations & Production Management*, *31*(4), 452–476. DOI: 10.1108/01443571111119551

Braun, V., & Clarke, V. (2019). Reflecting on reflexive thematic analysis. *Qualitative Research in Sport, Exercise and Health*, *11*(4), 589–597. DOI: 10.1080/2159676X.2019.1628806

Breusch, T. S., & Pagan, A. R. (1979). A simple test for heteroscedasticity and random coefficient variation. *Econometrica*, *47*(5), 1287–1294. DOI: 10.2307/1911963

Bruggeman, B., Hidding, K., Struyven, K., Pynoo, B., Garone, A., & Tondeur, J. (2022). Negotiating teacher educators' beliefs about blended learning: Using stimulated recall to explore design choices. *Australasian Journal of Educational Technology*, •••, 100–114. DOI: 10.14742/ajet.7175

Brunei Darussalam Department of Economic Planning and Statistics (DEPS). (2023). Brunei Darussalam. *Statistical Yearbook. Annuaire Statistique, 2022*. https://deps.mofe.gov.bn

Brunei Darussalam, S. D. G. https://www.sdgbrunei.gov.bn/

Brunei Times. (2016). *HM Consents to Adoption of Accounting Standards*. Retrieved on 1 March 2020, from https://btarchive.org/news/national/2016/10/27/hm-consents-adoption-accounting-standards

Brunnée, J. (2006). International environmental law: Rising to the challenge of common concern? [American Society of International Law]. *Proceedings of the Annual Meeting (American Association of University Teachers of Insurance)*, *100*, 307–310.

Budescu, D. V. (1993). Dominance analysis: A new approach to the problem of relative importance of predictors in multiple regression. *Psychological Bulletin*, *114*(3), 542–551. DOI: 10.1037/0033-2909.114.3.542

Budiana, M. (2024). Regional Organizations and Global Governance: The Case of ASEAN. *Journal of Student Collaboration Research*, *1*(2), 1–15.

Bukko, D., & Liu, K. (2021). Developing preservice teachers' equity consciousness and equity literacy. *Frontiers in Education*, *6*, 586708. Advance online publication. DOI: 10.3389/feduc.2021.586708

Bulfin, S., Henderson, M., Johnson, N. F., & Selwyn, N. (2014). Methodological capacity within the field of "educational technology" research: An initial investigation. *British Journal of Educational Technology*, *45*(3), 403–414. DOI: 10.1111/bjet.12145

Bunkše, E. V. (2012). Sensescapes: Or a Paradigm Shift from Words and Images to All Human Senses in Creating Feelings of Home in Landscapes. *Proceedings of the Latvia University of Agriculture: Landscape Architecture & Art*, *1*(1).

Bunting, P., Rosenqvist, A., Lucas, R., Rebelo, L. M., Hilarides, L., Thomas, N., Hardy, A., Itoh, T., Shimada, M., & Finlayson, C. M. (2018). The Global Mangrove Watch—A new 2010 global baseline of mangrove extent. *Remote Sensing (Basel)*, *10*(10), 1669. DOI: 10.3390/rs10101669

Byanyima, F., Edison, M., Kadengye, D. T., Arineitwe, S., Mijwil, M. M., Gaballa, M., & Turyasingura, B. (2025). Foreign direct investment and environmental challenges: A case study of Uganda with analytical perspective. In *Community Resilience and Climate Change Challenges: Pursuit of Sustainable Development Goals (SDGs)* (pp. 283–296). IGI Global Scientific Publishing.

Caiado, R., Filho, W., Quelhas, O., Nascimento, D., & Ávila, L. (2018). A literature-based review on potentials and constraints in the implementation of the sustainable development goals. *Journal of Cleaner Production*, *198*, 1276–1288. Advance online publication. DOI: 10.1016/j.jclepro.2018.07.102

Cai, W., & Li, G. (2022). The drivers of green innovation: The role of governmental support and environmental regulations. *Journal of Cleaner Production*, *233*, 117–130. DOI: 10.1016/j.jclepro.2022.05.023

Cai, Z., Liu, H., Huang, Q., & Liang, L. (2019). Developing organizational agility in product innovation: The roles of IT capability, KM capability, and innovative climate. *R & D Management*, *49*(4), 421–438. DOI: 10.1111/radm.12305

Callao, S., & Jarne, I. (2010). Have IFRS Affected Earnings Management in the European Union? *Accounting in Europe*, *2*(2), 159–189. DOI: 10.1080/17449480.2010.511896

Cambra-Fierro, J., López-Pérez, M. E., Perez, L., & Tejada-Tejada, M. (2025). Managing public-private partnerships for urban design and regeneration: Lessons learned from the Hermitage Museum Barcelona odyssey. *Cities (London, England)*, *158*, 105714. DOI: 10.1016/j.cities.2025.105714

Camphuijsen, M. K. (2020). Coping with performance expectations: Towards a deeper understanding of variation in school principals' responses to accountability demands. *Educational Assessment, Evaluation and Accountability*, *33*(3), 427–453. DOI: 10.1007/s11092-020-09344-6

Campos, N. F., & Kinoshita, Y. (2010). Structural reforms, financial liberalization, and foreign direct investment. *IMF Staff Papers*, *57*(2), 326–365. https://www.jstor.org/stable/25681049. DOI: 10.1057/imfsp.2009.17

Cantú Rivera, H. (2018). The UN Human Rights Council: Achievements and challenges in its first decade. In Oberleitner, G. (Ed.), *International human rights institutions, tribunals, and courts* (pp. 1–20). Springer., DOI: 10.1007/978-981-10-5206-4_3

Cao, M., Chang, L., Ma, S., Zhao, Z., Wu, K., Hu, X., Gu, Q., Lu, G., & Chen, M. (2022). Multi-Scenario Simulation of Land Use for Sustainable Development Goals. *IEEE Journal of Selected Topics in Applied Earth Observations and Remote Sensing*, *15*, 2119–2127. DOI: 10.1109/JSTARS.2022.3152904

Capkun, V., Collins, D., & Jeanjean, T. (2016). The Effect of IAS/IFRS Adoption on Earnings Management (Smoothing): A Closer Look at Competing Explanations. *Journal of Accounting and Public Policy*, *35*(4), 352–394. DOI: 10.1016/j.jaccpubpol.2016.04.002

Caponio, T., & Pettrachin, A. (2025). Neither multilevel governance nor battleground. Understanding the politics of immigrant integration in small and medium European localities. *Governance: An International Journal of Policy, Administration and Institutions*, *38*(1), e12833. DOI: 10.1111/gove.12833

Carter, C. R., & Rogers, D. S. (2008). A framework of sustainable supply chain management: Moving toward new theory. *International Journal of Physical Distribution & Logistics Management, 38*(5), 360–387. DOI: 10.1108/09600030810882816

Catala, B., Savall, T., & Chaves-Ávila, R. (2025, January). Social Economy and Public Private Partnerships: Analysis, Drivers, and Prospects from the Local Level for Territorial Development. In *Nonprofit Policy Forum* (No. 0). De Gruyter.

Cattaneo, M., Malighetti, P., & Paleari, S. (2018). The Italian brain drain: Cream and milk. *Higher Education, 77*(4), 603–622. DOI: 10.1007/s10734-018-0292-8

Central Pollution Control Board (CPCB). (2022). *Annual Report on Environmental Status in India.* Government of India.

Cepeda, J., & Arias-Pérez, J. (2019). Information technology capabilities and organizational agility: The mediating effects of open innovation capabilities. *Multinational Business Review, 27*(2), 198–216. DOI: 10.1108/MBR-11-2017-0088

Ceva, E., & Ferretti, M. P. (2021). Upholding public institutions in the midst of conflicts: The threat of political corruption. *Ethics & Global Politics, 14*(3), 1961379. DOI: 10.1080/16544951.2021.1961379

Chalarhena, N., & Hendayani, R. (2022). Pengaruh green supply chain management terhadap performa ekonomi dan organisasi pada UMKM industri makanan di Wonosobo Jawa Tengah. *Fair Value: Jurnal Ilmiah Akuntansi Dan Keuangan, 5*(3), 1696–1713. DOI: 10.32670/fairvalue.v5i3.2420

Chan, Y. E., Denford, J. S., & Wang, J. (jane). (2019). The Co-Evolution of IT, Knowledge, and Agility in Micro and Small Enterprises. *Journal of Information and Knowledge Management.* DOI: 10.1142/S0219649219500278

Chaney, P. (2022). Exploring civil society perspectives on the situation of human rights defenders in the Commonwealth of Independent States. *Central Asian Survey, 42*(2), 293–318. DOI: 10.1080/02634937.2022.2113034

Chankseliani, M., & McCowan, T. (2020). Higher education and the sustainable development goals. *Higher Education, 81*(1), 1–8. DOI: 10.1007/s10734-020-00652-w

Charitonos, K., . . . Aristorenas, A. (2022). Researching communicative approaches to English language teaching using peer ethnographic methods in Jordan's refugee settings. https://inee.org/sites/default/files/resources/TiCC%20Case%20Studies%20v2%20-%20TPD%20v1.0%20EN%20LowRes.pdf

Chatterjee, S. (2023). Climate refugees: Arriving at a possible framework. In P. Singh, B. Ao, & A. Yadav (Eds.), *Global climate change and environmental refugees* (pp. Chapter 10). Springer, Cham. https://doi.org/DOI: 10.1007/978-3-031-24833-7_10

Chen, A., Li, W., & Fu, W. (2024). Unleashing digital superheroes: Unravelling the empathy factor in digital competence and online teacher autonomy support. *British Journal of Educational Technology*, *55*(4), 1790–1810. DOI: 10.1111/bjet.13433

Chen, C., Young, D., & Zhuang, Z. (2013). Externalities of Mandatory IFRS Adoption: Evidence from Cross-Border Spillover Effects of Financial Information on Investment Efficiency. *The Accounting Review*, *88*(3), 881–914. DOI: 10.2308/accr-50384

Chen, L., Zhao, X., Tang, O., Price, L., Zhang, S., & Zhu, W. (2017). Supply chain collaboration for sustainability: A literature review and future research agenda. *International Journal of Production Economics*, *194*, 73–87. DOI: 10.1016/j.ijpe.2017.04.005

Chen, S., & Yi, Y. (2021). The manufacturer decision analysis for corporate social responsibility under government subsidy. *Mathematical Problems in Engineering*, Article ID 6617625, 1–15. Advance online publication. DOI: 10.1155/2021/6617625

Chen, Y. S., Lai, S. B., & Wen, C. T. (2006). The influence of green innovation performance on corporate advantage in Taiwan. *Journal of Business Ethics*, *67*(4), 331–339. DOI: 10.1007/s10551-006-9025-5

Chen, Y., Cao, L., Guo, L., & Cheng, J. (2022). Driving is believing: Using telepresence robots to access makerspace for teachers in rural areas. *British Journal of Educational Technology*, *53*(6), 1956–1975. DOI: 10.1111/bjet.13225

Chen, Y., Wang, Y., & Zhou, L. (2021). Green innovation and sustainable development: The moderating effect of environmental regulatio. *Technological Forecasting and Social Change*, *166*, 120–130. DOI: 10.1016/j.techfore.2021.120653

Chesalina, O. (2020). Platform work as a new form of employment: Implications for labour and social law. In Wratny, J., & Ludera-Ruszel, A. (Eds.), *New forms of employment. Prekarisierung und soziale Entkopplung – transdisziplinäre Studien* (pp. 153–167). Springer VS., DOI: 10.1007/978-3-658-28511-1_9

Chesler, A. (2024). Environmental displacement and political instability: Evidence from Africa. *Journal of Peace Research*, *0*(0). Advance online publication. DOI: 10.1177/00223433241274979

Chewpreecha, U., Pollitt, H., & Mercure, J. F. (2023). Combining short-term economic recovery with long-term sustainability. In Arestis, P., & Sawyer, M. (Eds.), *Prospects and policies for global sustainable recovery* (pp. 215–263). Palgrave Macmillan., DOI: 10.1007/978-3-031-19256-2_6

Chin, S. W. L., Hassan, N. H., & Yong, G. Y. (2023). The new ecotourists of the 21st century: Brunei as a case study. *Cogent Social Sciences*, *9*(1), 2191444. DOI: 10.1080/23311886.2023.2191444

Chishti, M. Z., Salam, M., Xaisongkham, S., & Du, A. M. (2025). Influence of green ICT and socioeconomic factors on sustainable development: Evidence from Chinese provinces. *Research in International Business and Finance*, *73*, 102624. DOI: 10.1016/j.ribaf.2024.102624

Choirunisa, U., Permadia, S., & Gunawan, A. R. (2020). The Role of Zakat in Reducing the Poverty Gap: Case Study in Desa Berdaya Rumah Zakat. 4th International Conference of Zakat Proceedings.

Chopra, T., & Isser, D. (2012). Access to justice and legal pluralism in fragile states: The case of women's rights. *Hague Journal on the Rule of Law*, *4*(2), 337–358. DOI: 10.1017/S187640451200019X

Cohen, J. (1988). *Statistical power analysis for the behavioral sciences* (2nd ed.). Lawrence Erlbaum Associates.

Compen, B., Witte, K. D., & Schelfhout, W. (2020). The impact of teacher engagement in an interactive webinar series on the effectiveness of financial literacy education. *British Journal of Educational Technology*, *52*(1), 411–425. DOI: 10.1111/bjet.13013

Conroy, A., & Blackie, M. (2006). Breaking out of economic isolation and the poverty trap. In *Poverty, AIDS and Hunger*. Palgrave Macmillan, London. https://doi.org/ DOI: 10.1057/9780230627703_11

Cook, R. D. (1977). Detection of influential observation in linear regression. *Technometrics*, *19*(1), 15–18. DOI: 10.1080/00401706.1977.10489493

Costanza, R., de Groot, R., Braat, L., Kubiszewski, I., Fioramonti, L., Sutton, P., & Grasso, M. (2020). Twenty years of ecosystem services: How far have we come and how far do we still need to go? *Ecosystem Services*, *28*, 1–16. DOI: 10.1016/j.ecoser.2017.09.008

Council on Foreign Relations. (2016). *Human Rights and Duterte's War on Drugs*. https://www.cfr.org/interview/human-rights-and-dutertes-war-drugs

Cox, R., & Pezzullo, P. C. (2018). *Environmental communication and the public sphere* (5th ed.). SAGE Publications.

Crick, C., & Chew, E. K. (2017). Business processes in the agile organisation: A socio-technical perspective. *Software & Systems Modeling*, *16*(3), 631–648. DOI: 10.1007/s10270-015-0506-9

Crosling, G., Nair, M., & Vaithilingam, S. (2014). A creative learning ecosystem, quality of education and innovative capacity: A perspective from higher education. *Studies in Higher Education*, *40*(7), 1147–1163. DOI: 10.1080/03075079.2014.881342

D'Adamo, I., Gastaldi, M., Imbriani, C., & Morone, P. (2021). Assessing regional performance for the Sustainable Development Goals in Italy. *Scientific Reports*, *11*(1), 22069. DOI: 10.1038/s41598-021-03635-8 PMID: 34916565

Dai, C., Ke, F., Dai, Z., & Pachman, M. (2022). Improving teaching practices via virtual reality-supported simulation-based learning: Scenario design and the duration of implementation. *British Journal of Educational Technology*, *54*(4), 836–856. DOI: 10.1111/bjet.13296

Dalal, M., Archambault, L., & Shelton, C. (2020). Fostering the growth of TPACK among international teachers of developing nations through a cultural exchange program. *Australasian Journal of Educational Technology*, •••, 43–56. DOI: 10.14742/ajet.5964

Daly, H. E. (1996). *Beyond Growth: The Economics of Sustainable Development*. Beacon Press.

Damane, M., & Ho, S. Y. (2024). The impact of financial inclusion on financial stability: Review of theories and international evidence. *Development Studies Research*, *11*(1), 2373459. Advance online publication. DOI: 10.1080/21665095.2024.2373459

Dangelico, R. M., & Pujari, D. (2019). Mainstreaming green product innovation: Why and how companies integrate environmental sustainability. *Journal of Business Ethics*, *95*(3), 471–486. DOI: 10.1007/s10551-010-0434-0

Darling-Hammond, L. (2023). Response to how teacher education matters. *Journal of Teacher Education*, *74*(2), 157–159. DOI: 10.1177/00224871231160378

Darvishmotevali, M., Altinay, L., & Köseoglu, M. A. (2020). The link between environmental uncertainty, organizational agility, and organizational creativity in the hotel industry. *International Journal of Hospitality Management, 87(August*. DOI: 10.1016/j.ijhm.2020.102499 de Pablos, P. O. (2002). Evidence of intellectual capital measurement from Asia, Europe and the Middle East. *Journal of Intellectual Capital.*

de Pablos, P. O., & Lytras, M. D. (2008). Competencies and human resource management: Implications for organizational competitive advantage. *Journal of Knowledge Management.*

de Souza Barbosa, A., da Silva, M. C. B. C., da Silva, L. B., Morioka, S. N., & de Souza, V. F. (2023). Integration of environmental, social, and governance (ESG) criteria: Their impacts on corporate sustainability performance. *Humanities & Social Sciences Communications, 10*(1), 410. DOI: 10.1057/s41599-023-01919-0

Dean, M. (2015). Foucault must not be defended. *History and Theory, 54*(3), 389–403. DOI: 10.1111/hith.10767

Del Bonifro, F., Gabbrielli, M., Lisanti, G., & Zingaro, S. P. (2020). Student dropout prediction. *Lecture Notes in Computer Science, 12163*, 129–140. DOI: 10.1007/978-3-030-52237-7_11

den Boer, M. (Ed.). (2018). *Comparative policing from a legal perspective*. Edward Elgar Publishing., DOI: 10.4337/9781785369117

Deng, R., & Gao, Y. (2024). Practices and barriers to promoting educational equity for migrant children in china from the perspective of a local government. *SAGE Open, 14*(3), 21582440241269979. Advance online publication. DOI: 10.1177/21582440241269979

Department of International Organizations Ministry of Foreign Affairs. (2021). Thailand's Voluntary National Review On The Implementation Of The 2030 Agenda For. *Sustainable Development*, •••, 2021.

Destiana, A. R., & Hamzah, M. Z. (2021). Fiscal policy design in Malaysia and Indonesia about Zakat and Taxes. 5th International Conference of Zakat Proceedings.

Dhar, S., & Dhar, A. (2025). Integrating Machine Learning Into Wildlife and Forest Conservation: Ethical Considerations and Sustainable Practices. In *Machine Learning and Internet of Things in Fire Ecology* (pp. 129-156). IGI Global Scientific Publishing.

Dhar, B. K., Sarkar, S. M., & Ayittey, F. K. (2022). Impact of social responsibility disclosure between implementation of green accounting and sustainable development: A study on heavily polluting companies in Bangladesh. *Corporate Social Responsibility and Environmental Management, 29*(1), 71–78. DOI: 10.1002/csr.2174

Dhyani, S., Shukla, J., Kadaverugu, R., Dasgupta, R., Panda, M., Kundu, S. K., Santhanam, H., Pujari, P. R., Kumar, P., & Hashimoto, S. (2023). Participatory stakeholder assessment for drivers of mangrove loss to prioritize evidence-based conservation and restoration in Bhitarkanika and Mahanadi Delta, India. *Sustainability, 15*(2), 963. DOI: 10.3390/su15020963

Di Baldassarre, G., Sivapalan, M., Rusca, M., Cudennec, C., Garcia, M., Kreibich, H., Konar, M., Mondino, E., Mård, J., Pande, S., Sanderson, M. R., Tian, F., Viglione, A., Wei, J., Wei, Y., Yu, D. J., Srinivasan, V., & Blöschl, G. (2019). Sociohydrology: Scientific challenges in addressing the sustainable development goals. *Water Resources Research, 55*(8), 6327–6355. DOI: 10.1029/2018WR023901 PMID: 32742038

Di Gregorio, M., Massarella, K., Schroeder, H., Brockhaus, M., & Pham, T. T. (2020). Building authority and legitimacy in transnational climate change governance: Evidence from the Governors' Climate and Forests Task Force. *Global Environmental Change, 64*, 102126. DOI: 10.1016/j.gloenvcha.2020.102126

Dienelt, A. (2022). The environment and international environmental law. In *Armed conflicts and the environment* (pp. 199–256). Springer., DOI: 10.1007/978-3-030-99339-9_4

Dinh, L. Q., Oanh, T. T. K., & Ha, N. T. H. (2024). Financial stability and sustainable development: Perspectives from fiscal and monetary policy. *International Journal of Finance & Economics*, 1–18. DOI: 10.1002/ijfe.2981

Dirikgil, N. (2023). Addressing the prevention of internal displacement: The right not to be arbitrarily displaced. *Journal of International Migration and Integration, 24*(1), 113–138. DOI: 10.1007/s12134-022-00935-4

Distefano, T., Isaza, A. S., Morlin, G. S., Carmona, V. P., Palacio, C. V., & Arango-Aramburo, S. (2025). The roads towards complex water governance: The Colombian case study. *World Development, 188*, 106874. DOI: 10.1016/j.worlddev.2024.106874

Dixon, T., Thompson, B., & McAllister, P. (2002). *The Value of ICT for SMEs in the UK: A Critical Literature Review*. A Report for Small Business Service Research Programme. Small Business Service, Reading: College of Estate Management.

Drèze, J., & Sen, A. (2017). *An Uncertain Glory: India and its Contradictions*. Princeton University Press.

Dubey, R., Gunasekaran, A., Childe, S. J., Papadopoulos, T., & Helo, P. (2019). Supplier relationship management for circular economy: Influence of external pressures and top management commitment. *Management Decision, 57*(4), 767–790. DOI: 10.1108/MD-04-2018-0396

Durbin, J., & Watson, G. S. (1951). Testing for serial correlation in least squares regression. II. *Biometrika*, *38*(1/2), 159–177. DOI: 10.2307/2332325 PMID: 14848121

Easterly, W. (2015). "The SDGs Should Stand for Senseless, Dreamy, Garbled", Foreign Policy. https://foreignpolicy.com/2015/09/28/the-sdgs-are-utopian-and-worthless-mdgs-development-rise-of-the-rest/

Ebekozien, A., Aigbavboa, C. O., Samsurijan, M. S., Radin Firdaus, R. B., & Rohayati, M. I. (2025). Expanded corporate social responsibility framework: Companies' role in improving higher education institutions infrastructure to Sustainable Development Goal 4. *Journal of Facilities Management*, *23*(1), 1–18. DOI: 10.1108/JFM-11-2022-0120

Economic Community of West African States (ECOWAS). (https://au.int/en/recs/ecowas)

Edralin, D. M., & Pastrana, R. (2022). Advancing the pursuit of the United Nations Sustainable Development Goals: Initiatives of selected publicly listed companies in the Philippines. *Bedan Research Journal*, *7*(1), 1–47. DOI: 10.58870/berj.v7i1.31

Egger, P., Syropoulos, C., & Yotov, Y. V. (2024). Analyzing the effects of economic sanctions: Recent theory, data, and quantification. *Review of International Economics*, *32*(1), 1–11. DOI: 10.1111/roie.12724

El Moll, A. (2023). Water resources and climate change: Regional, national and international perspective. In Prasad, M. N. V., & Smol, M. (Eds.), *Sustainable and circular management of resources and waste towards a green deal* (pp. 309–336). Elsevier., DOI: 10.1016/B978-0-323-95278-1.00010-3

Elkington, J. (2019). 25 Years Ago I Coined the Phrase "Triple Bottom Line". Here's Why It's Time to Rethink It. *Harvard Business Review*. https://hbr.org/2019/06/25-years-ago-i-coined-the-phrase-triple-bottom-line-heres-why-im-giving-up-on-it

Emerson, K., & Nabatchi, T. (2015). *Collaborative governance regimes*. Georgetown University Press.

Enayati, M., Lukambagire, I., & Manianga, A. (2024). Navigating sustainability and resilience: A collective case study of four Indian communities. *Environment, Development and Sustainability*. Advance online publication. DOI: 10.1007/s10668-024-05445-0

Eng, R., & Lim, S. (2025). The Influence of the Informal Economy on the Growth Rate of Real GDP within the Association of Southeast Asian Nations. *International Journal of Economics and Financial Issues*, *15*(3), 59–65. DOI: 10.32479/ijefi.18312

Epstein, B. J., & Jermakowicz, E. K. (2010). *Interpretation and Application of International Financial Reporting Standards*. John Wiley & Sons, Inc.

Escano-Arias, E. A., Abarquez, R. A. D., Cruz, R. V., Espeleta, R., Ong, M. M., Loreche, A. M., Pepito, V. C. F., Gomez, V., & Dayrit, M. M. (2025). Strengthening local health systems and governance for Universal Health Coverage: Experiences and lessons from the COVID-19 pandemic response in Quezon City, Philippines. *Health Policy and Planning*, *40*(3), czaf002. DOI: 10.1093/heapol/czaf002 PMID: 39801295

Escobar, A. (2015). Degrowth, postdevelopment, and transitions: A preliminary conversation. *Sustainability Science*, *10*(3), 451–462. DOI: 10.1007/s11625-015-0297-5

Escobar, A. (2018). *Designs for the Pluriverse: Radical Interdependence, Autonomy, and the Making of Worlds*. Duke University Press., DOI: 10.1215/9780822371816

ESCWA. (2017). *Intergenerational Justice: Meeting the Needs of the Future Arab Generations*. United Nations.

Ezugwu, A. E.-S. (2024). Metaheuristic Optimization for Sustainable Unrelated Parallel Machine Scheduling: A Concise Overview With a Proof-of-Concept Study. *IEEE Access : Practical Innovations, Open Solutions*, *12*, 3386–3416. DOI: 10.1109/ACCESS.2023.3347047

FAO. (2021). *Empowering farmers through sustainable practices*. Food and Agriculture Organization.

Fatima, T., & Elbanna, S. (2023). Corporate social responsibility (CSR) implementation: A review and a research agenda towards an integrative framework. *Journal of Business Ethics*, *183*(1), 105–121. DOI: 10.1007/s10551-022-05047-8

Faul, F., Erdfelder, E., Buchner, A., & Lang, A. G. (2009). Statistical power analyses using G*Power 3.1: Tests for correlation and regression analyses.* Behavior Research Methods, 41*(4), 1149-1160.

Federičová, M. (2020). Teacher turnover: What can we learn from europe? *European Journal of Education*, *56*(1), 102–116. DOI: 10.1111/ejed.12429

Feiock, R. C. (2007). Rational choice and regional governance. *Journal of Urban Affairs*, *29*(1), 47–63. DOI: 10.1111/j.1467-9906.2007.00322.x

Fernández Ugalde, R. (2024). Media articulation and the stabilisation of teacher policies in Chile. *Critical Studies in Education*, •••, 1–20. DOI: 10.1080/17508487.2024.2428291

Filho, W., Wolf, F., Lange Salvia, A., Beynaghi, A., Shulla, K., Kovaleva, M., & Vasconcelos, C. R. P. (2020). Heading towards an unsustainable world: Some of the implications of not achieving the SDGs. *Discover Sustainability*, *1*(1), 2. DOI: 10.1007/s43621-020-00002-x PMID: 38624553

Florian, L., & Camedda, D. (2019). Enhancing teacher education for inclusion. *European Journal of Teacher Education*, *43*(1), 4–8. DOI: 10.1080/02619768.2020.1707579

Florini, A., & Pauli, M. (2018). Collaborative governance for the sustainable development goals. *Asia & the Pacific Policy Studies*, *5*(3), 583–598. DOI: 10.1002/app5.252

Foley, J. A., DeFries, R., Asner, G. P., Barford, C., Bonan, G., Carpenter, S. R., & Snyder, P. K. (2011). Solutions for a cultivated planet. *Nature*, *478*(7369), 337–342. DOI: 10.1038/nature10452 PMID: 21993620

Foroudi, P., Marvi, R., Cuomo, M. T., & D'Amato, A. (2025). Sustainable Development Goals in a regional context: Conceptualising, measuring and managing residents' perceptions. *Regional Studies*, *59*(1), 93–109. DOI: 10.1080/00343404.2024.2373871

Foster, M., Mark, S. L., & Baize, J. (2023). Introduction special issue: educational equity: cultural and ethnic diversity in schools. *Education Sciences*, *14*(1), 38. DOI: 10.3390/educsci14010038

Fraisl, D., Campbell, J., See, L., Wehn, U., Wardlaw, J., Gold, M., Moorthy, I., Arias, R., Piera, J., Oliver, J. L., Masó, J., Penker, M., & Fritz, S. (2020). Mapping citizen science contributions to the UN sustainable development goals. *Sustainability Science*, *15*(6), 1735–1751. DOI: 10.1007/s11625-020-00833-7

Freeman, R. E., Harrison, J. S., Wicks, A. C., Parmar, B. L., & De Colle, S. (2010). *Stakeholder theory: The state of the art*. Cambridge University Press. DOI: 10.1017/CBO9780511815768

French, D., & Kotzé, L. (2018). *Sustainable Development Goals: Law, Theory and Implementation*. Edward Elgar Publishing., DOI: 10.4337/9781786438768

Friess, D. A., Rogers, K., Lovelock, C. E., Krauss, K. W., Hamilton, S. E., Lee, S. Y., Lucas, R., Primavera, J., Rajkaran, A., & Shi, S. (2019). The state of the world's mangrove forests: Past, present, and future. *Annual Review of Environment and Resources*, *44*(1), 89–115. DOI: 10.1146/annurev-environ-101718-033302

Friess, D. A., Yando, E. S., Abuchahla, G. M., Adams, J. B., Cannicci, S., Canty, S. W., Cavanaugh, K. C., Connolly, R. M., Cormier, N., Dahdouh-Guebas, F., Diele, K., Feller, I. C., Fratini, S., Jennerjahn, T. C., Lee, S. Y., Ogurcak, D. E., Ouyang, X., Rogers, K., Rowntree, J. K., & Wee, A. K. S. (2020). Mangroves give cause for conservation optimism, for now. *Current Biology, 30*(4), R153–R154. DOI: 10.1016/j.cub.2019.12.054 PMID: 32097637

Fu, B., Wang, S., Zhang, J., Hou, Z., & Li, J. (2019). Unravelling the complexity in achieving the 17 sustainable-development goals. *National Science Review, 6*(3), 386–388. DOI: 10.1093/nsr/nwz038

Fukuda-Parr, S. (2019). Keeping out extreme inequality from the SDG agenda: The politics of indicators. *Global Policy, 10*(S1), 61–69. DOI: 10.1111/1758-5899.12602

Fuller, E. J. (2013). Shaky methods, shaky motives. *Journal of Teacher Education, 65*(1), 63–77. DOI: 10.1177/0022487113503872

Fürst, D. (2001). Regional governance: A new paradigm in the regional sciences? *Raumforschung und Raumordnung, 59*, 370–380. DOI: 10.1007/BF03183038

Furukawa, T., Shirakawa, N., & Okuwada, K. (2012). An empirical study of graduate student mobility underpinning research universities. *Higher Education, 66*(1), 17–37. DOI: 10.1007/s10734-012-9586-4

Gaikhorst, L., Beishuizen, J., Zijlstra, B. J. H., & Volman, M. (2014). Contribution of a professional development programme to the quality and retention of teachers in an urban environment. *European Journal of Teacher Education, 38*(1), 41–57. DOI: 10.1080/02619768.2014.902439

Gaikhorst, L., Post, J., März, V., & Soeterik, I. (2019). Teacher preparation for urban teaching: A multiple case study of three primary teacher education programmes. *European Journal of Teacher Education, 43*(3), 301–317. DOI: 10.1080/02619768.2019.1695772

Galileo, G. (1610). *Sidereus nuncius*. Thomas Baglioni. DOI: 10.5479/sil.95438.39088015628597

Gao, C., Khalid, S., & Tadesse, E. (2023). Understanding school enrollment in the free education era: Roadblocks to meeting the sustainability development goal. *European Journal of Education, 59*(2), e12600. Advance online publication. DOI: 10.1111/ejed.12600

Gao, P., Zhang, J., Gong, Y., & Li, H. (2020). Effects of technical IT capabilities on organizational agility: The moderating role of IT business spanning capability. *Industrial Management & Data Systems*, *120*(5), 941–961. DOI: 10.1108/IMDS-08-2019-0433

Garg, S., Mittal, S., & Garg, A. (2024). Navigating the path for economic resilience and sustainable economic growth: Empirical insight from G-20 nations. *Sustainable Futures : An Applied Journal of Technology, Environment and Society*, *8*, 100338. DOI: 10.1016/j.sftr.2024.100338

Garretsen, H., Kitson, M., & Yang, C. (2025). Global forces and local impacts: Megatrends in regional development. *Cambridge Journal of Regions, Economy and Society*, *18*(1), 1–16. DOI: 10.1093/cjres/rsae047

Gavin, M. C., McCarter, J., Mead, A., Berkes, F., Stepp, J. R., Peterson, D., & Tang, R. (2018). Defining biocultural approaches to conservation. *Trends in Ecology & Evolution*, *33*(2), 140–149. PMID: 25622889

Gavsker, K. K. (2021). Regional environmental governance: Analytical framework for resource conservation in the Eastern Ghats, India. In Thakur, B., Thakur, R. R., Chattopadhyay, S., & Abhay, R. K. (Eds.), *Resource management, sustainable development and governance* (pp. 179–191). Springer., DOI: 10.1007/978-3-030-85839-1_11

GEF. (2021). *Expanded constituency workshops: Bridging global and local knowledge*. Global Environmental Facility.

Ge, J., & Rios, J. (2025). The effects of corruption in natural resources on the achievement of sustainable development goals in Bolivia and Peru. *SN Social Sciences*, *5*(3), 23. DOI: 10.1007/s43545-025-01056-2

Gereffi, G., Humphrey, J., & Sturgeon, T. (2001). Introduction: Globalisation, value chains and development. *IDS Bulletin*, *32*(3), 1–8. DOI: 10.1111/j.1759-5436.2001.mp32003001.x

Ghasemi, A., & Zahediasl, S. (2012). Normality tests for statistical analysis: A guide for non-statisticians. *International Journal of Endocrinology and Metabolism*, *10*(2), 486–489. DOI: 10.5812/ijem.3505 PMID: 23843808

Giangrande, N., White, R., East, M., Jackson, R., Clarke, T., Saloff Coste, M., & Penha-Lopes, G. (2019). A competency framework to assess and activate education for sustainable development: Addressing the UN Sustainable Development Goals 4.7 challenge. *Sustainability (Basel)*, *11*(10), 2832. Advance online publication. DOI: 10.3390/su11102832

Gilbert, J., Macpherson, E., Jones, E., & Dehm, J. (2023). The rights of nature as a legal response to the global environmental crisis? A critical review of international law's 'greening' agenda. In D. Dam-de Jong & F. Amtenbrink (Eds.), *Netherlands Yearbook of International Law 2021* (Vol. 52). T.M.C. Asser Press. https://doi.org/ DOI: 10.1007/978-94-6265-587-4_3

Gisselquist, R. M. (2018). Legal empowerment and group-based inequality. *The Journal of Development Studies*, *55*(3), 333–347. DOI: 10.1080/00220388.2018.1451636

Global Alliance to Monitor Learning (GAML). (n.d.). *Regional framework for Sustainable Development Goals (SDGs)*. UNESCO Institute for Statistics. https://gaml.uis.unesco.org/

Góes, C., & Bekkers, E. (2023). The impact of geopolitical conflicts on trade, growth, and innovation. *arXiv*. https://arxiv.org/abs/2203.12173

Gogoi, B., & Sarmah, J. K. (2024). Climate change and regional cooperation in South-East Asian countries. *India Quarterly a Journal of International Affairs*, *80*(2), 252–268. DOI: 10.1177/09749284241241597

Goldberg, L., Lagomasino, D., Thomas, N., & Fatoyinbo, T. (2020). Global declines in human-driven mangrove loss. *Global Change Biology*, *26*(10), 5844–5855. DOI: 10.1111/gcb.15275 PMID: 32654309

Golini, R., & Boffelli, A. (2018). The role of manufacturing within industrial districts. *Local Clusters in Global Value Chains, October 2017*, 193–212. DOI: 10.4324/9781315182049-11

Gong, X., & Huybers, T. (2015). Chinese students and higher education destinations: Findings from a choice experiment. *Australian Journal of Education*, *59*(2), 196–218. DOI: 10.1177/0004944115584482

Goodwin, A. L., Smith, L., Souto-Manning, M., Cheruvu, R., Tan, M. Y., Reed, R., & Taveras, L. (2014). What should teacher educators know and be able to do? perspectives from practicing teacher educators. *Journal of Teacher Education*, *65*(4), 284–302. DOI: 10.1177/0022487114535266

Gordon, L. A., Loeb, M. P., & Zhu, W. (2012). The Impact of IFRS Adoption on Foreign Direct Investment. *Journal of Accounting and Public Policy*, *31*(4), 374–398. DOI: 10.1016/j.jaccpubpol.2012.06.001

Gould, M., & Rablen, M. D. (2017). Reform of the United Nations Security Council: Equity and efficiency. *Public Choice*, *173*(1-2), 145–168. DOI: 10.1007/s11127-017-0468-2

Govindan, K., Khodaverdi, R., & Jafarian, A. (2021). A fuzzy multi criteria approach for measuring sustainability performance of a supplier based on triple bottom line approach. *Journal of Cleaner Production*, *240*, 118–129. DOI: 10.1016/j.jclepro.2021.118129

Govuzela, S., & Mafini, C. (2019). Organisational agility, business best practices and the performance of small to medium enterprises in South Africa. *South African Journal of Business Management*, *50*(1). Advance online publication. DOI: 10.4102/sajbm.v50i1.1417

Griggs, D., Smith, M., & Rockström, J.. (2014). An integrated framework for sustainable development goals. *Ecology and Society*, *19*(4), 49. DOI: 10.5751/ES-07082-190449

Grissom, J. A., & Strunk, K. O. (2011). How should school districts shape teacher salary schedules? linking school performance to pay structure in traditional compensation schemes. *Educational Policy*, *26*(5), 663–695. DOI: 10.1177/0895904811417583

Guang-Wen, Z., Murshed, M., Siddik, A. B., Alam, M. S., Balsalobre-Lorente, D., & Mahmood, H. (2023). Achieving the objectives of the 2030 sustainable development goals agenda: Causalities between economic growth, environmental sustainability, financial development, and renewable energy consumption. *Sustainable Development (Bradford)*, *31*(2), 680–697. DOI: 10.1002/sd.2411

Guha, J., & Chakrabarti, B. (2019). Achieving the Sustainable Development Goals (SDGs) through decentralisation and the role of local governments: A systematic review. *Commonwealth Journal of Local Governance*, (22), 1–21. DOI: 10.5130/cjlg.v0i22.6855

Guthridge, M., Kirkman, M., Penovic, T., & Giummarra, M. J. (2022). Promoting gender equality: A systematic review of interventions. *Social Justice Research*, *35*(3), 318–343. DOI: 10.1007/s11211-022-00398-z

Gutmann, J., Neuenkirch, M., & Neumeier, F. (2023). The economic effects of international sanctions: An event study. *Journal of Comparative Economics*, *51*(4), 1214–1231. DOI: 10.1016/j.jce.2023.05.005

Güven, A. B. (2012). The IMF, the World Bank, and the global economic crisis: Exploring paradigm continuity. *Development and Change*, *43*(4), 869–898. DOI: 10.1111/j.1467-7660.2012.01781.x

Gwambene, B., Liwenga, E., & Mung'ong'o, C. (2023). Climate change and variability impacts on agricultural production and food security for the smallholder farmers in Rungwe, Tanzania. *Environmental Management, 71*(1), 3–14. DOI: 10.1007/s00267-022-01628-5

Gweshengwe, B. (2019). A critique of the income poverty line and global multidimensional poverty index. *Southeast Asia: A Multidisciplinary Journal, 19*, 37-46.

Gweshengwe, B. (2020). *Understanding poverty in Brunei Darussalam* (Doctoral dissertation, Universiti Brunei Darussalam).

Gweshengwe, B., & Hassan, N. H. (2020). Defining the characteristics of poverty and their implications for poverty analysis. *Cogent Social Sciences, 6*(1), 1768669. DOI: 10.1080/23311886.2020.1768669

Gweshengwe, B., Hassan, N. H., & Maricar, H. A. (2020). Understanding quality of life in Brunei Darussalam. *Cogent Social Sciences, 6*(1), 1–13. DOI: 10.1080/23311886.2020.1838705

Gyasi, A. K. (2009). *Adoption of International Financial Reporting Standards in Developing Countries: The Case of Ghana.* Published Thesis. Vaasa Amma Tikorkeakoulu, University of Applied Sciences. Retrieved on 16 March 2020, from https://publications.theseus.fi/bitstream/handle/10024/15526/ANTWI_K

Hair, J. F., Black, W. C., Babin, B. J., & Anderson, R. E. (2018). *Multivariate data analysis* (8th ed.). Cengage Learning.

Hák, T., Janoušková, S., & Moldan, B. (2016). Sustainable Development Goals: A need for relevant indicators. *Ecological Indicators, 60*, 565–573. DOI: 10.1016/j.ecolind.2015.08.003

Halász, G., & Looney, J. (2019). Teacher professional competences and standards. concepts and implementation. *European Journal of Education, 54*(3), 311–314. DOI: 10.1111/ejed.12351

Hambari., Arif, A. A., & Zaim, M. A. (2020). The Role of Zakat Institution in Facing Covid-19: A Case Study of the Federal Territory Islamic Council (MAIWP) of Malaysia. The 4th International Conference of Zakat (ICONZ), Surabaya, Indonesia.

Hamdan, M., & Low, K. C. P. (2013). Is Brunei Ready to Adopt the International Accounting System (IAS)? *Business Journal for Entrepreneurs, 1*, 12–48.

Hamsal, M. (2019). The effect of sales-marketing collaboration, market-oriented product development, and nimble network structure on organizational agility. *International Journal of Advanced Science and Technology, 28*(8), 696–713.

Hannoudi, O. A. (2015). *Comparative Analysis between the IFRS and the AAOIFI Accounting Standards*. Published (Master) thesis. Universite De Strasbourg.

Han, S. W., & Borgonovi, F. (2020). Students' teaching career expectations by gender and ability levels in science and math: The role of salary and numeracy skills. *European Journal of Education*, *55*(3), 405–427. DOI: 10.1111/ejed.12407

Han, S. W., Borgonovi, F., & Guerriero, S. (2017). What motivates high school students to want to be teachers? the role of salary, working conditions, and societal evaluations about occupations in a comparative perspective. *American Educational Research Journal*, *55*(1), 3–39. DOI: 10.3102/0002831217729875

Hansen, M. C., Potapov, P. V., Moore, R., Hancher, M., Turubanova, S. A., Tyukavina, A., Thau, D., Stehman, S. V., Goetz, S. J., Loveland, T. R., Kommareddy, A., Egorov, A., Chini, L., Justice, C. O., & Townshend, J. R. G. (2013). High-resolution global maps of 21st-century forest cover change. *Science*, *342*(6160), 850–853. DOI: 10.1126/science.1244693 PMID: 24233722

Hanushek, E. A., & Rivkin, S. G. (2010). Generalizations about Using Value-Added Measures of Teacher Quality. *The American Economic Review*, *100*(2), 267–271. DOI: 10.1257/aer.100.2.267

Hanušová, S., Píšová, M., Kohoutek, T., Minaříková, E., Ježek, S., Janík, T., Mareš, J., & Janík, M. (2020). Novice teachers in the czech republic and their drop-out intentions. *European Journal of Education*, *55*(2), 275–291. DOI: 10.1111/ejed.12373

Hardin, D., Dewi, I. K., Alzarliani, W. O. D., Azaludin, , Wardana, , Andara, D., Ramadhan, F. M., & Huda, A. M. (2019). The role of communities in conserving mangrove forests to achieve sustainable development. *IOP Conference Series. Earth and Environmental Science*, *343*(1), 012143. DOI: 10.1088/1755-1315/343/1/012143

Hardman, F., Hardman, J., Dachi, H. A., Elliott, L., Ihebuzor, N., Ntekim, M., & Tibuhinda, A. (2015). Implementing school-based teacher development in tanzania. *Professional Development in Education*, *41*(4), 602–623. DOI: 10.1080/19415257.2015.1026453

Harhoff, F. S., & Freeland, S. (2017). Addressing the intentional destruction of the environment during warfare under the Rome Statute of the International Criminal Court. *Netherlands International Law Review*, *64*(2), 343–347. DOI: 10.1007/s40802-017-0088-z

Harraf, A., Wanasika, I., Tate, K., & Talbott, K. (2015). Organizational Agility. *Journal of Applied Business Research*, *31*(2), 675. DOI: 10.19030/jabr.v31i2.9160

Hartshorne, R. (1939). *The nature of geography: A critical survey of current thought in the light of the past*. Association of American Geographers.

Hasan, M. T., & Abdul Rahman, A. (2017). IFRS Adoption and Earnings Management: A Review and Justification of Earnings Management Model for Developing Countries. *ELK Asia Pacific Journal of Finance and Risk Management*, *8*(3), 1–17.

Hasan, Z. (2020). Distribution Of Zakat Funds To Achieve SDGs Through Poverty Alleviation In Baznas Republic Of Indonesia. *Azka International Journal of Zakat & Social Finance*, *1*(1), 25–43. DOI: 10.51377/azjaf.vol1no01.7

Hassan, N. H., & Yong, G. Y. (2018). A vision in which every family has basic shelter. *Sustainable development goals in Southeast Asia and ASEAN: National and regional approaches*, 190-209.

Hassan, N. H. P. H. (2017). *Housing matters: the value of home ownership in Brunei Darussalam*. Institute of Asian Studies, Universiti Brunei Darussalam.

Hassan, N. H., Rigg, J., Azalie, I. A., Yong, G. Y., Zainuddin, N. H. H., & Shamsul, M. A. S. M. (2022). Making do and staying poor: The poverty context of Urban Brunei. *Geoforum*, *136*, 132–141. DOI: 10.1016/j.geoforum.2022.09.012

Hassan, N. H., Rigg, J., Yong, G. Y., Azalie, I. A., Muhammad Shamsul, M. A. S., & Zainuddin, N. H. (2024). Administering and encountering the poor: Poverty from above and below in Brunei Darussalam. *Singapore Journal of Tropical Geography*, *45*(2), 311–331. DOI: 10.1111/sjtg.12545

Hassan, N. H., & Yong, G. Y. (2023). Climate Change and Sustainable Development of Kampong Ayer: Its History, People, and Environment. In Hamdan, M., Anshari, M., Ahmad, N., & Ali, E. (Eds.), *Public Policy's Role in Achieving Sustainable Development Goals* (pp. 96–114). IGI Global Scientific Publishing., DOI: 10.4018/978-1-6684-8903-1.ch006

Hassan, N. H., Yong, G. Y. V., Azalie, I. A. N., & Kamarulzaman, N. (2022). Urban Tapestry: A Place-Sensitive Approach to Sustainable Urban Design. *International Journal of Environment, Architecture, and Societies*, *2*(02), 109–121. DOI: 10.26418/ijeas.2022.2.02.109-121

Haufler, V. (2009). The Kimberley Process Certification Scheme: An innovation in global governance and conflict prevention. *Journal of Business Ethics*, *89*(S4), 403–416. https://www.jstor.org/stable/40605378. DOI: 10.1007/s10551-010-0401-9

Haworth, B., Bruce, E., & Whittaker, J. (2019). Participatory mapping: Diverse applications in the era of crowdsourcing and geospatial big data. *Cartography and Geographic Information Science*, *46*(4), 277–290.

Hayes, A. F. (2018). *Introduction to mediation, moderation, and conditional process analysis: A regression-based approach* (2nd ed.). Guilford Press.

Hay, J., & Mimura, N. (2010). The changing nature of extreme weather and climate events: Risks to sustainable development. *Geomatics, Natural Hazards & Risk, 1*(1), 3–18. DOI: 10.1080/19475701003643433

Haynes, J. D., Marsh, L. T. S., & Anderson, K. (2023). Planting the seeds of culturally responsive, equity-centered, and trauma-informed attitudes among urban educators. *Urban Education, 60*(3), 821–853. DOI: 10.1177/00420859231175663

Heffernan, A., Bright, D., Kim, M., Longmuir, F., & Magyar, B. (2022). 'I cannot sustain the workload and the emotional toll': Reasons behind Australian teachers' intentions to leave the profession' intentions to leave the profession. *Australian Journal of Education, 66*(2), 196–209. DOI: 10.1177/00049441221086654

Heinrich, C. J., Darling-Aduana, J., & Martin, C. (2019). The potential and prerequisites of effective tablet integration in rural kenya. *British Journal of Educational Technology, 51*(2), 498–514. DOI: 10.1111/bjet.12870

Helms, L. (2021). Introduction: The nature of political opposition in contemporary electoral democracies and autocracies. *European Political Science, 20*(4), 569–579. DOI: 10.1057/s41304-021-00323-z

Henderson, K., & Loreau, M. (2023). A model of Sustainable Development Goals: Challenges and opportunities in promoting human well-being and environmental sustainability. *Ecological Modelling, 475*, 110164. DOI: 10.1016/j.ecolmodel.2022.110164

Herath, S. K., & Alsulmi, F. H. (2017). International Financial Reporting Standards (IFRS): The Benefits, Obstacles and Opportunities for Implementation in Saudi Arabia. *International Journal of Social Science and Business, 2*(1), 1–18.

Hertz, B., Clemson, H. G., Hansen, D. T., Laurillard, D., Murray, M., Fernandes, L., & Rutkauskiene, D. (2022). A pedagogical model for effective online teacher professional development—Findings from the teacher academy initiative of the European commission. *European Journal of Education, 57*(1), 142–159. DOI: 10.1111/ejed.12486

Hickel, J. (2017). *The Divide: A Brief Guide to Global Inequality and its Solutions*. Penguin Random House UK.

Hickel, J. (2020). *Less is More: How Degrowth Will Save the World*. Random House.

Hinnebusch, R. A. (1981). Egypt under Sadat: Elites, power structure, and political change in a post-populist state. *Social Problems*, *28*(4), 442–464. https://www.jstor.org/stable/800057. DOI: 10.2307/800057

Hinz, J., & Leromain, E. (2020). Critically important: The heterogeneous effect of diplomatic tensions on trade. *Review of Industrial Organization*, *57*(3), 309–331. DOI: 10.1007/s11151-020-09769-9

Hodgdon, C., Hughes, S. B., & Street, D. L. (2011). Framework-based Teaching of IFRS Judgements. *Accounting Education*, *20*(4), 415–439. DOI: 10.1080/09639284.2011.589587

Hojnik, J., & Ruzzier, M. (2016). What drives eco-innovation? A review of an emerging literature. *Environmental Innovation and Societal Transitions*, *19*, 31–41. DOI: 10.1016/j.eist.2015.09.006

HOK. (2011). *Bandar Seri Begawan Development Master Plan* (Vol. 1-4). BSB Municipal Department, Government of Brunei Darussalam.

Horbach, J., & Rammer, C. (2020). Environmental innovation and employment dynamics in different technology fields: An analysis based on the German Community Innovation Survey. *Research Policy*, *51*(1), 103–115. DOI: 10.1016/j.respol.2020.103115

Huang, L., Li, S., Poitras, E., & Lajoie, S. P. (2020). Latent profiles of self-regulated learning and their impacts on teachers' technology integration. *British Journal of Educational Technology*, *52*(2), 695–713. DOI: 10.1111/bjet.13050

Hubbard, P. (2017). *City*. Routledge. DOI: 10.4324/9781315471259

Hulme, D., & Limcaoco, J. A. A. (2012). Planning integrated rural development projects in the Philippines: From blueprint to process? *Project Appraisal*, *06*(04), 223–233. DOI: 10.1080/02688867.1991.9726824

Human Rights Watch. (2017, March 2). *"License to Kill": Philippine Police Killings in Duterte's "War on Drugs" | HRW*. Human Rights Watch. Retrieved April, 2025, from https://www.hrw.org/report/2017/03/02/license-kill/philippine-police-killings-dutertes-war-drugs

Hussain, S., Hussain, E., Saxena, P., Sharma, A., Thathola, P., & Sonwani, S. (2024). Navigating the impact of climate change in India: A perspective on climate action (SDG13) and sustainable cities and communities (SDG11). *Frontiers in Sustainable Cities*, *5*, 1308684. Advance online publication. DOI: 10.3389/frsc.2023.1308684

Huttunen, M. (2024). A tale of two sovereigns: The responsibility to protect and the competing notions of responsible sovereignty. *International Journal of Human Rights*, •••, 1–22. DOI: 10.1080/13642987.2024.2429467

Hveem, H. (2003). The regional project in global governance. In *Theories of new regionalism: A Palgrave reader* (pp. 81–98). Palgrave Macmillan UK. DOI: 10.1057/9781403938794_5

IASPlus. (2002). Europe adopts regulation requiring IAS by 2005 (Deloitte). Retrieved on 3 March 2019, from https://www.iasplus.com/en/news/2002/June/news504

Iatridis, G. E., & Dalla, K. (2011). The Impact of IFRS Implementation on Greek Listed Companies: An industrial Sector and Stock Market Index Analysis. *International Journal of Managerial Finance*, *7*(3), 284–303. DOI: 10.1108/17439131111144478

IDB. (2018). Mexico: Integrating stakeholders in mangrove assessments and management planning: A model based on collaborative planning in Veracruz, Mexico. Washington, DC

IFRS. (2017). *IFRS Application around the World. Jurisdictional Profile: Brunei Darussalam.* Retrieved on 3 March 2019, from https://cdn.ifrs.org/-/media/feature/around-the-world/jurisdiction-profiles/brunei-darussalam-ifrs-profile.pdf

IGAD Climate Prediction and Applications Centre (ICPAC) (https://www.icpac.net/)

Inter-American Court of Human RightsInter-American Court of Human Rights Recognizes the Right to a Healthy Environment. (https://en.wikipedia.org/wiki/Inter-American_Court_of_Human_Rights; https://www.asil.org/insights/volume/24/issue/14/inter-american-court-human-rights-recognizes-right-healthy-environment)

Intergovernmental Authority for Development (IGAD). (https://igad.int/)

International Covenant on Economic, Social and Cultural Rights (https://www.ohchr.org/en/instruments-mechanisms/instruments/international-covenant-economic-social-and-cultural-rights)

International Labour Organization (ILO). (2019). *World Employment and Social Outlook: Trends 2019*. ILO Publications.

International Labour Organization. (n.d). *Thailand Decent Work Country Programme 2019-2021*.

IPCC. (2021). *Climate Change: The Physical Science Basis. Contribution of Working Group I to the Sixth Assessment Report of the Intergovernmental Panel on Climate Change, edited by Valérie Masson- Delmotte, Panmao Zhai, Anna Pirani, Sarah L. Connors, Clotilde Péan, Sophie Berger, Nada Caud, Yang Chen, Leah Goldfarb, Melissa I. Gomis, Mengtian Huang, Katherine Leitzell, Elisabeth Lonnoy, J.B. Robin Matthews, Thomas K. Maycock, Tim Waterfield, Ozge Yelekçi, Rong Yu, and Baiquan Zhou*. Cambridge University Press.

Ip, E. C. (2010). Globalization and the future of the law of the sovereign state. *International Journal of Constitutional Law*, *8*(3), 636–655. DOI: 10.1093/icon/moq033

Irfan, M., Wang, M., & Akhtar, N. (2019). Impact of IT capabilities on supply chain capabilities and organizational agility: A dynamic capability view. *Operations Management Research : Advancing Practice Through Research*, *12*(3–4), 113–128. DOI: 10.1007/s12063-019-00142-y

Irtyshcheva, I., Pavlenko, O., Boiko, Y., Stehnei, M., Kramarenko, I., Hryshyna, N., & Ishchenko, O. (2022). Evaluation of efficiency of regional public governance in the context of achieving goals of sustainable development. *Management Theory and Studies for Rural Business and Infrastructure Development*, *44*(4), 497–505. DOI: 10.15544/mts.2022.49

Ishiyama, J. (2019). Is democracy necessary for good governance? *Social Science Quarterly*, *100*(6), 2188–2208. DOI: 10.1111/ssqu.12712

Islam, M. A. (2017). *Future of Accounting Profession: Three Major Changes and Implications for Teaching and Research*. International Federation of Accountants. Retrieved on 7 January 2020, from https://www.ifac.org/knowledge-gateway/preparing-future-ready-professionals/discussion/future-accounting-profession-three

Islam, H. (2025). Nexus of economic, social, and environmental factors on sustainable development goals: The moderating role of technological advancement and green innovation. *Innovation and Green Development*, *4*(1), 100183. DOI: 10.1016/j.igd.2024.100183

Islam, M. M., Hasan, M., Mia, M. S., Masud, A. A., & Islam, A. R. M. T. (2025). *Early Warning systems in climate Risk Management: Roles and implementations in eradicating barriers and overcoming challenges*. Natural Hazards Research., DOI: 10.1016/j.nhres.2025.01.007

Islam, M. R., & Khan, N. A. (2018b). Threats, vulnerability, resilience and displacement among the climate change and natural disaster-affected people in South-East Asia: An overview. *Journal of the Asia Pacific Economy*, *23*(2), 297–323. DOI: 10.1080/13547860.2018.1442153

Issar, R. (2018). Cooperation in Disaster and Climate Risk Management in AAGC. AAGC (Asia-Africa Growth Corridor) (https://aagc.ris.org.in/en/about-aagc); RIS (Research and Information Systems for Developing Countries).

Ivanov, D., Dolgui, A., & Sokolov, B. (2019). The impact of digital technology and Industry 4.0 on the ripple effect and supply chain risk analytics. *International Journal of Production Research, 57*(3), 829–846. DOI: 10.1080/00207543.2018.1488086

Jabbour, A. B. L. S., & Jabbour, C. J. C. (2020). Green supply chain management and firms' performance: Understanding potential relationships and the role of green sourcing and organizational learning. *International Journal of Production Research, 58*(14), 4291–4306. DOI: 10.1080/00207543.2020.1769377

Jabeen, F. (2024). The Alignment of Universities With Sustainable Development Goals: How Do Academics Perceive the Progress (Not) Made? *IEEE Transactions on Engineering Management, 71*, 13545–13557. DOI: 10.1109/TEM.2022.3183016

Jaccard, J., & Turrisi, R. (2003). *Interaction effects in multiple regression* (2nd ed.). Sage. DOI: 10.4135/9781412984522

Jahangir, A. (2025). Governance Innovation in South Asia: The Pakistan Governance Journey. In *Mapping Governance Innovations* (pp. 147-166). Routledge India.

Jahn, D. (2024). The stringency and potential impact of climate laws and policies in the European Union and the 21 OECD countries. *Npj Climate Action, 3*(1), 90. DOI: 10.1038/s44168-024-00175-5

Jain, P. (2011). IFRS Implementation in India: Opportunities and Challenges. *World Journal of Social Sciences, 1*(1), 125–136.

Jakulevičienė, L. (2023). Principle of non-refoulement in the context of recent ECtHR case law and the EU migration and asylum pact. *ERA Forum, 24*, 379–396. https://doi.org/DOI: 10.1007/s12027-023-00761-w

James, G., Witten, D., Hastie, T., & Tibshirani, R. (2013). *An introduction to statistical learning: With applications in R*. Springer. DOI: 10.1007/978-1-4614-7138-7

Jayaram, D., & Sethi, G. (2022). Geopolitics of climate change and water security in South Asia: Conflict and cooperation. In N. Khare (Ed.), *Science, policies and conflicts of climate change* (pp. XX–XX). Springer Climate. Springer, Cham. https://doi.org/DOI: 10.1007/978-3-031-16254-1_4

Jennex. M. E. (January, 2017). What is Knowledge Management? San Diego State University. Retrieved on 12 September 2021 from https://www.researchgate.net/publication/314500732

Jermakowicz, E. K., & Epstein, B. J. (2010). IFRS for SMEs - An Option for U.S. Private Entities? *Review of Business, 30*(2), 72–79.

Jermakowicz, E. K., & Tomaszewski, S. G. (2006). Implementing IFRS from the Perspective of EU Publicly Traded Companies. *Journal of International Accounting, Auditing & Taxation, 15*(2), 170–196. DOI: 10.1016/j.intaccaudtax.2006.08.003

Jessen, H. (2021). Multilateral and unilateral sanctions: Compliance and challenges. In Leal Filho, W., Azul, A. M., Brandli, L., Lange Salvia, A., Özuyar, P. G., & Wall, T. (Eds.), *Peace, justice and strong institutions. Encyclopedia of the UN Sustainable Development Goals* (pp. 1–11). Springer., DOI: 10.1007/978-3-319-71066-2_51-1

Jetter, I., & Molina, T. (2022). Persuasive agenda-setting: Rodrigo Duterte's inauguration speech and drugs in the Philippines. *Journal of Development Economics, 156*. DOI: 10.1016/j.jdeveco.2022.102843

Jiang, S., Jakobsen, K., Bueie, J., Li, J., & Haro, P. H. (2022). A Tertiary Review on Blockchain and Sustainability With Focus on Sustainable Development Goals. *IEEE Access : Practical Innovations, Open Solutions, 10*, 114975–115006. DOI: 10.1109/ACCESS.2022.3217683

Jiménez-Aceituno, A., Peterson, G. D., Norström, A. V., Wong, G. Y., & Downing, A. S. (2020). Local lens for SDG implementation: Lessons from bottom-up approaches in Africa. *Sustainability Science, 15*(3), 729–743. DOI: 10.1007/s11625-019-00746-0

Johari, A. H., Ahad, A. D., & Anshari, M. (2024). Examining the Impact of Gender Differences, Sustainable Urbanism, and Digital Transformation on Online Consumer Buying Behavior: A Perspective Towards SDG Alignment. In *Harnessing Green and Circular Skills for Digital Transformation* (pp. 194-212). IGI Global.

Johnson, S. (2008). *IFRS: No Longer IF, but When.* Retrieved on 15 March 2020, from https://www.cfo.com/accounting-tax/2008/02/ifrs-no-longer-if-but-when/

Jones, C., & Pimdee, P. (2017). Innovative ideas: Thailand 4.0 and the fourth industrial revolution. In Asian International Journal of Social Sciences (Vol. 17, pp. 4–35). https://doi.org/DOI: 10.29139/aijss.20170101

Joseph, J., & Pandey, A. (2025). Prospective Role of "BRICS Plus" Countries in Global Governance: An Exploratory Study. In *Changing the Global Political Economy: BRICS Countries and Alternative Relations Strategies* (pp. 93-118). IGI Global Scientific Publishing.

Joshi, J., Magal, A., Limaye, V. S., Madan, P., Jaiswal, A., Mavalankar, D., & Knowlton, K. (2022). Climate change and 2030 cooling demand in Ahmedabad, India: Opportunities for expansion of renewable energy and cool roofs. *Mitigation and Adaptation Strategies for Global Change*, *27*(7), 44. Advance online publication. DOI: 10.1007/s11027-022-10019-4 PMID: 35967931

Jugović, I., Marušić, I., Ivanec, T. P., & Vidović, V. V. (2012). Motivation and personality of preservice teachers in croatia. *Asia-Pacific Journal of Teacher Education*, *40*(3), 271–287. DOI: 10.1080/1359866X.2012.700044

Kabeer, N. (2019). Gender equality, economic growth, and women's empowerment: A critical review. *Development and Change*, *50*(5), 1299–1328.

Kacowicz, A. M. (2018). Regional governance and global governance: Links and explanations. *Global Governance*, *24*(1), 61–80. DOI: 10.1163/19426720-02401005

Kafyulilo, A., Fisser, P., Pieters, J. M., & Voogt, J. (2015). Ict use in science and mathematics teacher education in tanzania: Developing technological pedagogical content knowledge. *Australasian Journal of Educational Technology*, *31*(4). Advance online publication. DOI: 10.14742/ajet.1240

Kahler, M. (2017). Regional challenges to global governance. *Global Policy*, *8*(1), 97–100. DOI: 10.1111/1758-5899.12392

Kai, Z., Sharaf, M., Wei, S.-Y., Al Shraah, A., Le, L. T., Arvind Bedekar, D. A., & Bani Ahmad, A. Y. A. (2024). Exploring the asymmetric relationship between natural resources, fintech, remittance and environmental pollution for BRICS nations: New insights from MMQR approach. *Resources Policy*, *90*, 104693. Advance online publication. DOI: 10.1016/j.resourpol.2024.104693

Kakabadse, N. K., Kakabadse, A., & Kouzmin, A. (2003). Reviewing the knowledge management literature: towards a taxonomy. Retrieved on 15 September 2021 from https://www.researchgate.net/publication/220363071

Kalfagianni, A., & Young, O. R. (2022). The politics of multilateral environmental agreements: Lessons from 20 years of INEA. *International Environmental Agreement: Politics, Law and Economics*, *22*(2), 245–262. DOI: 10.1007/s10784-022-09567-6

Kalliojärvi, S. (2020). Age of changes: Threat of climate change and its meaning for security. In Heininen, L., & Exner-Pirot, H. (Eds.), *Climate change and Arctic security* (pp. 9–32). Palgrave Pivot., DOI: 10.1007/978-3-030-20230-9_2

Kanie, N., & Biermann, F. (Eds.). (2017). *Governing through goals: Sustainable development goals as governance innovation.* mit Press. DOI: 10.7551/mitpress/10894.001.0001

Kant, I. (1784). *An Answer to the Question: What is Enlightenment?* Penguin Classics.

Kantorowicz-Reznichenko, E. (2020). The enforcement of international law. In Hosli, M. O., & Selleslaghs, J. (Eds.), *The changing global order* (Vol. 17). Springer., DOI: 10.1007/978-3-030-21603-0_18

Karakhanyan, S., & Veen, K. (2011). Teachers' voices in the context of higher education reforms in armenia. *European Journal of Education*, *46*(4), 508–523. DOI: 10.1111/j.1465-3435.2011.01497.x

Kasatpibal, N., Viseskul, N., Untong, A., Thummathai, K., Kamnon, K., Sangkampang, S., Tokilay, R., Assawapalanggool, S., & Apisarnthanarak, A. (2023). Impact of the COVID-19 pandemic on the Thai population: Delineating the effects of the pandemic and policy measures. *Antimicrobial Stewardship & Healthcare Epidemiology : ASHE*, *3*(1), e241. Advance online publication. DOI: 10.1017/ash.2023.523 PMID: 38156233

Kasych, A., Suler, P., & Rowland, Z. (2020). Corporate Environmental Responsibility through the prism of Strategic Management. *Sustainability (Basel)*, *12*(22), 9589. DOI: 10.3390/su12229589

Katila, P., Colfer, C. J. P., de Jong, W., Galloway, G., Pacheco, P., & Winkel, G. (Eds.). (2019). *Sustainable Development Goals: Their Impacts on Forests and People*. Cambridge University Press. DOI: 10.1017/9781108765015

Katsaliaki, K., Galetsi, P., & Kumar, S. (2022). Supply chain disruptions and resilience: A major review and future research agenda. *Annals of Operations Research*, *319*(1), 965–1002. DOI: 10.1007/s10479-020-03912-1

Kavaratzis, M. (2004). From city marketing to city branding: Towards a theoretical framework for developing city brands. *Place Branding*, *1*(1), 58–73. DOI: 10.1057/palgrave.pb.5990005

Kaya, D., & Koch, M. (2015). Countries' Adoption of the International Financial Reporting Standard for Small and Medium-sized Entities (IFRS for SMEs) – Early Empirical Evidence. *Accounting and Business Research, 45* (1), p. 93.120.

Ke, L., Liu, D., Tan, Q., Wang, S., Wang, Q., & Yang, J. (2024). Spatial-Temporal Pattern of Land Use and SDG15 Assessment in the Bohai Rim Region Based on GEE and RF Algorithms. *IEEE Journal of Selected Topics in Applied Earth Observations and Remote Sensing*, *17*, 7541–7553. DOI: 10.1109/JSTARS.2024.3380580

Kenny, M. E., Horne, A. M., Orpinas, P., & Reese, L. E. (Eds.). (2009). *Realizing social justice: The challenge of preventive interventions.* American Psychological Association., https://www.jstor.org/stable/j.ctv1chs94p DOI: 10.1037/11870-000

Keohane, R. O., & Nye, J. S. (2005). *Power and Interdependence*. Longman.

Kerswell, T. (2015). *Development and dispossession: The crisis of forced displacement and resettlement* by Anthony Oliver-Smith (Ed.). *Human Rights Review, 16*, 417–418. https://doi.org/DOI: 10.1007/s12142-015-0386-y

Kešeljević, A., & Spruk, R. (2024). Estimating the effects of the Syrian civil war. *Empirical Economics*, *66*(2), 671–703. DOI: 10.1007/s00181-023-02470-2

Khandker, S. R. (2005). Microfinance and poverty: Evidence using panel data from Bangladesh. *The World Bank Economic Review*, *19*(2), 263–286. DOI: 10.1093/wber/lhi008

Khan, J. (2021). European academic brain drain: A meta-synthesis. *European Journal of Education*, *56*(2), 265–278. DOI: 10.1111/ejed.12449

Khan, M. T. I., Anwar, S., Sarkodie, S. A., Yaseen, M. R., Nadeem, A. M., & Ali, Q. (2023). Natural disasters, resilience-building, and risk: Achieving sustainable cities and human settlements. *Natural Hazards*, *118*(1), 611–640. DOI: 10.1007/s11069-023-06021-x

Khan, P. A., Johl, S. K., & Akhtar, S. A. (2022). Vinculum of Sustainable Development Goal Practices and Firms' Financial Performance: A Moderation Role of Green Innovation. *Journal of Risk and Financial Management*, *15*(3), 96. https://doi.org/https://doi.org/10.3390/jrfm15030096. DOI: 10.3390/jrfm15030096

Kharrazi, A., Qin, H., & Zhang, Y. (2016). Urban big data and sustainable development goals: Challenges and opportunities. *Sustainability (Basel)*, *8*(12), 1293. DOI: 10.3390/su8121293

Khatter, A. (2023). Strategic management and stakeholder engagement: A case for environmental sustainability. *Sustainable Development*. Advance online publication. DOI: 10.5772/intechopen.112065

Khayer, A., Jahan, N., Hossain, M. N., & Hossain, M. Y. (2020). The adoption of cloud computing in small and medium enterprises: A developing country perspective. *VINE Journal of Information and Knowledge Management Systems*. Advance online publication. DOI: 10.1108/VJIKMS-05-2019-0064

Khoo, Y. (2021). Becoming globally competent through inter-school reciprocal learning partnerships: An inquiry into canadian and chinese teachers' narratives. *Journal of Teacher Education*, *73*(1), 110–122. DOI: 10.1177/00224871211042306

Kilic, M., & Uyar, A. (2017). Adoption Process of IFRS for SMEs in Turkey: Insights from Academics and Accountants. *Accounting and Management Information Systems, 16*(2), 313–339. DOI: 10.24818/jamis.2017.02005

Kılınç, A., Watt, H. M. G., & Richardson, P. (2012). Factors influencing teaching choice in turkey. *Asia-Pacific Journal of Teacher Education, 40*(3), 199–226. DOI: 10.1080/1359866X.2012.700048

Kimeli, E. K. (2017). IFRS Adoption and Capital Markets. *Journal of Finance and Accounting, 5*(1), 19–30.

Kinnard, J. N. (2014). *Places in motion: The fluid identities of temples, images, and pilgrims*. Oxford University Press. DOI: 10.1093/acprof:oso/9780199359653.001.0001

Kioupi, V., & Voulvoulis, N. (2019). Education for Sustainable Development: A systemic framework for connecting the SDGs to educational outcomes. *Sustainability (Basel), 11*(21), 6104. Advance online publication. DOI: 10.3390/su11216104

Kirksey, J. J. (2021). Preparedness and experiences of novice teachers in the sociopolitical context of heightened immigration enforcement: Evidence from a survey of california teachers. *Journal of Teacher Education, 73*(1), 37–51. DOI: 10.1177/00224871211045980

Kiron, D., Unruh, G., Reeves, M., Kruschwitz, N., & Rubel, H. (2021). Corporate Sustainability at a Crossroads. *MIT Sloan Management Review*. https://sloanreview.mit.edu/article/corporate-sustainability-at-a-crossroads/

Klewitz, J., & Hansen, E. G. (2021). Sustainability-oriented innovation of SMEs: A systematic review. *Journal of Cleaner Production, 65*, 57–75. DOI: 10.1016/j.jclepro.2013.07.017

Kocot, D., Kocot, M., & Soboń, J. (2024). Competitive Advantage of an Agile Organization as a Determinant of The Functioning of Global Market. *Globalization, the State and the Individual.* https://doi.org/https://doi.org/10.5604/01.3001.0054.9309

Kon, J. (2025). Brunei joins regional effort to combat human trafficking. *Borneo Bulletin.* https://borneobulletin.com.bn/brunei-joins-regional-effort-to-combat-human-trafficking/?

Kostoska, O., & Kocarev, L. (2019). A novel ICT framework for Sustainable Development Goals. *Sustainability (Basel), 11*(7), 1961. Advance online publication. DOI: 10.3390/su11071961

Kothari, A., Salleh, A., Escobar, A., Demaria, F., & Acosta, A. (Eds.). (2019). *Pluriverse: A Post-Development Dictionary*. Authors Up Front.

KPMG. (2024). *Thailand's Ageing Society: The opportunities for businesses in Thailand.* https://assets.kpmg.com/content/dam/kpmg/th/pdf/2024/07/thailand-ageing-society-the-opportunities-for-businesses-in-thailand.pdf

Kraft, M. A., & Papay, J. P. (2014). Can Professional Environments in Schools Promote Teacher Development? Explaining Heterogeneity in Returns to Teaching Experience. *Educational Evaluation and Policy Analysis*, *36*(4), 476–500. DOI: 10.3102/0162373713519496

Kranke, M. (2020). IMF-World Bank cooperation before and after the global financial crisis. *Global Policy*, *11*(1, S1), 15–25. DOI: 10.1111/1758-5899.12743

Krantz, V., & Gustafsson, S. (2023). Regional collaboration for the sustainable development goals: Experiences from developing a multi-actor platform in Sweden. *Sustainable Development (Bradford)*, *31*(6), 4007–4018. DOI: 10.1002/sd.2580

Kumagai, S. (2019). Thailand's Efforts to Cope with a Rapidly Aging Population — Can Thailand Become a Trailblazer for Other Middle-Income Countries Struggling with Demographic Aging? *Pacific Business and Industries, XIX*(71). https://www.jri.co.jp/MediaLibrary/file/english/periodical/rim/2019/71.pdf

Kumar, A., Soti, N., Gupta, S., & Deepa, . (2024). Addressing energy poverty in BRICS economies: Insights from panel data analysis and policy implications for sustainable development goals. *Environment, Development and Sustainability*. Advance online publication. DOI: 10.1007/s10668-024-05652-9

Kumar, S., & Anbanandam, R. (2020). Impact of risk management culture on supply chain resilience: An empirical study from Indian manufacturing industry. *Proceedings of the Institution of Mechanical Engineers. Part O, Journal of Risk and Reliability*, *234*(2), 246–259. DOI: 10.1177/1748006X19886718

Kumar, S., & Chandrakar, R. (2022). Green supply chain management: A study of recent practices, trends, and issues. *Journal of Cleaner Production*, *234*, 1243–1257. DOI: 10.1016/j.jclepro.2022.124387

Lakkala, M., & Ilomäki, L. (2015). A case study of developing ict-supported pedagogy through a collegial practice transfer process. Computers &Amp. *Computers & Education*, *90*, 1–12. DOI: 10.1016/j.compedu.2015.09.001

Laurien, F., Martin, J. G., & Mehryar, S. (2022). Climate and disaster resilience measurement: Persistent gaps in multiple hazards, methods, and practicability. *Climate Risk Management*, *37*, 100443. DOI: 10.1016/j.crm.2022.100443

Le Billon, P. (2001). The political ecology of war: Natural resources and armed conflicts. *Political Geography*, *20*(5), 561–584. DOI: 10.1016/S0962-6298(01)00015-4

Le, A. N. N., Pham, H., Pham, D. T. N., & Duong, K. D. (2023). Political stability and foreign direct investment inflows in 25 Asia-Pacific countries: The moderating role of trade openness. *Humanities & Social Sciences Communications*, *10*(1), 606. DOI: 10.1057/s41599-023-02075-1

Leach, M. (2016). *Gender equality and sustainable development: A pathways approach*. Routledge.

Leal Filho, W., Tripathi, S. K., Andrade Guerra, J. B. S. O. D., Giné-Garriga, R., Orlovic Lovren, V., & Willats, J. (2019). Using the sustainable development goals towards a better understanding of sustainability challenges. *International Journal of Sustainable Development and World Ecology*, *26*(2), 179–190. DOI: 10.1080/13504509.2018.1505674

Lee, J., & Fuller, B. (2020). Does progressive finance alter school organizations and raise achievement? the case of Los Angeles. *Educational Policy*, *36*(3), 587–623. DOI: 10.1177/0895904820901472

Lee, K. H., Noh, J., & Khim, J. S. (2020). The Blue Economy and the United Nations' sustainable development goals: Challenges and opportunities. *Environment International*, *137*, 105528. DOI: 10.1016/j.envint.2020.105528 PMID: 32014791

Le, T.-H., Bui, M.-T., & Uddin, G. S. (2022). Economic and social impacts of conflict: A cross-country analysis. *Economic Modelling*, *115*, 105980. DOI: 10.1016/j.econmod.2022.105980

Lhosupasirirat, P., Dahdouh-Guebas, F., Hugé, J., Wodehouse, D., & Enright, J. (2023). Stakeholder perceptions on community-based Ecological Mangrove Restoration (CBEMR): A case study in Thailand. *Restoration Ecology*, *31*(5), e13894. Advance online publication. DOI: 10.1111/rec.13894

Liapis, K., & Thalassinos, I. E. (2013). A Comparative Analysis for the Accounting Reporting of Employee Benefits between IFRS and other Accounting Standards: A Case Study for the Biggest Listed Entities in Greece. *International Journal of Economics and Business Administration*, *1*(1), 91–116. DOI: 10.35808/ijeba/5

Lillo Lobos, R. (2022). The right to access to justice as a key to understand the right to a fair trial in civil matters. In *Understanding due process in non-criminal matters* (Vol. 97, pp. 251–281). Springer., DOI: 10.1007/978-3-030-95534-2_13

Lin, J., Li, L., Luo, X., & Benitez, J. (2020, August). How do agribusinesses thrive through complexity? The pivotal role of e-commerce capability and business agility. *Decision Support Systems*, *135*(113342), 113342. Advance online publication. DOI: 10.1016/j.dss.2020.113342 PMID: 32834263

Lin, K., Wang, L., Ranjan, R., & Zhou, H. (2024). The changing contexts of international aid_ examining Indian experiences for a BRICS way. *India Review*, *23*(3), 245–265. DOI: 10.1080/14736489.2024.2341570

Li, Q. (2020). The impact of green innovation on firm performance: Evidence from China. *Technology in Society*, *62*, 101–111. DOI: 10.1016/j.techsoc.2020.101342

Liu, J., Hull, V., Godfray, H. C. J., Tilman, D., Gleick, P., Hoff, H., Pahl-Wostl, C., Xu, Z., Chung, M. G., Sun, J., & Li, S. (2018). Nexus approaches to global sustainable development. *Nature Sustainability*, *1*(9), 466–476. DOI: 10.1038/s41893-018-0135-8

Li, Y., Wang, Z., Lin, Z., & Gao, Y. (2025). Revisiting regional governance and regional development: Measurements, linkages and coupling effect. *World Development*, *185*, 106816. DOI: 10.1016/j.worlddev.2024.106816

Ljung, G. M., & Box, G. E. P. (1978). On a measure of lack of fit in time series models. *Biometrika*, *65*(2), 297–303. DOI: 10.1093/biomet/65.2.297

Lorthanavanich, D. (2021), 'Introduction', in Lorthanavanich, D. and O. Komazawa (eds.), Population Ageing in Thailand Long-term Care Model: Review of Population Ageing Practices and Policies, Vol. 2. ERIA Research Project Report FY2021 No. 06b, Jakarta: ERIA, pp.1-11.

Lovelock, C. E., & Brown, B. M. (2019). Land tenure considerations are key to successful mangrove restoration. *Nature Ecology & Evolution*, *3*(8), 1135. DOI: 10.1038/s41559-019-0942-y PMID: 31285575

Lu, J., Lin, W., & Dabić, M. (2024). Coevolution Analysis of the Sustainable Development Systems of Energy Firms From a Strategic Philanthropy Perspective. *IEEE Transactions on Engineering Management*, *71*, 6887–6902. DOI: 10.1109/TEM.2023.3244500

Lülf, C. (2016). Non-refoulement in international refugee law, human rights law, and asylum laws. In Heintze, H. J., & Thielbörger, P. (Eds.), *From Cold War to Cyber War*. Springer., DOI: 10.1007/978-3-319-19087-7_12

Luthra, S., Garg, D., & Haleem, A. (2015). Critical success factors of green supply chain management for achieving sustainability in Indian automobile industry. *Production Planning and Control*, *26*(5), 339–362. DOI: 10.1080/09537287.2014.904532

Lütkehermöller, K., Smit, S., & Kuramochi, T. (2021). Non-state and subnational climate action in India – An overview of the current landscape, emission reduction potential and implementation. Working paper. Project number 319041. NewClimate Institute (https://newclimate.org/resources/publications)

Mabokela, R. O., & Mlambo, Y. A. (2014). "the older women are men:" navigating the academic terrain, perspectives from ghana. *Higher Education*, *69*(5), 759–778. DOI: 10.1007/s10734-014-9804-3

Mackinder, H. J. (2004). The geographical pivot of history (1904). *The Geographical Journal*, *170*(4), 298–321. DOI: 10.1111/j.0016-7398.2004.00132.x

MacLeod, G., & Goodwin, M. (1999). Space, scale and state strategy: Rethinking urban and regional governance. *Progress in Human Geography*, *23*(4), 503–527. DOI: 10.1191/030913299669861026

MacPhail, A., Ulvik, M., Guberman, A., Czerniawski, G., Oolbekkink-Marchand, H., & Bain, Y. (2018). The professional development of higher education-based teacher educators: Needs and realities. *Professional Development in Education*, *45*(5), 848–861. DOI: 10.1080/19415257.2018.1529610

Maina, M. F., Ortiz, L. G., Mancini, F., & Melo, M. M. (2022). A micro-credentialing methodology for improved recognition of HE employability skills. *International Journal of Educational Technology in Higher Education*, *19*(1), 10. Advance online publication. DOI: 10.1186/s41239-021-00315-5

Main, S., & Slater, E. (2021). Online continuous professional learning: A model for improving reading outcomes in regional and remote schools? *Journal of Teacher Education*, *73*(2), 201–214. DOI: 10.1177/00224871211009110

MAIWP. (n.d.). https://www.maiwp.gov.my/i/index.php/en/mengenai-kami/objektif-visi-misi

Makinda, S. M. (1998). The United Nations and state sovereignty: Mechanism for managing international security. *Australian Journal of Political Science*, *33*(1), 101–115. DOI: 10.1080/10361149850750

Maksum, . (2018). *Fikih Sustainable Development Goals*. UIN Jakarta Press.

Malpass, A., Le Poidevin, A., Bamford, A., Lowndes, S., Grant, G., & Selman, L. (2025). From trusted intruder to real collaborator: Mapping the ecosystem of stakeholders and community assets in a coastal region in England, with a focus on end-of-life care and bereavement support. *Social Science & Medicine*, *365*, 117555. DOI: 10.1016/j.socscimed.2024.117555 PMID: 39642586

Mamabolo, M.A. (2018). Positioning stakeholder engagement theory on governance of communal farms: a proposed framework for land governance in South Africa.

Ma, N., Du, L., & Lu, Y. (2022). A model of factors influencing in-service teachers' social network prestige in online peer assessment. *Australasian Journal of Educational Technology*, •••, 100–118. DOI: 10.14742/ajet.7622

Mangan, J., Hughes, A., & Slack, K. (2010). Student finance, information and decision making. *Higher Education*, *60*(5), 459–472. DOI: 10.1007/s10734-010-9309-7

Mansoor, S. I. U. (2025). Renewable energy and international law: Managing risk in relation to economic volatility and geopolitical instability. In *Geopolitical landscapes of renewable energy and urban growth* (pp. 97–124). IGI Global., DOI: 10.4018/979-8-3693-8814-3.ch004

Manulak, M. W. (2024). The sources of influence in multilateral diplomacy: Replaceability and intergovernmental networks in international organizations. *The Review of International Organizations*, *19*(3), 579–610. DOI: 10.1007/s11558-024-09536-5

Marcovecchio, I., Thinyane, M., Estevez, E., Fillottrani, P., Marcovecchio, I., & Estevez, E. (2018). Capability Maturity Models as a Means to Standardize Sustainable Development Goals Indicators Data Production. *Journal of ICT Standardisation*, *6*(3), 217–244. DOI: 10.13052/jicts2245-800X.633

Marquez-Ramos, L. (2008). *The Effect of IFRS Adoption on Trade and Foreign Direct Investments*. International Trade and Finance Association. Working paper. 19 p.

Martell, C. C. (2019). Barriers to inquiry-based instruction: A longitudinal study of history teachers. *Journal of Teacher Education*, *71*(3), 279–291. DOI: 10.1177/0022487119841880

Martínez-Caro, E., Cepeda-Carrión, G., Cegarra-Navarro, J. G., & Garcia-Perez, A. (2020). The effect of information technology assimilation on firm performance in B2B scenarios. *Industrial Management & Data Systems*, *120*(12), 2269–2296. Advance online publication. DOI: 10.1108/IMDS-10-2019-0554

Martínez, J. G., Martí, M. C., & Cervera, M. G. (2019). Inside a 3d simulation: Realism, dramatism and challenge in the development of students' teacher digital competence. *Australasian Journal of Educational Technology*. Advance online publication. DOI: 10.14742/ajet.3885

Marx, A., Pertiwi, S. B., Depoorter, C., Hoornick, M., Mursitama, T. N., Otteburn, K., & Arnakim, L. Y. (2021). What role for regional organizations in goal-setting global governance? An analysis of the role of the European Union and ASEAN in the Sustainable Development Goals. *Global Public Policy and Governance*, *1*(4), 421–445. DOI: 10.1007/s43508-021-00027-6

Matta, S., Bleaney, M., & Appleton, S. (2022). The economic impact of political instability and mass civil protest. *Economics and Politics*, *34*(1), 253–270. DOI: 10.1111/ecpo.12197

McAleavy, T., Hall-Chen, A., Horrocks, S., & Riggall, A. (2022). Technology-supported professional development for teachers: lessons from developing countries. https://edtlive.b-cdn.net/live/media/egshqxa0/technology-supported-professional-development-for-teachers.pdf

McChesney, K., & Aldridge, J. M. (2018). A review of practitioner-led evaluation of teacher professional development. *Professional Development in Education*, *45*(2), 307–324. DOI: 10.1080/19415257.2018.1452782

McMenamin, R. (2023). Advisory opinion on obligations of states in respect of climate change: Potential contribution of human rights bodies. *Climate Law*, *13*(3–4), 213–223. DOI: 10.1163/18786561-bja10045

Meadowcroft, J. (2020). Sustainable development: A new (ish) idea for a post-pandemic world. *Journal of Environmental Policy and Planning*, *22*(4), 482–495.

Meaza, H., Ghebreyohannes, T., Nyssen, J., Tesfamariam, Z., Demissie, B., Poesen, J., Gebrehiwot, M., Weldemichel, T. G., Deckers, S., Gidey, D. G., & Vanmaercke, M. (2024). Managing the environmental impacts of war: What can be learned from conflict-vulnerable communities? *The Science of the Total Environment*, *927*, 171974. DOI: 10.1016/j.scitotenv.2024.171974

Mehmood, S. (2023). *The rule of law approach for more resilient institutions: Judicial accountability and independence, and global economic activities*. Asian Development Bank Institute. DOI: 10.56506/FGHQ1674

Melo, R. H., Kusmana, C., Eriyatno, , & Nurrochmat, D. R. (2019). A stakeholder analysis of sustainable mangrove management in Kwandang, sub-district of north, Gorontalo District. *IOP Conference Series. Earth and Environmental Science*, *399*(1), 012071. DOI: 10.1088/1755-1315/399/1/012071

Menon, M., & Paretti, M. C. (2024, September). Faculty Perspectives on Integrating Sustainable Development Into Engineering Education. *IEEE Transactions on Technology and Society*, *5*(3), 316–324. DOI: 10.1109/TTS.2024.3403482

Mensah, F. M. (2019). Finding voice and passion: Critical race theory methodology in science teacher education. *American Educational Research Journal*, *56*(4), 1412–1456. DOI: 10.3102/0002831218818093

Miao, S., Chen, B., & Jiang, N. (2025). Collaboration among governments, agribusinesses, and rural households for improving the effectiveness of conservation tillage technology adoption. *Scientific Reports*, *15*(1), 45. DOI: 10.1038/s41598-024-83827-0 PMID: 39747408

Miljkovic, D., & Rimal, A. (2008). The impact of socio-economic factors on political instability: A cross-country analysis. *Journal of Socio-Economics*, *37*(6), 2454–2463. DOI: 10.1016/j.socec.2008.04.007

Miller, M. A., Rigg, J., & Taylor, D. (2022). Transboundary environmental governance: Emerging themes and lessons from Southeast Asia. *Environmental Policy and Governance*, *32*(4), 275–280. DOI: 10.1002/eet.2012

Mill, J. S. (1863). *Utilitarianism*. Oxford University Press.

Ministry of External Affairs (MEA). (2023). *India's Role in Global Sustainable Development*. Government of India.

Mirjafari, S. H., & Ebrahimi, F. (2016) The Role of Regional Organizations in Environmental Protection. *SSRN*. https://papers.ssrn.com/sol3/papers.cfm?abstract_id=2818942 DOI: 10.2139/ssrn.2818942

Mitchell, D. (2000). *Cultural geography: A critical introduction*. Wiley-Blackwell.

Mlachila, M., Tapsoba, R., & Tapsoba, S. J. A. (2017). A quality of growth index for developing countries: A proposal. *Social Indicators Research*, *134*(2), 675–710. DOI: 10.1007/s11205-016-1439-6

Moallemi, E., Malekpour, S., Hadjikakou, M., Raven, R., Szetey, K., Ningrum, D., Dhiaulhaq, A., & Bryan, B. A. (2020). Achieving the Sustainable Development Goals requires transdisciplinary innovation at the local scale. *One Earth*, *3*(3), 300–313. Advance online publication. DOI: 10.1016/j.oneear.2020.08.006

Mohamad, L. (2018). Briefing Raises Awareness on Accounting Standards. *Borneo Bulletin*. Retrieved on 1 March 2020, from https://borneobulletin.com.bn/briefing-raises-awareness-on-accounting-standards/

Mohammed, A., Al-Mahdi, N., & Ghaithan, A. (2024). The role of environmental social governance on achieving sustainable development goals in Gulf Cooperation Council countries. Environment, Development and Sustainability. https://www.nber.org/system/files/chapters/c14917/c14917.pdfDOI: 10.1007/s10668-024-05662-7

Moka-Mubelo, W. (2017). Civil society and human rights. In *Reconciling law and morality in human rights discourse* (Vol. 3, pp. 147–167). Springer., DOI: 10.1007/978-3-319-49496-8_6

Moosa, I. A., & Moosa, N. (2019). Conditionality and structural adjustment programmes. In *Eliminating the IMF* (pp. 55–87). Palgrave Macmillan., DOI: 10.1007/978-3-030-05761-9_3

Morgan, K. (2004). Sustainable regions: Governance, innovation and scale. *European Planning Studies, 12*(6), 871–889. DOI: 10.1080/0965431042000251909

Morgan, T. C., & Schwebach, V. L. (1995). Economic sanctions as an instrument of foreign policy: The role of domestic politics. *International Interactions, 21*(3), 247–263. DOI: 10.1080/03050629508434868

Morita, K., Okitasari, M., & Masuda, H. (2020). Analysis of national and local governance systems to achieve the sustainable development goals: Case studies of Japan and Indonesia. *Sustainability Science, 15*(1), 179–202. DOI: 10.1007/s11625-019-00739-z

Morton, S., Pencheon, D., & Squires, N. (2017). Sustainable Development Goals (SDGs), and their implementation: A national global framework for health, development and equity needs a systems approach at every level. *British Medical Bulletin, 124*(1), 81–90. DOI: 10.1093/bmb/ldx031

Morvaridi, B. (2008). Introduction: Social justice and development. In *Social justice and development* (pp. 1–32). Palgrave Macmillan, London. https://doi.org/DOI: 10.1057/9780230581999_1

Mouboua, P. D., Atobatele, F. A., & Akintayo, O. T. (2024). Multilingual education and social equity: a comparative study of integration policies in multicultural societies. GSC Advanced Research and Reviews, 19(2), 032-042. DOI: 10.30574/gscarr.2024.19.2.0165

Mousavi, S., Bossink, B., & Van Vliet, M. (2020). Dynamic capabilities and organizational routines for managing innovation towards sustainability. *Journal of Cleaner Production, 247*, 119–135. DOI: 10.1016/j.jclepro.2019.119648

Moyer, J. D., & Bohl, D. K. (2019). Alternative pathways to human development: Assessing trade-offs and synergies in achieving the Sustainable Development Goals. *Futures, 105*, 199–210. DOI: 10.1016/j.futures.2018.10.007

Mufungizi, I. (2024). Mineral potential facing socio-economic development challenges: Case study of the Democratic Republic of Congo, a 'geological scandal.'. *International Geology Review*, 1–23. DOI: 10.1080/00206814.2024.2401575

Murdiyarso, D., Purbopuspito, J., Kauffman, J. B., Warren, M. W., Sasmito, S. D., Donato, D. C., Manuri, S., Krisnawati, H., Taberima, S., & Kurnianto, S. (2015). The potential of Indonesian mangrove forests for global climate change mitigation. *Nature Climate Change*, *5*(12), 1089–1092. DOI: 10.1038/nclimate2734

Muremela, M. G., Kutame, A. P., Kapueja, I. S., & Ajani, O. A.Matodzi Grace MuremelaAzwidohwi Phillip KutameIngrid Sibongile KapuejaOluwatoyin Ayodele Ajani. (2024). Optimising the recruitment of teachers for scarce skills subjects in rural secondary schools: A study of practices in african countries. *International Journal of Innovative Technologies in Social Science*, (4(44)). Advance online publication. DOI: 10.31435/ijitss.4(44).2024.3116

Mutawa, A. M. (2025). Enforcing the Ethics of Artificial Intelligence in Education: To Prevent or Brace AI Tools in the Educational System. In *Impacts of Generative AI on the Future of Research and Education* (pp. 143-168). IGI Global.

Nagel, A. S. (2023). Unilateral extraterritorial sanctions: The search for a jurisdictional justification under international law. *LSE Law Review*, *8*(3), 368–395. DOI: 10.61315/lselr.478

Nakamura, J., Diz, D., & Morgera, E. (2022). International legal requirements for environmental and socio-cultural assessments for large-scale industrial fisheries. *Review of European, Comparative & International Environmental Law*, *31*(3), 336–348. DOI: 10.1111/reel.12462

Narasimham, R. (2010). *Enhancing Financial Disclosure Standards in Transitional Economies*. Asian Development Bank.

National Economic and Development Authority. (2017). *Philippine Development Plan (PDP) 2017-2022*. https://pdp.neda.gov.ph/wp-content/uploads/2021/02/Pre -publication-copy-Updated-PD P-2017-2022.pdf

National Economic and Social Development Council. (n.d.). The Thirteenth National Economic And Social Development Plan (2023-2027).

Nayan, H. M. B. H., & Ullah, A. K. M. (2021). Indian migrants in Brunei: The role of social networks. *Migration Letters : An International Journal of Migration Studies*, *18*(6), 697–709.

NewClimate Institute. (https://newclimate.org/)

Newig, J., Challies, E., Jager, N. W., Kochskaemper, E., & Adzersen, A. (2018). The environmental performance of participatory and collaborative governance: A framework of causal mechanisms. *Policy Studies Journal: the Journal of the Policy Studies Organization*, *46*(2), 269–297. DOI: 10.1111/psj.12209 PMID: 30034065

Newig, J., & Fritsch, O. (2009). Environmental governance: Participatory, multi-level—and effective? *Environmental Policy and Governance*, *19*(3), 197–214. DOI: 10.1002/eet.509

Newman, P. (2000). Changing patterns of regional governance in the EU. *Urban Studies (Edinburgh, Scotland)*, *37*(5-6), 895–908. DOI: 10.1080/00420980050011145

Newton, I. (1687). *Philosophiæ naturalis principia mathematica*. Royal Society. DOI: 10.5479/sil.52126.39088015628399

Nguyen, D. T., Le, T. H., Do, D. D., & Nguyen, H. N. (2023). Does geopolitical risk hinder sustainable development goals? Evidence from a panel analysis. *Journal of Environmental Management*, *347*, 119204. DOI: 10.1016/j.jenvman.2023.119204

Nguyen, D., & Ng, D. (2020). Teacher collaboration for change:sharing, improving, and spreading. *Professional Development in Education*, *46*(4), 638–651. DOI: 10.1080/19415257.2020.1787206

Nguyen, X. H., Nguyen, L. T., & Pham, M. A. (2023). Influence of Green Supply Chain Management on Business Performance of Vietnamese Electronic Firms. *International Journal of Research and Review*, *10*(2), 753–771. DOI: 10.52403/ijrr.20230291

Niaz, M. U. (2021). Socio-economic development and sustainable development goals: A roadmap from vulnerability to sustainability through financial inclusion. *Ekonomska Istrazivanja*, *35*(1), 3243–3275. DOI: 10.1080/1331677X.2021.1989319

Nilsson, M., Chisholm, E., Griggs, D., Howden-Chapman, P., McCollum, D., Messerli, P., Neumann, B., Stevance, A. S., Visbeck, M., & Stafford-Smith, M. (2018). Mapping interactions between the sustainable development goals: Lessons learned and ways forward. *Sustainability Science*, *13*(6), 1489–1503. DOI: 10.1007/s11625-018-0604-z

Ningsih, F. O. (2024). *Analyzing the Implementation of Green Industries Practice at PT . Semen Indonesia (Persero) Tbk . Tuban Plant towards Achieving Sustainable Development Goals. 9*(5).

Nogueiro, T., Saraiva, M., Jorge, F., & Chaleta, E. (2022). The Erasmus+ Programme and Sustainable Development Goals—Contribution of mobility actions in higher education. *Sustainability (Basel)*, *14*(3), 1628. DOI: 10.3390/su14031628

Nonaka, I., & Takeuchi, H. (1995). The knowledge-creating company. New York: Oxford University Press. Retrieved on 23 September 2021 from https://memberfiles.freewebs.com/84/90/65819084/documents/The%20Knowledge-Creating%20Company.pdf

Nonaka, I., & Takeuchi, H. (1995). *The knowledge-creating company: How Japanese companies create the dynamics of innovation*. Oxford University Press. DOI: 10.1093/oso/9780195092691.001.0001

Nonet, G. A. H., Gössling, T., Van Tulder, R., & Bryson, J. M. (2022). Multi-stakeholder engagement for the Sustainable Development Goals: Introduction to the special issue. *Journal of Business Ethics, 180*(4), 945–957. DOI: 10.1007/s10551-022-05192-0

Noor, Z., & Pickup, F. (2017). The role of zakat in supporting the Sustainable Development Goals. Brief. May 2017.

Norabidin, R. N. R. (2016). Zakāh Collection and Distribution Framework in Federal Territory of Kuala Lumpur: A Study on Selected Islamic Banking Institutions. International Journal of Management and Applied Research, 4(1). DOI: 10.18646/2056.41.17-002

Norjidi, D. (2018). SME Bank Receives 134 Applications for Finance. *Borneo Bulletin*. Retrieved on 3 March 2020, from https://borneobulletin.com.bn/sme-bank-receives-134-applications-for-finance/

Novakovic, A., Michel, R. E., & Ockerman, M. S. (2020). Teaching school counselors to use evidence-based practice to advance social justice: A case study. *Professional School Counseling, 23*(1_part_3), 2156759X20904472. Advance online publication. DOI: 10.1177/2156759X20904472

Núñez, J. E. (2024). State sovereignty: Concept and conceptions. *International Journal for the Semiotics of Law, 37*(7), 2131–2150. DOI: 10.1007/s11196-024-10170-y

Nurasyiah, A., Adam, F., & Pertiwi, R. S. (2019). Efficiency and Productivity of Zakat Institution in Malaysia and Indonesia: The Comparative Study. International Conference of Zakat 2019 Proceedings.

Nurfitriana. (2008). Analisis perlakuan zakat dalam perhitungan pajak penghasilan orang pribadi: "studi komparasi" antara peraturan perpajakan Indonesia dengan Malaysia. Fakultas Ekonomi dan Bisnis Universitas Indonesia.

Nurhapsari, F., Dwijayanti, R., Ramadhani, R., & Al Azizah, U. S. (2021). The Role of Zakat for Poverty Alleviation in Indonesia. 5th International Conference of Zakat Proceedings

Nurkhozin. (2021). Manajemen Pendayagunaan Zakat Tinjauan Tujuan Pembangunan Berkelanjutan (Sustainable Development Goals). Jurnal An-Nahl, 8(1), 52-62.

Nussbaum, M. C. (2000). *Women and Human Development: The Capabilities Approach*. Cambridge University Press. DOI: 10.1017/CBO9780511841286

Nwankpa, J. K., & Merhout, J. W. (2020). Exploring the effect of digital investment on IT innovation. *Sustainability (Basel)*, *12*(18), 7374. Advance online publication. DOI: 10.3390/su12187374

O'Brien, R. M. (2007). A caution regarding rules of thumb for variance inflation factors. *Quality & Quantity*, *41*(5), 673–690. DOI: 10.1007/s11135-006-9018-6

O'Flaherty, M., & O'Brien, C. (2007). Reform of UN human rights treaty monitoring bodies: A critique of the concept paper on the High Commissioner's proposal for a unified standing treaty body. *Human Rights Law Review*, *7*(1), 141–172. DOI: 10.1093/hrlr/ngl035

OECD. (2020). *How Islamic Finance Contributes To Achieving The Sustainable Development Goals. OECD Development. Policy Papers. June 2020 No. 30*. OECD Publishing.

Office of the National Economic and Social Development Board Office of the Prime Minister Bangkok. (2017). *The Twelfth National Economic And Social Development Plan*.

Ofori, E. K., & Appiah-Opoku, S. (2024). Sustainable Development Goals in BRICS and G7 countries: Increasing accomplishments through policy synergies in four dimensions. *Sustainable Development (Bradford)*, *32*(1), 287–306. DOI: 10.1002/sd.2653

Okitasari, M., & Katramiz, T. (2022). The national development plans after the SDGs: Steering implications of the global goals towards national development planning. *Earth System Governance*, *12*, 100136. DOI: 10.1016/j.esg.2022.100136

Okombi, I. F., & Ndoum Babouama, V. B. D. (2024). Environmental taxation and inclusive green growth in developing countries: Does the quality of institutions matter? *Environmental Science and Pollution Research International*, *31*(21), 30633–30662. DOI: 10.1007/s11356-024-33245-6

Oliveira, S. V., Gurgel, H. C., & Oliveira, M. A. B. (2021). Geographical information systems in public health: An overview of the Brazilian experience. *Global Health Action*, *14*(1), 1840285.

Olkin, I., & Finn, J. D. (1995). Correlations redux. *Psychological Bulletin*, *118*(1), 155–164. DOI: 10.1037/0033-2909.118.1.155

Ono, Y., & Ferreira, J. (2010). A case study of continuing teacher professional development through lesson study in South Africa. *South African Journal of Education, 30*(1), 59–75. DOI: 10.15700/saje.v30n1a320

Oore-Ofe, O. R., & Ojakorotu, V. (2023). Conflict in Democratic Republic of Congo: Framework for resolution. In Ani, K. J. (Ed.), *Resource conflict and environmental relations in Africa* (pp. 121–137). Palgrave Macmillan., DOI: 10.1007/978-981-19-7343-7_8

Ordóñez de Pablos, P., Anshari, M., & Almunawar, M. N. (Eds.). (2024). *Harnessing Green and Circular Skills for Digital Transformation*. IGI Global. DOI: 10.4018/979-8-3693-2865-1

Orefice, G. (2017). Non-tariff measures, specific trade concerns and tariff reduction. *World Economy, 40*(8), 1807–1835. DOI: 10.1111/twec.12447

Orford, A. (2021). Regional orders, geopolitics, and the future of international law. *Current Legal Problems, 74*(1), 149–194. DOI: 10.1093/clp/cuab005

Organisation for Economic Co-operation and Development (OECD). (2020). *A territorial approach to the Sustainable Development Goals: Synthesis report*. OECD Publishing. https://www.oecd.org/en/publications/a-territorial-approach-to-the-sustainable-development-goals_e86fa715-en.html

Organisation for Economic Co-operation and Development (OECD). (2020). *Blended Finance for Sustainable Development*. OECD Publishing.

Osnoss, J. (2010). *Some Points to Ponder in the IFRS Debate*. Retrieved on 5 February 2020, from https://www.accountingtoday.com/news/some-points-to-ponder-in-the-ifrs-debate

Othman, A. (2017). SMEs Must Embrace Technology, Develop Tech-Advanced Businesses. *Borneo Bulletin*. Retrieved on 23 March 2020, from https://borneobulletin.com.bn/smes-must-embrace-technology-develop-tech-advanced-businesses/page/1936/

Oulu, M. (2015). Climate change governance: Emerging legal and institutional frameworks for developing countries. In W. Leal Filho (Ed.), *Handbook of climate change adaptation* (pp. [insert page numbers if available]). Springer, Berlin, Heidelberg. https://doi.org/DOI: 10.1007/978-3-642-38670-1_9

Our World in Data Team. (2023). SDG Tracker: Measuring progress towards the Sustainable Development Goals. OurWorldinData.org. URL: https://ourworldindata.org/sdgs

Ovcharuk, O., Ivaniuk, I., & Leshchenko, M. (2023). Impact of school lockdown on access to online instruction during the war in ukraine. *European Journal of Education, 58*(4), 561–574. DOI: 10.1111/ejed.12589

Overland, I., Sagbakken, H. F., Chan, H., Merdekawati, M., Suryadi, B., Utama, N. A., & Vakulchuk, R. (2020). The ASEAN climate and energy paradox. *Energy and Climate Change, 2*, 100019. DOI: 10.1016/j.egycc.2020.100019

Özdamar, Ö., & Shahin, E. (2021). Consequences of economic sanctions: The state of the art and paths forward. *International Studies Review, 23*(4), 1646–1671. DOI: 10.1093/isr/viab029

Özkan, H., & Salepçioğlu, M. A. (2022). Does Organizational Agility Affect Sustainable Quality Perception? The Mediating Role of Innovation Orientation. *International Journal of Business, 27*(4). Advance online publication. DOI: 10.55802/IJB.027(4).002

Paasi, A. (1986). The institutionalisation of regions. *Fennia, 164*(1), 105–146. DOI: 10.11143/9052

Pacific Disaster Center. (2020). *DisasterAWARE platform: Enhancing disaster resilience.*

Paes, F., & Geraldes, A. (2019). Convention on the Elimination of All Forms of Discrimination Against Women. In Leal Filho, W., Azul, A., Brandli, L., Özuyar, P., & Wall, T. (Eds.), *Gender equality. Encyclopedia of the UN Sustainable Development Goals*. Springer., DOI: 10.1007/978-3-319-70060-1_39-1

Pagiola, S. (2008). Payments for environmental services in Costa Rica. *Ecological Economics, 65*(4), 712–724. DOI: 10.1016/j.ecolecon.2007.07.033

Paizin, M. N. (2017). *Usage of Islamic Alms's Online Portal: Case Study on Pusat Pungutan Zakat (PPZ)*. Federal Territories of Malaysia. DOI: 10.4172/2375-4516.1000175

Palea, V. (2013). IAS/IFRS and Financial Reporting Quality: Lessons from the European experience. *China Journal of Accounting Research, 6*(4), 247–263. DOI: 10.1016/j.cjar.2013.08.003

Palei, T. (2015). Assessing the impact of infrastructure on economic growth and global competitiveness. *Procedia Economics and Finance, 23*, 168–175. DOI: 10.1016/S2212-5671(15)00322-6

Parashar, S., & Schulz, M. (2021). Colonial legacies, postcolonial 'selfhood' and the (un)doing of Africa. *Third World Quarterly*, *42*(5), 867–881. DOI: 10.1080/01436597.2021.1903313

Paris Agreement. (https://unfccc.int/process-and-meetings/the-paris-agreement)

Parker, L., Xu, S., & Chi, C. (2022). Chinese preservice teachers' perspectives of mentoring relationships in an international learning partnership. *Journal of Teacher Education*, *73*(5), 525–537. DOI: 10.1177/00224871221108655

Patro, A., & Gupta, V. K. (2014). Impact of International Financial Reporting Standards on Cost of equity Capital for Asian countries. *International Journal of Accounting and Financial Reporting*, *4*(2), 148–170. DOI: 10.5296/ijafr.v4i2.6462

Pattberg, P., Kaiser, C., Widerberg, O., & Stripple, J. (2022). 20 years of global climate change governance research: Taking stock and moving forward. *International Environmental Agreement: Politics, Law and Economics*, *22*(2), 295–315. DOI: 10.1007/s10784-022-09568-5

Pattiroei, A. A., Anshari, M., Hamdan, M., & Ali, E. (2023). Radiological Detector Aircraft for Marine Life's Sustainability in Supporting Sustainable Development Goals (SDGs). In *Public Policy's Role in Achieving Sustainable Development Goals* (pp. 261-278). IGI Global.

Paul, T. V., He, K., & Wivel, A. (2025). Soft balancing in the regions: Causes, characteristics and consequences. *International Affairs*, *101*(1), 3–15. DOI: 10.1093/ia/iiae286

PDP NEDA Government Philippines. (2024). *Development Plans - Philippine Development Plan*. Philippine Development Plan. Retrieved July 19, 2024, from https://pdp.neda.gov.ph/devt-plans/

Peeraer, J., & Petegem, P. V. (2012). The limits of programmed professional development on integration of information and communication technology in education. *Australasian Journal of Educational Technology*, *28*(6). Advance online publication. DOI: 10.14742/ajet.809

Pena, H. F. P., & Franco, J. B. (2017). Impact of IFRS on the Quality of Financial Information in the United Kingdom and France: Evidence from a New Perspective. *Intangible Capital*, *13*(4), 850–878. DOI: 10.3926/ic.939

Perera, D., & Chand, P. (2015). Issues in the Adoption of International Financial Reporting Standards (IFRS) for Small and Medium-Sized Enterprises (SMES). *Advances in Accounting*, *31*(1), 165–178. DOI: 10.1016/j.adiac.2015.03.012

Perez, M. F., Brada, J. C., & Drabek, Z. (2012). Illicit money flows as motives for FDI. *Journal of Comparative Economics*, *40*(1), 108–126. DOI: 10.1016/j.jce.2011.03.007

Perifanou, M., Tzafilkou, K., & Economides, A. A. (2022). Teacher intention to transfer ict training when integrating digital technologies in education: The teacher transfer of ict training model (tetra-ict). *European Journal of Education*, *58*(1), 111–129. DOI: 10.1111/ejed.12534

Permadi, A., Mursitama, T. N., Hamsal, M., & Bandur, A. (2020). The impact of IT capability and supply chain agility in the electronic home appliance retail at Indonesia. *International Journal of Advanced Trends in Computer Science and Engineering*, *9*(1), 833–844. DOI: 10.30534/ijatcse/2020/120912020

Perna, L. W., Walsh, E., & Fester, R. (2010). Promoting the educational attainment of adults. *Educational Policy*, *25*(6), 935–963. DOI: 10.1177/0895904810386599

Perry, E., & Bevins, S. (2018). Building capacity for professional development: The development of teachers as facilitators in ghana. *Professional Development in Education*, *45*(3), 390–404. DOI: 10.1080/19415257.2018.1474489

Peterson, L., & Skovgaard, J. (2019). Bureaucratic politics and the allocation of climate finance. *World Development*, *117*, 72–97. DOI: 10.1016/j.worlddev.2018.12.011

Philip, T. M. (2011). Moving beyond our progressive lenses. *Journal of Teacher Education*, *62*(4), 356–366. DOI: 10.1177/0022487111409414

Pierre, J. (2016). Urban and regional governance. In *Handbook on theories of governance* (pp. 477–485). Edward Elgar Publishing. DOI: 10.4337/9781782548508.00051

Pieterse, J. N. (2009). *Globalization and culture: Global mélange*. Rowman & Littlefield.

Piketty, T. (2014). *Capital in the Twenty-First Century*. Harvard University Press. DOI: 10.4159/9780674369542

Pisillo Mazzeschi, R., & Viviani, A. (2018). General principles of international law: From rules to values? In Pisillo Mazzeschi, R., & De Sena, P. (Eds.), *Global justice, human rights and the modernization of international law* (pp. 113–161). Springer., DOI: 10.1007/978-3-319-90227-2_6

Pogge, T., & Sengupta, M. (2016). Assessing the Sustainable Development Goals from a human rights perspective. *Journal of International and Comparative Social Policy*, *32*(2), 83–97. DOI: 10.1080/21699763.2016.1198268

Porter, M. E., & Kramer, M. R. (2011). Creating shared value. *SSRN*, *89*(1–2). Advance online publication. DOI: 10.2139/ssrn.3683975

Prime Minister's Office (PMO). Brunei. (2021). *National Climate Change Policy: Strategies and Action Plan*. Bandar Seri Begawan: Prime Minister's Office.

Pundhir, S., Marwah, T., & Gupta, V. (2025). Strengthening Rural Women's Entrepreneurship for Sustainable Development: The Impact of Digital Tools, Government Policies, and NGO Support. In *Empowering Women Through Rural Sustainable Development and Entrepreneurship* (pp. 287-308). IGI Global Scientific Publishing.

Purba Rao, D. H. (2005). Do green supply chains lead to competitiveness and economic performance? *International Journal of Operations & Production Management*, *25*(9), 898–916. https://doi.org/https://doi.org/10.1108/01443570510613956. DOI: 10.1108/01443570510613956

Purvis, B., Mao, Y., & Robinson, D. (2019). Three pillars of sustainability: In search of conceptual origins. *Sustainability Science*, *14*(3), 681–695. DOI: 10.1007/s11625-018-0627-5

Putra, A. A., Hasibuan, H. S., Tambunan, R. P., & Lautetu, L. M. (2024). Integration of the Sustainable Development Goals into a Regional Development Plan in Indonesia. *Sustainability (Basel)*, *16*(23), 10235. DOI: 10.3390/su162310235

Putra, B. A. (2024). The politics of countering climate change in Southeast Asia. *Frontiers in Environmental Science*, *12*, 1486796. Advance online publication. DOI: 10.3389/fenvs.2024.1486796

Qosasi, A., Permana, E., Muftiadi, A., Purnomo, M., & Maulina, E. (2019). Building SMEs' competitive advantage and the organizational agility of apparel retailers in indonesia: The role of ICT as an initial trigger. *Gadjah Mada International Journal of Business*, *21*(1), 69–90. DOI: 10.22146/gamaijb.39001

Qurroh, A., Pramanik, A. H., Saad, N. M., & Ariffin, M. I. (2017). The Comparison between Consumption and Production-based Zakat Distribution Programs for Poverty Alleviation and Income Inequality Reduction. *International Journal of Zakat*, *2*(2), 11–28. DOI: 10.37706/ijaz.v2i2.22

Rabal Conesa, J., Jiménez Jiménez, D., & Martínez Costa, M. (2024). Performing in green: Towards eco-innovation through agility and customer involvement. *Journal of Organizational Change Management*, *37*(3), 641–658. Advance online publication. DOI: 10.1108/JOCM-09-2023-0359

Rabal-Conesa, J., Jiménez-Jiménez, D., & Martínez-Costa, M. (2022). Organisational agility, environmental knowledge and green product succes. *Journal of Knowledge Management, 26*(9), 2440–2462. DOI: 10.1108/JKM-06-2021-0486

Raikes, A. (2017). Measuring child development and learning. *European Journal of Education, 52*(4), 511–522. DOI: 10.1111/ejed.12249

Ramadan, A. B., & Hussain, M. (2019). Zakat Institutions and Universities Strategic Partnership for Poverty Alleviation. *International Journal of Zakat, 4*(2), 73–86. DOI: 10.37706/ijaz.v4i2.172

Rawls, J. (1971). *A Theory of Justice*. Harvard University Press. DOI: 10.4159/9780674042605

Reed, M. S. (2008). Stakeholder participation for environmental management: A literature review. *Biological Conservation, 141*(10), 2417–2431. DOI: 10.1016/j.biocon.2008.07.014

Reed, M. S., Vella, S., Challies, E., de Vente, J., Frewer, L., Hohenwallner-Ries, D., Huber, T., Neumann, R. K., Oughton, E. A., Sidoli del Ceno, J., & van Delden, H. (2017). A theory of participation: What makes stakeholder and public engagement in environmental management work? *Restoration Ecology, 26*(S1). Advance online publication. DOI: 10.1111/rec.12541

Regional Centre for Training and Application in Agrometeorology and Operational Hydrology (AGRHYMET). AGRHYMET Regional Centre, A capacity assessment, 2022 (https://www.agrisource.org/en/7_108/5c332f1507c805cd14cf5232/agrhymet-regional-centre.html)

Regions4 (https://regions4.org/)

Regoli, N. (2019). 19 Advantages and Disadvantages of Adopting IFRS. Connect Us. Retrieved on 22 March 2020, from https://connectusfund.org/6-advantages-and-disadvantages-of-adopting-ifrs

Rehman, A. A., & Pickup, F. (2018). Zakat for the SDGs. *Islamic Finance News, 15*(34).

Relph, E. (1976). *Place and placelessness*. Pion.

Reserve Bank of India (RBI). (2023). *Report on State Finances*. RBI Publications.

Reuters. (2024). How AI can supercharge the Sustainable Development Goals. https://www.reuters.com/sustainability/society-equity/comment-heres-how-we-can-use-ai-supercharge-sustainable-development-goals-2024-07-23

Rice, J. K., Malen, B., Baumann, P., Chen, E., Dougherty, A., Hyde, L., Jackson, C., Jacobson, R., & McKithen, C. (2012). The persistent problems and confounding challenges of educator incentives. *Educational Policy*, *26*(6), 892–933. DOI: 10.1177/0895904812465708

Richards, D. R., & Friess, D. A. (2016). Rates and drivers of mangrove deforestation in Southeast Asia, 2000–2012. *Proceedings of the National Academy of Sciences of the United States of America*, *113*(2), 344–349. DOI: 10.1073/pnas.1510272113 PMID: 26712025

Robertson, R. (1995). Glocalization: Time-space and homogeneity-heterogeneity. In M.

Robinson, J., Harrison, P., Croese, S., Sheburah Essien, R., Kombe, W., Lane, M., Mwathunga, E., Owusu, G., & Yang, Y. (2025). Reframing urban development politics: Transcalarity in sovereign, developmental and private circuits. *Urban Studies (Edinburgh, Scotland)*, *62*(1), 00420980241284763. DOI: 10.1177/00420980241284763

Rockström, J., Steffen, W., Noone, K., Persson, Å., Chapin, F. S.III, Lambin, E. F., Lenton, T. M., Scheffer, M., Folke, C., Schellnhuber, H. J., Nykvist, B., de Wit, C. A., Hughes, T., van der Leeuw, S., Rodhe, H., Sörlin, S., Snyder, P. K., Costanza, R., Svedin, U., & Foley, J. A. (2009). A safe operating space for humanity. *Nature*, *461*(7263), 472–475. DOI: 10.1038/461472a

Roldán-álvarez, D., Martínez-Martínez, F., Martín, E., & Haya, P. A. (2021). Understanding Discussions of Citizen Science Around Sustainable Development Goals in Twitter. *IEEE Access : Practical Innovations, Open Solutions*, *9*, 144106–144120. DOI: 10.1109/ACCESS.2021.3122086

Roy, A. (2017). The Barefoot College: Women's empowerment through solar energy. *Renewable Energy Journal*, *25*(3), 223–230.

Rudall, J. (2021). The Obligation to Cooperate in the Fight against Climate Change. *International Community Law Review*, *23*(2–3), 184–196. DOI: 10.1163/18719732-12341469

Rus, H. A. (2014). Corruption, conflict, and the management of natural resources. *Economics of Governance*, *15*(4), 355–386. DOI: 10.1007/s10101-014-0148-3

Sa'a, E. A., & Asplund, F. (2025). Unpacking social capital in University–Industry Collaborations: Pathways to cross-industry knowledge sharing. *Technovation*, *140*, 103160. DOI: 10.1016/j.technovation.2024.103160

Saba, C. S., & Ngepah, N. (2024). The impact of artificial intelligence (AI) on employment and economic growth in BRICS: Does the moderating role of governance Matter? *Research in Globalization, 8*, 100213. Advance online publication. DOI: 10.1016/j.resglo.2024.100213

Sachs, J. D., Schmidt-Traub, G., Kroll, C., Lafortune, G., & Fuller, G. (2019). SDG Index and Dashboards Report 2019. *New York: Bertelsmann Stiftung and Sustainable Development Solutions Network (SDSN).*

Sachs, J. D. (2012). From Millennium Development Goals to Sustainable Development Goals. *Lancet, 379*(9832), 2206–2211. DOI: 10.1016/S0140-6736(12)60685-0

Sachs, J. D. (2015). *The Age of Sustainable Development*. Columbia University Press. DOI: 10.7312/sach17314

Sachs, J. D., Schmidt-Traub, G., Kroll, C., Lafortune, G., & Fuller, G. (2019). *The sustainable development goals and Covid-19: Sustainable development report 2019*. Cambridge University Press.

Sachs, J. D., Schmidt-Traub, G., Kroll, C., Lafortune, G., & Fuller, G. (2021). *The Decade of Action for the Sustainable Development Goals: Sustainable Development Report 2021*. Cambridge University Press., DOI: 10.1017/9781108646967

Saeed, U. F., Kamil, R., & Wiredu, I. (2025). The roles of ICT and governance quality in the finance-growth nexus of developing countries: A dynamic GMM approach. *Cogent Economics & Finance, 13*(1), 2448228. DOI: 10.1080/23322039.2024.2448228

Sahoo, M., Bhujabal, P., Gupta, M., & Islam, M. K. (2024). Empowering BRICS economies: The crucial role of green finance, information and communication technology and innovation in sustainable development. *Sustainable Development (Bradford), 32*(6), 7292–7308. DOI: 10.1002/sd.3083

Saigal, A., Kogekar, U., & Kong, P. (2022). Sustaining student learning in India through quality teacher support during the COVID-19 pandemic. https://inee.org/sites/default/files/resources/TiCC%20Case%20Studies%20v2%20-%20TPD%20v1.0%20EN%20LowRes.pdf

Sandage, J. (2014). The universal approach of the United Nations Convention against Corruption. In S. Manacorda, F. Centonze, & G. Forti (Eds.), *Preventing corporate corruption* (pp. [insert page numbers if available]). Springer, Cham. https://doi.org/ DOI: 10.1007/978-3-319-04480-4_2

Sankaran, K. (2022). Transition from the informal to the formal economy: The need for a multi-faceted approach. *The Indian Journal of Labour Economics : the Quarterly Journal of the Indian Society of Labour Economics*, *65*(3), 625–642. DOI: 10.1007/s41027-022-00398-2

Santosus, M., & Surmacz, J. (2001). The ABC's of Knowledge Management. Retrieved on 23 September 2021 from https://www.globalbioenergy.org/uploads/media/ABC_of_KM.pdf

Sapovadia, V. K. (2017). Importance of aligning IFRS with SDGs. *Available at SSRN* 3042562.

Sari, F. (2022). Gandeng Baznas, Bank BJB Mudahkan Masyarakat Salurkan Zakat dan Infak. Kontan. https://pressrelease.kontan.co.id/news/gandeng-baznas-bank-bjb-mudahkan-masyarakat-salurkan-zakat-dan-infak

Sauer, C. (1925). The morphology of landscape. *University of California Publications in Geography*, *2*(2), 19–53.

Schaefer, F. K. (1953). Exceptionalism in geography: A methodological examination. *Annals of the Association of American Geographers*, *43*(3), 226–249. DOI: 10.1080/00045605309352114

Schiederig, T., Tietze, F., & Herstatt, C. (2012). Green innovation in technology and innovation management - an exploratory literature review. *R & D Management*, *42*(2), 180–192. DOI: 10.1111/j.1467-9310.2011.00672.x

Schmidt-Traub, G. (2015). *Investment needs to achieve the Sustainable Development Goals*. Sustainable Development Solutions Network.

Schrijver, N. (2016). Managing the global commons: Common good or common sink? *Third World Quarterly*, *37*(7), 1252–1267. DOI: 10.1080/01436597.2016.1154441

Sciaccaluga, G. (2020). Why states should protect forced climate migrants. In *International law and the protection of*. In *climate refugees"*. Palgrave Macmillan., DOI: 10.1007/978-3-030-52402-9_9

SDSN. (2021). *Sustainable development solutions network: Knowledge for a sustainable future*.

Sen, A. (1999). *Development as Freedom*. Oxford University Press.

Sengupta, A. (2004). The human right to development. *Oxford Development Studies*, *32*(2), 179–203. DOI: 10.1080/1360081041000169948

Setzer, J., De Murieta, E. S., Galarraga, I., Rei, F., & Pinho, M. M. L. (2020). Transnationalization of climate adaptation by regional governments and the RegionsAdapt initiative. *Global Sustainability*, *3*, e10. Advance online publication. DOI: 10.1017/sus.2020.6

Shaffer, G., & Sandholtz, W. (2024). The rule of law under pressure: A transnational perspective. *Hague Journal on the Rule of Law*, *16*(2), 393–438. DOI: 10.1007/s40803-024-00210-x

Shah, S. A. R., Abbas, N., Serbanescu, L., Niu, R., & Nassani, A. A. (2025). The key challenges and best alternatives to environmental sustainability: A comprehensive study. *Scientific Reports*, *15*(1), 7042. DOI: 10.1038/s41598-025-90187-w

Shaikh, S. A., & Hassan, M. K. (2020). *Where Do Muslim Countries Stand On Sustainable Development Goals. Islamic Perspective For Sustainable Financial System 87-108*. Istanbul University Press.

Sharairi, M. H. (2020). Factors that Influenced the Adoption of IFRS by Islamic Banks in the UAE. *Accounting Research Journal*, *33*(1), 75–91. DOI: 10.1108/ARJ-11-2017-0185

Sharairi, M. H., & Islam, J. (2019). Comparison Study between IFRS and AAOIFI Accounting Standards: A Case Study of Islamic Banks in the United Arab Emirates. *Communications on Stochastic Analysis*, *12*(5), 23–32.

Shaw, R., Luo, Y., Cheong, T. S., Abdul Halim, S., Chaturvedi, S., Hashizume, M., Insarov, G. E., Ishikawa, Y., Jafari, M., Kitoh, A., Pulhin, J., Singh, C., Vasant, K., & Zhang, Z. (2022). Asia. In: *Climate Change 2022: Impacts, Adaptation and Vulnerability.* Contribution of Working Group II to the Sixth Assessment Report of the Intergovernmental Panel on Climate Change [H.-O. Pörtner, D.C. Roberts, M. Tignor, E.S. Poloczanska, K. Mintenbeck, A. Alegría, M. Craig, S. Langsdorf, S. Löschke, V. Möller, A. Okem, B. Rama (eds.)]. Cambridge University Press, Cambridge, UK and New York, NY, USA, pp. 1457–1579, DOI: 10.1017/9781009325844.012

Shayan, N. F., Mohabbati-Kalejahi, N., Alavi, S., & Zahed, M. A. (2022). Sustainable Development Goals (SDGs) as a framework for Corporate Social Responsibility (CSR). *Sustainability (Basel)*, *14*(3), 1222. DOI: 10.3390/su14031222

Shiva, V. (2016). Who Really Feeds the World? North Atlantic Books., ISBN-10.

Singh, S., Sharma, G. D., Parihar, J. S., Dev, D., & Bansal, P. (2024). Nexus between environmental degradation and climate change during the times of global conflict: Evidence from CS-ARDL model. *Environmental and Sustainability Indicators*, *22*, 100368. DOI: 10.1016/j.indic.2024.100368

Singh, Z. (2016). Sustainable development goals: Challenges and opportunities. *Indian Journal of Public Health*, *60*(4), 247–250. DOI: 10.4103/0019-557X.195862 PMID: 27976644

Smith, K., & Flores, M. A. (2019). The janus faced teacher educator. *European Journal of Teacher Education*, *42*(4), 433–446. DOI: 10.1080/02619768.2019.1646242

Snoek, M., Dengerink, J., & Wit, B. (2019). Reframing the teacher profession as a dynamic multifaceted profession: A wider perspective on teacher quality and teacher competence frameworks. *European Journal of Education*, *54*(3), 413–425. DOI: 10.1111/ejed.12350

Snoek, M., Swennen, A., & Klink, M. (2011). The quality of teacher educators in the European policy debate: Actions and measures to improve the professionalism of teacher educators. *Professional Development in Education*, *37*(5), 651–664. DOI: 10.1080/19415257.2011.616095

Soares, F. V., Ribas, R. P., & Osório, R. G. (2011). Evaluating the impact of Brazil's Bolsa Família: Cash transfer programmes in comparative perspective. *Latin American Research Review*, *45*(2), 173–190. DOI: 10.1017/S0023879100009390

Socio-Cultural Community, A. S. E. A. N. (ASSC) (https://asean.org/our-communities/asean-socio-cultural-community/)

Soleri, P. (1969). *Arcology, City in the Image of Man*. MIT Press.

Soo-Chen, K., & McCoy, D. (2022). *Climate Change in South-East Asia: Where Are We and What Are We Bound For?* United Nations University.

Sorooshian, S. (2024). The sustainable development goals of the United Nations: A comparative midterm research review. *Journal of Cleaner Production*, *453*, 142272. DOI: 10.1016/j.jclepro.2024.142272

South Asian Association for Regional Cooperation (SAARC). (https://www.saarc-sec.org/)

Spalding, M., & Parrett, C. L. (2019). Global patterns in mangrove recreation and tourism. *Marine Policy*, *110*, 103540. DOI: 10.1016/j.marpol.2019.103540

Spash, C. L. (2019). Time for a Paradigm Shift: From Economic Growth and Price-Making Markets to Social Ecological Economics. WU Vienna University of Economics and Business. SRE - Discussion Papers No. 2019/07 DOI: 10.57938/36774aa4-5549-4b97-89cc-ca64006577f2

Spash, C. L. (2020). Apologists for growth: Passive revolutionaries in a passive revolution. *Globalizations*, *18*(7), 1123–1148. DOI: 10.1080/14747731.2020.1824864

Spyromitros, E., & Panagiotidis, M. (2022). The impact of corruption on economic growth in developing countries and a comparative analysis of corruption measurement indicators. *Cogent Economics & Finance*, *10*(1), 2129368. DOI: 10.1080/23322039.2022.2129368

Staněk, M. (2017). The humanitarian crisis and civil war in Syria: Its impact and influence on the migration crisis in Europe. *Kontakt*, *19*(4), e270–e275. DOI: 10.1016/j.kontakt.2017.09.013

Sterling, E. J., Betley, E., Sigouin, A., Gomez, A., Toomey, A., Cullman, G., Malone, C., Pekor, A., Arengo, F., Blair, M., Filardi, C., Landrigan, K., & Porzecanski, A. L. (2017). Assessing the evidence for stakeholder engagement in biodiversity conservation. *Biological Conservation*, *209*, 159–171. DOI: 10.1016/j.biocon.2017.02.008

Stone, C. D. (2004). Common but differentiated responsibilities in international law. *The American Journal of International Law*, *98*(2), 276–301. DOI: 10.2307/3176729

Storper, M. (1998). *The regional world*. The Guildford Press.

Street, R., Buontempo, C., Mysiak, J., Karali, E., Pulquério, M., Murray, V., & Swart, R. (2018). How could climate services support disaster risk reduction in the 21st century. *International Journal of Disaster Risk Reduction*, *34*, 28–33. DOI: 10.1016/j.ijdrr.2018.12.001

Šturma, P. (2024). The notion of "sanctions" and "countermeasures" in international law. In Šturma, P. (Ed.), *International sanctions and human rights* (pp. xx–xx). Springer., DOI: 10.1007/978-3-031-69019-8_2

Subrahmanyam, S., Aishwaryalaxmi, N. S., Khalife, D., Shaikh, I. A. K., Faldu, R., & Asthana, N. (2024). Impact of Knowledge Management and Big Data Analytics Capabilities on Firm Performance. *2024 Ninth International Conference on Science Technology Engineering and Mathematics (ICONSTEM)*, Chennai, India, 2024, pp. 1-5. DOI: 10.1109/ICONSTEM60960.2024.10568874

Subramani, R. V. (2009). Accounting for Investments: Vol. 1. *Equities, Futures and Options*. John Wiley & Sons.

Suliman, F., Khwanda, H., & Murthy, R. V. R. (2023). An analysis of the Syrian economy in the era of military conflict, 2011–2020: The perspective of government and economics. *Journal of Government and Economics*, *11*, 100082. DOI: 10.1016/j.jge.2023.100082

Sun, J., Sarfraz, M., Turi, J., & Iva cu, L. (2022). Organizational Agility and Sustainable Manufacturing Practices in the Context of Emerging Economy: A Mediated Moderation Model. *Processes (Basel, Switzerland), 10*(12), 2567. Advance online publication. DOI: 10.3390/pr10122567

Sun, X., Meng, Z., Zhang, X., & Wu, J. (2025). The role of institutional quality in the nexus between green financing and sustainable development. *Research in International Business and Finance, 73*, 102531. DOI: 10.1016/j.ribaf.2024.102531

Sun, Y., & Page, N. (2024). Climate-biodiversity nexus in transnational climate governance: Variation across net zero initiatives. *Carbon Management, 15*(1), 2306895. Advance online publication. DOI: 10.1080/17583004.2024.2306895

Suprayitno, E., Aslam, M., & Harun, A. (2017). Zakat and SDGs: Impact Zakat on Human Development in the Five States of Malaysia. *International Journal of Zakat, 2*(1), 61–69. DOI: 10.37706/ijaz.v2i1.15

Suriani, Nurdin, R., & Riyaldi, M. H. (2020). Optimization of Zakat for Sustainable Development Goals: Evidence from Baitul Mal Aceh. 4th international Conference of Zakat Proceedings.

Syed, M. S. B., Kelly, P., Stacey, P., & Berry, D. (2024). Towards an IoT-Enabled Digital Earth for SDGs: The Data Quality Challenge. *IEEE Access : Practical Innovations, Open Solutions, 12*, 153421–153438. DOI: 10.1109/ACCESS.2024.3478813

Syna, H. D. (2020). Understanding the interface of diversity and political tensions in the context of divided societies: A multifocal perspective of social psychology and critical theory. In *Diversity management in places and times of tensions*. Palgrave Pivot., DOI: 10.1007/978-3-030-37723-6_3

Tabachnick, B. G., & Fidell, L. S. (2019). *Using multivariate statistics* (7th ed.). Pearson.

Taddese, E. T., & Rao, C. (2022). School-based continuous professional development of teachers: a case study of primary school teachers in Ethiopia. Education 3-13, 50(8), 1059-1071. https://doi.org/DOI: 10.1080/03004279.2021.1929382

Talwar, S., Dhir, A., Madanaguli, A. T., & Ractham, P. (2024). SDG-Embeddedness in Business School Operations: State-of-the-Art Literature, Ground Realities, and Measurement Parameters. *IEEE Transactions on Engineering Management, 71*, 13428–13445. DOI: 10.1109/TEM.2023.3236609

Tan, J. (2009). Multinational corporations and social responsibility in emerging markets: Opportunities and challenges for research and practice. *Journal of Business Ethics, 86*(S2), 151–153. https://www.jstor.org/stable/27749681. DOI: 10.1007/s10551-009-0189-7

Teece, D., Peteraf, M., & Leih, S. (2016). Dynamic Capabilities and Organizational Agility. *California Management Review, 58*(4), 13–35. DOI: 10.1525/cmr.2016.58.4.13

Telecom Review. (2024, May 31). Thailand's Digital Economy: Utilizing AI Strategies and Transformation Initiatives - Telecom Review Asia Pacific. Retrieved from https://telecomreviewasia.com/news/featured-articles/4276-thailand-s-digital-economy-utilizing-ai-strategies-and-transformation-initiatives/

Thacker, S., Adshead, D., Fay, M., Hallegatte, S., Harvey, M., Meller, H., O'Regan, N., Rozenberg, J., Watkins, G., & Hall, J. W. (2019). Infrastructure for sustainable development. *Nature Sustainability, 2*(4), 324–331. DOI: 10.1038/s41893-019-0256-8

The 2019 Voluntary National Review of the Philippines. (2019). https://hlpf.un.org/sites/default/files/vnrs/2022/06222022 VNR of the Philippines.pdf

The Lancet. (2021). India's COVID-19 crisis: A call for international action. *Lancet, 397*(10286).

Theparat, C. (2019, November 7). Bangkok Post - New ministry takes on human factor. Retrieved from https://www.bangkokpost.com/business/1789159/new-ministry-takes-on-human-factor

Thomas, N., Lucas, R., Bunting, P., Hardy, A., Rosenqvist, A., & Simard, M. (2017). Distribution and drivers of global mangrove forest change, 1996–2010. *PLoS One, 12*(6), e0179302. DOI: 10.1371/journal.pone.0179302 PMID: 28594908

Thomason, K. (2019). *The Difference between Principles and Rules-Based Accounting Standards*. Retrieved on 1 March 2019, from https://bizfluent.com/info-8333888-difference-principles-rulesbased-accounting-standards.html

Thomassen, W. E., & Munthe, E. (2020). Educating norwegian preservice teachers for the multicultural classroom – what knowledge do student teachers and mentor teachers express? *European Journal of Teacher Education, 44*(2), 234–248. DOI: 10.1080/02619768.2020.1758661

Thompson, S. C. (2016). *Accounting for a Developing World: A look at International Standards on Developing Countries*. Honors Thesis Project. University of Tennessee, Knoxville. Retrieved on 16 March 2020, from https://trace.tennessee.edu/cgi/viewcontent.cgi?article=2938&context=utk_chanhonoproj

Thompson, B. S., Clubbe, C. P., Primavera, J. H., Curnick, D., & Koldewey, H. J. (2018). Locally assessing the economic viability of blue carbon: A case study from Panay Island, the Philippines. *Ecosystem Services, 34*, 24–36. DOI: 10.1016/j.ecoser.2018.09.008

Thompson, F. C. (2009). Nomenclature and classification, principles of. In *Encyclopedia of insects* (pp. 707–714). Academic Press. DOI: 10.1016/B978-0-12-374144-8.00191-0

Thorbecke, E., & Charumilind, C. (2002). Economic inequality and its socioeconomic impact. *World Development, 30*(9), 1477–1495. DOI: 10.1016/S0305-750X(02)00052-9

Thwaites, K., & Simkins, I. (2006). *Experiential landscape: an approach to people, place and space*. Routledge. DOI: 10.4324/9780203462096

Torkayesh, A. E., Alizadeh, R., & Soltanisehat, L. (2024). Assessing Entrepreneurial Performance of G7 Countries Under Context of Social Sustainable Development Goals: A Multicriteria Decision Analysis Approach. *IEEE Transactions on Engineering Management, 71*, 4918–4929. DOI: 10.1109/TEM.2022.3227240

Trahan, J. (2020). Questioning the legality of veto use in the face of genocide, crimes against humanity, and/or war crimes. In *Existing legal limits to Security Council veto power in the face of atrocity crimes* (pp. 142–259). Cambridge University Press. DOI: 10.1017/9781108765251.006

Tran, H., & Smith, D. A. (2021). How hard-to-staff rural school districts use state funds to address teacher shortages. *Journal of Education Finance, 47*(2), 130–156. DOI: 10.1353/jef.2021.a846294

Traore, S. B., Ali, A., Tinni, S. H., Samake, M., Garba, I., Maigari, I., Alhassane, A., Samba, A., Diao, M. B., Atta, S., Dieye, P. O., Nacro, H. B., & Bouafou, K. G. (2014). AGRHYMET: A drought monitoring and capacity building center in the West Africa Region. *Weather and Climate Extremes, 3*, 22–30. DOI: 10.1016/j.wace.2014.03.008

Trends, G. (2017). *Challenges and opportunities in the implementation of the Sustainable Development Goals*. United Nations Development Programme& United Nations Research Institute for Social Development.

Tripathi, S., & Roepstorff, K. (2020). Decentering peace and conflict studies: Conceptualisations of peace in India. *Zeitschrift für Friedens- und Konfliktforschung, 9*, 79–99. https://doi.org/DOI: 10.1007/s42597-019-00014-z

Tseng, M. L., Tan, R. R., & Sirikrai, S. B. (2021). Implementation of green supply chain management in Thailand: A case study of the manufacturing sector. *Journal of Cleaner Production*, *203*, 709–717. DOI: 10.1016/j.jclepro.2021.09.113

Tuan, Y. F. (1979). Space and place: humanistic perspective. In *Philosophy in geography* (pp. 387–427). Springer Netherlands. DOI: 10.1007/978-94-009-9394-5_19

Tuan, Y. F. (2017). *A historical geography of China*. Routledge. DOI: 10.4324/9781315083537

Tuan, Y.-F. (1977). *Space and place: The perspective of experience*. University of Minnesota Press.

Tuffour, P., Chen, G., Opoku-Mensah, E., Darko, P. A., & Adu Agyapong, R. (2023). Toward sustainable development: Developing a decision-making framework for cross-sectoral engagement in Green Procurement. *Sustainable Development*. Advance online publication. DOI: 10.1002/sd.2772

Tuori, K. (2023). Pacta sunt servanda. *Suomalainen Tiedeakatemia Toimituksia. Sar. A.4: Biologica*, *2*(1), 44–57. DOI: 10.57048/aasf.130107

Turner, M. G., Collins, S. L., Lugo, A. L., Magnuson, J. J., Rupp, T. S., & Swanson, F. J. (2003). Disturbance dynamics and ecological response: The contribution of long-term ecological research. *Bioscience*, *53*(1), 46–56. DOI: 10.1641/0006-3568(2003)053[0046:DDAERT]2.0.CO;2

Ullah, A. A., & Chattoraj, D. (2022). *COVID-19 pandemic and the migrant population in Southeast Asia: Vaccine, diplomacy and disparity* (Vol. 2). World Scientific. DOI: 10.1142/12761

Ullah, A. A., Hassan, N. H., Mohamad, S. M., & Chattoraj, D. (2021). Privileged migrants and their sense of belonging: Insider or outsider? *Asian Journal of Social Science*, *49*(3), 161–169.

Ullah, A., Nobanee, H., Ullah, S., & Khan, S. (2025). Addressing climate change and advancing green development: Regional transition mechanisms for a sustainable future in Belt and Road Initiative nations. *Sustainable Development (Bradford)*, *2025*(4), 1–29. DOI: 10.1002/sd.3446

Ullah, M. H. (2013). Compliance of AAOIFI Guidelines in General Presentation and Disclosure in the Financial Statements of Islamic Banks in Bangladesh. *International Journal of Social Science Research*, *1*(2), 111–123. DOI: 10.14239/IJSSR.2013.01205

UN. (2020). *Goal 17: Revitalize the global partnership for sustainable development.* Retrieved 14 September 2022 from https://www.un.org/sustainabledevelopment/globalpartnerships/

UNCTAD. (2014). *World Investment Report 2014: Investing in the SDGs. United Nations Conference on Trade and Development.*

Under2 Coalition (https://www.theclimategroup.org/under2-coalition)

UNDP. (2021). *Human Development Report 2021.* United Nations Development Programme. https://hdr.undp.org/en/2021-report

UNDP. (n.d.). The SDGs in action. Retrieved from https://www.undp.org/sustainable-development-goals

UNESCO. (2019). *Water management and capacity building in sub-Saharan Africa: A comprehensive approach.* United Nations Educational, Scientific, and Cultural Organization.

UNESCO. (2020). *The future of education: Shaping the future of learning.* United Nations Educational, Scientific and Cultural Organization.

UNESCO. (2021). World Atlas of Languages: Towards a global online language repository.

UNESCO. (2023). Central Asian Experts Discuss Reforms of Teacher Policies to Improve Educational Outcomes in the Region. https://www.unesco.org/en/articles/central-asian-experts-discuss-reforms-teacher-policies-improve-educational-outcomes-region

UNESCO. (2023). *Global Education Monitoring Report 2023: Technology in Education – A Tool on Whose Terms?* UNESCO.

UNFPA. (n.d.). *A Life-Cycle Approach to Ageing in Thailand.*

United Nation. (2019). Sustainable Development Goal 8: Promote sustained, inclusive and sustainable economic growth, full and productive employment and decent work for all. https://www.globalgoals.org/goals/8-decent-work-and-economic-growth/

United Nations Development Programme (UNDP). (2018). *Capacity development: UNDP primer.* UNDP.

United Nations Environment Management Group. (n.d.). *Implementing the environment dimension of Agenda 2030 for sustainable development: Synergies between the implementation of the SDGs and international environmental objectives – Briefing paper.* UNITAR. https://www.unep.org/explore-topics/sustainable-development-goals

United Nations Statistics Division. (n.d.). SDG Indicators — SDG Indicators. https://unstats.un.org/sdgs/metadata/?Text=&Goal=8

United Nations. (1987). *Our Common Future: Report of the World Commission on Environment and Development*. United Nations.

United Nations. (2015). *Transforming our world: The 2030 Agenda for Sustainable Development*. United Nations General Assembly.

United Nations. (2015). *Transforming our world: The 2030 agenda for sustainable development*. United Nations.

United Nations. (2016). Report of the Inter-Agency and Expert Group on Sustainable Development Goal Indicators (E/CN.3/2016/2/Rev.1).

United Nations. (2020). *Webinars for sustainable development: Adapting to COVID-19 challenges*. United Nations Development Programme.

United Nations. (2023). *Goal 16 | Department of Economic and Social Affairs*. Sustainable Development Goals. Retrieved April, 2025, from https://sdgs.un.org/goals/goal16

United Nations. (2024). *SDG Progress Report 2024*. United Nations Publications.

United Nations. (2024). *The Sustainable Development Goals Report*. URL https://unstats.un.org/sdgs/report/2024/The-Sustainable-Development-Goals-Report-2024.pdf

Urama, K., Ozor, N., & Acheampong, E. (2014). *Achieving Sustainable Development Goals (SDGs) through transformative governance practices and vertical alignment at the national and subnational levels in Africa*. SDplanNet–Africa, African Technology Policy Studies Network.

Urzedo, D., Westerlaken, M., & Gabrys, J. (2022). Digitalizing Forest Landscape Restoration: A social and political analysis of emerging technological practices. *Environmental Politics*, *32*(3), 485–510. DOI: 10.1080/09644016.2022.2091417 PMID: 37207120

Valensisi, G., & Karingi, S. (2017). From global goals to regional strategies: Towards an African approach to SDGs. *African Geographical Review*, *36*(1), 45–60. DOI: 10.1080/19376812.2016.1185738

Valipour, E., Ketabchi, H., Shali, R. S., & Morid, S. (2024). Water resources allocation: Interactions between equity/justice and allocation strategies. *Water Resources Management*, *38*(2), 505–535. DOI: 10.1007/s11269-023-03682-y

Van Thao, N., Bańka, M. S., & Tien, N. H. (2025). Completing Dried Coconut Value Chain of Vietnam Towards Sustainable Development Goals. *Journal of Lifestyle and SDGs Review*, *5*(1), e03321–e03321. DOI: 10.47172/2965-730X.SDGsReview.v5.n01.pe03321

Varadharajan, M., Buchanan, J., & Schuck, S. (2019). Navigating and negotiating: Career changers in teacher education programmes. *Asia-Pacific Journal of Teacher Education*, *48*(5), 477–490. DOI: 10.1080/1359866X.2019.1669136

Vento, E. (2024). The global institutionalization of human rights discourse: A cross-national analysis of the language used in the International Labour Conference during the Cold War. *Nordic Journal of Human Rights*, *42*(2), 135–157. DOI: 10.1080/18918131.2024.2313907

Verdejo Espinosa, Á., Lendínez, A. M., Melguizo, F. J., & Estévez, M. E. (2023). Engineering and Technology Education in University Studies: Driving Digital, Sustainable, and Resilient Development—A Case Study in Andalusia, Spain. *IEEE Access : Practical Innovations, Open Solutions*, *11*, 108967–108981. DOI: 10.1109/ACCESS.2023.3321135

Viberg, O., Kizilcec, R. F., Wise, A. F., Jivet, I., & Nixon, N. (2024). Advancing equity and inclusion in educational practices with ai-powered educational decision support systems (ai-edss). *British Journal of Educational Technology*, *55*(5), 1974–1981. DOI: 10.1111/bjet.13507

Visser, F. (1988). The principle of permanent sovereignty over natural resources and the nationalisation of foreign interests. *The Comparative and International Law Journal of Southern Africa*, *21*(1), 76–91. https://www.jstor.org/stable/23247647

Wahlen, J. M., Jones, J. P., & Pagach, D. P. (2013). *Intermediate Accounting: Reporting and Analysis*. South-Western Cencage Learning.

Waltz, L. A., Muñoz, L., Johnson, H. W., & Rodriguez, T. (2020). Exploring job satisfaction and workplace engagement in millennial nurses. *Journal of Nursing Management*, *28*(3), 673–681. DOI: 10.1111/jonm.12981 PMID: 32068932

Wang, K. (2016). Reforming a university during political transformation: A case study of Yangon University in myanmar. *Studies in Higher Education*, *43*(7), 1184–1195. DOI: 10.1080/03075079.2016.1239250

Wang, W., Bai, Y., Xu, J., & Li, T. (2025). Collaborative governance of China-ASEAN maritime drug control: Mechanisms, challenges, and improvements. *Ocean and Coastal Management*, *261*, 107548. DOI: 10.1016/j.ocecoaman.2025.107548

Wasima, S. (2025). *Enhancing Resilience through Policy Frameworks: The Role of Regional Cooperation in Disaster Risk Management in South Asia*. BIPSS Commentary.

WCED. (1987). *Brundtland Report*. World Commission on Environment and Development.

Wei, C., Xu, J., Wang, Z., Wu, H., & Wang, J. (2025). How to implement pairing assistance during fighting COVID-19 in China: Collaborative governance between local governments under the authoritative regulation. *Frontiers in Public Health, 12*, 1417832. DOI: 10.3389/fpubh.2024.1417832 PMID: 39850856

Wewerinke-Singh, M. (2023). The rising tide of rights: Addressing climate loss and damage through rights-based litigation. *Transnational Environmental Law, 12*(3), 537–566. DOI: 10.1017/S2047102523000183

White, E. L., Dickerson, C., & Weston, K. M. (2015). Developing an appreciation of what it means to be a school-based teacher educator. *European Journal of Teacher Education, 38*(4), 445–459. DOI: 10.1080/02619768.2015.1077514

White, H. (1980). A heteroskedasticity-consistent covariance matrix estimator and a direct test for heteroskedasticity. *Econometrica, 48*(4), 817–838. DOI: 10.2307/1912934

Widegren, P., & Doherty, C. (2010). Is the world their oyster? the global imagination of pre-service teachers. *Asia-Pacific Journal of Teacher Education, 38*(1), 5–22. DOI: 10.1080/13598660903474155

Wieland, A., & Wallenburg, C. M. (2013). The influence of relational competencies on supply chain resilience: A relational view. *International Journal of Physical Distribution & Logistics Management, 43*(4), 300–320. DOI: 10.1108/IJPDLM-08-2012-0243

Willgerodt, M. A., Abu-Rish Blakeney, E., Summerside, N., Vogel, M. T., Liner, D. A., & Zierler, B. (2019). Impact of leadership development workshops in facilitating team-based practice transformation. *Journal of Interprofessional Care, 34*(1), 76–86. DOI: 10.1080/13561820.2019.1604496 PMID: 31039647

Winkler, I. T., & Williams, C. (2017). The Sustainable Development Goals and human rights: A critical early review. *International Journal of Human Rights, 21*(8), 1023–1028. DOI: 10.1080/13642987.2017.1348695

Wolfenden, F. (2023). Adapting and scaling teacher professional development approaches in Ghana, Honduras and Uzbekistan. Technical Report, http://hdl.handle.net/10625/61824

Wolf, J., Casado-Muñoz, R., & Pedone, F. (2022). Partnership of schools and civil society organisations to support education of students of varied linguistic backgrounds—The situation in the Czech republic, Italy and Spain. *European Journal of Education, 57*(4), 674–691. DOI: 10.1111/ejed.12524

Wolf, S., Aber, J. L., Behrman, J. R., & Tsinigo, E. (2018). Experimental impacts of the "quality preschool for ghana" interventions on teacher professional well-being, classroom quality, and children's school readiness. *Journal of Research on Educational Effectiveness, 12*(1), 10–37. DOI: 10.1080/19345747.2018.1517199

World Bank Data. (n.d.). World Bank Open Data. https://data.worldbank.org/indicator/SP.POP.DPND?locations=TH

World Bank Group. (n.d.). *GENDER AND INFORMAL WORK IN THAILAND*. https://documents1.worldbank.org/curated/en/099506009112325206/pdf/IDU044678b42097e804da90a85008d7c4c45b4b9.pdf

World Bank. (2020). *Digital economy for Africa (DE4A) initiative*. World Bank.

World Bank. (2020). *Global Partnership for Education (GPE) Annual Report*. World Bank.

World Bank. (n.d.). Decent work & economic growth | SDG 8: Decent work & economic growth. Atlas of Sustainable Development Goals 2023. Retrieved July 17, 2024, from https://datatopics.worldbank.org/sdgatlas/goal-8-decent-work-and-economic-growth?lang=en

World Commission on Environment and Development. (1987). *Our Commo Future*. Oxford University Press.

World Health Organization. (2023, February 9). Thailand's leadership and innovations towards healthy ageing. Retrieved from https://www.who.int/southeastasia/news/feature-stories/detail/thailands-leadership-and-innovation-towards-healthy-ageing

Worthington, T. A., Andradi-Brown, D. A., Bhargava, R., Buelow, C., Bunting, P., Duncan, C., Fatoyinbo, L., Friess, D. A., Goldberg, L., Hilarides, L., Lagomasino, D., Landis, E., Longley-Wood, K., Lovelock, C. E., Murray, N. J., Narayan, S., Rosenqvist, A., Sievers, M., Simard, M., & Spalding, M. D. (2020). Harnessing big data to support the conservation and rehabilitation of mangrove forests globally. *One Earth, 2*(5), 429–443. DOI: 10.1016/j.oneear.2020.04.018

Worthington, T., & Spalding, M. (2018). *Mangrove restoration potential: A global map highlighting a critical opportunity. The Nature Conservancy, IUCN*. University of Cambridge.

Wu, J., Guo, S., Huang, H., Liu, W., & Xiang, Y. "Information and Communications Technologies for Sustainable Development Goals: State-of-the-Art, Needs and Perspectives," in *IEEE Communications Surveys & Tutorials*, vol. 20, no. 3, pp. 2389-2406, thirdquarter 2018, DOI: 10.1109/COMST.2018.2812301

Wu, G., Ding, Q., & Chen, X. (2021). The effects of environmental regulation on green innovation: A panel study based on 31 provinces in China. *Journal of Cleaner Production*, *202*, 1200–1211. DOI: 10.1016/j.jclepro.2021.122356

Wu, G., Zhou, K., Hu, Z., Wang, G., Liu, B., & Zhang, W. (2025). The Symphony of Evolution: Unraveling Infrastructure Public–Private Partnerships Collaboration Networks Through Participant Characteristics. *Governance: An International Journal of Policy, Administration and Institutions*, *38*(1), e12911. DOI: 10.1111/gove.12911

Wu, L., Adebayo, T., Yue, X.-G., & Umut, A. (2022). The role of renewable energy consumption and financial development in environmental sustainability: Implications for the Nordic Countries. *International Journal of Sustainable Development and World Ecology*, *30*(1), 21–36. https://doi.org/https://doi.org/10.1080/13504509.2022.2115577. DOI: 10.1080/13504509.2022.2115577

Xie, L., & Jia, S. (2017). Diplomatic water cooperation: The case of the Sino-India dispute over Brahmaputra. *International Environmental Agreement: Politics, Law and Economics*, *17*(5), 677–694. DOI: 10.1007/s10784-016-9339-4

Xu, D., Glick, D., Rodríguez, F., Cung, B., Li, Q., & Warschauer, M. (2019). Does blended instruction enhance English language learning in developing countries? evidence from mexico. *British Journal of Educational Technology*, *51*(1), 211–227. DOI: 10.1111/bjet.12797

Yapa, P. W. S., & Joshi, M. (2012). The Socio-Economic Impacts of the Adoption of IFRS: A Comparative Study between the ASEAN countries of Singapore, Malaysia and Indonesia. Socio-Econ-IFRS, RMIT University (p. 1-26).

Yeh, Y. F., Hsu, Y. S., Wu, H. K., Hwang, F. K., & Lin, T. (2013). Developing and validating technological pedagogical content knowledge-practical (tpack-practical) through the delphi survey technique. *British Journal of Educational Technology*, *45*(4), 707–722. DOI: 10.1111/bjet.12078

Yin, R. K. (2017). Case study research and applications: Design and methods. Sage publications. Arifanti, V. B., Sidik, F., Mulyanto, B., Susilowati, A., Wahyuni, T., Subarno, S., Yulianti, Y., Yuniarti, N., Aminah, A., Suita, E., Karlina, E., Suharti, S., Pratiwi, P., Turjaman, M., Hidayat, A., Rachmat, H. H., Imanuddin, R., Yeny, I., Darwiati, W.,Novita, N. (2022). Challenges and strategies for Sustainable Mangrove Management in Indonesia: A Review. *Forests*, *13*(5), 695. 5DOI: 10.3390/f1305069

Yi, Y., Luo, J., & Wübbenhorst, M. (2020). Research on political instability, uncertainty, and risk during 1953–2019: A scientometric review. *Scientometrics, 123*(2), 1051–1076. DOI: 10.1007/s11192-020-03416-6

Yong, G. Y. V. (2021). Identifying Cultural Traits of the Historic Kampong Ayer of Brunei Darussalam using Biomimetic Analysis. *International Journal of Environment, Architecture, and Societies, 1*(2), 83–94. DOI: 10.26418/ijeas.2021.1.02.83-94

Yong, G. Y. V. (2022). Climate-Resilient Urban Design from a Biomimicry-Arcology Perspective. *International Journal of Environment, Architecture, and Societies, 2*(1), 1–15. DOI: 10.26418/ijeas.2022.2.01.1-15

Yong, G. Y. V., & Hassan, N. H. (2024). Blue Urbanism: An Insight of Navigating Climate Change in Humid Tropical Cities in Brunei Darussalam. *International Journal of Environment, Architecture, and Societies, 4*(01), 23–36. DOI: 10.26418/ijeas.2024.4.01.23-36

Yonpiam, C. (2024, August 17). Ageing society a boost for care industry. https://www.bangkokpost.com. https://www.bangkokpost.com/thailand/general/2848708/ageing-society-a-boost-for-care-industry

Young, J. C., Rose, D. C., Mumby, H. S., Benitez-Capistros, F., Derrick, C. J., Finch, T., Garcia, C., Home, C., Marwaha, E., Morgans, C., Parkinson, S., Shah, J., Wilson, K. A., & Mukherjee, N. (2020). A methodological guide to using and reporting on interviews in conservation science research. *Methods in Ecology and Evolution, 11*(1), 22–32. DOI: 10.1111/2041-210X.13342

Yudha, A. T. R. C., Awwaliah, H., & Pertiwi, E. M. (2021). SDGs Value and Islamic Philanthropy Through Zakat Institutions During the Covid-19. *Journal of Islamic Economics, Finance, and Banking, 4*(1), 31–44.

Yu, J., & Ye, C. (2025). A new conceptual framework and case analysis of rural environmental cross-boundary governance in megacities of China. *Habitat International, 156*, 103258. DOI: 10.1016/j.habitatint.2024.103258

Yusof, B., & Azalie, I. A. (2025). Jawi and cultural heritage conservation: A case study on Brunei's Yayasan Sultan Haji Hassanal Bolkiah (YSHHB) Islamic Calligraphy and Arts Study Centre. In Knight, D., & Truong, V. D. (Eds.), *Heritage tourism in contemporary Southeast Asia*. Springer. DOI: 10.1007/978-981-96-5427-7_12

Zahoor, N., Wu, J., Khan, H., & Khan, Z. (2023). De-globalization, international trade protectionism, and the reconfigurations of global value chains. *MIR. Management International Review, 63*(5), 823–859. DOI: 10.1007/s11575-023-00522-4

Zaini, M. K., Masrek, M. N., & Abdullah Sani, M. K. J. (2020). The impact of information security management practices on organisational agility. *Information and Computer Security*, *28*(5), 681–700. DOI: 10.1108/ICS-02-2020-0020

Zapp, M. (2021). International organisations and the proliferation of scientised global reporting, 1947–2019. *Global Society*, *36*(4), 455–474. DOI: 10.1080/13600826.2021.1902284

Zarghami, S. A. (2025). The role of economic policies in achieving sustainable development goal 7: Insights from OECD and European countries. *Applied Energy*, *377*, 124558. DOI: 10.1016/j.apenergy.2024.124558

Zhang, M. (2025). Deepening Webs of Functional Trilateralism and Challenges: A Case of Transboundary Air Pollution Cooperation. In *China–Japan–South Korea Trilateral Cooperation: Institution Building and Power Politics* (pp. 33–49). Springer Nature Singapore.

Zhang, M., Zhang, H., Zhang, L., Peng, X., Zhu, J., Liu, D., & You, S. (2023). Corruption, anti-corruption, and economic development. *Humanities & Social Sciences Communications*, *10*(1), 434. DOI: 10.1057/s41599-023-01930-5

Zhao, L., & Ye, J. (2025). Can civilized city construction facilitate green total factor productivity? A quasi-natural experiment based on China's pilot civilized city. *Journal of Environmental Planning and Management*, *68*(2), 437–462. DOI: 10.1080/09640568.2023.2259602

Zheng, N., & Xu, Y. Q. (2025). Spatial correlation and urban collaborative governance of air pollution in Anhui province based on complex networks. *Atmospheric Pollution Research*, *16*(2), 102369. DOI: 10.1016/j.apr.2024.102369

Zhu, Q., Sarkis, J., & Lai, K. (2007). Green supply chain management: Pressures, practices and performance within the Chinese automobile industry. *Journal of Cleaner Production*, *15*(11–12), 1041–1052. DOI: 10.1016/j.jclepro.2006.05.021

Zhu, Q., Sarkis, J., & Lai, K. (2008). Confirmation of a measurement model for green supply chain management practices implementation. *International Journal of Production Economics*, *111*(2), 261–273. DOI: 10.1016/j.ijpe.2006.11.029

Zhu, R., Xu, Q., Xiqiang, X., Sibt-e-Ali, M., Waqas, M., Ullah, I., & Anwar, A. (2024). Role of resources rent, research and development, and information and communication technologies on CO2 emissions in BRICS economies. *Resources Policy*, *93*, 105072. Advance online publication. DOI: 10.1016/j.resourpol.2024.105072

Zikargae, M. H., Woldearegay, A. G., & Skjerdal, T. (2022). Assessing the roles of stakeholders in community projects on environmental security and livelihood of Impoverished Rural Society: A nongovernmental organization implementation strategy in focus. *Heliyon*, *8*(10), e10987. Advance online publication. DOI: 10.1016/j.heliyon.2022.e10987 PMID: 36276717

Zipline. (2022). *Revolutionizing Healthcare Delivery in Rwanda*. Zipline Publications.

Zuccaro, G., & Leone, M. F. (2021). Climate services to support disaster risk reduction and climate change adaptation in urban areas: The CLARITY Project and the Napoli case study. *Frontiers in Environmental Science*, *9*, 693319. Advance online publication. DOI: 10.3389/fenvs.2021.693319

Zulhendra, Nurwahidin, Fahruroji, & Hamzah, M. A. (2020). Zakat Fund Management by Amil Zakat Institutions: Case Study on LAZNAS BSM Umat. 4th Internationa Conference of Zakat Proceedings.

About the Contributors

Lili Yulyadi Arnakim is currently an Associate Professor at the Faculty of Humanities and attached at Department of International Relations, BINUS University, Jakarta as Lecturer Specialist, and was a visiting professor at the Asian-Europe Institute (AEI), University of Malaya. He was attached at the Department of Southeast Asian Studies, University Malaya for more than ten years. His current research projects are on Integrating SDGs into the National Development Planning in Southeast Asian Countries (2024), Financing for Development in Southeast Asia (2023), BRI's Influence towards Chinese Entrepreneurs in Southeast Asia (2022), China's State Capitalism and Economic Development in Southeast Asia (2021), Gender Representation and Development in Southeast Asia (2020), Regional developmentalism in Southeast Asia (2019), funded by BINUS University, and a team member of H2020 Global governance research funded by European Commission. His latest works have published in Journal of ASEAN Studies (JAS), Intellectual Discourse, AEI Insights: An International Journal of Asia-Europe Relations, Transylvanian Review, Revista De Cercetare Si Interventie Sociala, Sejarah, CCPS, and Jati (Journal of Southeast Asian Studies). Lili Yulyadi Arnakim holds Ph.D from University of Malaya.

Miranda P. Tahalele is a Subject Content Coordinator / Lecturer in the Department of International Relations, Faculty of Humanities, Bina Nusantara (BINUS) University and independent development consultant. Her main research topics are the politics of international development cooperation, international political economy of development, gender and development, and the globalized world of foreign aid, with a focus on Indonesia's South–South Cooperation and North – South Relations. She holds a PhD in Anthropology (Development Studies) from the Australian National University (ANU) and a recipient of the Australian Awards Scholarship (AAS). She has a Master of Arts in Peace Studies (Conflict Resolution) from University of Bradford – UK, and a Bachelor of Political Science in

International Relations from Parahyangan Catholic University. She has also carried out worked as a consultant / development practitioners at the Ministry of Women Empowerment and Child Protection (for G20 and ASEAN), Ministry of National Development Planning (BAPPENAS), KFW, UNDP, UNESCAP, the Search for Common Ground (SFCG), and the Centre for East Indonesian Affairs (CEIA).

Muhammad Anshari is an Associate Professor in Business Information Systems at the Universiti Brunei Darussalam School of Business and Economics (UBDSBE), and currently serves as Deputy Director for the Institute of Policy Studies. He has been named on the Stanford University List of Top 2% World Researchers since 2021 for the categories of ICT, Business & Management. Together with colleagues from UBDSBE and IPS, winning Outstanding Paper awards at the Emerald Literati Awards 2023. Patent Impact: His research has also been referenced in two patents, US Patent 10,999,416, 2021, and US Patent 11,489,906, 2022. Following the completion of his PhD at FBEPS UBD, Dr. Anshari worked as a research fellow at National Taiwan University. Before that, he was a research fellow at King Saud University, Saudi Arabia. He has also completed eight research grant projects and three research industry consultancies. His research focuses on Business Information Systems, Knowledge Management Systems, Digital Business, Digital Humanities, and Digital Health, contributing primarily to SDG No. 9 on Industry, Innovation, and Infrastructure.

* * *

Sofiul Ahmed is an Assistant Professor of Law at the Alliance School of Law, Alliance University, Bangalore. With a strong academic background and a commitment to excellence in legal education, he specializes in areas such as energy laws, constitutional law, human rights, and international law. Mr. Ahmed is known for his engaging teaching style, dedication to student development, and contributions to academic research. He actively participates in seminars, workshops, and conferences, enriching the academic environment with his insights. His passion lies in fostering critical thinking and promoting a deeper understanding of law among his students.

Mesut Aydemir is an assistant professor at Open Education Faculty, Distance Education Department, Anadolu University. Asst. Prof. Mesut Aydemir gained his Ph.D. in Distance Education Department in 2019. His academic interest areas are online learning, recognition of prior learning, e-learning technology, enhanced learning, blended learning, distance learning, Instructional Technology, ICT in Education, mobile learning, teaching materials, social media in education, computer supported collaborative learning, andragogy,virtual collaboration. He has articles on e-learning, open and distance education in national and international indexes, papers submitted to international meetings. He has many supported projects by Universtiy and The Scientific and Technological Research Council of Turkiye, Erasmus+.

Susanta Das is a seasoned educator with a proven record of teaching and mentoring diverse students, working at multiple institutions, and providing service to the various initiatives of organizations. He received his Ph.D. and M.A. degrees from Western Michigan University (WMU), USA, and M.Sc. degree from Banaras Hindu University (BHU), India all in Physics. He continued his research as a Marie-Curie post-doctoral fellow at Stockholm University, Stockholm, Sweden on beam diagnostics for the DESIREE (Double ElectroStatic Ion Ring ExpEriment) in the project DITANET (Diagnostic Techniques for particle Accelerators – a Marie-Curie initial training NETwork), at the Indian Institute of Science Education and Research-Kolkata, India on high-pressure physics as a Project Scientist-B, and at the University of Electro-Communications, Tokyo, Japan on ion-surface interactions as a Post-doctoral fellow. Throughout his career, Dr. Das worked and collaborated with many researchers, Ph.D., master, and visiting students from different countries. He further visited many countries to discuss research and an international conference participant (UK, Italy, Germany, Greece, Belgium, Bulgaria, Romania, Brazil etc.). He co-authored several research articles in WoS/Scopus indexed international journals, conference proceedings, and scholarly book chapters published by renowned international publishing houses. Beyond his research endeavors, Dr. Das has a storied history of service to academic and administrative committees, exhibiting his commitment to the growth and development of educational institutions. His experience includes tenure at Central University South Bihar, Sri Sri University, and P.K. University before assuming his current role at Ajeenkya D Y Patil University. He received the Marie-Curie post-doctoral fellowship in Sweden, Science Academies' Summer Research Fellowship in India, Gwen Frostic Doctoral Fellowship, Department Graduate Research and Creative Scholar Award by WMU, and Leo R. Parpart Doctoral Fellowship by Dept. of Physics, WMU, among many others, throughout his academic journey. At present, Dr. Das continues to delve into cutting-edge fields, with a keen interest in nanotechnology, UN SDGs, and data science. His multifaceted contributions, spanning teaching, research, and administrative leadership, showcase a dedicated professional who is instrumental in advancing the vision and mission of the institutions he serves.

Anton Dwi Fitriyanto is a lecturer with a focus on digitalization, Bina Nusantara University. He holds a Doctor of Management degree and a Master's degree in Information and Communication Technology. He is also an expert in digitalization strategies, innovation development, and new technologies. He frequently collaborates with industry, universities, and government on research projects involving Big Data, Artificial Intelligence, Digital Strategic Management, Sustainability and Carbon Economy.

S. Ida Evangeline is an Assistant Professor in the Department of Electrical and Electronics Engineering at Government College of Engineering, Tirunelveli, Tamil Nadu, India. She holds a Bachelor's degree in Electrical and Electronics Engineering from Manonmaniam Sundaranar University (2003) and a Master's degree in Power Systems Engineering from Anna University, Chennai (2005). She earned her Ph.D. from Anna University, Chennai, in 2023. She has published six research papers in international journals. Her areas of interest include optimal power flow, renewable energy, soft computing, artificial intelligence, machine learning, and deep learning.

Asiyah Kumpoh is the Director of the Institute for Leadership, Innovation and Advancement (ILIA), Universiti Brunei Darussalam (UBD). She also holds an academic position as Senior Assistant Professor of History at the Faculty of Arts and Social Sciences (FASS), UBD.

Bharathi M. was born in Renigunta, India, in 1982. She received B.Tech. degree in Electronics and Communication Engineering from the KakatiyaUnversity, India, in 2004, and the M.Tech in Embedded Systems from the JNTUA, Ananthapuramu, India, in 2015, and currently doing her Ph.D in VTU. In 2024, she joined the Department of ECE, SKIT, as a Lecturer, and SreeVidyanikethan Engineering College as an Assistant Professor in 2006. She has 15 years of teaching experience Her current research interests include Low Power Design techniques, VLSI systems for next generations, Digital Arithmetic, Signal Processing Algorithm and VLSI Architectures. She has published 36 technical papers in various reputed International Journals & conferences She is a Life Member of the Indian Society for Technical Education (MISTE), International Association of Engineers, and the Institute of Research Engineers and Doctors.

Tanishka Prashant Madke is currently pursuing her B.Tech. in Computer Engineering at the School of Engineering, Ajeenkya D Y Patil University, Pune, Maharashtra, India. She has a strong passion for research and has authored multiple articles in NOVA Science Publishers and IGI-Global, USA. Her research interests encompass AI (Artificial Intelligence), Big Data, climate change, and environmental sustainability. She is particularly focused on contributing to the United Nations Sustainable Development Goals (SDGs), especially those related to climate action, innovation in technology, and sustainable cities. Through her work, she aims to explore how AI and emerging technologies can drive sustainable solutions, bridging the gap between innovation and global challenges.

Sheikh Inam Ul Mansoor (BA.LLB Hon's, LLM, CPCT WIPO, Ph.D. International Trade Law) is a highly accomplished academician presently working as an Assistant Professor at Symbiosis Law School, Hyderabad Campus, Symbiosis International Deemed University, Pune, India. With a strong background in law and a passion for academia, Dr. Sheikh has made significant contributions to the field through his extensive research and publications. Pursued his BA. LLB (Hon's) from University of Kashmir, LLM from JNU Jaipur and Ph.D. International Trade Law from LPU Punjab. Prior to his academic career, Dr. Sheikh served as an Advocate and Solicitor at the Jammu and Kashmir High Court and other subordinate courts in Jammu and Kashmir. This experience as a legal practitioner has provided him with valuable insights into the practical application of law and enhanced his understanding of legal dynamics.

Mulyono as a permanent lecturer majoring in management with specialization in economics and management, Bina Nusantara University, West Jakarta, Indonesia, since 2013. He is also an expert in financial management, capital market, and business economics. He graduated from a Doctor of Research in Management, Master's degree in Financial Management, and Bachelor degree in Economics Development Study.

Izni Azrein Noor Azalie is a human geographer specialising in industrial development and its impact on Brunei. He graduated with an M.Sc. in Global Transformations from Loughborough University UK and a Ph.D in Development Policy at University College London, UK, researching halal certification governance and its effects on local food producers. With more than 15 years of experience as a lecturer at Universiti Brunei Darussalam (UBD), Dr Izni Azalie has also consulted on environmental and social issues related to oil and gas projects, urban development, and coastal communities. His research interests range from industrial production networks to creative industries and heritage conservation within Brunei's urban centers. As Brunei explores the blue economy, he is now focused on the relationship between policy, governance strategies, and their real-world implementation from a multi-analytical perspectives combining Integrated Coastal Management (ICM) and Marine Spatial Planning (MSP) as part of developing a Sustainable blue industries.

Safira Puspitarani is a graduating student from the Department of International Relations, Faculty of Humanities, Bina Nusantara University, Jakarta. She is currently a research assistant at the Centre for Business and Diplomatic Studies, at Binus University, Jakarta.

Golla Sandhya Kumari, M.Tech, Ph.D, Currently Working as an Assistant Professor in the Department of Electronics and Communication Engineering at Siddharth Institute of Engineering & Technology, Puttur, Andhra Pradesh, India. She has completed her Ph.D in Koneru Lakshmaiah Education Foundation, Vijayawada, Andhra Pradesh, India. She has completed a Master of Technology in Digital Electronics and Communication Systems at Sri Padmavathi Mahila Visvavidyalayam, Tirupati, Andhra Pradesh and a Bachelor of Technology in Electronics and Communication Engineering at Sri Venkateswara Engineering College for Women, Tirupati under Jawaharlal Nehru Technological University, Anantapur. She has been worked as a research associate for the DST-CSRI project. Her research interests include Bio-Medical Signal Processing, Image Processing, Machine Learning and Deep Learning. She has published research papers in International journals and conferences. She is addressing as a reviewer for peer-reviewed journals as well as Web of Science, Scopus.

Namira Shabrina is a freelance researcher with a Master's degree in Southeast Asian Studies from the University of Malaya, Kuala Lumpur.

Mumtaz Ahmad Shah is an assistant professor of International relations and politics with experience in both teaching and research. His area of research includes international politics, great power politics and regional and global dynamics.

Amisha Shaikh is currently doing her B.Tech. in Computer Engineering at the School of Engineering, Ajeenkya D Y Patil University, Pune, Maharashtra, India. She has broad research interests from understanding the properties of biogenic gold nanoparticles to their application in various fields. In addition to nanoparticles, her interests extend into AI (Artificial Intelligence), Biotechnology, and Cybersecurity. She strives to contribute significantly in the possibilities of nanoparticles and AI, focusing on their vast applications. She aims to conduct research in these domains and make them more applied.

Rizwan Shaikh is a distinguished academician, researcher, and software consultant with over 30 years of rich experience. He holds a Ph.D. in Computer Applications from Pune University, India. His doctoral work focused on the evaluation and analysis of e-Governance systems. Dr. Rizwan currently serves as Professor (Computer Applications) in prestigious D Y Patil Group of Institutes, Pune. He was Director (R&D, Science and Technology) at Anam Green Revolution (AGR) IT Solutions LLP. As a Board of Studies (BoS) member and syllabus designer, he contributed to curriculum reforms at RTM University and SPPU. He has authored and co-authored over 25 research papers on e-Governance, CRM, semantic web, and cloud computing. He has received research grants from BCUD, University of

Pune, and holds a patent on electronic assessment systems. He has presented research at international conferences organized by IIM Bangalore, CSIR-IPR, and London Business School. Dr. Rizwan is an active member of professional bodies including INETA, CSI, ISTE, and the National Entrepreneurship Network (NEN). He played a pioneering role in establishing EDP cells in collaboration with IIM Bangalore and IIT Mumbai. He has served as coordinator for MBA online exams across 153 institutions affiliated with SPPU since 2013. As an educator, he has taught Computer Science, Management, Physics, and Engineering subjects. His contributions span academic leadership, accreditation processes (NAAC, NBA), and institutional development. He is a passionate advocate for value-based education and digital transformation in governance. With his cross-disciplinary expertise, he mentors students, faculty, and professionals in research and innovation. His vision is to create impactful digital ecosystems through governance, education, and technology.

Saleha Tabassum, M.Tech, is working as an assistant professor in the dept of Electrical and Electronics Engineering in KSRM College of Engineering. She is pursuing Ph.D in the dept of Electrical and Electronics Engineering.

Showkat Ahmad Wani is a Senior Scale Assistant Professor at, the School of Law, Alliance University Bengaluru, (India). Earlier, Dr. Wani was associated with the School of Law, Lovely Professional University (India). Dr. Wani has so far written 30 research papers in various reputed journals. He has also presented over 40 research papers at various National and International Conferences. His key research areas are Constitutional law and Human rights. Dr Wani has also contributed to valuable legal assistance to the Under- Trail Prisoners in collaboration with Tata Institute of Social Science (India) to the venerable classes. Dr. Wani has supervised eight students in the LLM dissertation. In addition, three Ph.D. Scholars have completed their Thesis under his supervision. Dr. Wani is a chief editor of one book, Legal Metamorphosis in Contemporary India. Dr. Wani is very keen on Research and Human Rights & a strong believer in the power of positive thinking in the workplace.

Putri Sekar Diasti Wibowo is a graduating student from the Department of International Relations, Faculty of Humanities, Binus University, Jakarta. She is currently attached at the Centre for Business and Diplomatic Studies as research assistant, Binus University.

Govind Yadav is an Assistant Professor of Law at the School of Law, Manipal University, Jaipur. He holds a doctorate in law and brings a rich blend of academic rigor and practical insight to his teaching. His areas of expertise include constitutional law, administrative law, and legal theory. Dr. Yadav is committed to nurturing analytical and research skills among students and has contributed to various academic publications and conferences. His approach to legal education emphasizes clarity, critical analysis, and a strong foundation in both theory and practice.

Gabriel Yit Vui Yong is a lecturer in the Geography, Environment & Development Programme at Universiti Brunei Darussalam (UBD), specializing in sustainable development, environmental systems, and coastal management. With a BSc in Geology/Geophysics (University of Reading) and an MSc in Environmental Analysis & Dynamics (University of Hull), his research integrates environmental science with sociocultural contexts, focusing on Brunei's unique challenges. He has led or contributed to key UBD projects, including coastal erosion assessments, mangrove conservation, urban poverty studies (e.g., Kampong Ayer), and climate-resilient urban design. His work critiques standardized sustainability frameworks (e.g., SDGs) for overlooking local realities, advocating place-sensitive methodologies that blend quantitative data with qualitative insights. Yong's interdisciplinary research spans environmental impact assessment, blue urbanism, and biomimicry-arcology, addressing issues like spatial inequality, hydrocarbon dependency, and climate adaptation. Notable publications explore Kampong Ayer's cultural heritage, coastal erosion dynamics, and Brunei's urban poverty. He collaborates extensively with scholars, policymakers, and communities, emphasizing participatory approaches in projects such as Teraja Ecotourism Development and Brunei's Voluntary National Reviews. His consultancy roles include environmental policy reviews (UNDP, ASEAN) and contributions to Brunei's State of the Environment Report. Yong's scholarship advances context-specific strategies for sustainable development, bridging academic rigour and practical policy implementation.

Index

Symbols

A

B

C

D

E

G

H

I

K

M

N

P

R

S

T

U

V

W

Z

www.ingramcontent.com/pod-product-compliance
Lightning Source LLC
LaVergne TN
LVHW060007031026
849161LV00025B/2244